Nanodroplets

LECTURE NOTES IN NANOSCALE SCIENCE AND TECHNOLOGY

Volumes Published in this Series:
Volume 1: Self-Assembled Quantum Dots, Wang, Z.M., 2008
Volume 2: Nanoscale Phenomena: Basic Science to Device Applications, Tang, Z., and Sheng, P., 2008
Volume 3: One-Dimensional Nanostructures, Wang, Z.M., 2008

Forthcoming Titles:
B-C-N Nanotubes and Related Nanostructures, Yap, Y.K., 2008
Towards Functional Nanomaterials, Wang, Z.M., 2008
Epitaxial Semiconductor Nanostructures, Wang, Z.M., and Salamo, G., 2008

For further volumes:
http://www.springer.com/series/7544

Zhiming M. Wang
Editor

Nanodroplets

Springer

Editor
Zhiming M. Wang
State Key Laboratory of Electronic
University of Electronic Science
and Technology
Chengdu, People's Republic of China

ISSN 2195-2159 ISSN 2195-2167 (electronic)
ISBN 978-1-4614-9471-3 ISBN 978-1-4614-9472-0 (eBook)
DOI 10.1007/978-1-4614-9472-0
Springer New York Heidelberg Dordrecht London

Library of Congress Control Number: 2013958214

Printed on acid-free paper

Springer is part of Springer Science+Business Media (www.springer.com)

Preface

There are many kinds of nanostructures, including the extensively studied quantum dots, nanowires, nanotubes, and graphenes. Most of them are solid-state nanomaterials. Nanodroplets in liquid form offer a unique platform for nanoscience and nanotechnology. Their liquid drop dynamics plays an important role in the formation, characterization, and application of related nanostructures. This book, with 15 chapters from 38 authors, combines experimental studies and theoretical analysis of nanosized droplets, with the aim of establishing collective effects as a promising research field of nanodroplets science and technology. Because nanodroplets may offer their own particular valuable properties, such as confined environment for nanomaterial synthesis, a venue to carry other nanomaterials, or subsequent novel nanostructures through liquid transition, nanodroplets science and technology finds a broad spectrum of applications in electronics and optoelectronics.

The first five chapters cover the formation of several kinds of nanodroplets. Charged nanodroplets and ion clusters are of great interest for the manipulation and fabrication of complex and adjustable nanostructures. In Chap. 1, a method to generate charged nanodroplets and its underlying mechanism based on the Thomson theory of ion-induced nucleation are discussed together with several successful applications of charged nanodroplets in nanotechnology and biotechnology. Chapter 2 shows that by thermally annealing electrospun fibers of immiscible polymer blends, the ribbon- or fiber-like dispersed phase can offer polymeric nanodroplets, which in turn can be used to study the fractionated crystallization and homogeneous nucleation of multifarious semicrystalline polymers. The latter include poly(ethylene oxide), poly(vinylidene fluoride), polyethylene, and polypropylene. Chapter 3 highlights dynamics studies of nanodroplet nucleation and growth using a wet scanning transmission electron microscope detector in environmental scanning electron microscopy. Chapter 4 reviews physical, chemical, and hybrid methods and mechanisms to assemble nanoscale metallic structures in suspensions, on two-dimensional substrates, and in 3-dimensional matrices, respectively. Chapter 5 provides a way to assemble ordered Ga metal nanodroplets by low-energy ion sputtering on GaAs surfaces.

The intrinsic properties of nanodroplets and their reactions with the surrounding environment are discussed in Chaps. 6–9. Chapter 6 investigates molecular dynamics simulations of the crystallization of a nanometer-sized Au droplet and the coexistence of liquid and solid phases in a nanometer-sized Ag droplet. Chapter 7 focuses on the static and dynamic peculiarities of nanodroplets on structured surfaces with special emphasis on theory. Chapter 8 reports the fragmentation behavior of pure nanodroplets and of polymer-loaded droplets using molecular dynamics simulation. Chapter 9 reveals light properties, such as light extinction, polarization, and phase, of polymer films containing nanosized liquid crystal droplets.

Chapters 10–15 review recent experimental and theoretical advances in various aspects of nanodroplet applications. Chapter 10 introduces fundamental properties of helium droplets and their applications to the study of clusters and nanoparticles, which shows new possibilities for using helium droplets as a powerful tool in nanoparticle synthesis. Chapter 11 focuses on reactive dynamics and confinement effects of Auramine O in confined water environment by reversed micelles. Methods and simulation of Brownian deposited nanodroplets and carbon nanofiber growth via the "vapor–liquid–solid" route are presented in Chap. 12. In Chap. 13, molecular dynamics simulations of water nanodroplets and drag phenomena of nanodroplets on carbon nanotubes by vibrations and by coupling to distantly solvated ions are presented. Chapter 14 then introduces atomistic pseudopotential theory and related investigations on nanodroplet epitaxial GaAs/AlGaAs quantum dots by molecular beam epitaxy using this methodology. The Ga nanodroplets are transformed into GaAs quantum dots in this case. Chapter 15 concentrates on the local droplet etching nanofabrication technique which is applied in situ during molecular beam epitaxy and shows great advantages in growing strain-free, highly uniform, and size adjustable nanostructures like quantum dots and quantum pillars.

Finally, the editor would like to express great appreciation for the chapter authors' thoughtful contributions and hopes that this book will encourage greater interest in our research community in the promising field of nanodroplets science and technology. The editor is also grateful to Mr. Lei Gao for providing helpful editorial assistance and is pleased to acknowledge financial support from 111 project No. B13042.

Chengdu, People's Republic of China Zhiming M. Wang

Contents

Contributors

Motoaki Adachi Department of Chemical Engineering, Graduate School of Engineering, Osaka Prefecture University, Sakai, Osaka, Japan, adachim@chemeng.osakafu-u.ac.jp

Zahava Barkay Wolfson Applied Materials Research Center, Tel-Aviv University, Tel-Aviv, Israel, barkay@post.tau.ac.il

Mourad Benamara Institute for Nanoscale Materials Science and Engineering, University of Arkansas, Fayetteville, AR, USA

Gabriel Bester Max-Planck-Institut für Festkörperforschung, Stuttgart, Germany, G.Bester@fkf.mpg.de

Francesco Delogu Dipartimento di Ingegneria Meccanica, Chimica, e dei Materiali, Università degli Studi di Cagliari, Cagliari, Italy, francesco.delogu@dimcm.unica.it

Andrew M. Ellis Department of Chemistry, University of Leicester, Leicester, UK, ame2@le.ac.uk

S.P. Fisenko A.V. Luikov Heat & Mass Transfer of National Academy of Sciences of Belarus, Minsk, Republic Belarus, fsp@hmti.ac.by

Hao Fong Department of Chemistry and Applied Biological Sciences, South Dakota School of Mines and Technology, Rapid City, SD, USA, Hao.Fong@sdsmt.edu

Lei Gao State Key Laboratory of Electronic Thin Film and Integrated Devices, University of Electronic Science and Technology of China, Chengdu, People's Republic of China, gaolei2010@gmail.com

Wolfgang Hansen Institute of Applied Physics, University of Hamburg, Hamburg, Germany, hansen@physnet.uni-hamburg.de

Ismael A. Heisler School of Chemistry, University of East Anglia, Norwich, UK, i.heisler@uea.ac.uk

Christian Heyn Institute of Applied Physics, University of Hamburg, Hamburg, Germany, heyn@physnet.uni-hamburg.de

Takuya Kinoshita Department of Chemical Engineering, Graduate School of Engineering, Osaka Prefecture University, Sakai, Osaka, Japan, t-kinoshita @chemeng.osakafu-u.ac.jp

Minako Kondo Molecular Photosciences Research Center, Kobe University, Kobe, Japan, mkondo@sapphire.kobe-u.ac.jp

Sabina Koukourinkova Institute for Nanoscale Materials Science and Engineering, University of Arkansas, Fayetteville, AR, USA, skoukourinkova@gmail.com

Valery A. Loiko B.I. Stepanov Institute of Physics of the National Academy of Sciences of Belarus, Minsk, Belarus, loiko@dragon.bas-net.by

Jun-Wei Luo National Renewable Energy Laboratory, Golden, CO, USA, jun-wei.luo@nrel.gov

Stephene R. Meech School of Chemistry, University of East Anglia, Norwich, UK, s.meech@uea.ac.uk

Peter Moeck Nano-Crystallography Group, Portland State University, Portland, OR, USA

Oregon Nanoscience and Microtechnologies Institute, Corvallis, OR, USA

N. Patra Department of Chemistry, University of Illinois at Chicago, Chicago, IL, USA, npatra2@uic.edu

Markus Rauscher Max-Planck-Institut für Intelligente Systeme, Stuttgart, Germany

Institut für Theoretische Physik IV, Stuttgart, Germany, rauscher@is.mpg.de

D. Keith Roper University of Arkansas, Fayetteville, AR, USA, dkroper@uark.edu

J. Russell Materials Science Division, Argonne National Laboratory, Argonne, IL, USA, jrussell@anl.gov

Gregory J. Salamo Institute for Nanoscale Materials Science and Engineering, University of Arkansas, Fayetteville, AR, USA, salamo@uark.edu

David Sonnenberg Institute of Applied Physics, University of Hamburg, Hamburg, Germany, dsonnenb@physnet.uni-hamburg.de

Si Neng Sun Fachbereich Physik und Forschungszentrum OPTIMAS, Universität Kaiserslautern, Kaiserslautern, Germany, sunsineng05@gmail.com

D.A. Takopulo A.V. Luikov Heat & Mass Transfer of National, National Academy of Sciences of Belarus, Minsk, Belarus, takopulo@mail.ru

Herbert M. Urbassek Fachbereich Physik und Forschungszentrum OPTIMAS, Universität Kaiserslautern, Kaiserslautern, Germany, urbassek@rhrk.uni-kl.de

B. Wang Institute of Chemistry, Chinese Academy of Science, Beijing, wangby@itp.ac.cn

Zhiming M. Wang Institute for Nanoscale Materials Science and Engineering, University of Arkansas, Fayetteville, USA, zhmwang@gmail.com

Jiang Wu State Key Laboratory of Electronic Thin Film and Integrated Devices, University of Electronic Science and Technology of China, Chengdu, People's Republic of China, jiangwu@uestc.edu.cn

Xingliang Xu State Key Laboratory of Electronic Thin Film and Integrated Devices, University of Electronic Science and Technology of China, Chengdu, People's Republic of China, xingliangxu@gmail.com

Shengfu Yang Department of Chemistry, University of Leicester, Leicester, UK, sfy1@le.ac.uk

Ganji Zhong College of Polymer Science and Engineering, Sichuan University, Chengdu, Sichuan, People's Republic of China, ganji.zhong@scu.edu.cn

Lei Zhu Department of Macromolecular Science and Engineering, Case Western Reserve University, Cleveland, OH, USA, lxz121@case.edu

Alex Zunger University of Colorado, Boulder, CO, USA, alex.Zunger@colorado.edu

Chapter 1
Generation of Nanodroplets and Its Applications

Motoaki Adachi and Takuya Kinoshita

Abstract The generation mechanism of charged nanodroplets is explained here using Thomson theory for ion-induced nucleation. Also described are the charged nanodroplet generator (CNDG) based on this theory and the method to measure the size of charged nanodroplets. The CNDG can generate charged nanodroplets with a geometric mean diameter $D_{pg} = 1.3$–1.8 nm for TEOS and $D_{pg} = 1.3$ nm for H_2O in O_2 gas and with $D_{pg} = 5.0$–5.5 nm for $Co(CO)_3NO$ in H_2 gas. Four successful applications of these charged nanodroplets are also described as the following: (1) synthesis of non-agglomerated SiO_2 nanoparticles with a diameter smaller than 10 nm, (2) patterning on a substrate by selectively depositing nanoparticles onto a charged line pattern, (3) fabrication of a high-performance magnetoresistance device, and (4) sterilization of bacteria (yeast fungi and *Escherichia coli*) by negatively charged H_2O nanodroplets.

1.1 Introduction

Nanodroplets are an underdeveloped research area. Some scientists even doubt their existence. The reason for such lack and misunderstanding is the difficulty in measuring the size of nanodroplets due to their instability. Their instability is caused by their large diffusivity and high evaporation rate. Recently, numerous atmospheric aerosol researchers have reported that the development of measurement instruments such as the differential mobility analyzer (DMA) now enables the observation of atmospheric nanoparticles a few nanometers in diameter (see Special

---

M. Adachi (✉) • T. Kinoshita
Department of Chemical Engineering, Graduate School of Engineering, Osaka Prefecture University, 1-1 Gakuen-cho, Naka-ku, Sakai, Osaka 599-8531, Japan
e-mail: adachim@chemeng.osakafu-u.ac.jp; t-kinoshita@chemeng.osakafu-u.ac.jp

Z.M. Wang (ed.), *Nanodroplets*, Lecture Notes in Nanoscale Science and Technology 18,
DOI 10.1007/978-1-4614-9472-0_1,

Issue of *Aerosol Sci. Technol.* Vol. 45, No. 3, 2011). These observed nanoparticles are actually nanodroplets because their main component is water. Such observations confirm the existence of nanodroplets in nature.

Nanodroplets potentially have widespread applications in biotechnology and nanotechnology. However, such applications require equipment or techniques by which nanodroplets at high concentration can be stably generated. In this chapter, we focus on nanodroplets that have a charge, which we call "charged nanodroplets," because charged nanodroplets are more stable and their high concentration generation is easier than uncharged nanodroplets. Here, "charged nanodroplets" comprise cluster ions in which molecules agglomerate because one does not define the limit of both numbers of agglomerate molecules in "charged nanodroplets" and "cluster ions." We describe (1) the theory behind their generation, (2) a charged nanodroplet generator (CNDG) based on this theory, (3) the method to measure the size distribution of the charged nanodroplets, (4) measurement results, and (5) four successful applications of charged nanodroplets in nanotechnology and biotechnology.

1.2 Theory of Charged Nanodroplet Generation [1]

Figure 1.1 shows a schematic of a charged nanodroplet in which dozens or possibly hundreds of vapor molecules have condensed around a primary ion generated by the ionization of gas. The size of a charged nanodroplet can range from 1 to several tens of nanometers in diameter, although gaseous ions are smaller than 1 nm. The generation mechanism of nanodroplets can be explained as follows based on the ion-induced nucleation theory proposed by Thomson [2]. Gibbs free energy ΔG induced by ion-induced nucleation can be expressed as

$$\Delta G = -\frac{4}{3}\pi {r_\mathrm{p}}^3 \frac{kT}{v_\mathrm{l}} \ln S + 4\pi {r_\mathrm{p}}^2 \sigma + \frac{q^2}{2}\left(1 - \frac{1}{\varepsilon}\right)\left(\frac{1}{r_\mathrm{p}} - \frac{1}{r_0}\right), \quad (1.1)$$

where r_p is the radius of a droplet generated by the condensation of vapor molecules on an gaseous ion, k is Boltzmann constant, T is temperature, v_l is volume of a liquid molecule, S is saturation ratio, σ is surface tension of the liquid, q is charge of a droplet, ε is specific dielectric constant of the liquid, and r_0 is radius of a gaseous ion

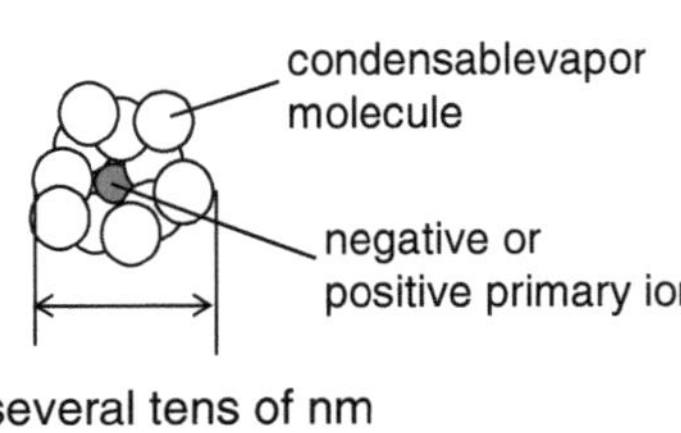

Fig. 1.1 Schematic of a charged nanodroplet

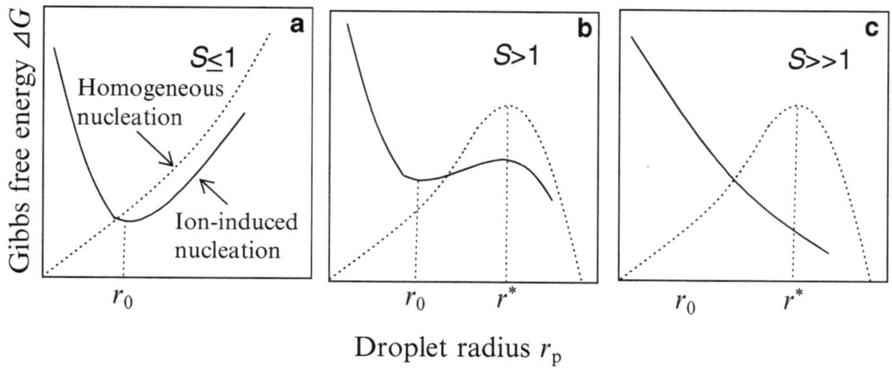

Fig. 1.2 Change in Gibbs free energy ΔG by ion-induced nucleation and homogeneous nucleation at different saturation ratio S as a function of droplet radius r_p [1]

(a primary ion). The first term in the right-hand side of Eq. (1.1) shows the energy change due to the phase change from gas to liquid, the second term is the surface energy, and the third term is the electrostatic energy. Figure 1.2a–c show changes in ΔG with respect to r_p by ion-induced nucleation at different S [solid line; Eq. (1.1)] and by homogeneous nucleation [dotted line; at $q=0$ in Eq. (1.1)]. At $S \leq 1$ (Fig. 1.2a), ΔG for the homogeneous nucleation increases with increasing r_p, indicating that a nanodroplet shrinks due to evaporation of the liquid. In contrast, ΔG for the ion-induced nucleation has a minimum at $r_p = r_0$, indicating that a nanodroplet reaches the size of gaseous ions r_0 because gaseous ions do not grow to nanodroplets when $r_p > r_0$ and do not evaporate when $r_p < r_0$. Therefore, gaseous ions can exist steadily in an atmospheric environment. At $S > 1$ (Fig. 1.2b), ΔG for homogeneous nucleation has a maximum at $r_p = r^*$, where r^* is called the critical radius or Kelvin radius. This maximum indicates that droplets with $r_p > r^*$ grow by condensation, whereas nanodroplets with $r_p < r^*$ shrink by evaporation. Also at $S > 1$, the ΔG for ion-induced nucleation has a minimum at $r_p = r_0$ and a maximum at $r_p = r^*$, indicating that further growth of nanodroplets with $r_p < r^*$ will require that the nanodroplet exceed this maximum ΔG. Furthermore, at $S > 1$, the maximum ΔG for ion-induced nucleation is much smaller than that for homogeneous nucleation, suggesting that growth of nanodroplets by ion-induced nucleation occurs easier than that by homogeneous nucleation. At $S \gg 1$ (Fig. 1.2c), ΔG for the ion-induced nucleation decreases with increasing r_p, and all gaseous ions grow to charged droplets.

1.3 Charged Nanodroplet Generator [3]

Figure 1.3 shows the charged nanodroplet generator (CNDG), which consists of an evaporator, a high-pressure ionizer (HPI), and a nucleation tube [4]. The meaning of high pressure used in this chapter is the pressure higher than an atmospheric pressure

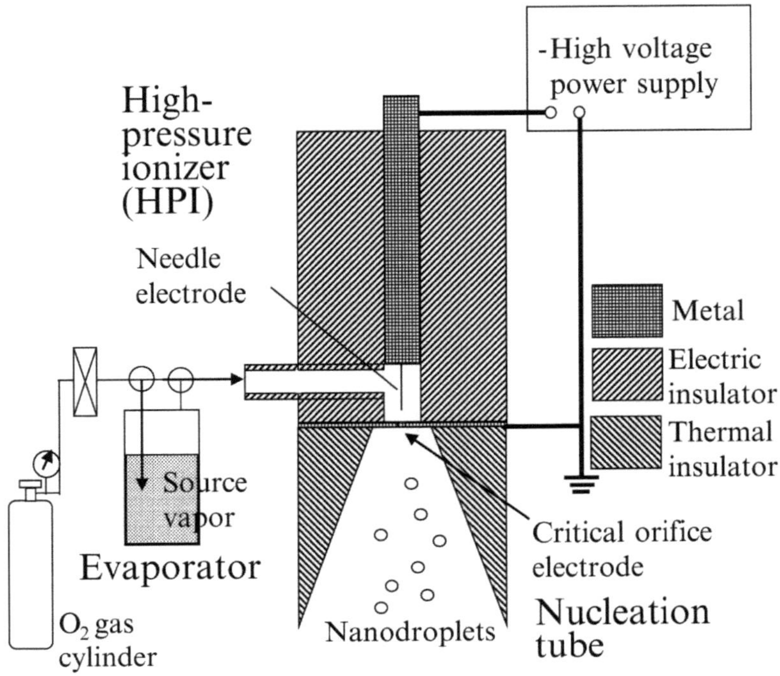

Fig. 1.3 Charged nanodroplet generator (CNDG) [3]

$P > 0.1$ MPa. The source vapor such as water and organic metals generated in the evaporator is transported to the ionizer by a carrier gas (e.g., oxygen) compressed to $P = 0.1$–0.3 MPa. The ionizer is kept at $P > 0.1$ MPa and thus creates a corona discharge between a tungsten needle electrode and a stainless steel critical orifice electrode. Free electrons generated by a negative corona discharge attach to the source molecules and produce negative source-molecule ions. The sonic jet formed by the critical orifice ejects source-molecule ions to the nucleation tube, which is covered with a thermal insulator. In the nucleation tube, source-molecule ions grow to nanodroplets because a supersaturation occurs due to adiabatic expansion.

The HPI has three advantageous characteristics: (1) stable ion generation in the mixture containing vapor, (2) high ion-density production, and (3) low deposition loss of ionized vapor in the ionizer. Advantages (1) and (2) are due to the discharge at high pressure, and (3) is due to the sonic jet exiting the critical orifice.

The corona discharge current in O_2 gas between the needle electrode and the orifice electrode has been measured by connecting an electrometer to the orifice electrode. Figure 1.4 shows this measured corona discharge current versus applied discharge voltage (I–V) curves at different discharge distance (defined as the distance between the needle and orifice electrodes) and O_2 gas pressure P, revealing that I increases with increasing V. When $V < -12$ kV, I does not arc when the discharge distance is either 4 or 6 mm, but does when the distance is 2 mm,

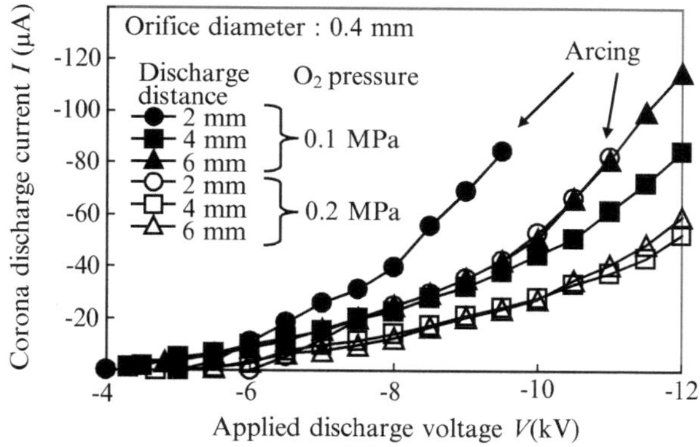

Fig. 1.4 *I*–*V* curves for corona discharge in HPI in CNDG [3]

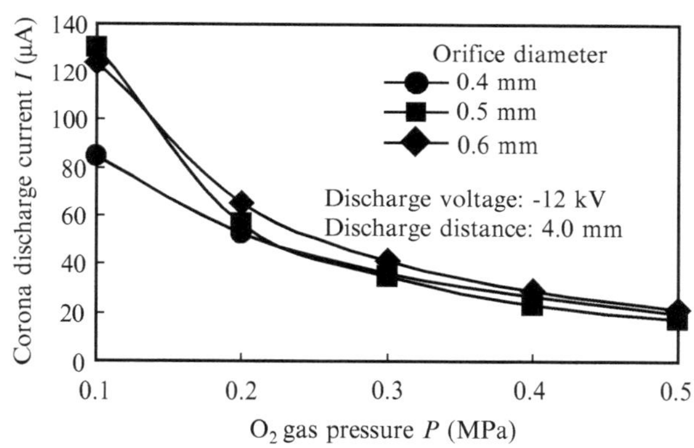

Fig. 1.5 Effect of O_2 gas pressure on corona discharge current *I* in HPI in CNDG [3]

independent of O_2 gas pressure. The transfer voltage from corona to arc increases with increasing O_2 gas pressure.

Figure 1.5 shows the measured effect of gas pressure and orifice size on *I*, revealing that *I* decreases with *P* because ion mobility is slower under higher gas pressure. *I* is affected by the orifice size at $P = 0.1$ MPa, but independent of the gas pressure at $P > 0.2$ MPa. The cause of differences is that the gas flow in the orifice does not reach sonic velocity ($P = 0.1$ MPa) or does it ($P > 0.2$ MPa).

Figure 1.6 shows the system to measure the current of ions I_{ion} ejected from the HPI. The system consists of two parallel electrodes working as an electro condenser, a dc voltage power supply and an electrometer. The two parallel electrodes are set inside a stainless steel chamber, which acts as an electrical shield. In the measurements, when the dc voltage was increased from 0 to 500 V, all of the I_{ion}–V curves reached saturation at 80 V. Therefore, I_{ion} was measured at 100 V.

Figure 1.7 shows the measured I_{ion} at various *V* of the HPI, revealing that I_{ion} increased with increasing *V* and finally reached saturation. The saturation I_{ion} increased with increasing *P*. The ion concentration n_{ion} is calculated from I_{ion} as

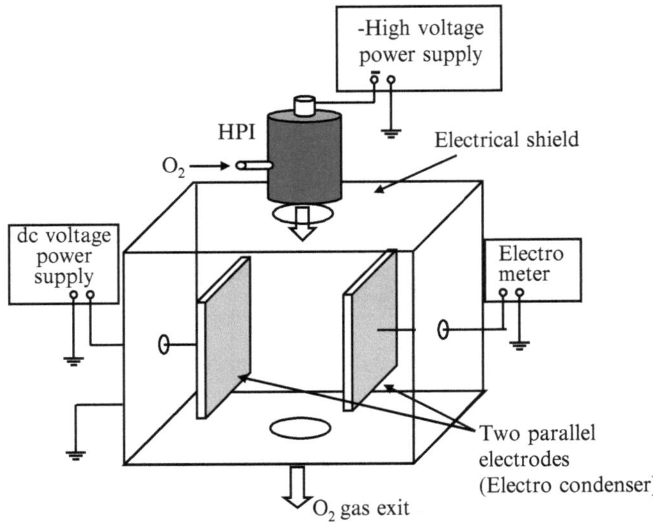

Fig. 1.6 System to measure current of ions (I_{ion}) ejected from HPI in CNDG [3]

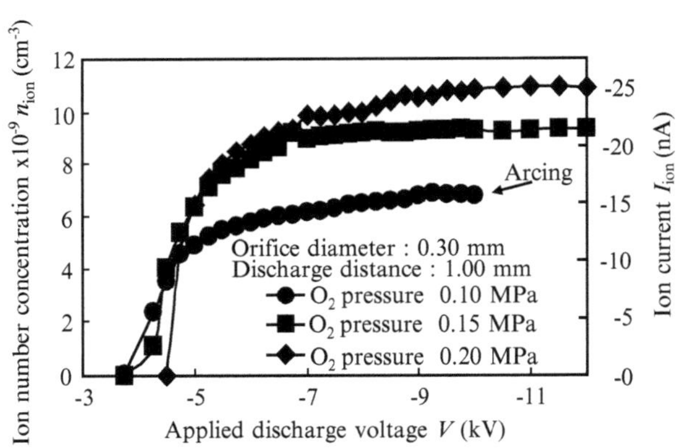

Fig. 1.7 I_{ion}–V curves of ions ejected from HPI [3]

$$n_{\text{ion}} = \frac{I_{\text{ion}}}{eQ_{\text{HPI}}}, \tag{1.2}$$

where e is elementary unit of a charge and Q_{HPI} is flow rate of carrier gas exited from HPI. The ion concentration at saturation was 6.8×10^9, 9.1×10^9, and 1.1×10^{10} cm^{-3} at $P = 0.1$ (circle key), 0.15 (square key), and 0.2 (diamond key) MPa, respectively. These concentration values are 100–1,000 times higher than a common corona discharge at an atmospheric pressure and were maintained longer than 100 h of operation of the CNDG.

1.4 Measurement of Charged Nanodroplets

1.4.1 Size Measurement Method

The size distribution of nanoparticles whose diameter is smaller than several tens of nanometers suspended in the gas phase has been measured using a differential mobility analyzer (DMA)/Faraday cup electrometer (FCE) system (Model DMA III, Wyckoff Co.) when they are solid or amorphous such as SiO_2 [5]. After here, we call nanometer-sized solid or amorphous particles "nanoparticles." Figure 1.8, respectively, shows (a) schematics of a DMA and (b) the size classification in the DMA. The DMA is composed of coaxial double-cylinder electrodes. A sampling slit and an entrance slit are set on the inner and outer electrodes, respectively. A laminar flow of charged nanoparticles travels down along the outer electrode, and a laminar sheath flow (i.e., "clean" air in which particles have been removed by an air filter) travels down along the inner electrode to prohibit the entry of nanoparticles into the sampling slit. An external electric field is applied at a set voltage level V (dc voltage is supplied to the inner electrode and the outer electrode is connected to ground) so that only the charged nanoparticles with a specific electrical mobility Z_p can cross the sheath flow from the outer electrode to the inner electrode and thus reach the sampling slit. The current I_p of these nanoparticles is then measured by the FCE. The V is then changed to sample charged particles of different Z_p. The Z_p of nanoparticles that reach the sampling slit at a set V is expressed as

$$Z_p = \left[\frac{Q_{sh} + 0.5\,(Q_{nano} + Q_s)}{2\pi L V}\right] \ln\left(\frac{R_1}{R_2}\right)' \tag{1.3}$$

where R_1 is the inner radius of the outer cylinder electrode, R_2 is the outer radius of the inner cylinder electrode, L is the distance from the entrance slit of nanoparticle flow to the sampling slit, and Q_{sh}, Q_{nano}, and Q_s are flow rates of the sheath air, carrier gas of nanoparticles, and sampling flow, respectively.

The number concentration of charged nanoparticles classified by the DMA, n_p, is calculated from the measured I_p as

$$n_p = \frac{I_p}{eQ_s}. \tag{1.4}$$

Then, the Z_p distribution based on n_p can then be calculated from the I_p–V curve measured by the DMA/FCE system.

Furthermore, Z_p can then be expressed as a function of particle diameter D_p as

$$Z_p = \frac{C_c p e}{3\pi\mu D_p}' \tag{1.5}$$

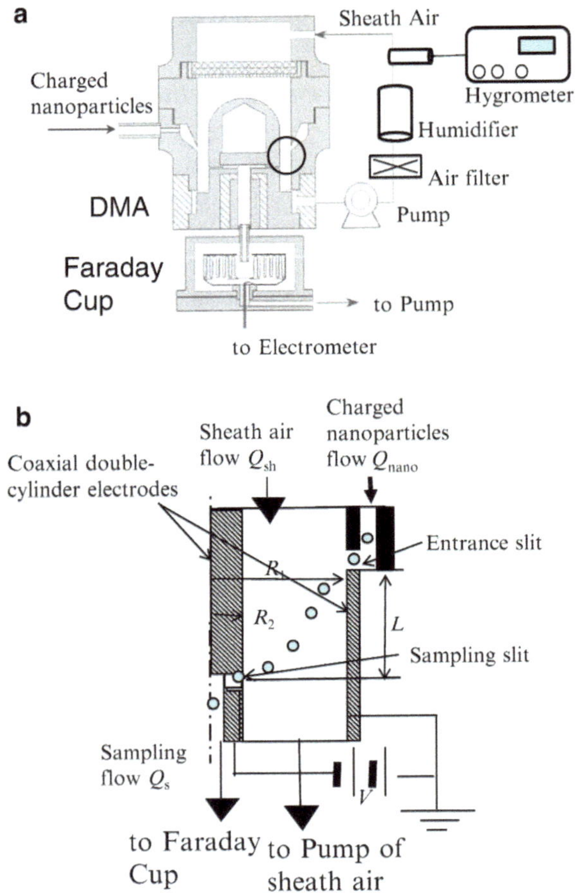

Fig. 1.8 (**a**) Differential mobility analyzer (DMA)/Faraday cup electrometer (FCE) system and (**b**) size clarification of charged nanoparticles in the DMA (schematic in *circle* on Fig. 1.8a is magnified)

where C_c is Cunningham coefficient, p is the number of charges on a particle, and μ is the gas viscosity. Finally, the D_p distribution of the nanoparticles can then be obtained from the Z_p distribution when the particle charge is known.

If size changes of charged nanodroplets in the DMA can be ignored, above Eqs. (1.3), (1.4), and (1.5) are used as the size classification theory for nanodroplets. The size distribution of nanodroplets suspended in the gas phase can therefore be measured by the DMA/FCE system. It should be noted that the humidity in the DMA must be controlled by a humidifier and hygrometer (see Fig. 1.8a) to prevent the size change of nanodroplets because D_p of nanodroplets is changed by the humidity [1].

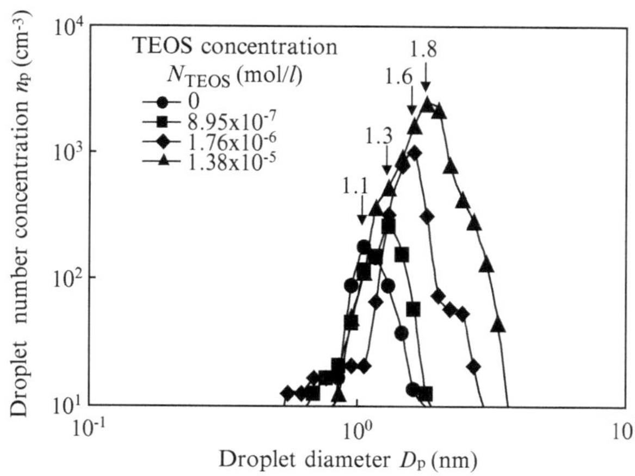

Fig. 1.9 Size distributions of negatively charged TEOS nanodroplets at different TEOS concentration N_{TEOS}

1.4.2 Size Distributions [3, 6]

Figure 1.9 shows the D_p distributions of charged nanodroplets when tetraethylorthosilicate (TEOS) is ionized by a negative discharge. In the absence of TEOS, the peak size is 1.1 nm, whose $Z_p = 1.65$ cm^2 V^{-1} s^{-1} and is due to O_2^- and/or $O_2^-(H_2O)_n$ ions [7, 8]. For TEOS concentration of 8.95×10^{-7} mol l^{-1}, the peak shifted to 1.3 nm, suggesting that the ion species changes to heavier molecules. As the TEOS concentration increases from 8.95×10^{-7} to 1.38×10^{-5} mol l^{-1}, the peak intensity increased and the peak shifted to larger D_p. These results indicate that negative TEOS and/or TEOS cluster ions produce and/or grow to nanodroplets. TEOS molecules apparently are ionized by attached electrons because the mean electron energy in the discharge field is lower than 3.2 eV [9, 10].

Figure 1.10 shows D_p distributions of negatively charged nanodroplets for H_2O vapor. The peak size of 1.1 nm when H_2O vapor was not added (note that H_2O concentration was actually 100 ppm due to contamination as discussed in the next paragraph) is due to $O_2^-(H_2O)_n$ ions as discussed for Fig. 1.9. When H_2O vapor concentration of either 6,000 or 12,000 ppm was added, the peak shifted to 1.3 nm.

For both D_p measurements (Figs. 1.9 and 1.10), O_2 gas was the carrier gas. The H_2O concentration in the O_2 gas cylinder was less than 5 ppm. The hygrometer, however, showed that the carrier gas contained H_2O vapor at a concentration of 100 ppm. The cause of the H_2O contamination is that H_2O molecules passed through the wall of the vinyl tube used in the gas line. The result of this H_2O contamination was the generation of $O_2^-(H_2O)_n$ ions even when H_2O vapor was not added during these measurements [3]. In Figs. 1.9 and 1.10, the peak height increased with increasing D_p. This increase has two possible explanations: (1) an increase in the number of nanodroplets and/or (2) a decrease in particle losses by Brownian diffusion. The I_{ion}–V curves for the O_2 gas to which either TEOS or H_2O vapor was added are similar to those in which neither was added, as shown in Fig. 1.7. The total concentrations of ion and charged nanodroplets determined from

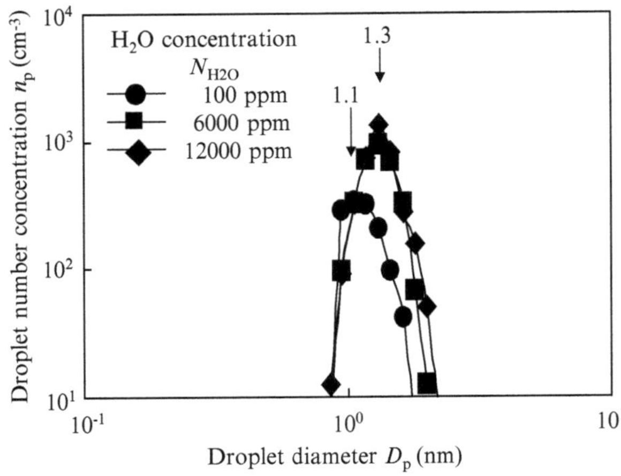

Fig. 1.10 Size distributions for negatively charged H_2O nanodroplets at different H_2O concentration N_{H2O}

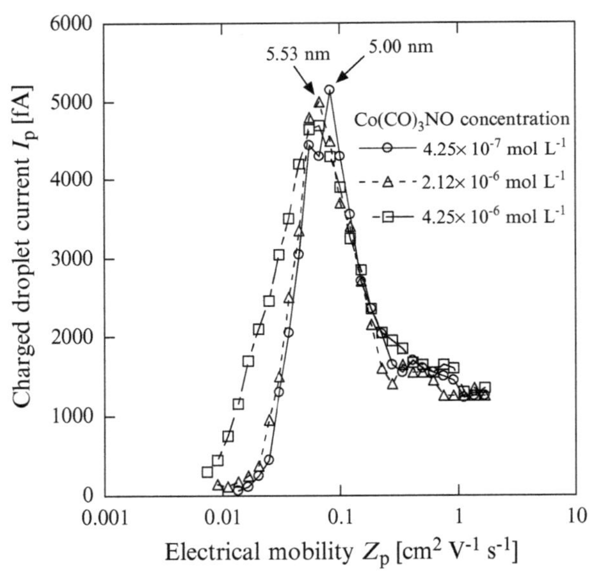

Fig. 1.11 Electrical mobility distribution Z_p of positively charged $Co(CO)_3NO$ nanodroplets in H_2 gas [6]

the I_{ion}–V results are apparently independent of the vapor concentration. If droplet losses remain constant during the droplet growth from ions, the peak height should also remain constant due to the constancy of the total concentration. In contrast, the droplet losses by Brownian diffusion in the DMA and the sampling line decrease with increasing droplet size. Therefore, this increase in peak height with increasing size is apparently caused by the particle losses by Brownian diffusion.

Figure 1.11 shows the Z_p distribution of positively charged nanodroplets for $Co(CO)_3NO$ vapor in H_2 gas. All of the distributions show a single peak at around 6×10^{-2} cm^2 V^{-1} s^{-1}, which encloses the mobility of 5.5-nm-diameter nanoparticles. On the other hand, there is no peak due to water cluster ions which appear usually around 1.65 $cm^2\ V^{-1}\ s^{-1}$. This suggests that carrier gas ions grow to charged nanodroplets by condensation of the vapor. Although the D_p for the peaks slightly increased with increasing vapor concentration, the height of the peak remained unchanged. This constant peak height is apparently caused by a decrease in particle losses due to particle sizes larger than either the TEOS or H_2O nanodroplets.

1.5 Application of Charged Nanodroplets

1.5.1 Synthesis of Non-agglomerated Nanoparticles [3]

A single-layer film of non-agglomerated nanoparticles has generated considerable interest in electronic and optoelectronic applications ([11, 12], etc.). Figure 1.12 shows a non-agglomerated SiO_2-nanoparticle generator using charged nanodroplets. The generator consists of the CNDG, a tube reactor heated by an infrared furnace, and a substrate-holder electrode. A dc voltage with the opposite sign to that of the charged nanodroplets is applied to the substrate-holder electrode. TEOS nanodroplets generated by the CNDG have size distributions shown in Fig. 1.9. These nanodroplets are introduced into the tube reactor in which intermediate nanodroplets react with oxygen molecules to form charged SiO_2 nanoparticles. Repulsive Coulombic forces between the charged particles reduce the Brownian coagulation of these charged nanoparticles, thus maintaining the primary nanoparticles [13, 14]. An external electric field is then applied to attract and then deposit the charged nanoparticles onto a substrate.

Figure 1.13 shows scanning electron microscopy (SEM) images of SiO_2 nanoparticles produced at a TEOS concentration $N_{TEOS} = 3.51 \times 10^{-5}$ mol l^{-1} and a furnace temperature $T_f = 1{,}173$ K and then deposited on an Si wafer using the system shown

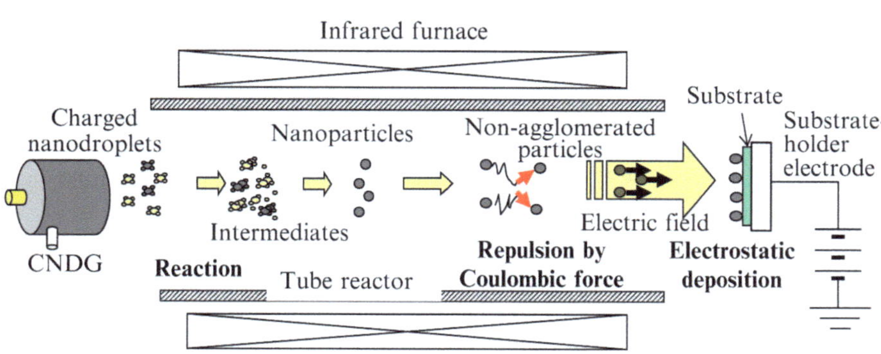

Fig. 1.12 System to generate non-agglomerated SiO_2 nanoparticles using charged nanodroplets [3]

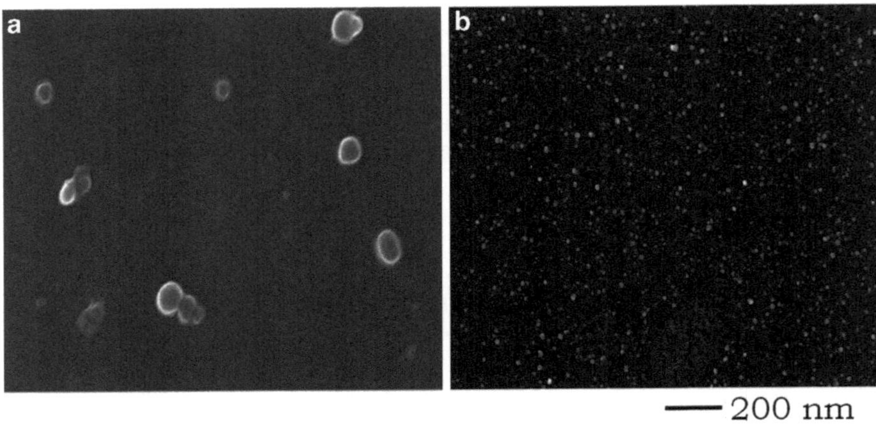

Fig. 1.13 SEM images of SiO_2 nanoparticles prepared using generator in Fig. 1.12 with ionization (**b**) and without ionization (**a**) at TEOS concentration of 3.51×10^{-5} mol l^{-1} and furnace temperature of 1,173 K. Particles were deposited on a Si wafer [3]

in Fig. 1.12. When TEOS vapor was ionized by the corona discharge in the CNDG (Fig. 1.13b), the nanoparticles produced by this generator were non-agglomerated and had a geometric mean diameter of $D_{pg} = 9.8$ nm, a geometric standard deviation of $\sigma_g = 1.35$, and a deposition density of 1.4×10^{10} cm^{-2}. When TEOS vapor was not ionized (Fig. 1.13a), the nanoparticles were agglomerated and had $D_p = 96$ nm and a deposition density of about 1×10^7 cm^{-2}. The electron diffraction patterns, Fourier transform infrared spectra, and thermal desorption spectra indicate that both sets of nanoparticles were amorphous silica that contained hydroxyl groups [3].

Figure 1.14 shows the D_p distribution of negatively charged nanodroplets and nanoparticles measured by the DMA/FCE system. They were generated at $293 \leq T_f \leq 1{,}173$ K and $N_{TEOS} = 3.51 \times 10^{-6}$ mol l^{-1}. At $T_f = 293$ and 473 K, the D_p distribution showed a single peak at around 1.6–2.0 nm, which is associated with charged TEOS nanodroplets. At $T_f = 673$ K, a new peak appeared at about 15 nm, which is associated with SiO_2 nanoparticles. At $T_f \geq 873$ K, the "charged nanodroplet peak" disappeared, while the "nanoparticle peak" remained. As T_f was increased from 873 to 1,173 K, the peak shifted to a larger D_p. The appearance of a new peak at $T_f = 673$ K indicates that TEOS molecules react and produce intermediates which have a lower vapor pressure at $T_f = 673$ K. The temperature of $T_f = 673$ K is similar to the particle generation temperature in a $TEOS/O_2$-atmospheric CVD film formation process [15]. The increase in D_p at $873 \leq T_f \leq 1{,}173$ K is caused by the large production of intermediates via the thermal reaction. The disappearance of charged nanodroplet peak at $T_f \geq 873$ K suggests that nanodroplets act as nuclei in the nanoparticle formation. A previous study on the particle formation mechanism in the ionization CVD process [5] reported that (1) particle formation with ionization occurs at a sufficiently low TEOS concentration at which particles are not generated without the ionization and (2) the

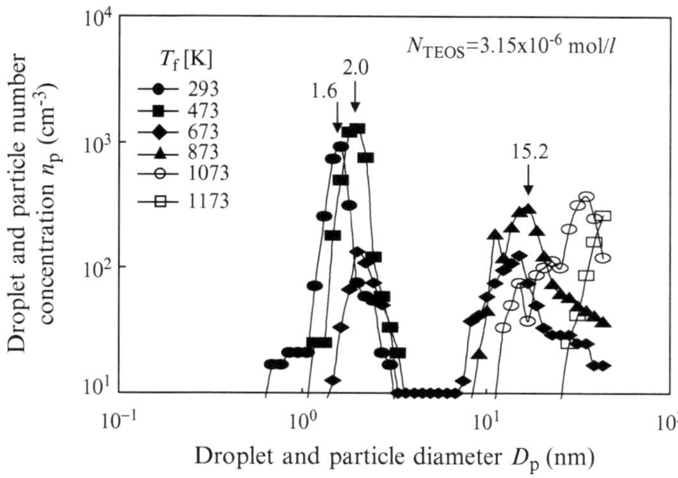

Fig. 1.14 Size distributions of SiO_2 nanoparticles and TEOS droplets prepared using generator in Fig. 1.12 with ionization at various furnace temperatures T_f [3]

nanodroplet peaks disappeared and the nanoparticle peaks appeared at $T_f = 873$ K when N_{TEOS} was increased. These results suggest that particles in ionization CVD are formed by condensation of gas-phase intermediates on charged nanodroplets (i.e., ion-induced nucleation).

1.5.2 Patterning of Nanoparticles on a Substrate [16]

Nanoparticles patterned on a substrate have generated considerable interest for electronic, optoelectronic, and magnetic device applications (e.g., [17]). Numerous methods for nanoparticle arrangement on a substrate by using a liquid-phase process have therefore been developed ([18, 19], etc.). In contrast, methods using a gas-phase process such as CVD have not yet been established because nanoparticles have large Brownian motion in a gas. One such method that we developed involves direct deposition of charged nanodroplets onto a substrate as follows [16].

Figure 1.15 shows the charged nanodroplet deposition reactor consisting of the CNDG and a substrate-holder electrode heated by an electric heater. Nanoparticle arrangement on a substrate using charged nanodroplets via this reactor was experimentally validated as follows using TEOS as the reactant gas. Charged TEOS nanodroplets with $D_p = 1.5$–2 nm and $n_p = 10^{10}$ cm^{-3} were generated by the CNDG at $N_{TEOS} = 3.75 \times 10^{-5}$ mol l^{-1}. The flow rate of O_2 gas containing the charged TEOS nanodroplets was kept constant at 1.4 l/min. In this reactor, the flux of charged nanodroplets was about 10^4 cm^{-2} s^{-1}. These charged nanodroplets were attracted to and then deposited onto the substrate-holder electrode by Coulombic

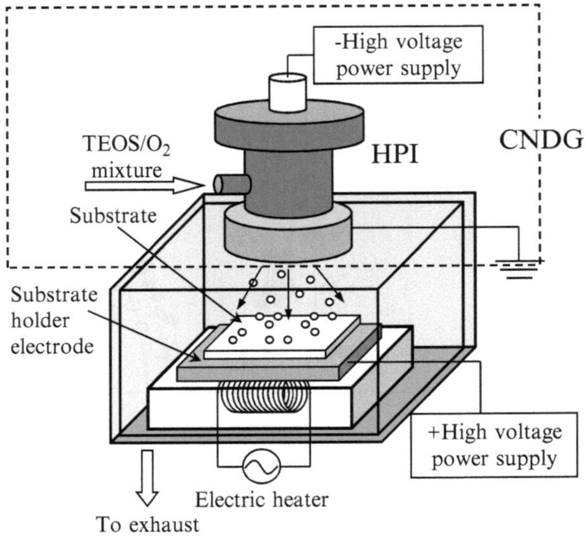

Fig. 1.15 Charged nanodroplet deposition reactor [16]

force by applying a dc voltage to the substrate on the electrode in which the sign of the applied voltage was opposite that of the charged nanodroplets. During a 10-min deposition, the holder electrode was kept at a deposition temperature ($T_{depo} = 491$ K) lower than the reaction temperature of $TEOS/O_2$ (723 K). Then, O_2 flow was used to anneal the substrate with the deposited TEOS charged nanodroplets at 873 K for 30 s. A Si (100) wafer (10–20 Ωcm, 1.4×1.4 cm^2) covered with a SiO_2 film was used as the substrate. In the arrangement, a line of induced surface charges was "drawn" on a SiO_2 film on a substrate by contact charging [20]. For this contact charging, a metal blade with a 5–10 μm edge and an applied voltage of 1 kV was brought into contact with the SiO_2 surface.

Figure 1.16 shows SEM images of nanodroplets and nanoparticles deposited selectively on a one-line (a) and three-line (b) charge pattern drawn by contact charging before and after annealing. The $N_{TEOS} = 3.75 \times 10^{-5}$ mol l^{-1}. Independent of annealing, nanoparticles were successfully deposited selectively along all of the lines, and the line widths (27 μm for the one-line pattern and 37 μm in the three-line pattern) were comparable to the edge width (5–10 μm) of the metal blade used in the contact charging.

1.5.3 Fabrication of a Magnetoresistance Device [6]

Magnetic nanoparticles have led to new fundamental and technological applications involving magnetoresistance response [21–24]. Methods to fabricate

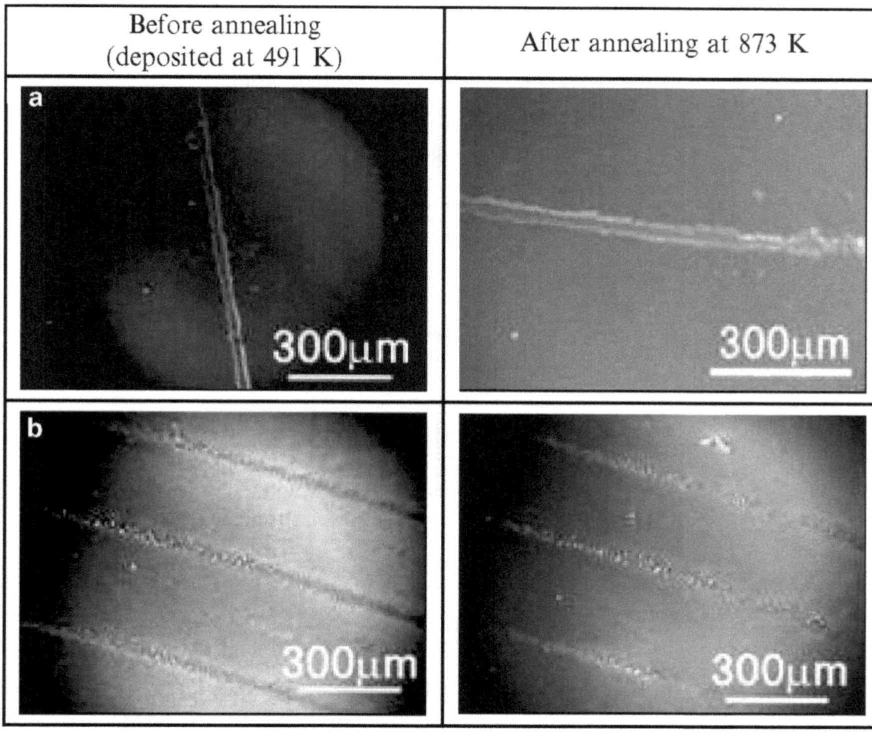

Fig. 1.16 SEM images of nanoparticles deposited selectively on a charge pattern before and after annealing using the reactor in Fig. 1.15 with TEOS vapor concentration N_{TEOS} of 3.75×10^{-7} mol l^{-1} [16]

magnetoresistance devices using ferromagnetic nanoparticles have been developed mainly involving a liquid-phase process and a nanoscale lithography process [11, 25–27].

Charged nanodroplets have been used to fabricate magnetoresistance devices. In this fabrication, first, charged nanodroplets of a source vapor formed by the CNDG are introduced into a furnace reactor, where they react and produce solid nanoparticles by thermal decomposition reaction. These solid nanoparticles then deposit electrostatically on an insulator substrate placed in the reactor. This insulator substrate has two Au electrodes, and the gap between these electrodes is filled by these nanoparticles. This nanoparticle device is then annealed under H_2 gas to deoxidize the deposited particles. After annealing, two Cu wires are then attached to the two Au electrodes to respectively connect to a dc voltage supplier and an ammeter.

Figure 1.17 shows the nanodroplet generation and deposition apparatus used in this fabrication of magnetoresistance devices. The apparatus consists of a CNDG, a quartz tube reactor heated by an infrared furnace, and a substrate-holder electrode to which dc high voltage is applied. In the device fabrication, cobalt tricarbonyl

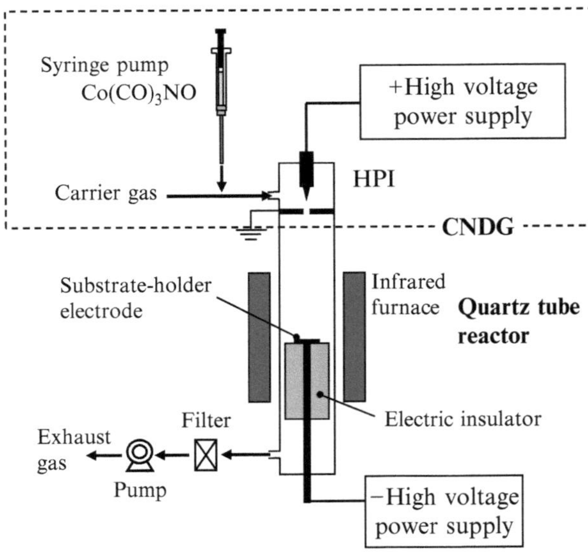

Fig. 1.17 System to fabricate a magnetoresistance device by Co-nanoparticle generation and deposition using charged nanodroplets [6]

nitrosyl [$Co(CO)_3NO$] vapor has been used as the reactant gas and injected into the HPI (via a syringe pump) using N_2 gas as the carrier gas. The CNDG generates charged $Co(CO)_3NO$ nanodroplets with Z_p distributions shown in Fig. 1.11. They resolve into a cobalt oxide (CoO), carbon oxides (CO_x)/monoxide/dioxide, and nitrogen oxides (NO_x)/monoxide/dioxide by thermal decomposition and form solid nanoparticles in the reactor. Due to their charge, these nanoparticles are attracted to and then deposited onto the substrate-holder electrode by Coulombic force by an applied dc voltage.

In an actual device fabrication, the conditions were as follows: The $Co(CO)_3NO$ concentration in the carrier gas was maintained at 3.04×10^{-6} mol l^{-1}. The gas pressure in the HPI was 0.3 MPa. The applied voltage to the needle electrode in the HPI was +7 kV. The flow rate of N_2 gas containing nanometer-sized charged droplets of $Co(CO)_3NO$ and the vapor exiting the HPI was 1.4 l min^{-1}. The furnace temperature was 623 K. The negative dc voltage applied to the substrate-holder electrode was −10 kV.

Figure 1.18 shows a schematic of this actual fabricated magnetoresistance device and the system used to measure its electro resistance *R*. A Si(100) wafer covered with a 350-nm-thick thermal oxide (SiO_2) film was used as an insulator substrate. To measure *R*, an Au-film electrode was deposited on the substrate by a physical vapor deposition method, and the area of the gap between the electrodes that was then filled with nanoparticles was 25 μm × 3 mm.

Figure 1.19 shows the XRD patterns of the nanoparticles before (i.e., as deposited) and after annealing in H_2 gas at 623 K for 20 min. Comparisons with

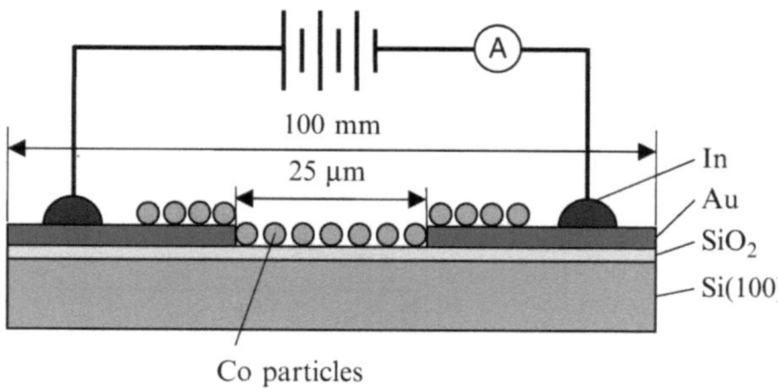

Fig. 1.18 Magnetoresistance device fabricated using the system in Fig. 1.17 and system to measure its electric resistance [6]

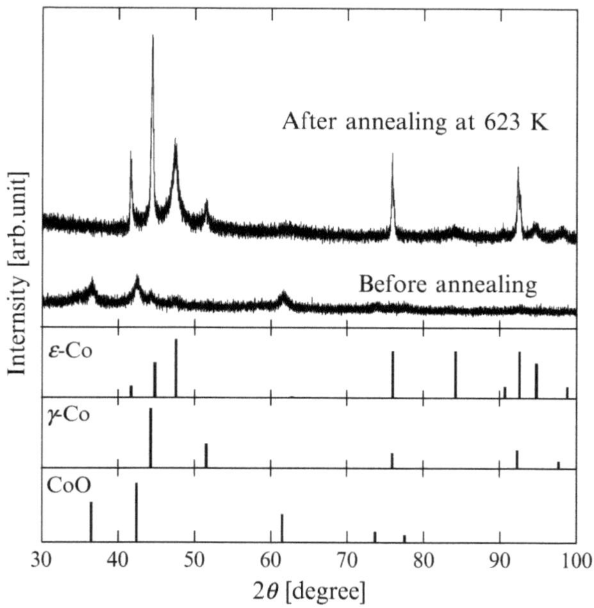

Fig. 1.19 X-ray diffraction patterns of nanoparticles before (as deposited) and after annealing in H_2 gas for 20 min [6]

ICDD (International Centre for Diffraction Data) data (ε-Co; ICDD 15-0806, γ-Co; ICDD 05-0727, CoO; ICDD 78-0431) reveal that the as-deposited particles (before annealing) were deoxidized to Co nanoparticles by the annealing and that most of the Co nanoparticles were hexagonal (γ-Co) and a portion was cubic (ε-Co).

The magnetoresistance of the device can be evaluated by measuring its electrical resistance using a superconducting quantum interference device (SQUID) magnetometer. Figure 1.20 shows the (a) magnetization electro resistance and (b) magnetoresistance ratio $\Delta R/R_0$ of a Co-nanoparticle device measured at 5 K fabricated using the system in Fig. 1.17. The magnetization M versus magnetic field H plot (Fig. 1.20a) shows a clear magnetic hysteresis, indicating that Co nanoparticles deposited on the device were also ferromagnetic. The electro resistance R versus

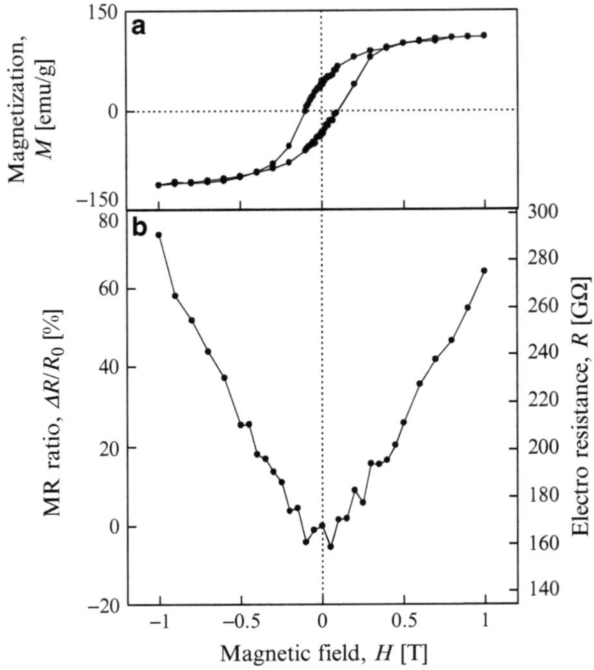

Fig. 1.20 Measured (**a**) magnetization M and (**b**) electric resistance R and magnetoresistance ratio $\Delta R/R_0$ for the device fabricated using the system in Fig. 1.17 [6]

H plot (Fig. 1.20b) shows that R increased with increasing H from 159 GΩ at $H=0$ T to 293 GΩ at $H=-1$ T, indicating that the Co-nanoparticle device had magnetoresistance. The magnetoresistance ratio $\Delta R/R_0$ of the ordinate in Fig. 1.20b was obtained as

$$\Delta R/R_0 = \frac{R - R_0}{R_0}, \quad (1.6)$$

where R_0 is the electric resistance absence of a magnetic field, respectively. The result shows that $\Delta R/R_0$ of this device reaches 73 % at $H=-1$ T.

1.5.4 *Sterilization of Bacteria [3, 28]*

A negative ion comprising H_2O molecules is called a "minus ion" in Japan and is believed to have beneficial health effects. The only effect confirmed scientifically is sterilization of bacteria. The effects of negatively charged H_2O nanodroplets on such sterilization have been examined as follows.

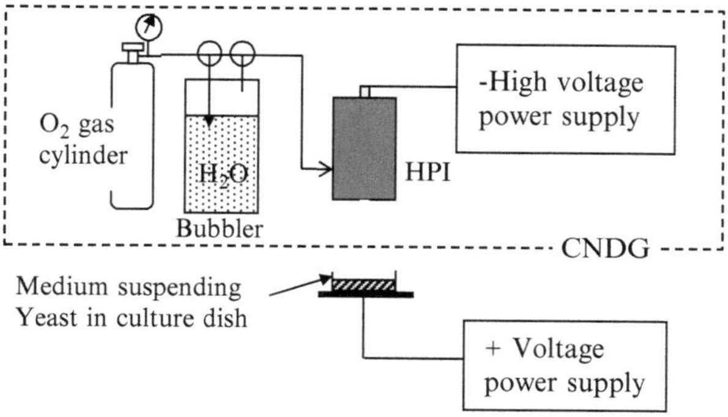

Fig. 1.21 Experimental apparatus for sterilization of bacteria by negatively charged H_2O nanodroplets [3]

Table 1.1 Effect of charged H_2O nanodroplets generated using H_2O concentration of 12,000 ppm on survival rate of yeast [3]

	Colony number		
Irradiation time (min)	With discharge	Without discharge	Survival rate (%)
10	198	374	53
20	45	374	12
30	36	374	9.6
40	14	265	5.3
60	6	265	2.2

Figure 1.21 shows the experimental apparatus for sterilization using negatively charged H_2O nanodroplets. The culture dish in which yeast was suspended in a medium was placed on an electrode to which positive voltage was applied. Negatively charged H_2O nanodroplets (see Fig. 1.10) produced by a CNDG were deposited onto the culture dish by Coulombic force. The culture dish was then exposed to the charged nanodroplets for an irradiation time of 10–60 min, and then cultivated for 24 h in an incubator. The colonies that appeared on the dish were then counted.

Table 1.1 shows the effects of negatively charged H_2O nanodroplets on the survival rates of yeast. As a reference for comparison, the colony number without discharge was also counted. In that reference experiment, voltage was not applied to the CNDG, and thus the carrier gas containing only the H_2O vapor was introduced onto the culture dish. The survival rate was defined as the colony number with discharge divided by that without discharge. Results reveal that the survival rate decreased with increasing irradiation time, reaching 2 % at 60 min. Similar results were obtained for the survival rate of *E. coli* [28]. These results indicate that yeast and *E. coli* can be killed by negatively charged H_2O nanodroplets and/or $O_2^-(H_2O)_n$ ions.

To clarify the sterilization mechanism by charged H_2O nanodroplets, metabolic heat during the growth stage of *E. coli* after nanodroplet irradiation was measured

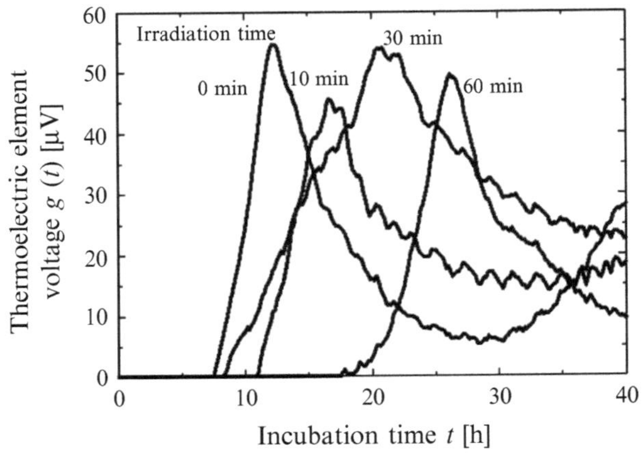

Fig. 1.22 Growth thermograms of *Escherichia coli* irradiated by negatively charged H_2O nanodroplets for different irradiation times using H_2O concentration of 12,000 ppm [28]

by calorimetry in which the heat was detected by a thermoelectric element. The measured growth thermograms of *E. coli* after different irradiation times are shown in Fig. 1.22 in which the abscissa is incubation time t and the ordinate is voltage $g(t)$ measured by the thermoelectric element. These thermograms reveal that an increase in $g(t)$ at $t = 8$ h, reaching a maximum at $t = 12$ h with no irradiation by charged H_2O nanodroplets (irradiation time was 0 min). This increase in $g(t)$ is due to the growth of bacterial cells. The growth stage is delayed by the nanodroplet irradiation (irradiation time of 10, 30, 60 min). The slope of each of the growth curves for irradiation times of 10, 30, and 60 min is more gradual than that for irradiation time of 0 min. This delay in the growth stage indicates that some of the bacterial cells are killed by the nanodroplets, and the gradual slope of the growth curve indicates that the activities of bacterial cells are weakened by the nanodroplets. These results suggest that charged H_2O nanodroplets have both bactericidal and bacteriostatic effects.

Now, sizes of disinfectant droplets which are used to sterilize in a hospital are a few millimeters because they are splayed by a duster or sprayer. A heavy use of disinfectants causes sometimes the generation of tolerance bacteria. Charged nanodroplets formed from disinfectsant vapor can be reduced by the disinfectant consumption. Charged nanodroplets formed from ethanol, acetic acid, and essential oil (Hiba) have been successfully generated using the CNDG. Electrical mobility distributions of these charged nanodroplets are shown in Fig. 1.23a–c. In these experiments, volatile disinfectant liquids were injected to the carrier gas by a syringe pump. Measured results showed that charged nanodroplets were produced from these disinfectant liquid by the CNDG.

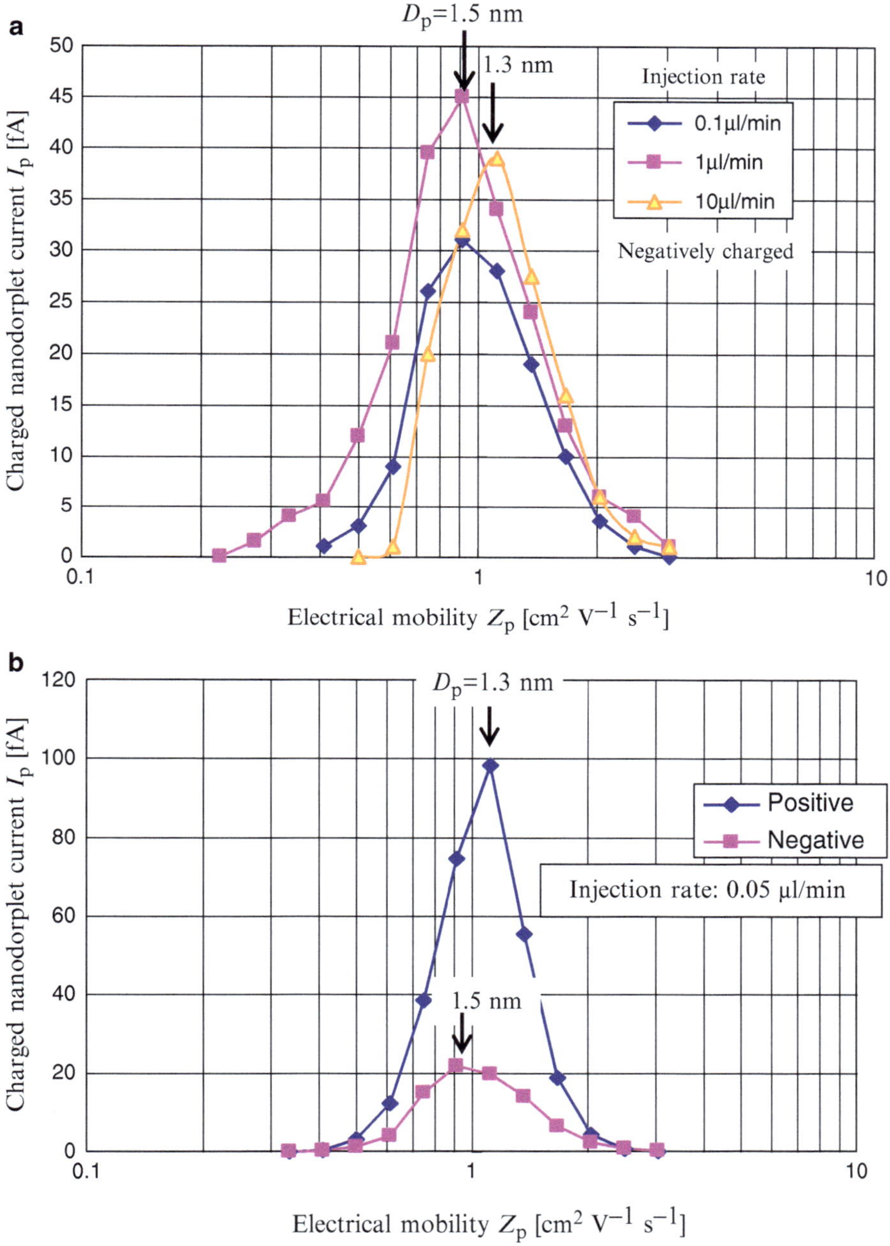

Fig. 1.23 (continued)

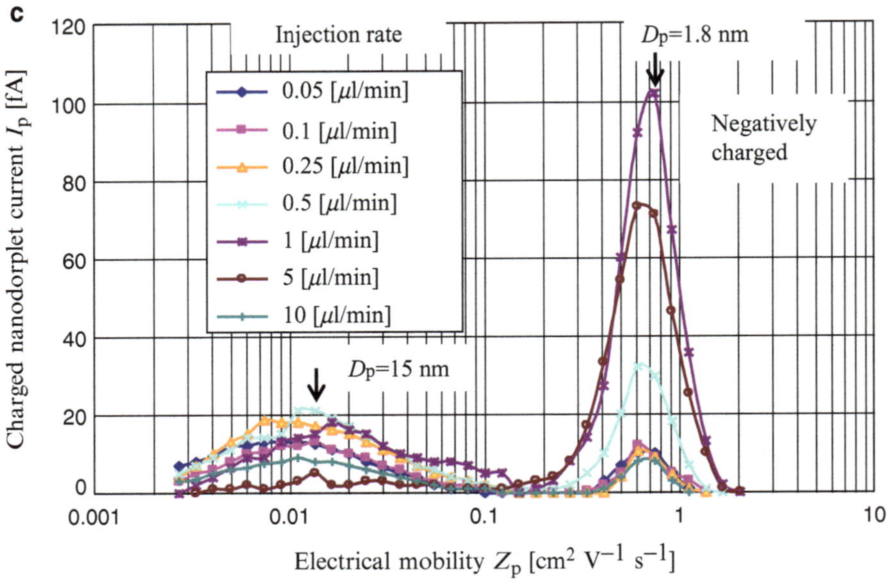

Fig. 1.23 Electrical mobility distributions of charged nanodroplets from (**a**) ethanol, (**b**) acetic acid, and (**c**) essential oil (Hiba)

1.6 Conclusions

The charged nanodroplet generator (CNDG) using high-pressure ionizer (HPI) was developed, and the *I–V* characteristics and size distributions of charged nanodroplets generated were measured. The CNDG can produce $O_2^-(H_2O)_n$ ions at a high concentration (10^9–10^{10} cm^{-3}) by a corona discharge at 0.1–0.3 MPa when a condensable vapor is not added to O_2 gas. When TEOS or H_2O vapor is added to O_2 gas, the CNDG can generate negatively charged nanodroplets that have an average diameter of 1.3–1.8 and 1.3 nm, respectively. When $Co(CO)_3NO$ vapor is added to H_2 gas, the CNDG can generate nanodroplets with average diameter of 5.0–5.5 nm.

These charged nanodroplets were successfully applied to the (1) synthesis of non-agglomerated nanoparticles, (2) patterning on a substrate, (3) fabrication of a high-performance magnetoresistance device, and (4) sterilization of bacteria.

In application (1), SiO_2 nanoparticles synthesized from charged TEOS nanodroplets were non-agglomerated and had a geometric mean diameter of 9.0 nm. The size distributions measured at various furnace temperatures indicate that ion-induced nucleation is the formation mechanism of the nanoparticles. In application (2), charged TEOS nanodroplets were selectively deposited onto a charged line pattern on a SiO_2 surface, and SiO_2-nanoparticle lines were formed on this pattern by annealing. In application (3), a fabricated magnetoresistance device had a high magnetoresistance ratio that reached 73 % at $H = -1$ T. In application (4), negatively charged H_2O nanodroplets clearly reduced the survival rates of yeast fungi and *Escherichia coli*.

References

1. Adachi, M.: Nanoparticle generation from air pollutant gases by ion-induced nucleation. Hoshasen Kagaku **86**, 26–35 (2008) (in Japanese)
2. Thomson, J.J.: Conduction of Electricity Through Gases. Dover, New York, NY (1969)
3. Adachi, M., Kusumi, M., Furuta, M., Tsukui, M.: Generation of nanodroplets and nanoparticles by ion-induced nucleation. Adv. Powder Technol. **16**, 549–559 (2005)
4. Adachi, M., Tsukui, S., Okuyama, K.: Nanoparticle synthesis by ionization source gas in chemical vapor deposition. Jpn. J. Appl. Phys. **42**, L77–L79 (2003)
5. Adachi, M., Kusumi, M., Tsukui, S.: Ion-induced nucleation in nanoparticle synthesis by ionization chemical vapor deposition. Aerosol Sci. Technol. **38**, 496–505 (2004)
6. Shirai, H., Kinoshita, T., Adachi, M.: Synthesis of cobalt nanoparticle and fabrication of magnetoresistance devices by ion-assisted aerosol generation method. Aerosol Sci. Technol. **45**, 1240–1244 (2011)
7. Morzzi, J.L., Phelps, A.V.: Survey of negative-ion-molecule reactions in O_2, CO_2, H_2O, CO, and mixtures of these gases at high pressures. J. Chem. Phys. **45**, 4617–4627 (1966)
8. Snuggs, R.M., Volz, D.J., Schummers, J.H., Martin, D.W., MacDaniel, E.W.: Mobilities and longitudinal diffusion coefficients of mass-identified potassium ions and positive and negative oxygen ions in oxygen. Phys. Rev. **A3**, 477–487 (1971)
9. Caledonia, G.E.: A survey of the gas-phase negative ion kinetics of inorganic molecules. Electron attachment reactions. Chem. Rev. **73**, 333–351 (1975)
10. Dutton, J.: A survey of electron swarm data. J. Phys. Chem. Ref. Data **4**, 577–857 (1975)
11. Black, C.T., Murray, C.B., Sandstrom, R.L., Sun, S.: Spin-dependent tunneling in self-assembled cobalt-nanocrystal superlattices. Science **290**, 1131–1134 (2000)
12. Coe, S., Woo, W.W., Bawendiand, M., Bulovic, V.: Electroluminescence from single monolayers of nanocrystals in molecular organic devices. Nature **420**, 800 (2002)
13. Fujimoto, T., Kuga, Y., Pratsinis, S.E., Okuyama, K.: Unipolar ion charging and coagulation during aerosol formation by chemical reaction. Powder Technol. **135–136**, 321–335 (2003)
14. Matsoukas, T.: The coagulation rate of charged aerosol in ionized gases. J. Colloid Interface Sci. **187**, 474–483 (1997)
15. Adachi, M., Okuyama, K., Thoge, N.: Particle generation and film formation in an atmospheric-pressure chemical vapor deposition process using tetraethylorthosilicate. J. Mater. Sci. **30**, 932–937 (1995)
16. Shirai, H., Kinoshita, T., Adachi, M.: Patterning and formation of SiO_2 nanoparticles on a substrate by electrically attracting of cluster ions. Jpn. J. Appl. Phys. **48**, 070216 (2009)
17. Huynh, W.U., Dittmer, J.J., Libby, W.C., Whiting, G.L., Alivisatos, A.P.: Controlling the morphology of nanocrystal-polymer composites for solar cells. Adv. Funct. Mater. **13**, 73–79 (2003)
18. Bethell, D., Brust, M., Schiffrin, D.J., Kiely, C.: From monolayers to nanostructured materials: an organic chemist's view of self-assembly. J. Electroanal. Chem. **409**, 137–143 (1996)
19. Daniel, M.C., Astruc, D.: Gold nanoparticles: assembly supramolecular chemistry, quantum-size-related properties, and applications toward biology, catalysis, and nanotechnology. Chem. Rev. **104**, 293–346 (2004)
20. Krinke, T.J., Fissan, H., Deppert, K., Magnusson, M.H., Samuelson, L.: Positioning of nanometer-sized particles on flat surface by direct deposition from the gas phase. Appl. Phys. Lett. **78**, 3708–3710 (2001)

21. Deshmukh, M.M., Kleff, S., Gueron, G., Bonet, E., Pasupathy, A.N., von Delft, J., Ralph, D.C.: Magnetic anisotropy variations and nonequilibrium tunneling in a cobalt nanoparticle. Phys. Rev. Lett. **87**, 226801 (2001)
22. Gerber, A., Milner, A., Groisman, B., Karpovsky, M., Gladkikh, A.: Magnetoresistance of granular ferromagnets. Phys. Rev. B **55**, 6446–6452 (1997)
23. Prinz, G.A.: Magnetoelectronics. Science **282**, 1660–1663 (1998)
24. Taub, N., Tsukernik, A., Markovich, G.: Inter-particle spin-polarized tunneling in arrays of magnetite nanocrystals. J. Magn. Magn. Mater. **321**, 1933–1938 (2009)
25. Marzin, J.Y., Gerard, J.M., Izael, A., Barrier, D., Bastard, G.: Photoluminescence of single InAs quantum dots obtained by self-organized growth on GaAs. Phys. Rev. Lett. **73**, 716–719 (1994)
26. Sun, S., Murray, C.B., Weller, D., Folks, L., Moser, A.: Monodisperse FePt nanoparticles and ferromagnetic FePt nanocrystal superlattices. Science **287**, 1989–1992 (2000)
27. Yang, J.Y., Yoon, K.S., Do, Y.H., Kim, C.O., Homg, J.P., Rho, Y.H., Kim, H.J.: Ferromagnetic quantum dots formed by external laser irradiation. J. Appl. Phys. **93**, 8766–8768 (2003)
28. Mizota, Y., Furuta, M., Adachi, M.: Generation of charged nanodroplets and their sterilization effects on Escherichia coli. Bokin Bobai **35**(10), 639–645 (2007) (in Japanese)

Chapter 2
Nanodroplet Formations in Electrospun Fibers of Immiscible Polymer Blends and Their Effects on Fractionated Crystallization

Ganji Zhong, Lei Zhu, and Hao Fong

Abstract A new and facile method to obtain polymeric nanodroplets was developed by thermally annealing of electrospun fibers of immiscible polymer blends. Through thermally annealing at a temperature slightly above the glass transition temperature of the matrix, the ribbon- or fiber-like dispersed phase broke up into nanodroplet with the diameter mainly in the range of 50–300 nm due to the Plateau-Rayleigh instability. Our study shows that these nanodroplets can be used to study the fractionated crystallization and homogeneous nucleation of almost any semicrystalline polymers, such as poly(ethylene oxide), poly(vinylidene fluoride), polyethylene, and polypropylene. We observed that the homogeneously nucleated crystallization of PVDF took place at 55–60°C for the first time. Additionally, this method can be utilized to investigate the effect of nanoconfinement on crystalline morphologies of semicrystalline polymers.

G. Zhong (✉)
College of Polymer Science and Engineering, Sichuan University, Chengdu 610065, Sichuan, People's Republic of China
e-mail: ganji.zhong@scu.edu.cn

L. Zhu
Department of Macromolecular Science and Engineering, Case Western Reserve University, Cleveland, OH 44106-7202, USA
e-mail: lxz121@case.edu

H. Fong
Department of Chemistry and Applied Biological Sciences, South Dakota School of Mines and Technology, 501 East St. Joseph Street, Rapid City, SD 57701-3995, USA
e-mail: Hao.Fong@sdsmt.edu

Z.M. Wang (ed.), *Nanodroplets*, Lecture Notes in Nanoscale Science and Technology 18,
DOI 10.1007/978-1-4614-9472-0_2, © Springer Science+Business Media New York 2013

2.1 Introduction

Nanostructured polymers have attracted growing attention because they are recognized as an efficient way to obtain novel polymeric materials with synergistic and better mechanical, electrical, and optical properties due to nanoconfinement effects. Recently, through electrospinning of immiscible polymer blends, various nanostructures, such as core-shell structure [1–3] and co-continuous structure [4], have been achieved, which provides a new approach to obtain nanostructured polymer blends even without addition of any compatibilizers. By thermal processing of these novel nanostructured polymer blends, other than common microstructured polymer blends, new types of devices with improved properties and potential electrical, optical, mechanical, and biomedical applications might be developed [1].

When polymers are confined in nanometer spaces in nanostructured polymers, new physics and unprecedented properties can emerge, which is significant for developing new applications and devices. For amorphous polymers with the absence of long-range or quasi-long-range order, their glass transition temperatures ($T_g s$) can be significantly altered when they are confined to the nanometer scales [5, 6]. This is believed to originate from the polymeric nature of the amorphous materials, because the confinement effect for glass-forming small molecules is much less significant than that for polymers, i.e., a vast variety of chain conformations may be obtained during the sample preparation processes [7], and a broad distribution of $T_g s$ can be resulted from the polymer interactions with the environment/substrate as far as several tens of nanometers into the bulk material [6].

For ordered polymers with long-range or quasi-long-range order, including crystalline polymers, liquid crystalline polymer, and block copolymers, novel ordered hierarchical superstructures are observed when they are confined in nanometer spaces [8]. First, when block copolymers are confined to nanometer spaces, single crystalline domains with specific orientations can be obtained [9–12]. These single crystalline domains have good potentials for next-generation nanolithography and microelectronics applications. Second, when liquid crystalline polymers are confined to nanometer spaces, the ordered mesogens adopt a specific orientation with respect to the confinement geometry [13]. In the above two kinds of ordered polymers, the commensurability between the polymer long period and the confinement dimension is key to the final domain (for block copolymers) and mesogen (for liquid crystals) orientations. Third, when crystalline polymers are confined to nanoscale spaces, crystallization behavior, crystal orientation, polymorphism, phase transformation, and even melting behavior can be drastically changed. Finely dispersed immiscible polymer blends [1], polymers in nanopores [14, 15], and microphase-separated crystalline block copolymers [16] have been used to achieve confined polymer crystallization with uniform crystal orientation.

Among the above nanoconfined effects on polymer crystallization, the most pronounced effect observed is the crystallization behavior. Homogeneous nucleation, which is rarely observed in bulk polymers and usually happens at a large supercooling (ca. 60–100°C) [17, 18], can occur when the confined geometry changes from three-dimensional (3D) to two-dimensional (2D), to one-dimensional (1D), and finally to zero-dimensional (0D), and confined space decreases from the bulk to nanometer scales. Closely related to the crystallization behavior, uniform crystal orientation can be achieved by confining the crystal growth in nanoscale lamellae and cylinders [8]. Moreover, polymorphic phases and phase transformation may also be affected by confined crystallization, although the effects are much less pronounced. For example, when nylon 6 is confined in the form of nanofibers (200 nm diameter), the Brill transition (i.e., from the monoclinic α-form to a high-temperature monoclinic form) is observed at 180–190°C, which is at least 20°C higher than that (160°C) in the bulk sample [19].

In this chapter, we illustrate the nanostructure development in electrospun polymer blend fibers during thermal treatment. Nanodroplets can be formed easily by thermally annealing the co-continuous polymer blend fibers via the Plateau-Rayleigh instability [20–22]. In Plateau-Rayleigh instability, a falling or flowing stream breaks up into smaller droplets with the same volume but less surface area as driven by the surface tension of the liquid. Subsequently, the confinement effects on the crystallization behavior as well as crystalline structure are introduced.

2.2 Nanodroplet Formation and Exclusive Homogenously Nucleated Crystallization in Confined Electrospun Immiscible Polymer Blend Fibers of Polystyrene and Poly(ethylene Oxide)

We chose the electrospun immiscible polymer blend fibers of polystyrene (PS) and poly(ethylene oxide) (PEO) as a model system for investigation of nanostructure development during thermal treatment, because the crystallization of PEO has been intensively studied [23] and PS has a higher T_g (~100°C) than melting point of PEO (~63°C) for good confinement of the PEO nanodroplets. Electrospinning of binary immiscible blend solutions using a single nozzle has been reported in the literature, including polyoxymethylene/thermoplastic polyurethane blends [2], polylactide/polyvinylpyrrolidone blends [24], poly(vinyl chloride)/polyurethane blend [25], and poly(methyl methacrylate)/polyacrylonitrile blends [26]. The solution viscosity, mainly depending on concentration of polymer solution (assuming the solvents are the same), is a key factor for the formation of bead-free and uniform electrospun fibers. To obtain well-defined electrospun fibers of PS/PEO bends, the mixed solution for electrospinning was obtained by dissolving polymers (15 wt%) of PS and PEO in a mixture solvent of tetrahydrofuran (THF)/*N,N*-dimethyl formamide (DMF). A uniform milk-like suspension was obtained by strong agitation at 80°C. Fibers of the PS/PEO (70/30 wt/wt) blend were prepared via electrospinning under optimum conditions. From the scanning electron microscopy (SEM) image (Fig. 2.1a), the as-spun PS/PEO blend fibers appeared uniform and bead-free, and the diameter ranged from 0.4 to 1.0 μm. The electrospun PS/PEO blend fibers were

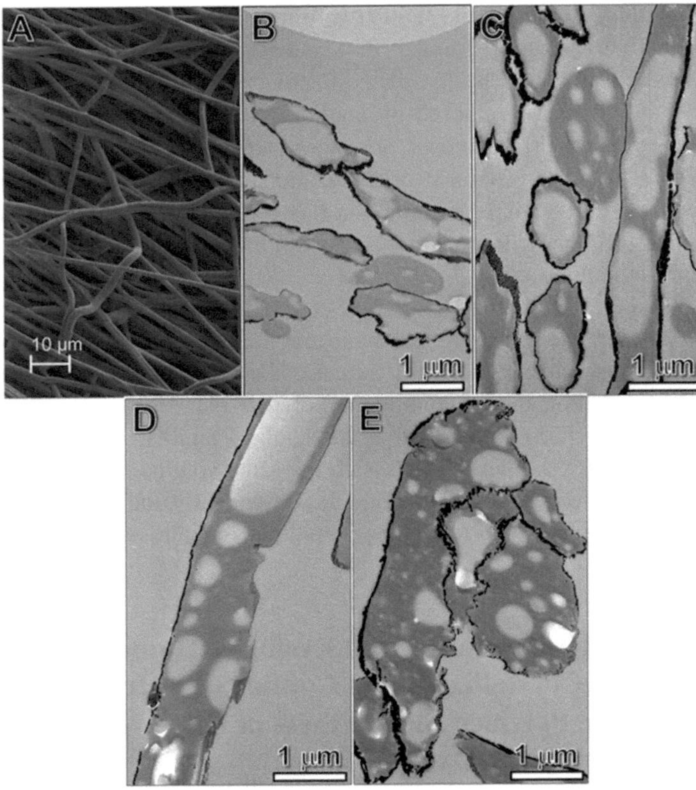

Fig. 2.1 (**a**) SEM and (**b**) TEM (microtomed thin section) micrograph of as-spun PS/PEO blend fibers. (**c–e**) TEM micrographs of thin sections of P4tBS-coated PS/PEO blend fibers annealed at different temperatures for 15 min, (**c**) 85°C, (**d**) 105°C, and (**e**) 150°C, respectively. The fibers were sputter-coated with gold before being embedded into standard epoxy

further coated with a high T_g polymer, poly(4-*t*-butylstyrene) (P4tBS) (T_g ~143°C), in order to study the confinement effect on the evolution of phase morphology during thermal treatment in details. The as-spun PS/PEO fibers were dipped into a 3 wt% P4tBS/hexane solution followed by being dried at ambient temperature.

2.2.1 Nanodroplet Formation During Thermal Annealing of Electrospun PS/PEO Blend Fibers

The evolution of phase morphology upon thermal annealing at different temperatures for the P4tBS-coated PS/PEO blend fibers is shown in Fig. 2.1. The phase morphology can be revealed clearly through the transmission electron microscopy (TEM) image on a microtomed thin section (Fig. 2.1b–e), since PEO was dissolved into water during microtoming, and floating the thin sections onto the water surface, the morphology of PEO phase appeared bright in TEM. As shown in Fig. 2.1b, PEO domains (or fibers) were encapsulated in the PS matrix. When the P4tBS-coated PS/PEO blend fibers were annealed below the T_g of the PS matrix (e.g., at 85°C for 15 min), there should not be any changes in the shape and size of the PEO domains confined within the PS matrix as compared with the as-spun blend fibers, although the PEO crystals had melted. From Fig. 2.1c, it was evident that the elongated PEO domains were mostly confined in the PS matrix, forming more or less core-shell morphology.

After annealing the P4tBS-coated PS/PEO blend fibers at 105°C, which was above the T_g of the PS matrix but below the T_g of P4tBS, the overall shape of the PS/PEO blend fibers was retained due to the P4tBS confinement. However, a portion of the continuous PEO domain in the PS fibers broke up into smaller droplets (ca. 150–300 nm, see the middle broken part of the PEO fiber in Fig. 2.1d) due to the Plateau-Rayleigh instability [20, 22]. When the P4tBS-coated PS/PEO blend fibers were annealed slightly above the T_g of the P4tBS coating (e.g., 150°C), as shown in Fig. 2.1e, they started to break into short pieces. Intriguingly, large PEO domains further broken into smaller domains (with sizes from slightly less than 50 nm to ca. 300 nm) inside the PS matrix.

2.2.2 *The Crystallization Behavior of Nanodroplets in Thermally Annealed Electrospun PS/PEO Blend Fibers*

After annealing the P4tBS-confined PS/PEO blend fibers at different temperatures, the nonisothermal crystallization results are shown in Fig. 2.2. When the annealing temperature was below the T_g of the PS matrix (e.g., at 85 and 95°C), a major crystallization peak was observed at ca. 40°C, and a small broad crystallization peak was observed at about −20°C. On the basis of previous studies, homogeneous nucleation of poly(ethylene oxide) (PEO) was observed at ca. −23°C in miniemulsions, which is much lower than that of heterogeneous nucleation in the bulk PEO (ca. 40°C) [27, 28]. Therefore, we assumed that the high- and low-temperature crystallization processes were induced by (unconfined) heterogeneous and homogeneous nucleation, respectively. When the annealing temperature increased to slightly above the T_g of PS (e.g., at 105°C), the heterogeneously nucleated crystallization peak at 40°C decreased its intensity, whereas the intensity of the homogeneously nucleated crystallization peak at −20°C significantly increased. With further increasing the annealing temperature, the crystallization peak at 40°C continuously decreased its intensity while the crystallization peak at −20°C gradually increased. Finally, after annealing at 150°C, which is slightly above the T_g of the confining P4tBS (∼143°C), the heterogeneously nucleated crystallization at 40°C completely disappeared, and the PEO crystallization became exclusively homogeneously nucleated.

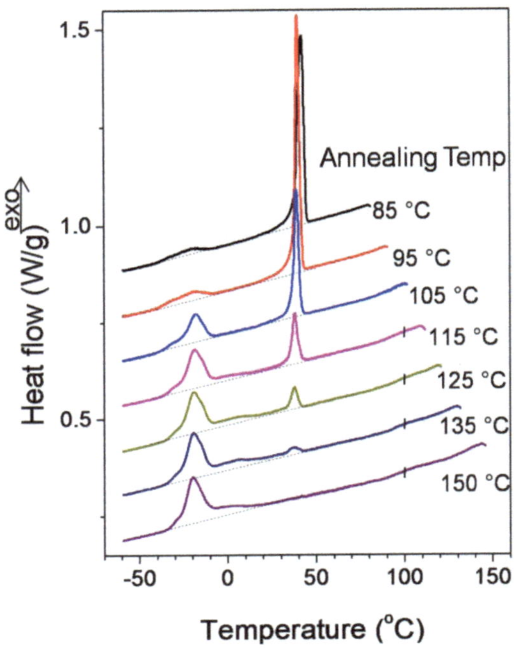

Fig. 2.2 DSC cooling curves for nonisothermal crystallization of P4tBS-coated electrospun PS/PEO blend fibers annealed at different temperatures. The cooling rate was −5°C/min

By combining TEM and differential scanning calorimetry (DSC) results, we can conclude that the homogeneous nucleation is directly related to the formation of PEO nanodroplets.

To confirm the above assumption for heterogeneous and homogeneous nucleation, we studied the crystallization kinetics for the P4tBS-coated PS/PEO blend fibers annealed at 85 and 150°C, respectively. For the study of crystallization kinetics by heterogeneous nucleation, samples were melted at 85°C for 15 min before fast cooling at −80°C/min to a preset crystallization temperature (T_c). The crystallization kinetics could be directly extracted from the exothermic peak as a function of the crystallization time because the crystallization rate was fast enough. For the study of crystallization kinetics by homogeneous nucleation, the crystallization was too slow to show a well-defined exothermic peak in DSC. Instead, the sample was melted at 150°C for 15 min and then quickly cooled at −80°C/min to a preset T_c for isothermal crystallization. The crystallinity (x_t) at different crystallization times was obtained from the melting enthalpy during a subsequent heating process to 100°C at a heating rate of 10°C/min. The final crystallinity (x_∞) was obtained by isothermally annealing the sample at the T_c for a long time. For comparison, the crystallization kinetics of pure PEO at T_c of 53°C was also studied (see Fig. 2.3), and the Avrami index (n) was 3.0, revealing the unconfined 2D crystal growth after heterogeneous nucleation. For the P4tBS-coated PS/PEO blend fibers annealed at 85°C, the Avrami index for isothermal crystallization at T_c of 48°C was 2.1, smaller than that (3.0) for the unconfined PEO isothermal crystallization. The decreased Avrami index could be attributed to

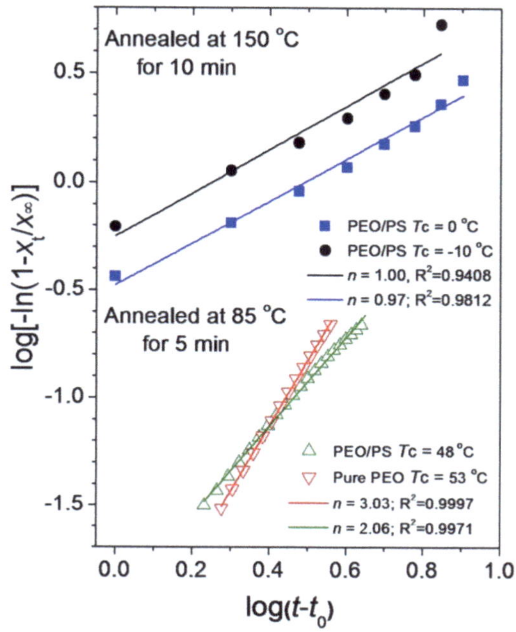

Fig. 2.3 Avrami plots for the isothermal crystallization of P4tBS-coated PS/PEO blend fibers annealed at different temperatures. Result for the pure PEO isothermal crystallization is also included for comparison

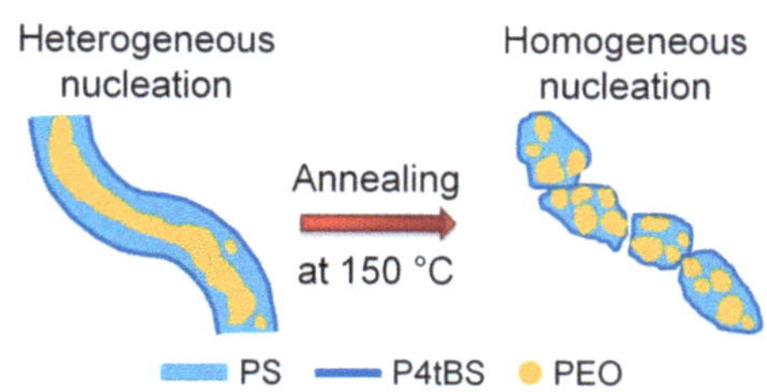

Fig. 2.4 Schematic representation showing the nanosized PEO droplet formation in P4tBS-coated electrospun PS/PEO blend fibers upon annealing at high temperatures (e.g., 150°C for 15 min)

the confinement effect because the annealing temperature of 85°C was below the T_g of the PS matrix (100°C). The Avrami index of 2.1 indicated the 1D crystal growth confined in the electrospun blend fibers after heterogeneous nucleation. When the P4tBS-coated PS/PEO blend fibers were annealed at 150°C for 15 min, the Avrami index for isothermal crystallization at T_c of -10 and 0°C decreased to 1.0, confirming the homogeneous nucleation mechanism [27, 29].

On the basis of the above TEM and DSC results, the nanostructure evolution in P4tBS-coated electrospun PS/PEO blend fibers and the subsequent crystallization are proposed, as schematically shown in Fig. 2.4. Submicron-sized fibers of immiscible PS/PEO blends are obtained via electrospinning with the minor component (PEO) forming elongated domains along the fiber axis. Through annealing at temperatures above the $T_g s$ of the PS matrix and the P4tBS coating, nanosized PEO droplets are formed as a result of the Rayleigh breakup of the elongated PEO domains. When there was no heterogeneity in the PEO droplets, homogeneous nucleation will occur at a large supercooling. Here, we consider that the viscoelasticity contrast between molten PEO and the PS/P4tBS matrix plays an important role for the Rayleigh breakup of the elongated PEO domains. For example, when the PS/PEO blend fibers are coated with P4tBS, the high viscosity of the P4tBS coating can effectively promote the breakup of the elongated PEO domains inside the PS matrix and prevented the coalescence of the small PEO droplets. As a result, small PEO domains (50–300 nm) can exist for a long enough time, and thus only homogeneously nucleated crystallization is observed at −20°C upon cooling from the melt.

It is envisioned this method can be generalized to other immiscible polymer blends in order to achieve nanodroplet. If the diameter of electrospun blend fibers can be further decreased to 100–200 nm, even smaller polymer nanodroplets can be obtained. It is noteworthy that small polymer droplets can also be obtained by adding a compatibilizer into the blend system [1]. However, the advantage of the reported method is the absence of a large amount of compatibilizers and the potential to achieve smaller droplet sizes.

2.3 Evolution of Nanodroplets and Crystallization in Thermally Annealed Electrospun Blend Fibers of Poly(vinylidene Fluoride) and Polysulfone

2.3.1 Blend Morphology of As-Spun PVDF/PSF Blend Fibers

For the Poly(vinylidene fluoride) (PVDF)/polysulfone (PSF) blend, we found that bead-free and uniform fibers could be obtained when the blend concentration was as high as 29 wt% in the mixture solvent of DMAc/THF 80/20 (wt/wt) [30]. Figure 2.5 shows electrospun blend fibers with various PVDF/PSF weight ratios of 5/95, 10/90, 20/80, and 30/70. The insets show the corresponding diameter distributions for these blend fibers. It was found that the diameter distributions did not change much with the weight fraction of PVDF, and the diameters of most fibers were in the range of 1.0–1.5 μm. In contrary to conventional polymer blends, whose phase morphology is either island and sea or co-continuous depending on the composition, the phase morphology in electrospun fibers of immiscible polymer blends is either co-continuous [24, 31] or sometimes core shell [2, 3, 26], nearly regardless of the blend composition. This is because of the tremendous elongation force during electrospinning and the small diameter of the resulting fibers. From our recent study [30], the electrospun PVDF/PSF blend fibers in Fig. 2.5 contained a co-continuous fibrillar morphology, despite that macrophase separation (the domain size is about tens of microns) occurred in the blend solution, from which the fibers were electrospun. Thermodynamically, the co-continuous fibrillar morphology in electrospun blend fibers should be far away from equilibrium because of its fast formation kinetics.

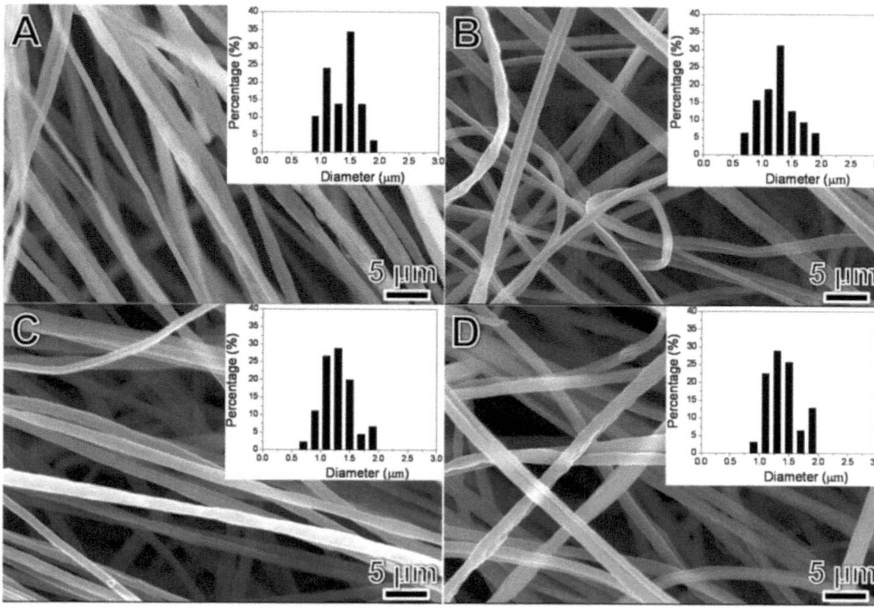

Fig. 2.5 SEM micrographs of electrospun blend fibers with the PVDF/PSF weight ratio of (**a**) 5/95, (**b**) 10/90, (**c**) 20/80, and (**d**) 30/70. Corresponding fiber diameter distributions are shown in the *insets*

2.3.2 Morphology Evolution of Electrospun PVDF/PSF Blend Fibers After Thermal Treatment

In this study, we investigated how thermal annealing above the melting temperature of PVDF would affect the blend morphology and subsequent crystallization of PVDF. First, the as-spun blend fibers were annealed at 180°C for 5 min, followed by a fast cooling (at 10°C/min) at room temperature. This annealing temperature is just below the T_g of PSF (185°C). Chloroform, a good solvent for PSF but a poor solvent for PVDF, was used to extract the PSF matrix to reveal the morphology of the PVDF phase. As shown in Fig. 2.6, despite different PVDF/PSF weight ratios, the leftover PVDF remained as ribbon-like fibers, similar to that in the as-spun samples etched by chloroform [30]. This result indicated that annealing slightly below the T_g of PSF for a short period of time (5 min) could not change the co-continuous morphology formed by electrospinning. In Fig. 2.6, the fibrillar PVDF appeared coarsened with rough surfaces when the PVDF content increased.

When the PVDF/PSF blend fibers were annealed at 220°C, the glassy PSF matrix became liquid, and the molten PVDF fibers in the PSF matrix started to break up due to the Plateau-Rayleigh instability [22, 32], leading to the formation of spherical nanosized PVDF domains, as shown in Fig. 2.7. Because the electrospun fiber diameter was about 1–1.5 μm, most droplets of PVDF were fairly small.

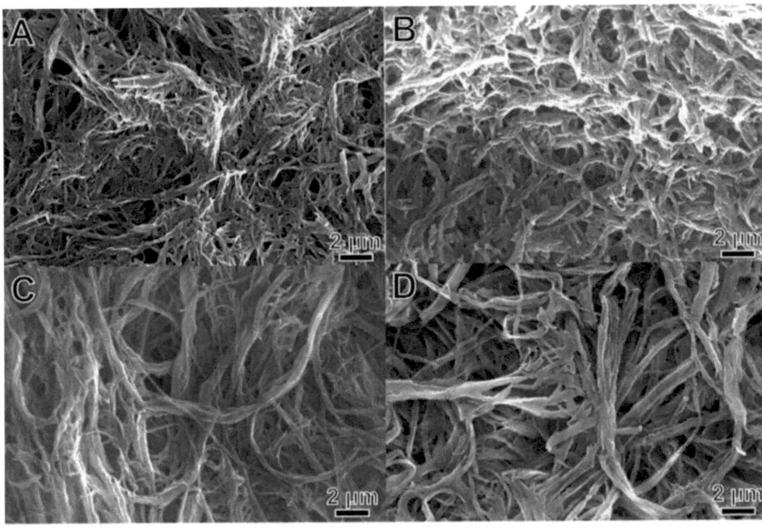

Fig. 2.6 SEM micrographs showing the representative morphology after electrospun PVDF/PSF blend fibers were annealed at 180°C for 5 min followed by etching with chloroform: (**a**) PVDF/PSF 5/95, (**b**) PVDF/PSF 10/90, (**c**) PVDF/PSF 20/80, and (**d**) PVDF/PSF 30/70

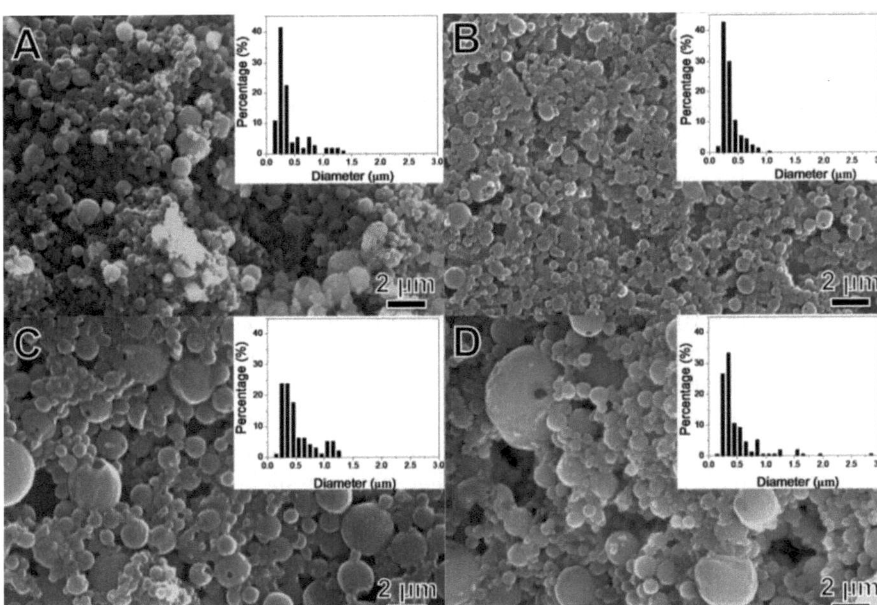

Fig. 2.7 SEM micrographs showing the morphology of the PVDF phase after electrospun PVDF/PSF blend fibers were annealed at 220°C for 5 min followed by etching with chloroform: (**a**) PVDF/PSF 5/95, (**b**) PVDF/PSF 10/90, (**c**) PVDF/PSF 20/80, and (**d**) PVDF/PSF 30/70. Corresponding droplet diameter distributions are shown in the *insets*

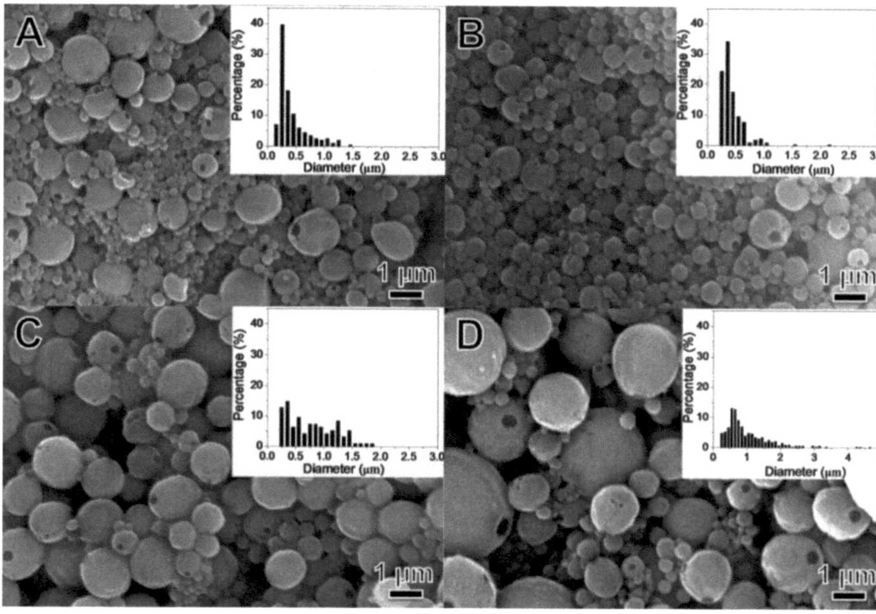

Fig. 2.8 SEM micrographs showing the morphology of the PVDF phase after electrospun PVDF/PSF blend fibers were annealed at 240°C for 5 min followed by etching with chloroform: (**a**) PVDF/PSF 5/95, (**b**) PVDF/PSF 10/90, (**c**) PVDF/PSF 20/80, and (**d**) PVDF/PSF 30/70. Corresponding droplet diameter distributions are shown in the *insets*

Occasionally, a hole was observed in a PVDF sphere, which might be attributed to a nanosized PSF droplet entrapped in the PVDF phase. The diameter distributions of PVDF domains formed in the annealed PVDF/PSF fibers with different PVDF contents were shown in the insets of Fig. 2.7. For all samples, the most frequent diameters fell nearly in the same range, i.e., between 200 and 300 nm. However, the population of domains larger than 500 nm increased as the PVDF content increases. Note that for the PVDF/PSF 30/70 fiber after annealing at 220°C for 5 min, some domains reached a diameter greater than 2.0 μm, which was even larger than the maximum diameter of as-spun fibers. This result suggested that breakup was dominant when the PVDF content was less than 20 wt% in the blend fibers annealed at 220°C for 5 min, whereas coalescence of small PVDF domains happened when the PVDF content reached 30 wt%. Our results showed that a prolonged annealing time at 220°C (over a few hours) could also enhance coalescence and thus resulted in larger domain sizes.

The effect of annealing temperature on the PVDF domain size was further investigated by annealing the as-spun PVDF/PSF blend fibers at 240°C for 5 min. Figure 2.8 shows the corresponding PVDF phase morphology and diameter distributions for blend fibers annealed at 240°C followed by etching with chloroform. When the PVDF contents were 5 and 10 wt%, the most frequent diameter still remained at 200–300 nm, indicating that breakup was dominant at 240°C because of the low

PVDF content in the blend fibers. However, when the PVDF contents were 20 and 30 wt%, the most frequent diameter obviously increased, indicating that significant coalescence of small PVDF domains took place due to the high PVDF content in the blend fibers. Annealing at even higher temperatures led to even more severe coalescence of PVDF domains.

2.3.3 Fractionated Crystallization in PVDF Nanodroplets

On the basis of the above studies, we have learned that the competition between the breakup of PVDF nanofibers into submicron-sized droplets and subsequent coalescence of these nanodroplets inside the PSF matrix (without any compatibilizers) determined the final size and size distribution of spherical PVDF domains. To avoid coalescence of small PVDF droplets, the annealing temperature should not exceed 220°C and the annealing time should not be long (e.g., ~5 min). In addition, electrospun fibers of immiscible blends containing a crystalline polymer as the minor component can provide a unique way to fabricate a smaller domain size (~200–300 nm) than the conventional melt blending without any compatibilizers. Actually, these nanosized domains are ideal for the study of homogeneous nucleation and fractionated crystallization for PVDF. Many interesting scientific questions are still outstanding. For example, where is the homogenous nucleation temperature for PVDF? How does homogeneous nucleation or confined (fractionated) crystallization affect the polymorphism formation in PVDF and subsequent electrical properties?

The first attempt to study fractionated PVDF crystallization was achieved in finely dispersed PVDF/nylon 6 blends. At a PVDF-to-nylon 6 volume ratio of 15/85, the PVDF crystallization split into two steps at 140 and 116°C, respectively [33]. The fractionated crystallization behavior was attributed to the small domain sizes of PVDF in the range of 0.1–0.6 μm after an intense melt mixing. The crystallization at 140°C was attributed to the heterogeneously nucleated crystallization [34], whereas the crystallization at 116°C was speculated to be the homogeneously nucleated crystallization. Similar T_c depression (down to the 110–120°C range) was also observed in PVDF/nylon 6 blends with a fine dispersion of the PVDF domains (0.2–0.7 μm) [35]. In addition to the polymer blend method, PVDF and its copolymers were also entrapped in the nanopores of ordered anodic aluminum oxide (AAO) [14, 15, 36, 37]. For example, when the pore diameter of the AAO decreased to 15 nm, the nonisothermal T_c of a poly(VDF-*co*-trifluoroethylene) [P(VDF-TrFE)] 70/30 (mol.%) sample decreased from 130°C for the bulk sample to 91°C. Meanwhile, the Curie temperature slightly increased from 45°C for the bulk sample to 60°C upon cooling [15, 37].

Figure 2.8a shows the nonisothermal crystallization behaviors for electrospun PVDF/PSF blend fibers after annealing at 180°C for 5 min. From the above results, PVDF still largely remained as co-continuous fibers inside the PSF matrix (see Fig. 2.6). In addition to the major crystallization peak around 121–133°C, a broad

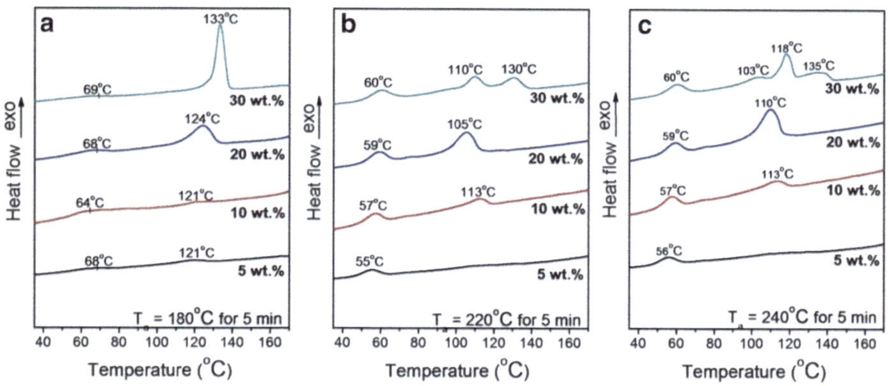

Fig. 2.9 DSC cooling curves for electrospun PVDF/PSF blend fibers after annealing at different temperatures: (**a**) 180 °C for 5 min, (**b**) 220°C for 5 min, and (**c**) 240°C for 5 min, respectively

and weak crystallization peak was observed around 64–69°C for all samples. The high-temperature crystallization could be attributed to the heterogeneously nucleated crystallization in fibrillar PVDF domains. Compared with the heterogeneous T_c (136°C) for the neat PVDF [30], the slightly decreased T_c (133°C) for the PVDF/PSF 30/70 blend fibers could be attributed to their small diameters confined in the PSF matrix. The further decrease in T_c (121–124°C) for PVDF/PSF blend fibers with the PVDF content less than 20 wt% could be attributed to even smaller diameters of the PVDF fibers confined in the PSF matrix. The low-temperature crystallization could be attributed to the homogeneously nucleated crystallization in small isolated PVDF domains. Observation of this low-temperature crystallization peak in both as-spun blend fibers and blend fibers annealed at 180°C indicated that a small fraction of isolated PVDF domains must coexist with the fibrillar domains in all electrospun fibers regardless of different PVDF contents. However, these small isolated PVDF domains might have been washed away by chloroform and were not observed in Fig. 2.6.

From the results in Fig. 2.9a, we for the first time observed that the homogeneous nucleation temperature for PVDF could be around 60°C. However, this crystallization peak was too weak for studying the crystallization kinetics to confirm the mechanism of homogeneous nucleation crystallization. After annealing the as-spun PVDF/PSF blend fibers above the T_g of PSF (i.e., at 220°C for 5 min), the PVDF fibers broke up into submicron-sized droplets and the corresponding crystallization behaviors are shown in Fig. 2.9b. For PVDF/PSF 5/95, the heterogeneously nucleated crystallization peak at 121°C became very weak and a clear crystallization peak took place at 55°C. For the annealed PVDF/PSF 10/90 and 20/80 blend fibers, two crystallization peaks were observed (i.e., fractionated crystallization) with a high-temperature peak at 105–113°C and a low-temperature peak at 57–59°C. For the PVDF/PSF 30/70 sample, three crystallization peaks were observed with two high-temperature peaks at 130 and 110°C and a low-temperature peak at 60°C. Further annealing the blend fibers at 240°C for 5 min showed similar low crystallization peaks around 56–60°C (see Fig. 2.8c). However, the high-temperature crystallization peak shifted to a higher temperature (110°C) for the PVDF/PSF 20/80 sample. For the PVDF/PSF 30/70 sample, three crystallization peaks were observed at 103, 118, and 135°C at high temperatures. These results indirectly suggested that annealing at 240°C caused significant coalescence of the broken PVDF nanodroplets. From the above nonisothermal crystallization studies for the annealed PVDF/PSF blend fibers, we attribute the low-temperature crystallization peak between 55 and 60°C to the homogeneously nucleated crystallization of PVDF. The temperatures of homogeneous nucleation for other semicrystalline polymers have also been reported for PEO at ca. −20°C [17, 18, 38, 39], isotactic polypropylene at ca. 45°C [40, 41], high-density polyethylene at ca. 74°C [42], and nylon 6 at ca. 90°C [43–46]. Generally speaking, the homogeneous nucleation temperature is about 60–100°C lower than the heterogeneous nucleation temperature for the bulk polymer. Therefore, it is reasonable to attribute the crystallization peak at 55–60°C to the homogeneously nucleated PVDF crystallization.

To determine the homogeneous nucleation mechanism of PVDF, PVDF crystallization kinetics in electrospun PVDF/PSF blend fibers was studied via isothermal crystallization at preset temperatures. The PVDF/PSF 20/80 blend fibers were chosen for the kinetics study because (1) it had two well-defined crystallization peaks upon cooling at −10°C/min (see Fig. 2.9b) and (2) the heat of transition for each crystallization peak was large enough to minimize experimental errors. Using DSC, the Avrami analysis was adopted to investigate both homogeneously and heterogeneously nucleated crystallization processes. Because the heterogeneously nucleated crystallization was relatively fast and the heat of transition was large, the crystallization kinetics could be recorded using the conventional isothermal crystallization procedure. For PVDF/PSF 20/80 blend fibers annealed at 180°C for 5 min, the preset isothermal T_c was chosen at 142–146°C. For the isothermal crystallization kinetics study on the homogeneously nucleated crystallization of the PVDF/PSF 20/80 blend fibers annealed at 220°C for 5 min, a different procedure was used because the crystallization processes were slow and the heat of transition was not high enough. First, the sample was melted at 180°C for 5 min and then fast cooled at −80°C/min to a preset temperature for isothermal crystallization. The preset temperature was chosen to be slightly (5–10°C) higher than the nonisothermal crystallization peak temperatures (59 and 105°C, respectively). Second, the x_t was obtained by the melting enthalpy during a subsequent heating process to 180°C at 10°C/min at different crystallization times. Finally, the x_∞ (5.25 J/g) was obtained by isothermally annealing the sample at the T_c for a long enough time (12 h).

For the isothermal crystallization of the PVDF/PSF 20/80 blend fiber annealed at 180°C, the Avrami index, n, was around 2.7–3.3 at $T_c = 142$–146°C (see Fig. 2.10a), indicating two-dimensional (2D) growth of crystallites due to the ribbon- or fiber-like PVDF morphology when the thermal nucleation mechanism was assumed. For the high-temperature (around 115–118°C) isothermal crystallization of the PVDF/PSF 20/80 blend fibers annealed at 220°C, the Avrami index n decreased to ca. 2.1 (see Fig. 2.10b) because of the confined crystallization in large

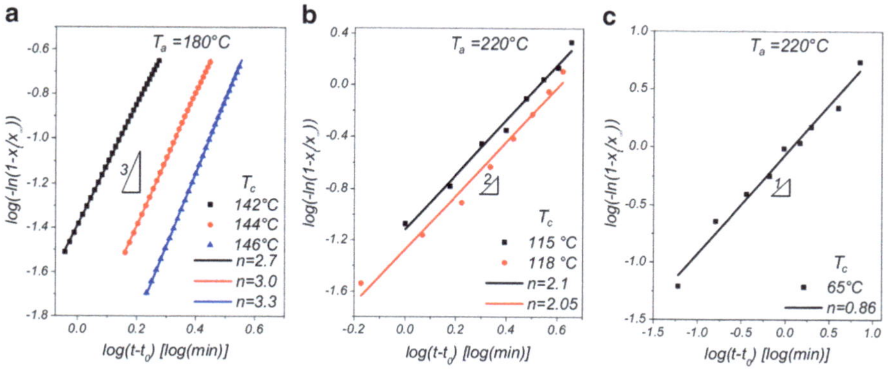

Fig. 2.10 (**a**) Avrami plots of isothermal crystallization at 142, 144, and 146°C for electrospun PVDF/PSF 20/80 fibers annealed at 180°C for 5 min. (**b**) Avrami plots of isothermal crystallization at 115 and 118°C for electrospun PVDF/PSF 20/80 fibers annealed at 220°C for 5 min. (**c**) Avrami plots of isothermal crystallization at 65 and 70°C for electrospun PVDF/PSF 20/80 fibers annealed at 220°C for 5 min

spherical domains. These high Avrami index values (~2.0) suggested that the high-temperature crystallization should be heterogeneously nucleated. However, when the same sample was isothermally crystallized at 65°C (close to the low-temperature crystallization peak in Fig. 2.9b), the Avrami index became 0.86 (see Fig. 2.10c). These low Avrami index values around 1.0 suggested that the low-temperature crystallization should be homogeneously nucleated, similar to the results observed for PEO homogeneous nucleation [29, 47]. From this crystallization kinetics study, we concluded that the high-temperature crystallization was heterogeneously nucleated in large PVDF droplets, while the low-temperature crystallization was homogeneously nucleated in PVDF nanodroplets.

2.3.4 Understanding Polymorphism Formation in Electrospun Fibers of Immiscible Poly(vinylidene Fluoride) Blends

It is well known that PVDF has polymorphic crystalline phases, namely, α, β, γ, and δ [48]. It would be interesting to understand the polymorphism formation induced by the homogeneous nucleation, which usually has a fast crystallization rate. Figure 2.11 shows the wide angle X-ray diffraction (WAXD) profiles of various PVDF/PSF blend fibers annealed at different temperatures (i.e., 180, 220, and 240°C) after etching with chloroform to remove the PSF matrix. Only α-crystal reflections, $(100)_\alpha$, $(020)_\alpha$, $(110)_\alpha$, and mixed $(021/111/120)_\alpha$, were seen at 12.65, 13.04, 14.25, and 18.9 nm^{-1}, respectively, for all samples. No obvious β-crystal reflections, $(110/200)_\beta$, were seen at 14.7 nm^{-1}. Therefore, homogeneously nucleated PVDF crystallization confined in nanosized droplets had no significant effect

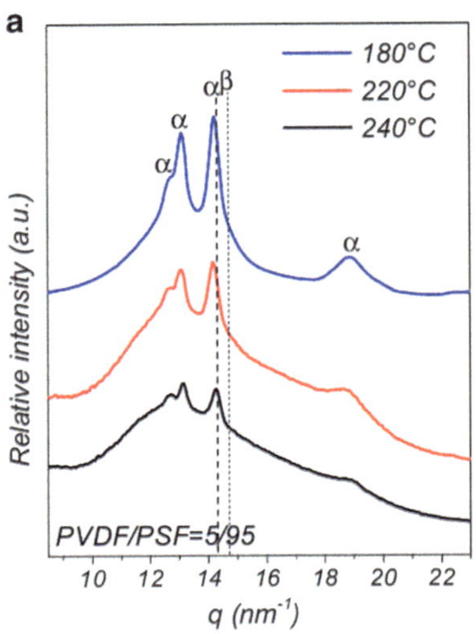

Fig. 2.11 WAXD profiles of electrospun PVDF/PSF blend fibers annealed at different temperatures (180, 220, 240°C) after etching with chloroform to remove the PSF matrix. (**a**) PVDF/PSF 5/95, (**b**) PVDF/PSF 20/80

on the formation of ferroelectric β or γ phases. However, it was reported that nearly a pure β phase was observed in electrospun PVDF composite fibers with nanoclays [49], magnetic nanoparticles [50], or even carbon nanotubes [51]. These phenomena really draw our attention to investigate polymorphism formation in electrospun fibers of immiscible PVDF blends.

To understand the mechanism of polymorphism formation in electrospun PVDF blend fibers, a comparative study in electrospun fibers of PVDF blends with nonpolar and polar matrices was carried out. The nonpolar polymer matrix was polysulfone (PSF) and the polar matrix was polyacrylonitrile (PAN). PAN has polar cyano side groups, which have a large dipole moment of ca. 13.4×10^{-30} C m or 4.0 Debye [52, 53]. Instead, PSF is a nonpolar polymer with a nearly net zero dipole moment perpendicular to the chains. The polarity of the matrix polymer had significant influence on the crystalline morphology of PVDF in its as-electrospun blend fibers, as shown in Fig. 2.12. For PVDF/PAN blend fibers, only the characteristic β phase reflection, $(110/200)_{\beta}$, was observed at 14.7 nm^{-1} in the WAXD profiles in Fig. 2.12A. Note that the $(100)_{PAN}$ reflection for the hexagonal PAN crystalline phase [54, 55] was observed at 11.9 nm^{-1}. The absence of the α phase in as-electrospun PVDF/PAN blend fibers was further confirmed by Fourier transform infrared spectroscopy (FTIR) results in Fig. 2.12B because only β absorption bands at 840 and 1,273 cm^{-1} and the γ absorption band at 1,233 cm^{-1} were observed. For as-electrospun PVDF/PSF blend fibers, as shown in the WAXD profiles (Fig. 2.12A, curve e and f), distinctive α-form reflections were observed. The existence of the β phase in as-electrospun PVDF/PSF fibers was evidenced by the absorption bands at 840 and 1,273 cm^{-1} in the FTIR spectra in Fig. 2.12B.

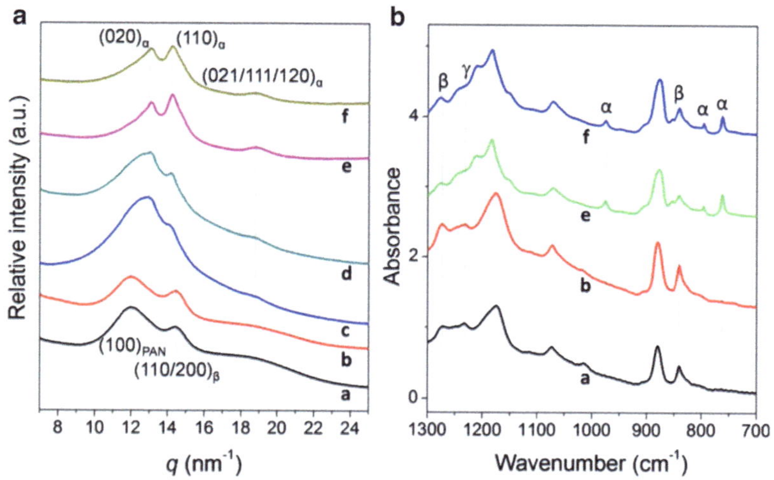

Fig. 2.12 (**A**) WAXD profiles and (**B**) FTIR spectra for electrospun fibers of PVDF/PAN blends and PVDF/PSF blends from 80/20 (wt/wt) DMAc/THF; (a) PVDF/PAN 20/80, (b) PVDF/PAN 30/70, (c) PVDF/PSF 20/80, (d) PVDF/PSF 30/70, (e) PVDF/PSF 20/80 washed by $CHCl_3$, and (f) PVDF/PSF 30/70 washed by $CHCl_3$

Using Eq. (2.1), $F(\beta)$ values of as-electrospun PVDF/PSF 20/80 and 30/70 fibers were calculated to be ca. 0.43 and 0.49, respectively, significantly lower than those ($\sim$1) of electrospun PVDF/PAN fibers:

$$F(\beta) = \frac{A_\beta}{1.3A_\alpha + A_\beta}, \tag{2.1}$$

where A_α and A_β are the absorbance at 765 and 840 cm^{-1}, respectively.

Crystalline polymorphism in the melt-recrystallized samples was studied by in situ temperature-dependent synchrotron WAXD studies during the nonisothermal crystallization, as shown in Fig. 2.13. For the melt-recrystallized PVDF/PSF blend fibers, only nonpolar α phase was observed at the initial stage (the green curves at 140°C) of crystallization (Fig. 2.13a,b). With continuously decreasing the temperature, the $(110)_\alpha$ reflection gradually increased its intensity (see insets in Fig. 2.13a,b), indicating the growth of the nonpolar α phase in the samples. For melt-recrystallized PVDF/PAN blend fibers, only the α phase [see the $(110)_\alpha$ reflection in the insets of Fig. 2.13c,d] was observed in the beginning of crystallization (the green curves at 130°C). Note that the sharpening of the $(100)_{PAN}$ reflection at 11.9 nm^{-1} was a direct result of thermal annealing at 180°C, which either enhanced the crystallinity or made the as-electrospun PAN crystals more perfect. With decreasing the temperature, ferroelectric (β and/or γ) phases started to appear. It was interesting to note that the temperatures, where the ferroelectric phases started to appear, were different for PVDF/PAN 20/80 and 30/70 blends. For PVDF/PAN 20/80, the ferroelectric phases appeared at 125°C (the blue curve in Fig. 2.13c), whereas they appeared at 120°C for PVDF/PAN 30/70 (the magenta

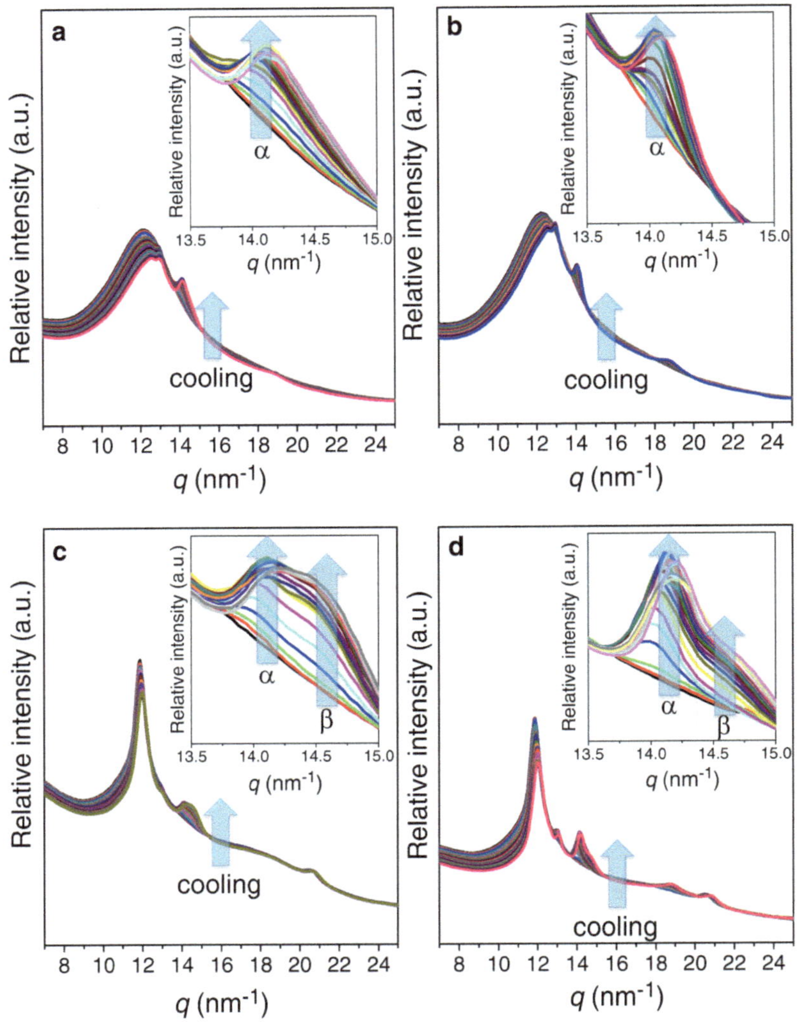

Fig. 2.13 Temperature-dependent WAXD profiles during the nonisothermal crystallization after melting at 180°C for 5 min for electrospun fibers of PVDF/PSF (**a**) 20/80 and (**b**) 30/70 (wt/wt) and PVDF/PAN (**c**) 20/80 and (**d**) 30/70 (wt/wt) blends. Insets show a magnified q range between 13.5 and 15 nm^{-1}. The cooling rate is 5°C/min. The start temperatures (the *first black curve*) for PVDF/PSF and PVDF/PAN are 150 and 140°C, respectively, and each curve is 5°C apart

curve in Fig. 2.13d). This phenomenon could be explained by the fact that during a moderate cooling from the melt, the α phase was kinetically favored and thus had a faster nucleation/growth rate than the β phase [56]. Meanwhile, the content of the ferroelectric phases was higher in PVDF/PAN 20/80 than in PVDF/PAN 30/70 (compare Fig. 2.13c,d), suggesting that the higher surface-to-volume ratio in PVDF/PAN 20/80 than that in PVDF/PAN 30/70 favored the nucleation of ferroelectric phases in PVDF near the PAN/PVDF interfaces.

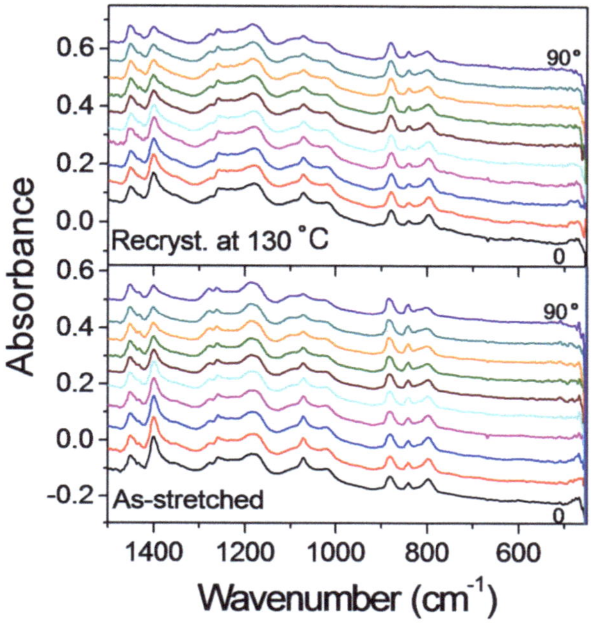

Fig. 2.14 Polarized FTIR spectra between 450 and 1,500 cm^{-1} for as-stretched and 130°C recrystallized PVDF/PAN 20/80 blend fibers at different polarization angles

Crystal orientation and polymorphism formation in electrospun PVDF/PAN blend fibers during melt-recrystallization can be further revealed by polarized FTIR, which can give more information on the interface-induced polymorphism mentioned above. In order to obtain aligned fibers for investigation of crystal orientation, a bundle of nonwoven PVDF/PAN 20/80 blend fibers was mechanically stretched at 110°C by fixing its one end and then hanging a constant weight (1–5 g depending on the amount of fibers) at the other end in a Blue M G01300A gravity oven. Figure 2.14 shows the polarized FTIR spectra for the as-stretched and 130°C recrystallized PVDF/PAN 20/80 blend fibers between 450 and 1,500 cm^{-1} when the polarization changed from 0° (parallel to the drawing direction) to 90° (perpendicular to the drawing direction). In this region, the absorption bands at 840 and 1,278 cm^{-1} were attributed to pure β-PVDF crystals, and the absorption bands at 1,070 and 1,400 cm^{-1} were attributed to both α- and β-PVDF crystals. With increasing the polarization angle, the absorbance in crystalline sensitive bands varied and detailed analyses are shown in Fig. 2.15. For the as-stretched PVDF/PAN blend fibers, the absorption in bands at 840 and 1,278 cm^{-1} increased with increasing polarization angle from 0° to 90°, whereas the absorption in bands at 1,070 and 1,400 cm^{-1} decreased with increasing polarization angle from 0° to 90°. The band at 840 cm^{-1} was assigned, in terms of potential energy distribution (PED), to $r(CH_2)(60)$—$\nu_a(CF_2)$ (31), and the band at 1,278 cm^{-1} was assigned to $\nu_s(CF_2)(40)$—$\nu_s(CC)(22) + \delta$ (CCC)(15) [57]. Here, ν_a stands for antisymmetric stretching, ν_s symmetric stretching, r rocking, and δ scissoring. The numbers in parenthesis represent the PED obtained by the normal coordinate treatment for a

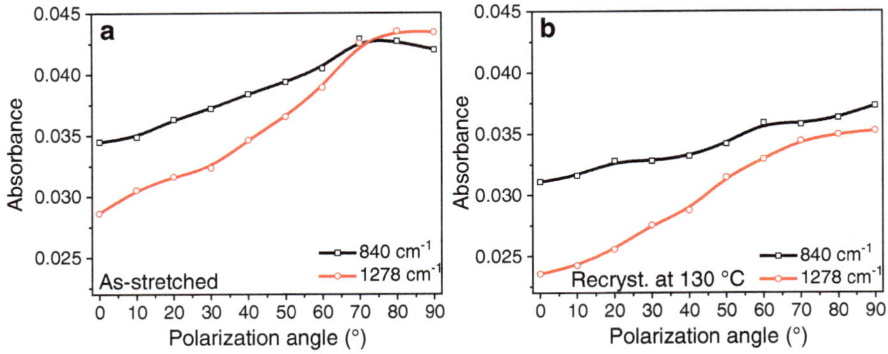

Fig. 2.15 Absorbance at 840 and 1,278 cm^{-1} as a function of polarization angle for (**a**) as-stretched and (**b**) 130°C recrystallized PVDF/PAN 20/80 blend fibers

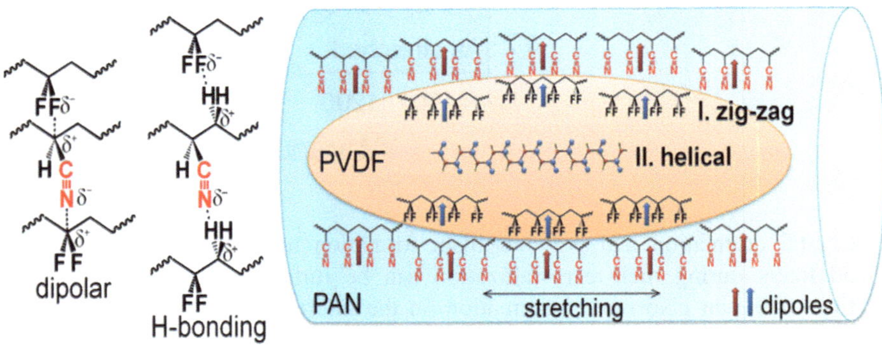

Fig. 2.16 Schematic representation of the polymorphism formation in melt-recrystallized PVDF/PAN blend fibers. The *left panel* shows possible dipolar and hydrogen bonding interactions between PAN and PVDF

single chain. Because the major contributions for the 840 cm^{-1} band [i.e., $r(CH_2)$] and the 1,278 cm^{-1} band [i.e., $\nu_s(CF_2)$] were perpendicular to the PVDF chains, polarized FTIR results indicate that the degree of orientation of ferroelectric β-PVDF crystals along the fiber axes was higher than that of nonpolar α-crystals. Therefore, it was concluded that the β-PVDF crystals should be formed at the PVDF/PAN interfaces (see the zigzag chain conformation at position I in Fig. 2.16) because of strong dipolar and hydrogen bonding interactions between VDF and AN units. The α-PVDF crystals should form in the center of the PVDF domain (or far away from the PVDF/PAN interface, see the helical chain conformation at position II in Fig. 2.15).

Finally, from the above results as well as studies in literature, we propose a possible universal mechanism for the formation of ferroelectric phases in the PVDF homopolymer, whose chain conformation is fairly adaptable to different crystalline forms. Note that PVDF random copolymers, including P(VDF-*co*-TrFE), P(VDF-*co*-tetrafluoroethylene), and P(VDF-*co*-vinylfluoride), should be excluded because their chain conformation favored the ferroelectric β phase [48]. First, nucleation of ferroelectric phases should be strongly influenced by the interactions between the local electric field and PVDF dipoles. Note that we refer the local electric field to the field among atoms and molecules, not the macroscopic electric field in the sample. These local field–dipole interactions can be in different forms, including ion–dipole, dipole–dipole, hydrogen bonding, and external electric field–dipole interactions. Generally speaking, the stronger the interactions between the local electric field and PVDF dipoles, the easier the nucleation of the ferroelectric phases. Second, the growth rate of ferroelectric phases after nucleation with respect to the rates of nucleation and growth of the nonpolar α phase determines the relative amount of ferroelectric phases in PVDF. If the nucleation of the α phase can be suppressed, a pure ferroelectric phase will be obtained. In the following, we will discuss the ferroelectric phase formation using this proposed mechanism in different situations:

1. *Ferroelectric Phase Formation from a Heterogeneous Phase.* This situation refers to PVDF crystallization in the presence of preexisting heterogeneous interfaces. These interfaces can be either nonpolar or polar. Nonpolar interfaces include glasses, metals, and (nonpolar) polymers. For example, transcrystallization of PVDF is observed from the surfaces of several nonpolar polymers, namely, poly(ethylene terephthalate), polytetrafluoroethylene, and polyimide; however, only the nonpolar α phase was observed [58]. This is consistent with our result for melt-recrystallized PVDF/PSF blend fibers. Polar interfaces can be either inorganic or polymeric. Inorganic polar interfaces include those from inorganic salts (e.g., (001) surfaces in KBr and NaCl [59–62]), charged nanoplatelets (e.g., mica and layered silicates [63–66]), or even electron-rich carbon nanotubes [67, 68]. The strong local electric field near these inorganic interfaces can effectively induce the nucleation of polar (ferroelectric) phases in PVDF. However, at locations far away from these polar interfaces, nonpolar α phase will be nucleated, leading to the coexistence of both nonpolar and ferroelectric phases in the composites. Therefore, the amount of induced ferroelectric phases depends on the content of these inorganic fillers or the surface-to-volume ratio in the system. For polar polymer interfaces such as in nylon 11 and PAN, ferroelectric phases will be nucleated with their *c*-axes parallel to the interface [69, 70]. Again, for locations away from the polar polymer interfaces, the α phase can be nucleated, leading to the coexistence of both nonpolar and ferroelectric phases in PVDF. Generally speaking, the lower the PVDF content (or the smaller PVDF domains) in these blends, the higher the surface-to-volume ratio and thus the higher the ferroelectric phase content. This is exactly observed in our melt-recrystallized PVDF/PAN blend fibers (see Fig. 2.12a,b). Intriguingly, only the α phase is observed when polar nylon 11 is replaced by nonpolar nylon 6 [35, 71, 72]. Therefore, hydrogen bonding may not be a necessary requirement for the formation of ferroelectric PVDF phases. Instead, we consider that the polar local crystalline phases in nylon 11 and PAN should be responsible for the nucleation of ferroelectric phases in PVDF.

2. *Ferroelectric Phase Formation from a Homogeneous Single Phase.* This situation includes PVDF crystallization from homogeneous solutions and miscible polymer blends. In most miscible blends with PVDF, hydrogen bonding between the carbonyl groups in the counterpart polymer and PVDF may induce local ordering in the single phase, as suggested by light-scattering and dielectric studies [73–75]. We speculate that the local ordering of polymer chains may effectively suppress the nucleation of the nonpolar phase and, meanwhile, induce the nucleation and growth of the ferroelectric phases in miscible PVDF blends. Currently, it is still not clear how the ferroelectric phases form in polar solvents, although we speculate that certain local ordering might also occur due to strong dipolar interactions between PVDF and the solvent molecules. More experiments should be dedicated to understand the underlying mechanism.
3. *Electrospinning Effect.* During electrospinning, both fast stretching and external electric field-dipole interaction will help the ferroelectric phase formation in PVDF. When polar inorganic or polymeric interfaces are present, the nucleation and growth of the nonpolar α phase will be completely suppressed, and pure ferroelectric phases will be obtained in PVDF blend fibers (e.g., PVDF/PAN). For nonpolar polymer interfaces (e.g., PVDF/PSF), ferroelectric phases will coexist with the α phase in electrospun PVDF blend fibers.

2.4 Conclusions

By taking advantage of the Plateau-Rayleigh instability in the electrospun co-continuous immiscible polymer blend fibers, nanosized droplets of 50–300 nm were successfully prepared via thermal annealing at an appropriate temperature above the T_g of the matrix for a short period of time without any help of compatibilizers. The competition between nanofiber breakup and subsequent droplet coalescence determined the final droplet size and size distribution. This technique is very simple and easy and, moreover, can be applicable to various polymeric materials. Our study shows that these nanodroplets can be used to study the fractionated crystallization and homogeneous nucleation of almost any semicrystalline polymers, such as PEO, PVDF, polyethylene, and polypropylene. By using this method, we observed the homogeneously nucleated crystallization of PVDF took place at 55–60°C for the first time. Additionally, this method can be utilized to investigate the effect of nanoconfinement on crystalline morphologies of semicrystalline polymers. Through this study, we obtained more insight into the polymorphism formation of PVDF through extensive investigation of crystalline structure formed in electrospun fibers of immiscible blends.

Acknowledgments The authors would acknowledge the Chinese Scholarship Council, startup fund of Sichuan University (Grant No. 2011SCU11072), and NSF (DMR-0907580) for financial support. The authors are also grateful to Mr. Linghui Meng and Professor Gary Wnek at Case

Western Reserve University for providing electrospinning instrument and technical assistance. The authors also acknowledge the assistance of synchrotron XRD experiments from Dr. Lixia Rong at Brookhaven National Laboratory and the work of Dr. Run Su.

References

1. Harrats, C., Thomas, S., Groeninckx, G.: Micro- and Nanostructured Multiphase Polymer Blend Systems: Phase Morphology and Interfaces. Taylor & Francis, New York (2006)
2. Peng, P., Chen, Y.Z., Gao, Y.F., Yu, J., Guo, Z.X.: Phase morphology and mechanical properties of the electrospun polyoxymethylene/polyurethane blend fiber mats. J. Polym. Sci. Part B: Polym. Phys. **47**, 1853–1859 (2009)
3. Wei, M., Lee, J., Kang, B., Mead, J.: Preparation of core-sheath nanofibers from conducting polymer blends. Macromol. Rapid Commun. **26**, 1127–1132 (2005)
4. Sun, X.-Y., Nobles, L.R., Börner, H.G., Spontak, R.J.: Field-driven surface segregation of biofunctional species on electrospun PMMA/PEO microfibers. Macromol. Rapid Commun. **29**, 1455–1460 (2008)
5. Forrest, J.A., Dalnoki-Veress, K., Dutcher, J.R.: Interface and chain confinement effects on the glass transition temperature of thin polymer films. Phys. Rev. E. **56**, 5705–5716 (1997)
6. Ellison, C.J., Torkelson, J.M.: The distribution of glass-transition temperatures in nanoscopically confined glass formers. Nat. Mater. **2**, 695–700 (2003)
7. Reiter, G., Napolitano, S.: Possible origin of thickness-dependent deviations from bulk properties of thin polymer films. J. Polym. Sci. Part B: Polym. Phys. **48**, 2544–2547 (2010)
8. Zhu, L.: Polymer ordering in nanoconfined spaces. In: Nalwa, H.S. (ed.) Polymeric Nanostructures and Their Applications, vol. 2. American Scientific Publishers, Stevenson Ranch (2007)
9. Huang, E., Rockford, L., Russell, T.P., Hawker, C.J.: Nanodomain control in copolymer thin films. Nature **395**, 757–758 (1998)
10. Kim, S.O., Solak, H.H., Stoykovich, M.P., Ferrier, N.J., de Pablo, J.J., Nealey, P.F.: Epitaxial self-assembly of block copolymers on lithographically defined nanopatterned substrates. Nature **424**, 411–414 (2003)
11. Cheng, J.Y., Mayes, A.M., Ross, C.A.: Nanostructure engineering by templated self-assembly of block copolymers. Nat. Mater. **3**, 823–828 (2004)
12. Park, S., Lee, D.H., Xu, J., Kim, B., Hong, S.W., Jeong, U., Xu, T., Russell, T.P.: Macroscopic 10-terabit-per-square- inch arrays from block copolymers with lateral order. Science **323**, 1030–1033 (2009)
13. Abeysekera, R., Bushby, R.J., Caillet, C., Hamley, I.W., Lozman, O.R., Lu, Z.B., Robards, A.W.: Discotic liquid crystalline triblock copolymers: interplay of liquid crystal architecture with microphase separation. Macromolecules **36**, 1526–1533 (2003)
14. Steinhart, M., Goring, P., Dernaika, H., Prabhukaran, M., Gosele, U., Hempel, E., Thurn-Albrecht, T.: Coherent kinetic control over crystal orientation in macroscopic ensembles of polymer nanorods and nanotubes. Phys. Rev. Lett. **97**, 027801 (2006)
15. Serghei, A., Lutkenhaus, J.L., Miranda, D.F., McEnnis, K., Kremer, F., Russell, T.P.: Density fluctuations and phase transitions of ferroelectric polymer nanowires. Small **6**, 1822–1826 (2010)
16. Loo, Y.-L., Register, R.A.: Crystallization within block copolymer mesophases. In: Hamley, I.W. (ed.) Developments in Block Copolymer Science and Technology. Wiley, Hoboken (2004)
17. Taden, A., Landfester, K.: Crystallization of poly(ethylene oxide) confined in miniemulsion droplets. Macromolecules **36**, 4037–4041 (2003)
18. Muller, A.J., Balsamo, V., Arnal, M.L., Jakob, T., Schmalz, H., Abetz, V.: Homogeneous nucleation and fractionated crystallization in block copolymers. Macromolecules **35**, 3048–3058 (2002)

19. Liu, Y., Cui, L., Guan, F.X., Gao, Y., Hedin, N.E., Zhu, L., Fong, H.: Crystalline morphology and polymorphic phase transitions in electrospun nylon-6 nanofibers. Macromolecules **40**, 6283–6290 (2007)
20. Rayleigh, L.: On the capillary phenomena of jets. Proc. R. Soc. Lond. **29**, 71–97 (1879)
21. Elemans, P.H.M., Janssen, J.M.H., Meijer, H.E.H.: The measurement of interfacial tension in polymer/polymer systems: the breaking thread method. J. Rheol. **34**, 1311–1325 (1990)
22. McGraw, J.D., Li, J.F., Tran, D.L., Shi, A.C., Dalnoki-Veress, K.: Plateau-Rayleigh instability in a torus: formation and breakup of a polymer ring. Soft Matter **6**, 1258–1262 (2010)
23. Zhong, G.J., Wang, K., Zhang, L.F., Li, Z.-M., Hao, F., Zhu, L.: Nanodroplet formation and exclusive homogeneously nucleated crystallization in confined electrospun immiscible polymer blend fibers of polystyrene and poly(ethylene oxide). Polymer **52**, 5397–5402 (2011)
24. Bognitzki, M., Frese, T., Steinhart, M., Greiner, A., Wendorff, J.: Preparation of fibers with nanoscaled morphologies: electrospinning of polymer blends. Polym. Eng. Sci. **41**, 982–989 (2001)
25. Lee, K.H., Kim, H.Y., Ryu, Y.J., Kim, K.W., Choi, S.W.: Mechanical behavior of electrospun fiber mats of poly(vinyl chloride)/polyurethane polyblends. J. Polym. Sci. Part B: Polym. Phys. **41**, 1256–1262 (2003)
26. Bazilevsky, A.V., Yarin, A.L., Megaridis, C.M.: Co-electrospinning of core-shell fibers using a single-nozzle technique. Langmuir **23**, 2311–2314 (2007)
27. Massa, M.V., Dalnoki-Veress, K.: Homogeneous crystallization of poly(ethylene oxide) confined to droplets: the dependence of the crystal nucleation rate on length scale and temperature. Phys. Rev. Lett. **92**, 255509 (2004)
28. He, Y., Zhu, B., Kai, W., Inoue, Y.: Nanoscale-confined and fractional crystallization of poly(ethylene oxide) in the interlamellar region of poly(butylene succinate). Macromolecules **37**, 3337–3345 (2004)
29. Massa, M.V., Lee, M.S.M., Dalnoki-Veress, K.: Crystal nucleation of polymers confined to droplets: memory effects. J. Polym. Sci. Part B: Polym. Phys. **43**, 3438–3443 (2005)
30. Zhong, G.J., Zhang, L.F., Su, R., Wang, K., Fong, H., Zhu, L.: Understanding polymorphism formation in electrospun fibers of immiscible poly(vinylidene fluoride) blends. Polymer **52**, 2228–2237 (2011)
31. You, Y., Youk, J.H., Lee, S.W., Min, B.-M., Lee, S.J., Park, W.H.: Preparation of porous ultrafine PGA fibers via selective dissolution of electrospun PGA/PLA blend fibers. Mater. Lett. **60**, 757–760 (2006)
32. Zhong, G.J., Su, R., Zhang, L.F., Wang, K., Li, Z.M., Hao, F., Zhu, L.: Evolution of nanodroplets and fractionated crystallization in thermally annealed electrospun blend fibers of poly(vinylidene fluoride) and polysulfone. Polymer **53**, 4472–4480 (2012)
33. Frensch, H., Jungnickel, B.J.: Some novel crystallization kinetic peculiarities in finely dispersing polymer blends. Colloid Polym. Sci. **267**, 16–27 (1989)
34. Liu, Z., Maréchal, P., Jérôme, R.: Melting and crystallization of poly(vinylidene fluoride) blended with polyamide 6. Polymer **38**, 5149–5153 (1997)
35. Na, B., Xu, W., Lv, R., Li, Z., Tian, N., Zou, S.: Toughening of nylon-6 by semicrystalline poly(vinylidene fluoride): role of phase transformation and fibrillation of dispersed particles. Macromolecules **43**(8), 3911–3915 (2010)
36. Steinhart, M., Senz, S., Wehrspohn, R.B., Gosele, U., Wendorff, J.H.: Curvature-directed crystallization of poly(vinylidene difluoride) in nanotube walls. Macromolecules **36**, 3646–3651 (2003)
37. Lutkenhaus, J.L., McEnnis, K., Serghei, A., Russell, T.P.: Confinement effects on crystallization and curie transitions of poly(vinylidene fluoride-co-trifluoroethylene). Macromolecules **43**, 3844–3850 (2010)
38. Muller, A.J., Arnal, M.L., Lopez-Carrasquero, F.: Nucleation and crystallization of PS-b-PEO-b-PCL triblock copolymers. Macromol. Symp. **183**, 199–204 (2002)

39. Hsiao, M.S., Zheng, J.X., Leng, S.W., Van Horn, R.M., Quirk, R.P., Thomas, E.L., Chen, H.L., Hsiao, B.S., Rong, L.X., Lotz, B., Cheng, S.Z.D.: Crystal orientation change and its origin in one-dimensional nanoconfinement constructed by polystyrene-block-poly(ethylene oxide) single crystal mats. Macromolecules **41**, 8114–8123 (2008)
40. Langhe, D.S., Keum, J.K., Hiltner, A., Baer, E.: Fractionated crystallization of alpha- and beta-nucleated polypropylene droplets. J. Polym. Sci. Part B: Polym. Phys. **49**, 159–171 (2011)
41. Arnal, M.L., Muller, A.J., Maiti, P., Hikosaka, M.: Nucleation and crystallization of isotactic poly(propylene) droplets in an immiscible polystyrene matrix. Macromol. Chem. Phys. **201**, 2493–2504 (2000)
42. Arnal, M.L., Matos, M.E., Morales, R.A., Santana, O.O., Muller, A.J.: Evaluation of the fractionated crystallization of dispersed polyolefins in a polystyrene matrix. Macromol. Chem. Phys. **199**, 2275–2288 (1998)
43. Tol, R.T., Mathot, V.B.F., Groeninckx, G.: Confined crystallization phenomena in immiscible polymer blends with dispersed micro- and nanometer sized PA6 droplets, part 3: crystallization kinetics and crystallinity of micro- and nanometer sized PA6 droplets crystallizing at high supercoolings. Polymer **46**, 2955–2965 (2005)
44. Tol, R.T., Mathot, V.B.F., Groeninckx, G.: Confined crystallization phenomena in immiscible polymer blends with dispersed micro- and nanometer sized PA6 droplets, part 1: uncompatibilized PS/PA6, (PPE/PS)/PA6 and PPE/PA6 blends. Polymer **46**, 369–382 (2005)
45. Tol, R.T., Mathot, V.B.F., Groeninckx, G.: Confined crystallization phenomena in immiscible polymer blends with dispersed micro- and nanometer sized PA6 droplets, part 2: reactively compatibilized PS/PA6 and (PPE/PS)/PA6 blends. Polymer **46**, 383–396 (2005)
46. Tol, R.T., Mathot, V.B.F., Reynaers, H., Goderis, B., Groeninckx, G.: Confined crystallization phenomena in immiscible polymer blends with dispersed micro-and nanometer sized PA6 droplets part 4: polymorphous structure and (meta)-stability of PA6 crystals formed in different temperature regions. Polymer **46**, 2966–2977 (2005)
47. Massa, M.V., Carvalho, J.L., Dalnoki-Veress, K.: Confinement effects in polymer crystal nucleation from the bulk to few-chain systems. Phys. Rev. Lett. **97**, 247802 (2006)
48. Tashiro, K.: Crystal structure and phase transition of PVDF and related copolymers. In: Nalwa, H.S. (ed.) Ferroelectric Polymers: Chemistry, Physics, and Applications, 1st edn, pp. 63–182. Dekker, New York (1995)
49. Yu, L., Cebe, P.: Crystal polymorphism in electrospun composite nanofibers of poly(vinylidene fluoride) with nanoclay. Polymer **50**, 2133–2141 (2009)
50. Andrew, J.S., Clarke, D.R.: Enhanced ferroelectric phase content of polyvinylidene difluoride fibers with the addition of magnetic nanoparticles. Langmuir **24**, 8435–8438 (2008)
51. Huang, S., Yee, W.A., Tjiu, W.C., Liu, Y., Kotaki, M., Boey, Y.C.F., Ma, J., Liu, T., Lu, X.: Electrospinning of polyvinylidene difluoride with carbon nanotubes: synergistic effects of extensional force and interfacial interaction on crystalline structures. Langmuir **24**, 13621–13626 (2008)
52. Blythe, A.R.: Electrical Properties of Polymers. Cambridge University Press, Cambridge (1979)
53. Tasaka, S.: Cyanopolymers. In: Nalwa, H.S. (ed.) Ferroelectric Polymers: Chemistry, Physics, and Applications, 1st edn, pp. 325–352. Dekker, New York (1995)
54. Colvin, B.G., Storr, P.: Crystal-structure of polyacrylonitrile. Eur. Polym. J. **10**, 337–340 (1974)
55. Liu, X.D., Ruland, W.: X-ray studies on the structure of polyacrylonitrile fibers. Macromolecules **26**, 3030–3036 (1993)
56. Gregorio, R., Cestari, M.: Effect of crystallization temperature on the crystalline phase content and morphology of poly(vinylidene fluoride). J. Polym. Sci. Part B: Polym. Phys. **32**, 859–870 (1994)
57. Kobayashi, M., Tashiro, K., Tadokoro, H.: Molecular vibrations of three crystal forms of poly(vinylidene fluoride). Macromolecules **8**, 158–171 (1975)
58. Benkhati, H., Tan, T.T.M., Jungnickel, B.J.: Transcrystallization kinetics of poly(vinylidene fluoride). J. Polym. Sci. Part B: Polym. Phys. **39**, 2130–2139 (2001)

59. Lovinger, A.J.: Crystallization of the beta-phase of poly(vinylidene fluoride) from the melt. Polymer **22**, 412–413 (1981)
60. Lovinger, A.J.: Unit-cell of the gamma-phase of poly(vinylidene fluoride). Macromolecules **14**, 322–325 (1981)
61. Lovinger, A.J.: Conformational defects and associated molecular motions in crystalline poly(vinylidene fluoride). J. Appl. Phys. **52**, 5934–5938 (1981)
62. Miyazaki, T., Takeda, Y., Akasaka, M., Sakai, M., Hoshiko, A.: Preparation of isothermally crystallized gamma-form poly(vinylidene fluoride) films by adding a KBr powder as a nucleating agent. Macromolecules **41**, 2749–2753 (2008)
63. Priya, L., Jog, J.P.: Poly(vinylidene fluoride)/clay nanocomposites prepared by melt intercalation: crystallization and dynamic mechanical behavior studies. J. Polym. Sci. Part B: Polym. Phys. **40**, 1682–1689 (2002)
64. Buckley, J., Cebe, P., Cherdack, D., Crawford, J., Ince, B.S., Jenkins, M., Pan, J., Reveley, M., Washington, N., Wolchover, N.: Nanocomposites of poly(vinylidene fluoride) with organically modified silicate. Polymer **47**, 2411–2422 (2006)
65. Dillon, D.R., Tenneti, K.K., Li, C.Y., Ko, F.K., Sics, I., Hsiao, B.S.: On the structure and morphology of polyvinylidene fluoride-nanoclay nanocomposites. Polymer **47**, 1678–1688 (2006)
66. Shah, D., Maiti, P., Gunn, E., Schmidt, D.F., Jiang, D.D., Batt, C.A., Giannelis, E.P.: Dramatic enhancements in toughness of polyvinylidene fluoride nanocomposites via nanoclay-directed crystal structure and morphology. Adv. Mater. **16**, 1173–1177 (2004)
67. He, L., Xu, Q., Hue, C., Song, R.: Effect of multi-walled carbon nanotubes on crystallization, thermal, and mechanical properties of poly(vinylidene fluoride). Polym. Comp. **31**, 921–927 (2010)
68. Kim, G.H., Hong, S.M., Seo, Y.: Piezoelectric properties of poly(vinylidene fluoride) and carbon nanotube blends: beta-phase development. Phys. Chem. Chem. Phys. **11**, 10506–10512 (2009)
69. Li, Y., Kaito, A.: Mechanistic investigation into the unique orientation textures of poly(vinylidene fluoride) in blends with nylon 11. Macromol. Rapid Comm. **24**, 603–608 (2003)
70. Li, Y., Kaito, A.: Crystallization and orientation behaviors of poly(vinylidene fluoride) in the oriented blend with nylon 11. Polymer **44**, 8167–8176 (2003)
71. Kim, K.J., Cho, H.W., Yoon, K.J.: Effect of P(MMA-co-MAA) compatibilizer on the miscibility of nylon 6/PVDF blends. Eur. Polym. J. **39**, 1249–1265 (2003)
72. Kaito, A., Iwakura, Y., Li, Y., Nakayama, K., Shimizu, H.: Unique orientation textures induced by confined crystal growth of poly(vinylidene fluoride) in oriented blends with polyamide 6. Macromol. Chem. Phys. **208**, 504–513 (2007)
73. Saito, H., Matsuura, M., Inoue, T.: Depolarized light-scattering-studies on single-phase mixtures of dissimilar polymers – evidence for local ordering. J. Polym. Sci. Part B: Polym. Phys. **29**, 1541–1546 (1991)
74. Saito, H., Matsuura, M., Okada, T., Inoue, T.: Short-range order in a miscible polymer blend. Polym. J. **21**, 357–360 (1989)
75. Aihara, T., Saito, H., Inoue, T., Wolff, H.P., Stuhn, B.: Dielectric studies of specific interaction and molecular motion in single-phase mixture of poly(methyl methacrylate) and poly(vinylidene fluoride). Polymer **39**, 129–134 (1998)

Chapter 3
Dynamic Study of Nanodroplet Nucleation and Growth Using Transmitted Electrons in ESEM

Zahava Barkay

Abstract Experimental methods for wettability research are reviewed. A novel method for quantitative wettability study at nanoscale is presented. It is based on measuring transmitted electrons through nanodroplets using wet scanning transmission electron microscope (wet-STEM) detector in environmental scanning electron microscope (ESEM). The quantitative information of the nanodroplet shape and contact angle is obtained by fitting Monte Carlo simulation results for transmitted electrons with experimental results. Dynamic in situ imaging has showed that irregularities at the water film boundaries constituted nucleation sites for both dropwise and filmwise condensation. The initial stages of nucleation, growth, and coalescence have been studied as well as the growth power law dependence.

List of Abbreviations

2D	Two dimensional
3D	Three dimensional
AFM	Atomic force microscope
BF	Bright field
EM	Electron microscope
ESEM	Environmental scanning electron microscope
ETEM	Environmental TEM
FEG	Field emission gun
MC	Monte Carlo
MEMS and NEMS	Micro/nano electromechanical systems
RH	Relative humidity
RTD	Resistive temperature device

Z. Barkay (✉)
Wolfson Applied Materials Research Center, Tel-Aviv University, Tel-Aviv 69978, Israel
e-mail: barkay@post.tau.ac.il

Z.M. Wang (ed.), *Nanodroplets*, Lecture Notes in Nanoscale Science and Technology 18,
DOI 10.1007/978-1-4614-9472-0_3,

SEM	Scanning electron microscope
TEM	Transmitted electron microscope
Wet-STEM	Wet scanning transmission electron microscope

3.1 Introduction

The nanoscale wettability research explores theoretical aspects such as static contact angle as well as nanodroplet dynamics including drop growth and coalescence, which have been previously intensively studied at the macroscale. The liquid drop dynamics has a major role in a range of scientific applications including emulsion preparation, metallurgy, phase transitions in liquids or polymers, and atmospheric science. The liquid drop macroscopic properties are averaged over nanoscale inhomogeneities, which play a significant role for nanodroplets. In addition, the drop macroscale physical properties change with reduction of size down to nanoscale.

The driving force for micro- and nanoscale wettability research stems from the request of physical understanding of nucleation and growth processes. Technological research refers to device miniaturization and nanofluidic requirements in biotechnology and materials sciences. Some examples are micro/nano lab-on-chip devices, referring to integrating a miniature "laboratory" into a single chemical-electronic device as well as micro/nanoelectromechanical systems (MEMS and NEMS). Other nano-wetting topics refer to biological components composed of nanofluidic ion channels, materials possessing by lithography or applications involving miniature quantities of liquids. The demand for wettability detection methods at the nanoscale poses technological challenges, which are here reviewed.

The introduction of this chapter is divided into two main subsections. The first (Sect. 3.1.1) provides a comparable background on wetting phenomena characterization at macro-, micro-, and nanoscales. The second (Sect. 3.1.2) reviews experimental microscopic methods with emphasis on electron microscopy and more particularly on the environmental scanning electron microscope (ESEM). The experimental part (Sect. 3.2) focuses on the novel method of wettability measurement, which uses wet-STEM detector in ESEM. The first part of results (Sect. 3.3.1) refers to the principle of quantifying a two-dimensional (2D) STEM image for obtaining the droplet three-dimensional (3D) shape and the corresponding contact angle. The droplet spatial projected dimensions are measured on the 2D STEM image. A Monte Carlo (MC) simulation for the interaction of electrons with the specimen is used for quantifying the height dimension using a calibration particle for image (offset/gain) setting. The characterization is demonstrated for initial stages of water droplet condensation. The second part of results (Sect. 3.3.2) refers to in situ wet-STEM imaging of droplet growth on a fluid self-supported interface (instead of commonly studied fluid–solid interface) with nano-lateral resolution and 1 s temporal resolution. The results demonstrate both dropwise and filmwise growth and the possible role of irregularities around the self-supported nano-thick water films. The droplet growth power law shows a distribution in values

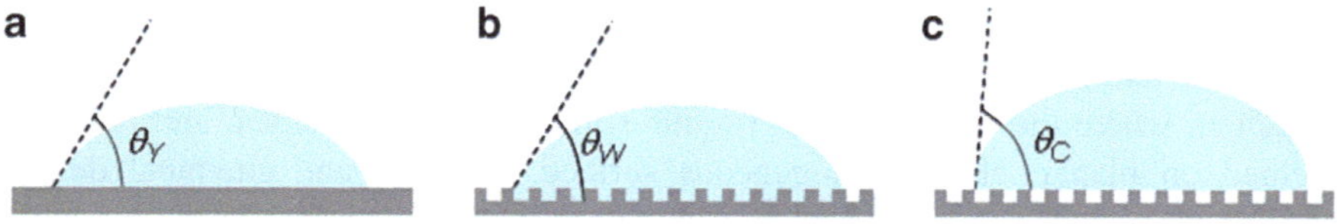

Fig. 3.1 (**a**) Young contact angle θ_Y, (**b**) Wenzel apparent contact angle θ_W, and (**c**) Cassie-Baxter apparent contact angle θ_C

and is compared with prior results for the micron and sub-mm scales. In situ imaging has showed interaction between droplets at tens nm separation, which is followed by coalescence events.

3.1.1 Wetting Phenomena at Macro-, Micro-, and Nanoscales

This section poses the following question, which has been already referred by D. Quéré almost a decade ago [1]: "how do the macroscopic factors that shape a drop change with the reduction in scale?" Macroscale drops are basically described by Young and Laplace equations, while Wenzel and Cassie-Baxter equations provide apparent contact angle on rough substrates [2]. On microscopic scale the contact angle is formulated by the modified Young equation, while the modified surface tension and the disjoining pressure are used for the nanoscale description. A comparable description will thus be given for wettability at macro-, micro-, and nanoscales. This enables the understanding of basic theoretical aspects as well as the technological challenges for inspection of these phenomena.

Basic Definitions for Macroscale: Young and Laplace Equations

Capillary and wetting phenomena at the macroscale have been studied since the nineteenth century as described in details elsewhere [2]. These phenomena are basically attributed to surface tension, which is related to excess of surface energy due to asymmetric liquid molecular interaction at the surface relative to the bulk liquid. The surface tension, γ, is defined by the energy required to increase a liquid unit area or by the force per unit length also named capillary force. These capillary forces are well known from everyday life being responsible for insects walking on water, capillary rise in a tube, or shower drops holding together. Surface tension is defined for any two interfaces such as liquid–vapor, solid–vapor, or solid–liquid by γ_{lv}, γ_{sv}, γ_{sl} correspondingly. The wettability phenomenon describes the tendency of liquid to spread on solid surface. It is quantitatively defined by the following Young contact angle θ_Y (Fig. 3.1a) using the surface energies [2]:

$$\gamma_{lv} \cos \theta_Y = \gamma_{sv} - \gamma_{sl} \tag{3.1}$$

A solid interface is defined as hydrophilic for $\theta_Y < 90°$, hydrophobic for $\theta_Y > 90°$, and superhydrophobic for $\theta_Y > 150°$. Equation (3.1) refers to a static case, i.e., where the droplet triplet (liquid–solid–vapor) line ceases, and is implemented on planar, clean homogeneous surface. Physical and chemical defects provide pinning of the triplet line, and therefore instead of the static Young angle, advancing and receding angles are defined. Contact angle hysteresis is defined as the difference between the advancing and receding angles and is relatively large for high pinning and relatively low under superhydrophobicity.

The Young contact angle of a liquid drop on a solid surface depends on the roughness and the chemical homogeneity of the surface, being correspondingly characterized by Wenzel and Cassie-Baxter apparent contact angles [2, 3]. For scale of inhomogeneity below the drop size, the Wenzel model applies to drop wetting the whole textured region, while the Cassie-Baxter model applies to drop wetting the top textured parts leaving below air pockets (Fig. 3.1b, c). According to Wenzel model, the roughness amplifies the intrinsic surface wettability properties, i.e., hydrophilic surfaces result with lower apparent contact angle and hydrophobic surfaces result with higher apparent contact angle. According to Cassie-Baxter model, the air pockets always strengthen water-repelling and increase the apparent contact angle, which could result with superhydrophobicity and low contact angle hysteresis. The transition between non-wetting Cassie-Baxter to wetting Wenzel situation has been further considered [4, 5].

Surface tension is the origin of overpressure existing in the interior of drops and bubbles. The pressure difference ΔP across the interface of a drop with radius r is given by the following Laplace equation [2]:

$$\Delta P = \frac{2\gamma_{lv}}{r} \tag{3.2}$$

Equation (3.2) explains the spherical shape of droplets in free space or in emulsion and the disappearance of small drops in favor of bigger ones while being in contact.

At the macroscale both capillary forces and long-range forces such as gravity are dominated. As a result, drop shape at the center tends to flatten under the force of gravity. However, gravity is neglected, and capillary effects dominate for drop size, r, below the capillary length κ^{-1}, which is defined by [2]

$$\kappa^{-1} = \sqrt{\frac{\gamma_{lv}}{\rho \cdot g}} \tag{3.3}$$

For water of $\gamma_{lv} = 72$ mN/m, density $\rho = 1$ g/cm^3, and Earth's gravity $g = 9.8$ m/s^2, Eq. (3.3) provides κ^{-1} 2 mm. For $r < \kappa^{-1}$, following the Laplace equation and the tendency to minimize the total surface area, the drop is of symmetrical spherical cap geometry.

Basic Definitions for Microscale: Modified Young Contact Angle

The physics of microdroplets has been described elsewhere [6]. The microscale regime is defined by the action of capillary forces while neglecting gravity. The drop shape on flat solid substrate is thus of spherical cap geometry as for $r < \kappa^{-1}$. However, the Young contact angle itself requires modification at the microscale taking into account the molecules in vicinity to the triple line. These molecules experience a different set of interactions than at the interface because their environment is different from that in the bulk liquid or at the surface. The modified Young's law is given by introducing the line tension τ, which for droplet of radius r obeys [2, 7]:

$$\gamma_{\mathrm{lv}} \cos \theta_{\mathrm{Y}} = \gamma_{\mathrm{sv}} - \gamma_{\mathrm{sl}} - \tau / r \tag{3.4}$$

Equation (3.4) was evaluated for hydrophobic and hydrophilic surfaces, each characterized by a typical line tension [7]. The correction term for cosine of Young contact angle for water drop of $r = 1$ μm is of the order of 10^{-3}–10^{-2} being dependent on the line tension value [2, 6].

Following Laplace equation (3.2), a spherical water drop of $r = 1$ μm yields ΔP comparable to the atmospheric pressure, and thus bubbles of this size would rapidly disappear in liquid. A significant lower contact angle has been measured for micro- and nanoscale drops relative to macro-drops on various solid substrates [8]. The reduction in contact angle and correspondingly increase of curvature explain the stabilization of droplets and bubbles at small scales. The consideration about the shape of immiscible liquid interfaces and drops is given in the framework of Neumann's construction [6].

Basic Definitions for Nanoscale: Modified Surface Tension

As the system size decreases, surface effects including capillary forces become dominant [2, 6]. The thermodynamic parameters in small nanoscale systems depend on the system size and geometry [9]. The theory of nanofluidic thin liquid films has been elsewhere described [10, 11]. Nanofluidics refers to movement of liquids in or around objects with at least one dimension below 10–50 nm. At nanoscale the molecular interaction range, thermal fluctuations, fluid channel walls, microscopic inhomogeneities, and electric charges become particularly important.

The most simplified nanofluidic system is a thin liquid film. A liquid film, typically below tens nm thick at equilibrium on a flat solid, is subject to forces acting perpendicular to the plane of the film named "disjoining pressure" as introduced by Derjaguin [12]. The disjoining pressure is also relevant to nanoscale drops or drops at nucleation state. The dependence of surface tension on the disjoining pressure and drop radius has been explicitly shown elsewhere for nanoscale drops [9, 12]. The disjoining pressure is also relevant for the precursor film theory, which refers to a nanometer thin film spreading ahead of a drop at the triple line. When one interface approaches the other, as for adjacent drops or bubbles, the surface tension and the disjoining pressure dividing the two surfaces change, which affects the drop or the bubble shape. The disjoining pressure is a sum of various types of interactions such as van der Waals, electrostatic, and structural. Thus the dependence of drop surface tension on radius is system specific at the nanoscale.

3.1.2 Experimental Methods for Micro- and Nanoscales

Macroscale imaging is usually carried out by optical microscopes including confocal microscopes and goniometric devices for contact angle derivation. Micro- and nanoscale imaging will require alternative methods, which overcome the submicron Rayleigh resolution limit of far-field optical microscopes.

What are then the basic technological challenges for dynamic wettability study at nanoscale? A suitable method should accomplish the following aspects:

- Nanoscale spatial resolution
- Fast recording for dynamic in situ imaging
- Study under environmental conditions (preserving sample humidity)
- A measurement which is free of detection perturbations

This section will be devoted to the atomic force microscope (AFM) method and to the electron microscope (EM) method with emphasis on ESEM, in which current experiments have been carried out.

Atomic Force Microscope

AFM can provide nanoscale lateral resolution but is rather slow with typical scan time of a few min per frame [13]. AFM measurements of liquids are usually conducted at non-contact or tapping modes by scanning with a tip on cantilever a few nm above the sample to reduce the direct contact with the liquid, which is unavoidable at contact mode. AFM measurements of contact angles for nanoscale drops on various solid substrates have provided values 2–10 times lower than for macro-drops on the same substrates [8]. Modification of the non-contact AFM method in terms of the scanning polarization force microscope (SPFM) has used the electrostatic forces to improve imaging of liquid films and the nanodroplet contact angle as function of droplet height [14].

Electron Microscope

The electron microscope (EM) is known since the invention of M. Knoll and E. Ruska at 1931. EM provides the required lateral resolution and fast imaging for wettability study. EM operation is traditionally categorized either in transmission mode based on the transmitted electron microscope (TEM) [15] or reflection mode based on the scanning electron microscope (SEM) [16]. However, electron microscopes are usually incompatible with imaging of liquids due to the high vacuum requirements for electron beam (e-beam) operation (the vacuum is usually below 10^{-5} torr). Thus for wet environment, two directions in EM technology have been evolved:

- Window technique
- Aperture-limited technique

In window technique a pair of electron transparent windows is placed above and below the specimen allowing inlet and outlet side horizontal flow while maintaining vacuum in the EM outside the window device. It is usually designed for TEM imaging using transmitted electrons of a hundred keV. Certain configurations are termed as "nanoaquarium" for micro-fabricated liquid cell that provides TEM imaging under liquid environment [17]. The "nanoaquarium" supports a thin liquid layer, from tens nanometers to a few microns, between two thin silicon nitride membranes, which altogether scatter only a small fraction of electrons allowing nanoscale imaging of suspended objects in liquid. Special window liquid capsules have been designed for SEM imaging at high vacuum using backscattered electrons of tens keV [18]. The alternative aperture-limited technique is referred to as environmental TEM (ETEM) [19] or environmental SEM (ESEM) [20]. It has the benefit of providing the study of open liquid systems. The apertures above and below the sample together with differential pumping allow the required pressures in the sample and at the same time keep the other parts of the EM at high vacuum.

Microscale and nanoscale applications of electron microscopes correspond to in situ imaging of hydration and dehydration experiments, oxidation and reduction experiments, materials growth (tubes, wires, films, and coatings), imaging dispersed particles, gas-mediated reactions, and beam-induced chemical reactions. The following section will be focused on ESEM as a microscopic method for dynamic wettability study.

ESEM for Wettability Study

The ESEM principles and properties were widely elsewhere described [20]. Wettability study is carried out in ESEM at wet-mode under conditions which mimic environmental conditions, i.e., by controlling the sample conditions according to the water vapor equilibrium curve of the water pressure-temperature phase diagram [20]. The water vapor is introduced into the sample chamber (up to usually 10–20 torr), and the sample temperature is controlled usually to a few degrees C (by Peltier cooling stage). Humidity sensors for accurate characterization of the conditions in ESEM have been elsewhere described [21]. Surface imaging in ESEM wet mode is obtained by a gaseous secondary electron detector. Varying the pressure and temperature and thus relative humidity (RH) in the sample chamber provides condensation and evaporation while in situ dynamic imaging.

ESEM wettability study involves several strategies for drop contact angle derivation including top view of planar structures [22] or side view of curved structures [23]. The various results in ESEM study include in situ imaging of liquid marbles [24], picoliter liquid flow through carbon nanotubes [25], and line tension derivation [7]. Contact angle hysteresis and Cassie-Baxter to Wenzel transition were also measured in ESEM for textured solid surfaces [6, 26].

The main limitations of EM, which affect wettability inspection, are [20, 27, 28]:

- E-beam contamination
- E-beam radiology
- E-beam electrical charging
- E-beam heating

E-beam contamination occurs in EM-based systems due to carbonization of the sample irradiated area. It results from dissociation of native sample or vacuum hydrocarbon under e-beam, which is typical under ordinary SEM sample vacuum conditions of 10^{-5} torr. Contamination could be also considered for low vacuum (usually up to about 1 torr) or ESEM conditions (usually up to about 10 torr) [20].

E-beam radiology is a result of inelastic e-beam scattering in water vapor environment, which forms ionized or excited water molecules. As explicitly has been explained [20, 28], the excited water molecules can decay into free radicals or ions; the latter could generally lead to sample radiation damage named sample radiology.

E-beam charging is due to accumulation of net charge in the sample due to electron-specimen interaction. The net accumulated charge is a result of nonzero balance between incoming and outgoing electrons. The sample charge could be either positive or negative, which depends on operation conditions such as beam voltage, sample conductivity, sample thickness, and environmental conditions in the EM. Charging effects are observed in the image in a way, which is also detector dependent.

E-beam heating is a result of inelastic electron-specimen interaction. Energy dissipation along the e-beam trajectory ($\mathrm{d}E/\mathrm{d}X$) named Bethe energy loss is inversely proportional to the energy E ($\mathrm{d}E/\mathrm{d}X \sim 1/E$) [16]. Therefore, e-beam in ESEM (typically below 30 keV) will yield more dissipation within the volume of the bulk samples in comparison with TEM (typically of 300 keV or below), which may require heat removal using a cooling device. Pioneer work has studied heating affects in SEM as a function of material thickness [29]. Recent work showed the effect of e-beam heating on droplets in ESEM [30].

Recently, wet-STEM detector in ESEM has been incorporated as a novel method for imaging nanoparticles in the liquid volume with a few nm resolution [31, 32]. Furthermore, it has been utilized for measuring the monomer miniemulsions [33] and magnetic nanovectors [34]. We here demonstrate quantitative study of nanodroplet nucleation and growth using wet-STEM.

3.2 Wet-STEM for Wettability Study

3.2.1 Wet-STEM Contrast

The wet-STEM detector uses two solid-state segments attached underneath the grid holder being used either in bright-field or dark-field modes. The grid sample is temperature controlled by a Peltier cooling device, while the humidity and the liquid film thickness could be controlled by both temperature and pressure according to the water vapor phase diagram. For simplicity, the contrast of STEM in SEM is shown here for nano-thick samples using single electron-atom scattering approximation. The dark-field STEM contrast is given by [16]

$$C = \frac{t}{\lambda_{\text{el}}}, \tag{3.5}$$

where t is the thickness of the material and λ_{el} is the elastic mean free path of the electrons in the material.

The elastic mean free path λ_{el} is inversely proportional to the scattering cross section for interaction and could be written using the unscreened Rutherford model [16] as

$$\lambda_{\text{el}} \sim \frac{E^2}{\rho \cdot Z^2}, \tag{3.6}$$

where E is the energy of the electron beam, Z is the atomic number of the material, and ρ is the density of the material. Substituting Eq. (3.6) in Eq. (3.5) gives an explicit expression for the dependence of contrast on E, ρ, Z, and t:

$$C \sim \frac{\rho \cdot t \cdot Z^2}{E^2}. \tag{3.7}$$

Wet-STEM in ESEM provides imaging with main beam ESEM energy, which typically is an order of magnitude lower than of conventional TEM, thus increasing the cross section for interaction relative to TEM and correspondingly improving the contrast as explicitly shown at Eq. (3.7). Low operation energy is particularly important for high STEM contrast of low atomic number materials including unstained biological samples [35], polymers or miniemulsion [33], carbon nanotubes [36], and water nanodroplets as described in this work.

3.2.2 Experimental Procedure

The samples for dynamic nucleation and growth were holey carbon transmission electron microscope grids being immersed in distilled water without any additives in order to rule out any external effect on the results. The holey carbon grid allowed measurement of primary beam signal through the empty holes. Several holes of a few μm in diameter supported overhanging thin water films and thus provided nanodroplet condensation over these areas and further examination of transmitted signal through condensed nanodroplets. The calibration sample for water droplet contact angle derivation was composed of polystyrene latex calibration spheres of 100 nm nominal diameter and mass density of 1.05 g/ml, which were diluted in water and randomly spread over the holey carbon grid.

Our experiments were carried out in FEI Quanta200 field emission gun (FEG) ESEM using the wet-STEM detector at bright-field (BF) mode. The condensation and evaporation conditions were provided by increasing or reducing the ESEM chamber pressure at constant sample temperature around dew point (5.3 torr and 2 °C). Prior to the ESEM pump down, the sample was cooled down to 2 °C by the Peltier cooling wet-STEM stage. The temperature of the Peltier cooling stage was maintained by a thermoelectric module, which was externally water cooled for removal of excess heat. A microprocessor controller provided accurate and stable automatic temperature control of the stage temperature. A resistive temperature device (RTD) in the stage measured the sample temperature. The temperature measurement accuracy was of ±0.5° as determined by the standard error limits of the RTD and the accuracy of the temperature measurement module.

In current experiments the substrate temperature T_s differed from the vapor temperature T_V, which provided a subcooling effect. The thermodynamic relation between the saturation vapor pressure and temperature for condensation to take place is given by [37]:

$$P_W = \left(\frac{T_S}{T_V}\right)^{0.5} \cdot P_{sat}(T_S), \tag{3.8}$$

where $P_{sat}(T_s)$ is saturation vapor pressure at temperature T_S. The pressure difference will be defined by $\delta P = P_W - P_{sat}$.

The vapor inside the ESEM chamber had originated from the water boiling in the ESEM glass bottle under low pressure, and the vapor temperature was equal to the environmental temperature at about 20 °C. For $T_V = 20$ °C, $T_S = 2$ °C, and $P_{sat} = 5.3$ torr, the pressure difference is $\delta P = 0.15$ torr.

The subcooling creates a non-equilibrium situation, according to which at the theoretical dew point (5.3 torr and 2 °C), more vapor molecules leave the substrate than condense, which reduces the number of nucleation sites. Thus the pressure can be elevated by 0.15 torr for condensation under saturated conditions. Adjustment of pressure with 0.1 torr resolution has been obtained during in situ imaging with condensation conditions in the range of 5.3–5.5 torr and 2 °C.

Primary e-beam of $E = 20$ keV provided suitable wet-STEM contrast (Eq. 3.7) for nanodroplet imaging in current experimental setup. It also resulted with suitable e-beam mean free path λ_{el} in the chamber vapor medium (Eq. 3.6), which minimized the primary electron beam collisions and thus provided a high signal to noise ratio at the detector. This energy provided low energy loss at the sample according to the Bethe equation and was suitable for high solid-state STEM detector efficiency. The e-beam scan size was of several microns, and scan time was 1 s per frame as determined by the detector response time. The carbon grid around the water film served as a local heat sink (typically 0.25 W/m K thermal conductance). An analytic solution, for the model of electron beam heating of cold stage specimens under beam power heating conditions similar to ours, resulted [29] with temperature rise less than 0.5 K for micron-thick specimens, i.e., within the Peltier cooling stage resolution. The thermal stabilization timescale could be estimated by [38] $\tau = d^2/\lambda$, where d is the film thickness and λ is the thermal diffusivity. Substituting the maximum thickness $d \sim 1$ μm and $\lambda \sim 1.4 \times 10^{-7}$ m^2/s for water [39] results in thermal stabilization timescale of $\tau = 10^{-5}$ s, i.e., much shorter than the acquisition frame time of 1 s, which provides imaging at isothermal conditions under dynamic acquisition.

3.3 Results and Discussion

3.3.1 Quantitative Approach for Droplet Shape Calculation

The quantitative approach for droplet shape and contact angle derivation is widely described elsewhere [40]. It is based on quantifying a two-dimensional (2D) STEM image for obtaining the droplet three-dimensional (3D) shape and the corresponding contact angle. The droplet spatial projected dimensions are measured on the 2D STEM image. A Monte Carlo (MC) simulation for the interaction of electrons with the specimen is used for quantifying the height dimension using a 100 nm polystyrene calibration particle for image (offset/gain) setting.

The MC simulation technique is suitable for studying interactions between a large number of electrons and the sample. The basic assumption is that although evaluation of parameters for a single incoming electron cannot represent the actual process, a large number of electrons will provide representative results. The interaction between incoming electron and solid is generally separated to elastic and inelastic interactions using a single electron-atom scattering model. The trajectory for the incoming electrons is calculated in discrete steps. Bethe energy loss and three random numbers (step distance, elastic, and azimuthal scattering angles) are calculated at each step for each incoming electron until complete energy loss. The experimental parameters are substituted in the expressions for probability distribution of the random numbers [41]. The MC simulation parameters have been 20 keV primary beam energy (of the experiment), density of 1.05 g/cm^3 for both spherical polystyrene calibration particles and water nanodroplets, and collection angle of 10° for the bright-field detector.

The STEM image calibration was demonstrated on 100 nm single particle at the center of Fig. 3.2a. Particles in simulation were assumed to have an ellipsoidal surface structure given by $(x/a)^2 + (y/b)^2 + (z/c)^2 = 1$, where parameters a and b

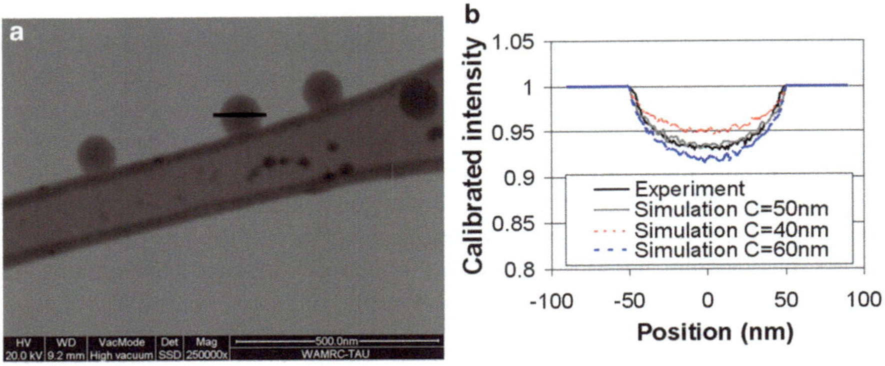

Fig. 3.2 100 nm polystyrene latex spheres: (**a**) BF-STEM image (scale bar is 500 nm) and (**b**) calibrated intensity profile with fit to an ellipsoidal structure of $a=b=50$ nm and $c=40$ nm, $c=50$ nm, $c=60$ nm

are the equatorial ellipsoid axes (along x and y axes) and parameter c is the polar axis corresponding to the height (along z axis). It was demonstrated in Fig. 3.2b that following a calibration for $a=b=c=50$ nm, elliptical particles of platelike structure or elongated structure could be clearly characterized relative to a spherical structure with better than ±10 nm accuracy in c value for the same projection dimensions.

Figure 3.3 shows polystyrene nanoparticles at water vapor equilibrium point of 2 °C and 5.3 torr. At elevated pressure (5.7 torr at 2 °C), condensation is shown over the carbon grid layer at the polystyrene nanoparticle position (Fig. 3.3b), while unsupported nanoparticles over the holey regions are almost unaffected. The continuous water condensation phenomenon over the carbon film is in situ studied as shown at a successive scan of the same sample region at 2 °C and 5.7 torr (Fig. 3.3c). The corresponding intensity profiles along the central line of the unsupported nanoparticle (indicated by an arrow at Fig. 3.3a) are shown in Fig. 3.3d after offset level adjustment for conditions (a)–(c). The intensity profiles are similar except for the right side, corresponding to initial water condensation at the nanoparticle side, which sticks to the carbon film. These results indicate that due to thermal isolation of the almost unsupported nanoparticle, there is no water condensation over it, which allows using it for calibration at first stages of condensation at wet-STEM mode.

In order to evaporate the water layer of the latex emulsion from the hydrated grid, the pressure was reduced to 1.1 torr at 2 °C (providing 20 % humidity). Then the pressure was elevated to 5.4 torr to initiate nanodroplet condensation (Fig. 3.4a). Droplets with typical lateral size of 100–400 nm were obtained on the carbon film and over the ultrathin water layers at the holes (of 0.5–1 μm) as indicated by an arrow at Fig. 3.4a. Figure 3.4b shows the calibrated measured intensity profile across the center of a typical nanodroplet and the calibration nanoparticle (marked at Fig. 3.4a) together with the MC simulation results. The droplet structure was fitted to the calibrated experimental results according to the spherical cap model (inset

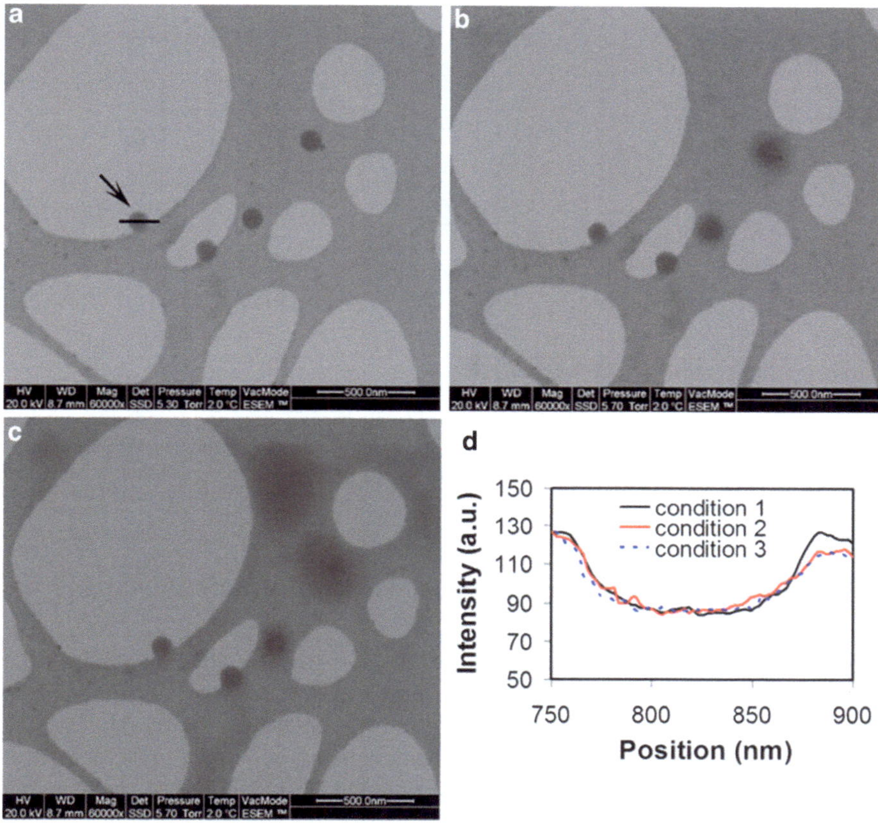

Fig. 3.3 Polystyrene water diluted spheres on grid (scale bar is 500 nm): (**a**) 2 °C and 5.3 torr, (**b**) 2 °C and 5.7 torr, (**c**) successive image at 2 °C and 5.7 torr, and (**d**) intensity profiles along the central line of the single particle for conditions (**a**)–(**c**)

of Fig. 3.4b), which yielded $c = 70$ nm, while the lateral droplet dimensions were measured as $a = 115$ nm and $b = 100$ nm on Fig. 3.4a. The contact angle θ_c for the spherical cap shape model obeys the geometrical relations:

$$\sin\theta_c = \frac{2\cdot a\cdot c}{a^2 + c^2}. \qquad (3.9)$$

Substituting $a = 115$ nm and $c = 70$ nm provided estimation of $\theta_c = 63 \pm 3°$. Previous work [7] showed an increase of contact angle from 30° to 45° for water droplet radius of about 0.2 μm, due to condensation of an ultrathin nanometer hydrocarbon layer on top of a bulk silicon dioxide. In our case, in addition to possible hydrocarbon contaminates, the droplet shape and contact angle might be affected by pinning at water–carbon interface defect site.

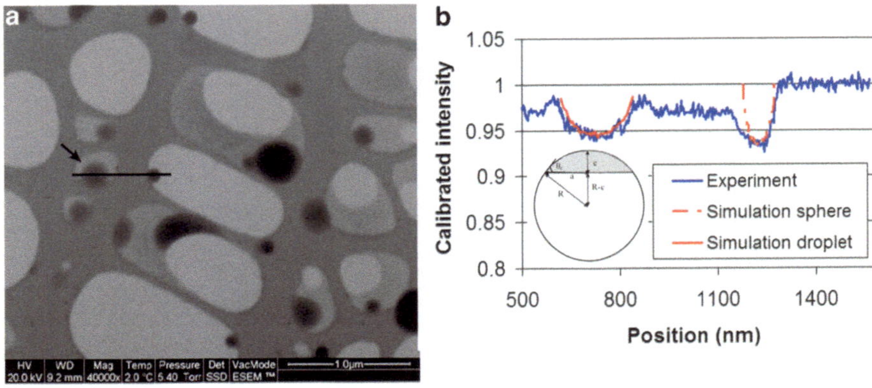

Fig. 3.4 Polystyrene water diluted spheres on grid (scale bar is 1 μm): (**a**) nanodroplet condensation at 2 °C and 5.4 torr and (**b**) calibrated profile along the center of the nanodroplet and the calibration nanoparticle together with the MC simulations (inset—schematic diagram of spherical cap droplet model for the *x*–*z* plan)

3.3.2 *Dynamic Nucleation and Growth upon Coalescence*

Dynamic nucleation and growth using wet-STEM was described elsewhere [42]. The holey carbon grid samples immersed in distilled water were cooled down to 2 °C by a Peltier cooling wet-STEM stage prior the ESEM pump down process. The pressure was in situ reduced well below the dew point, i.e., down to 20 % relative humidity, for thinning out water films and obtaining self-supported overhanging nano-thick films at the carbon grid holes. Condensation on these thin water films was further obtained by elevating the chamber pressure in increments of 0.1 torr up to about 5.3 torr, while in situ wet-STEM imaging at constant sample temperature of 2 °C. The dynamics was successively recorded using BF-STEM acquisition mode with 1 s intervals. The same order of magnifications and similar e-beam dose were previously [7] applied in ESEM for wettability study of solid interfaces, providing a clear distinction between hydrophobic and hydrophilic regions with corresponding contact angle measurement.

Complete and partial wetting phenomena (filmwise and dropwise growths correspondingly) were observed during in situ water vapor condensation over thin water films in the carbon grid holes. For the BF-STEM images (Figs. 3.5, 3.6, and 3.7), the gray scale levels were bright for empty holes, dark for the carbon grid, and slight gray for the thin water layers inside the carbon holes. In principle, the existence of the thin water layers could be destructively checked by condensing over it a heavy water layer (demonstrated below at Fig. 3.5c, d).

Figure 3.5 shows several stages of the filmwise condensation process over the thin water films. The initial stage shows a thin water film inside a carbon hole (arrow at Fig. 3.5a). Upon increase of pressure from 5.3 torr (Fig. 3.5a) to 5.5 torr (Fig. 3.5b) at constant temperature of 2 °C, a continuous film of water is formed

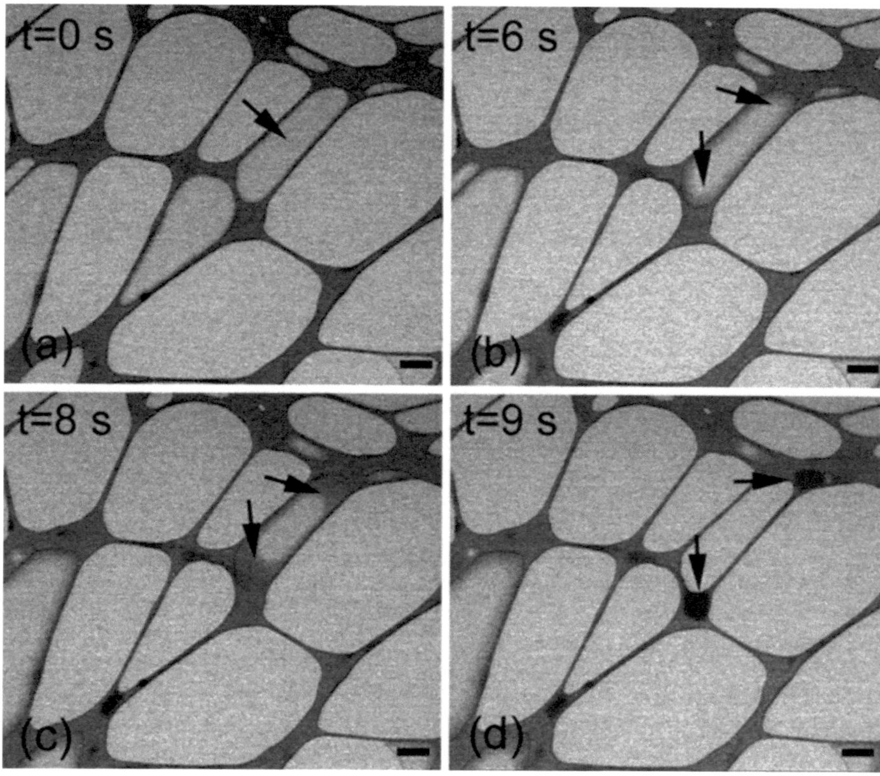

Fig. 3.5 (**a**)–(**d**) Wet-STEM in ESEM images of a sequence of filmwise growth stages on a self-supported water film (scale bar is 1 μm). (**a**) $t = 0$ s, (**b**) $t = 6$ s, (**c**) $t = 8$ s, and (**d**) $t = 9$ s

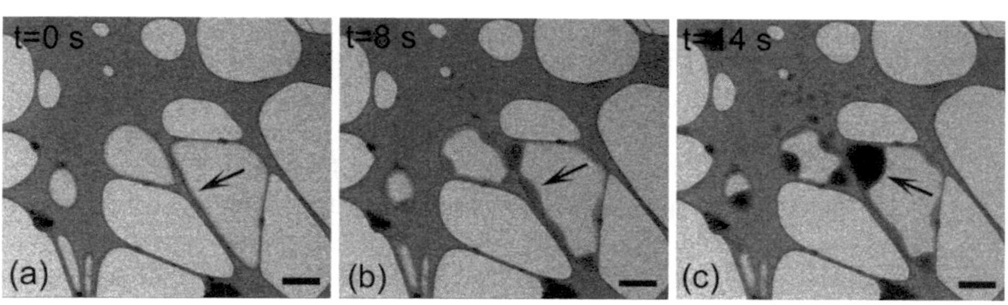

Fig. 3.6 (**a**)–(**c**) Wet-STEM in ESEM images of a sequence of filmwise growth stages on a self-supported water film (scale bar is 1 μm). (**a**) $t = 0$ s, (**b**) $t = 8$ s, and (**c**) $t = 14$ s

from the outer film boundaries (at the carbon interface) towards the center (arrows at Fig. 3.5b). As time proceeds, there is an increase in both the thickness and width of the condensed layer (Fig. 3.5c), and finally, it breaks up into drops (Fig. 3.5d) due to the capillary forces (for these scales the gravity is neglected).

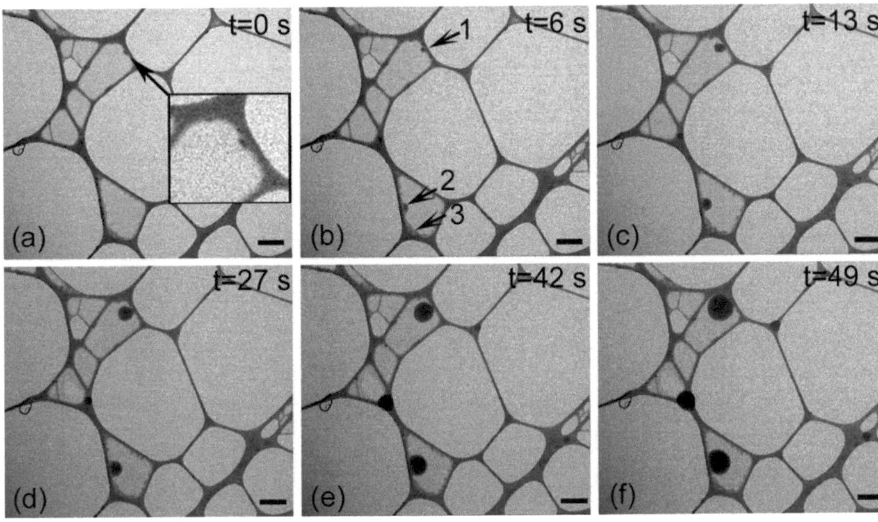

Fig. 3.7 (**a**)–(**f**) Wet-STEM in ESEM images of a sequence of dropwise nucleation and growth stages on a self-supported water film (scale bar is 1 μm). (**a**) $t = 0$ s, (**b**) $t = 6$ s, (**c**) $t = 13$ s, (**d**) $t = 27$ s, (**e**) $t = 42$ s, and (**f**) $t = 49$ s. ESEM environmental conditions are 2 °C and 4.4–5.3 torr

Figure 3.6b shows an asymmetric filmwise condensation with a wavelike structure (indicated by arrow) of about 1 μm periodicity and 200 nm deviation relative to the smooth initial state (Fig. 3.6a). Figure 3.6c shows a transition from film-like to droplet-like structure (pointed by arrow) upon further increase of RH in resemblance to Fig. 3.5d. The wavelike structure could be compared with the interfacial instability (known as Rayleigh–Taylor instability), which was reported [2] for perturbed liquid films on horizontal suspended substrates or cylindrical shapes, where the film wavelike structure kept on growing in amplitude until eventually it broke up into drops.

The water dropletwise condensation over water films could be possibly attributed to various effects as carbon contamination in ESEM, existence of air pockets, or possible deviation from equilibrium conditions. The dynamics of dropwise condensation and growth is here studied following the nucleation stage at the water–carbon boundary. A sequence of droplet condensation at selected time intervals is shown in Fig. 3.7.

A few nanodroplets ("1" and "2" in Fig. 3.7b) besides filmlike structure ("3" in Fig. 3.7b) have been preferably nucleated at the water interface in vicinity to the carbon boundary. The droplet signed as "1" (zoomed in at Fig. 3.7a) has a longitudinal diameter of 165 nm along the carbon interface and a perpendicular shorter diameter by a factor of 1.51. At the timescale between $t = 13$ s and $t = 49$ s, the droplet becomes of symmetrical circular shape, which is energetically favorable. The circularity has been defined [43] as

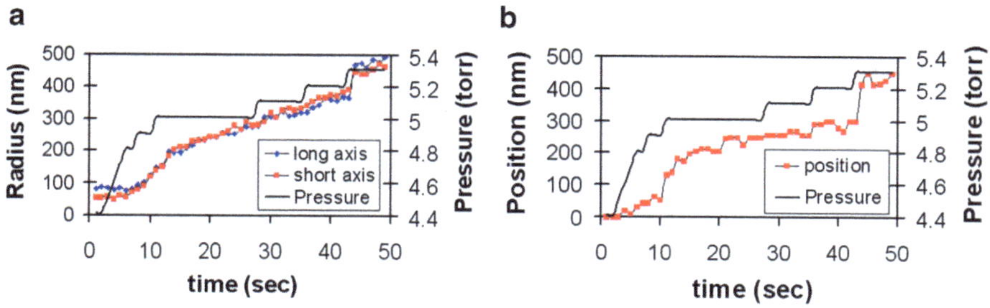

Fig. 3.8 (**a**) Droplet radius growth at two perpendicular directions versus time. (**b**) Droplet center of mass position versus time. The pressure variation within 4.4–5.3 torr is shown in both graphs

$$C = \frac{4 \cdot \pi \cdot A}{P^2} \tag{3.10}$$

where P is the perimeter and A is the area. The C parameter is 0.92 at the initial nucleation stage and approaches unity with accuracy of 1 % between 13 and 49 s. The time-dependent droplet growth radius at the two perpendicular directions is shown in Fig. 3.8a (for pressure range of 4.4–5.3 torr). The droplet eventually reaches the micron scale range, i.e., 960 nm in diameter (at $t = 49$ s) at dew point of 5.3 torr. The corresponding center of mass movement of the droplet is shown in Fig. 3.8b. The abrupt movements, upon increase of external pressure at $t = 11$ s and $t = 44$ s, are correlated with abrupt growth radius (Fig. 3.8a). The control of RH for this sample configuration could thus be used for nanodroplet movement over the liquid film at the vicinity of surface heterogeneities in similarity [7] to water manipulation over patterned hydrophilic and hydrophobic micro-channels on a solid substrate.

Figure 3.9 shows droplet growth radius as a function of time at constant RH conditions for duration of 14 s starting at $t = 13$ s (for droplets indicated as "1" and "2" at Fig. 3.7b). The droplet radius growth R is described [44] by power law dependence on time as

$$R \propto (t - t_0)^{\mu}, \tag{3.11}$$

where t is the time from droplet nucleation at t_0. The fitting provides a power law value of $\mu = 0.55 \pm 0.02$ for the upper droplet (indicated as "1" in Fig. 3.7b). Power law value of 0.30 ± 0.02 was measured for several asymmetric droplets, such as the one on the bottom of Fig. 3.7b (indicated as "2"). The accuracy of the estimation was improved by measuring for each droplet the radius at two perpendicular directions (in respect to the film boundary) and calculating the equivalent radius according to the droplet area. As elsewhere described [44], for small droplets, which have high perimeter to area ratios, the growth is dominated by diffusion of water from the surrounding substrate to the perimeter of the droplet and is expected to obey power law dependence on time with $\mu = 1/2$. According to this description, for large

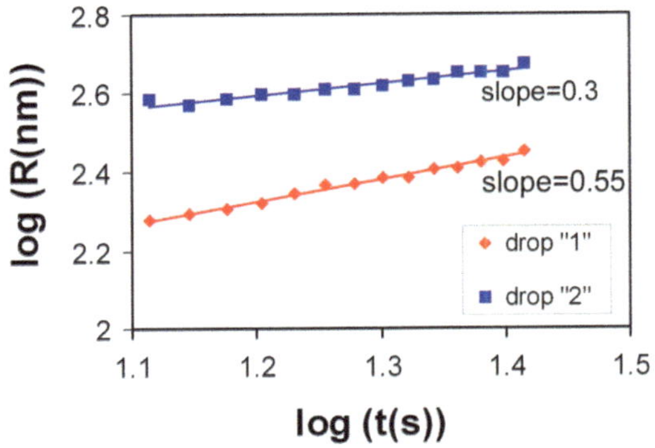

Fig. 3.9 The droplet growth radius at constant RH for two nanodroplets (signed as "1" and "2" in Fig. 3.7b)

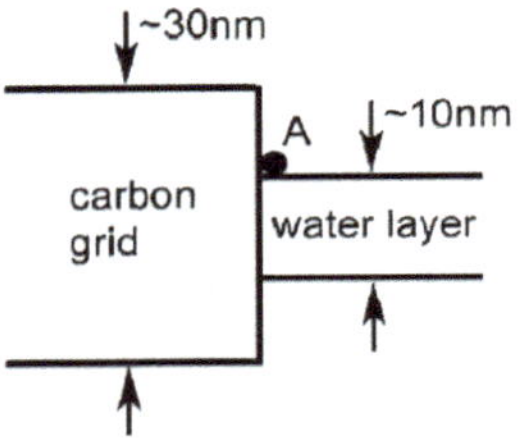

Fig. 3.10 Schematic cross section of the interface between the carbon grid and thin liquid layer with condensation site indicated at A

droplets, the vapor nucleates at the droplet surface with power law dependence of $\mu = 1$. However, lower power law values of $\mu = 1/3$ have been reported [45] as well. Our repeated experiments show a distribution of power law values corresponding to $\mu = 0.5 \pm 0.2$, i.e., with variation in μ value beyond the error of the experiment.

Following the previous work [40], estimation was done for the average thicknesses of the nano-water and carbon grid films, which yielded 10 nm and 30 nm correspondingly. Assuming a symmetric vertical geometry, the water film layer could be assumed surrounded by a carbon grid film with step height of approximately 10 nm at the water–carbon interface, which provided a condensation site (Fig. 3.10). The carbon film relative to the thin water film constituted a geometrical heterogeneity, which was possibly accompanied by chemical heterogeneities. The motion of nanodroplets near chemical heterogeneities was theoretically studied [46] assuming a smooth geometry. It was claimed that the dynamics of nanodroplets in the vicinity of chemical heterogeneities was determined by van der Waals long-range forces acting between the droplet and the different materials of the substrate rather by the short-range parts of the interaction potentials related to contact angles. In our case (Fig. 3.8b) the center of mass movement exceeded 440 nm towards the wet region, i.e., beyond the effectiveness of the van der Waals forces, while the droplet perimeter is still in touch with the water–carbon interface. We could thus assume that the dynamics of the droplet growth could be effected by structural heterogeneities although chemical heterogeneities should be considered as well at least for water films of a few μm size, i.e., comparable with the maximum droplet size.

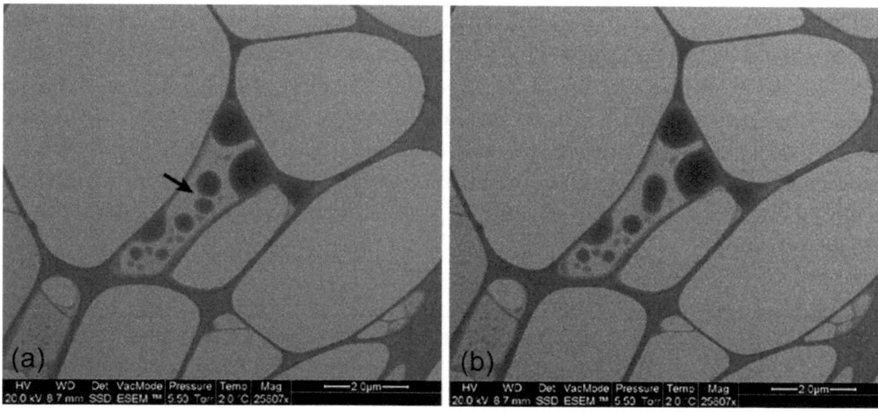

Fig. 3.11 Multi-droplets on a water film: (**a**) condensation and growth at 2 °C and 5.5 torr and (**b**) coalescence under the same conditions

Recent research [47] explored the interaction among nanodroplets. Multi-droplet condensation and growth over a self-supported thin liquid film were characterized for the first time for nanodroplet separation close to the resolution limit of 10 nm. Figure 3.11a shows the formation of flat facets of two nanodroplets (indicated by an arrow) upon interaction, while the coalescence occurs within the 1 s time resolution over 30 nm separation gap (Fig. 3.11b). The growth behavior of single nanodroplets is thus expected to be also affected by droplet-droplet interaction at the nanoscale.

3.4 Conclusions

Wettability characterization by transmitted electrons in ESEM using wet-STEM detector provided quantification of the initial stages of water vapor condensation over inhomogeneous water film areas. Both filmwise and dropwise phenomena were observed. MC simulation was fitted to experimental results using calibration procedure, which provided quantification of droplet shape and contact angle. Condensation and evaporation of nanoscale droplets over thin films could be dynamically in situ investigated by varying the relative humidity.

The kinetics of the nanodroplets supports previous power law growth description for the micron and sub-mm drops. However, the spread in power law values for a range of individual nanodroplets (even within the same scan area) might indicate sensitivity of the dynamics to structural heterogeneities at the film boundaries, initial nucleation stage (including nucleus shape asymmetry), and possible thermal fluctuations. The dependence of droplet growth power law values on relative humidity could be further explored as elsewhere described [48] as well as droplet interaction and coalescence events [49, 50].

The experimental part of wet-STEM acquisition, temperature, and humidity control could be further improved for better accuracy in nanoscale wettability study. The droplet growth study could be expanded for three dimensions for dynamic contact angle evaluation based on the calibration method. The whole study could be further applied to explore wettability properties of stand-alone polymer or biological films and for studying the interaction of nanomaterials with biofilms and within themselves. This method could be further implemented for nanoscale device characterization.

Acknowledgments The author thanks the "Center of Nanoscience and Nanotechnology" and the "Wolfson Applied Materials Research Center" at Tel-Aviv University for the ESEM facility.

References

1. Quéré, D.: Surface wetting: model droplets. Nat. Mater. **3**, 79–80 (2004)
2. De Gennes, P.G., Brochard-Wyart, F., Quéré, D.: Capillarity and Wetting Phenomena: Drops, Bubbles, Pearls, Waves. Springer, New York, NY (2004)
3. Bico, J., Thiele, U., Quéré, D.: Wetting of textured surfaces. Colloids Surf. A **206**, 41–46 (2002)
4. Bormashenko, E., Pogreb, R., Whyman, G., Erlich, M.: Cassie-Wenzel wetting transition in vibrating drops deposited on rough surfaces: is dynamic Cassie-Wenzel wetting transition a 2D or 1D affair? Langmuir **23**, 6501–6503 (2007)
5. Nosonovsky, M., Bhushan, B.: Biomimetic superhydrophobic surfaces: multiscale approach. Nano Lett. **7**, 2633–2637 (2007)
6. Berthier, J., Brakke, K.A.: The Physics of Microdroplets. Wiley, Hoboken, NJ (2012)
7. Aronov, D., Rosenman, G., Barkay, Z.: Wettability study of modified silicon dioxide surface using environmental scanning electron microscope. J. Appl. Phys. **101**, 084901–084905 (2007)
8. Mendez-Vilaz, A., Jodar-Reyes, A.B., Gonzalez-Martin, M.L.: Ultrasmall liquid droplets on solid surfaces: production, imaging, and relevance for current wetting research. Small **5**, 1366–1390 (2009)
9. Tsekov, R., Stockelhuber, K.W., Toshev, B.V.: Disjoining pressure and surface tension of small drops. Langmuir **16**, 3502–3505 (2000)
10. Satter, K.D.: Handbook of Nanophysics: Principles and Methods. CRC, Boca Raton, FL (2011)
11. Rauscher, M., Dietrich, S.: Wetting phenomena in nanofluidics. Annu. Rev. Mater. Res. **38**, 1–36 (2008)
12. Tsekov, R., Toshev, B.V.: Capillary pressure of van der Waals liquid nanodrops. Colloid J. **74**, 266–268 (2012)
13. Sarid, D.: Scanning Force Microscopy: With Applications to Electric, Magnetic, and Atomic Forces. Oxford University Press, New York, NY (1994)
14. Salmeron, M.: Scanning polarization force microscopy: a technique for studies of wetting phenomena at the nanometer scale. Oil Gas Sci. Technol. **56**, 63–75 (2001)
15. Williams, D.B., Carter, C.B.: Transmission Electron Microscopy: A Textbook for Materials Science. Plenum, New York, NY (1996)
16. Goldstein, J., Newbury, D., Joy, D., Lyman, C., Echlin, P., Lifshin, E., Sawyer, L., Michael, J.: Scanning Electron Microscopy and X-ray Microanalysis. Plenum, New York, NY (2003)
17. Grogan, J.M., Rotkina, L., Bau, H.: In situ liquid-cell electron microscopy of colloid aggregation and growth dynamics. Phys. Rev. E **83**, 061405/1–061405/5 (2011)
18. Ruach-Nir, I., Zrihan, O., Tzabari, Y.: A capsule for dynamic in-situ studies of hydration processes by conventional SEM. Microsc. Anal. **20**, 19–21 (2006)

19. Kishita, K., Sakai, H., Tanaka, H., Saka, H., Kuroda, K., Sakamoto, K., Watabe, H., Kamino, K.: Development of an analytical environmental TEM system and its application. J. Electron Microsc. **58**, 331–339 (2009)
20. Stokes, D.J.: Principles and Practice of Variable Pressure/Environmental Electron Microscopy. Wiley, Chichester (2008)
21. Leary, R., Brydson, R.: Characterisation of ESEM conditions for specimen hydration control. J. Phys. Conf. Ser. **241**, 012024/1–012024/4 (2010)
22. Stelmashenko, N.A., Craven, J.P., Donald, A.M., Terentjev, E.M., Thiel, B.L.: Topographic contrast of partially wetting water droplets in environmental scanning electron microscopy. Microscopy **204**, 172–183 (2001)
23. Brugnara, M., Volpe, C.D., Siboni, S., Zeni, D.: Contact angle analysis on polymethylmethacrylate and commercial wax by using environmental scanning electron microscope. Scanning **28**, 267–273 (2006)
24. Bormashenko, E., Pogreb, R., Whyman, G., Musin, A., Bormashenko, Y., Barkay, Z.: Shape, vibrations, and effective surface tension of water marbles. Langmuir **25**, 1893–1896 (2009)
25. Rossi, M.P., Ye, H., Gogotsi, Y., Babu, S., Ndungu, P., Bradley, J.C.: Environmental scanning electron microscope of water in carbon nanopipes. Nano Lett. **4**, 989–993 (2004)
26. Jung, Y.C., Bhushan, B.: Wetting behavior during condensation and evaporation of water microdroplets on superhydrophobic pattered surfaces. J. Microsc. **229**, 127–140 (2008)
27. Egerton, R.F., Li, P., Malac, M.: Radiation damage in the TEM and SEM. Micron **35**, 399–409 (2004)
28. Royall, C.P., Thiel, B.L., Donald, A.M.: Radiation damage of water in environmental scanning electron microscopy. J. Microsc. **204**, 185–195 (2001)
29. Lin, T.P.: Estimation of temperature rise in electron beam heating of thin films. IBM J. Res. Dev. **11**, 527–536 (1967)
30. Rykaczewski, K., Scott, J.H.J., Fedorov, A.G.: Electron beam heating effects during environmental scanning electron microscopy of water condensation on superhydrophobic surfaces. Appl. Phys. Lett. **98**, 093106–093109 (2011)
31. Bogner, A., Jouneau, P.H., Thollet, G., Basset, D., Gauthier, C.: A history of scanning electron microscopy developments: towards "wet-STEM" imaging. Micron **38**, 390–401 (2007)
32. Bogner, A., Thollet, G., Basset, D., Jouneau, P.H., Gauthier, C.: Wet-STEM: a new development in environmental SEM for imaging nano-objects included in a liquid phase. Ultramicroscopy **104**, 290–301 (2005)
33. Amaral, M., Borger, A., Gauthier, C., Thollet, G., Jouneau, P.H., Cavaille, J.Y., Asua, J.M.: Novel experimental technique for the determination of monomer drop size distribution in miniemulsion. Macromol. Rapid Commun. **26**, 365–368 (2005)
34. Maraloiu, V.A., Hamoudeh, M., Fessi, H., Blanchin, M.G.: Study of magnetic nanovectors by wet-STEM, a new ESEM mode in transmission. J. Colloid Interface Sci. **352**, 386–392 (2010)
35. Barkay, Z., Rivkin, I., Margalit, R.: Three-dimensional characterization of drug-encapsulating particles using STEM detector in FEG-SEM. Micron **40**, 480–485 (2009)
36. Probst, C., Gauvin, R., Drew, R.A.L.: Imaging of carbon nanotubes with tin–palladium particles using stem detector in a FE-SEM. Micron **38**, 402–408 (2007)
37. Anand, S., Son, S.Y.: Sub-micrometer dropwise condensation under superheated and rarefied vapor condition. Langmuir **26**, 17100–17110 (2010)
38. Bormashenko, E.: Correct values of Rayleigh and Marangoni numbers for liquid layers deposited on thin substrates. Ind. Eng. Chem. Res. **47**, 1726–1728 (2008)
39. James, D.W.: The thermal diffusivity of ice and water between −40 and +60°C. J. Mater. Sci. **3**, 540–543 (1968)
40. Barkay, Z.: Quantitative wettability study using transmission electrons in environmental scanning electron microscope. Appl. Phys. Lett. **96**, 183109/1–183109/3 (2010)
41. Joy, D.C.: An introduction to Monte Carlo simulations. Scanning Microsc. **5**, 329–337 (1991)
42. Barkay, Z.: Dynamic study of nanodroplet nucleation and growth on self-supported nanothick liquid films. Langmuir **26**, 18581–18584 (2010)

43. Zhao, H., Beysens, D.: From droplet growth to film growth on a heterogeneous surface: condensation associated with a wettability gradient. Langmuir **11**, 627–634 (1995)
44. Leach, R.N., Stevens, F., Langford, S.C., Dickinson, J.T.: Dropwise condensation: experiments and simulations of nucleation and growth of water drops in a cooling system. Langmuir **22**, 8864–8872 (2006)
45. Fritter, D., Knobler, C.M., Beysens, D.A.: Experiments and simulation of the growth of droplets on a surface (breath figures). Phys. Rev. A **43**, 2858–2869 (1991)
46. Moosavi, A., Rauscher, M., Dietrich, S.: Dynamics of nanodroplets on topographically structured substrates. Langmuir **24**, 734–742 (2008)
47. Barkay, Z.: Quantitative wettability study at nanometer scale based on WetSTEM in ESEM. Microsc. Microanal. **18**(Suppl 2), 1134–1135 (2012)
48. Hamill, P.J.: The time dependent growth of H_2O-H_2SO_4 aerosols by heteromolecular condensation. Aerosol Sci. **6**, 475–482 (1975)
49. Narhe, R., Beysens, D., Nikolayev, V.S.: Dynamics of drop coalescence on a surface: the role of initial conditions and surface properties. Int. J. Thermophys. **26**, 1743–1757 (2005)
50. Li, R., Ashgriz, N., Chandra, S., Andrews, J.R., Drappel, S.: Coalescence of two droplets impacting a solid surface. Exp. Fluids **48**, 1025–1035 (2010)

Chapter 4
Self-Assembly of Nanodroplets in Nanocomposite Materials in Nanodroplets Science and Technology

D. Keith Roper

Abstract Use of metal nanoarchitectures is increasing in electronics, diagnostics, therapeutics, sensing, and microelectromechanical systems due to their unique electromagnetic and physicochemical properties. This chapter examines physical, chemical, and hybrid methods to assemble metal nanodroplets in single- and multidimensional geometries and phases. Reductive self-assembly offers a route to economic, scale-able preparation of nanodroplets and stabilization on solid substrates that could lead to atom-level tune-ability. Enhanced control and real-time characterization have been used to uncover thermodynamic and transport mechanisms of nanodroplet self-assembly to enhance prediction and control of morphological features. Physicochemical principles of reductive nanodroplet self-assembly are examined to provide a framework to modulate local surface forces and control orderly self-assembly of metallic nanostructures.

4.1 Introduction

Nanoscale metal architectures are of growing interest in electronics, MEMS devices, diagnostics, biosensing, spectroscopy, and microscopy due to their unique electromagnetic and physicochemical properties. *Thin metal films* on semiconducting substrates are used in gateable electronic transistors [1, 2], and conductors [3], thermoplasmonic gratings to manipulate picoliter droplets [4], surface plasmon resonance (SPR) sensors (45–60-nm thickness), and SERS-active substrates [5–7]. Discontinuous *island films* are substrates for SERS [8], attenuated total reflection surface-enhanced IR absorption (SEIRA) [9] microspectroscopy [10], transmission localized surface plasmon sensors [11–13], and near-field scanning optical microscopy (NSOM) [14]. *Metal nanoparticles* (*NPs*) *assembled*

D.K. Roper (✉)
University of Arkansas, 3202 BELL, Fayetteville, AR 72701, USA
e-mail: dkroper@uark.edu

Z.M. Wang (ed.), *Nanodroplets*, Lecture Notes in Nanoscale Science and Technology 18,
DOI 10.1007/978-1-4614-9472-0_4, © Springer Science+Business Media New York 2013

on oxides find application in chemical and biosensing [15, 16], photovoltaics [17], thermal nanoprocessing [18, 19], nanoactuation [20], immunoassays [21], porphyrin-conjugated nanowires [22], waveguides [23], surface-enhanced Raman (SER) scattering [24–26], for single-molecule spectroscopy [27, 28], and SER nanosensors [29].

Such diverse and remarkable uses motivate developing a foundational understanding of, and technologies for, fabricating nanoscale metal architectures. Methods that allow enhancement of desirable features, predictable control, and economic scale-ability to manufacturing are highly desirable. This chapter is an overview of various physical, chemical, and hybrid methods to form nanoparticles in random and ordered arrangements and associate them with two- and three-dimensional solid substrates.

"Bottom-up" self-assembly of nanodroplets via electrochemical reduction offers a green, scale-able alternative to resource-intensive "top-down" strategies that use extreme, oft-toxic conditions. This chapter examines physicochemical and electromagnetic principles that underlie reductive metal NP self-assembly. Understanding these principles provides a framework within which modulating local surface forces to control orderly self-assembly of metallic nanostructures could provide atomic-level tune-ability of fine, nanoscale features.

4.2 Nanodroplets Self-Assembled in Suspensions

Colloidal suspensions of monodisperse gold (Au) NP are readily prepared using a method introduced by Turkevich [30] and later refined by Frens [31, 32],. Trivalent gold, Au(III) in hot, aqueous chlorauric acid (H[$AuCl_4$]), is reduced by adding trisodium citrate ($Na_3C_6H_5O_7$), which also acts as a capping agent. Reduction of Au(III) forms Au(0) atoms, which precipitate as red to red-violet, 12–20-nm AuNP in vigorously stirred suspension. Figure 4.1 illustrates AuNP made by this method. Brust [33] showed smaller, 5–6-nm AuNP could be prepared in similar fashion using sodium borohydride ($NaBH_4$) reductant in organic solutions containing an anticoagulant. Larger, 30–250-nm AuNP may be formed by hydroquinone reduction onto NP seeds [34]. Suspended AuNP may also be formed by block copolymer reduction and stabilization [35, 36].

4.2.1 Electrochemistry of Nanoparticle Reduction

Metal nanoparticles self-assemble from dissolved metal salts as a consequence of their electrochemical reduction potential. Reduction potentials for a number of metals at ambient conditions are listed in Table 4.1. The noble metal gold (Au) tops the list as a strong oxidizing agent with a low tendency for donating electrons. Its low reactivity and stability against oxidation paired with its high thermal and electrical

Fig. 4.1 Aqueous suspension of gold nanoparticles prepared by reduction of Turkevich and Frens

Table 4.1 Standard reduction potentials in aqueous solution, 25 °C

Ions	E^0(V)	Condition	Reference
Succinimidyl radical (*N*-bromosuccinimide)	>+1.84	vs. normal hydrogen electrode (NHE) from cyclic voltammetry (ACN)	[37]
$Au^+ + e^- \rightarrow Au(s)$	+1.68	vs. standard hydrogen electrode (SHE)	[38]
$Au^{3+} + 3e^- \rightarrow Au(s)$	+1.50	vs. SHE	[39]
$Au^{3+} + 2e^- \rightarrow Au^+$	+1.41	vs. SHE	[39]
$Ag^+ + e^- \rightarrow Ag(s)$	+0.80	vs. SHE	[40]
$Cu^+ + e^- \rightarrow Cu(s)$	+0.52	vs. SHE	[40]
$Cu^{2+} + 2e^- \rightarrow Cu(s)$	+0.34	vs. SHE	[40]
$Cu^+ + e^- \rightarrow Cu(s)$	+0.52	vs. SHE	[40]
$Cu^+ + e^- \rightarrow Cu(s)$	+0.52	vs. SHE	[40]
$Cu^{2+} + e^- \rightarrow Cu^+$	+0.15	vs. SHE	[40]
$Sn^{4+} + 2e^- \rightarrow Sn^{2+}$	+0.15	vs. SHE	[40]
$Sn^{2+} + 2e^- \rightarrow Sn(s)$	−0.14	vs. SHE	[40]
Formaldehyde (HCHO)	−1.07	vs. SHE	[41]
3,3′,5,5′-tetramethylbenzidine (TMB)	0.22	(Acetonitrile)	[42]

conductivity make plasmonic Au a metal of choice, particularly in reaction-sensitive bio/medical theranostics, optoelectronics, and corrosion-intolerant electronic and deep space applications. Interest in monovalent gold Au(I) compounds has grown recently due to their unique biological activity [43], their industrial utility as catalysts [44], and their role as oxidation agents in electroless reduction [45]. As examples, Au(I) phosphine complexes have been studied to treat breast cancer [46] and to catalyze oxidation of CO gas [47]. Solid-state islands, films, and particles of Au(0) may be formed by electroless reduction of Au(I) from aqueous solution [48–51]. Au(I) galvanically displaces the moderately oxidizing silver Ag(I) ion in electroless reduction onto oxide and polymer substrates.

Reactions of copper (Cu) and tin (Sn) cations occur under a broad range of conditions, affected by pH, conductivity, dissolved oxygen, and other solution conditions. Reduction potentials for these metal ions and the blue/yellow colorimetric reporter tetramethylbenzidine (TMB) lie intermediate between those of strong reducing formaldehyde (CH_2O) and oxidizing N-bromosuccinimide (NBS). The agents $Na_3C_6H_5O_7$, $NaBH_4$, and hydroquinone reduce Au(III) to Au(O) in AuNP suspensions. CH_2O reduces Au(I) [52] onto bare silica glass [53] and polymer [48] without electrical current. Glycerol ($C_3H_8O_3$) [54] and hydroxylamine (NH_2OH) [55] also reduce metal ions onto substrates like polymer-coated glass. Interestingly, titration of CH_2O with NBS has been used to interconvert Au between 0, I, and III oxidation states, effecting a sensitive new TMB-based spectrophotometric assay for Au(I) [56].

4.3 Nanodroplets Assembled on Two-Dimensional Substrates

Metal nanoparticles have been assembled in random and ordered configurations on two-dimensional substrates using a range of physical, chemical, and combined methodologies. These techniques and envisioned applications along with their corresponding strengths and limitations are summarized in Tables 4.2 and 4.3 [57]. Primary physical methods for assembling NP in random arrangements have been pulsed laser evaporation, sputtering, and microwave-assisted deposition by varying time of operation. Such mechanical means can introduce solid-state metal islands, films, and particles of interest (Au, Ag, Cu, Ti, etc.) onto nonconductive (Si and SiO_2) surfaces. However, this requires expensive and sophisticated equipment that operates at extreme temperature and pressure conditions. While NP formation is fast and easy, the size, morphology, and distribution of NP may be highly nonuniform. Alternatively, evaporated thin films may be thermally annealed to form NP with improved uniformity without effects of stabilizing molecules.

Primary chemical methods for assembling NP in random arrangements have been chemical functionalization; self-assembly on thiols, dendrimers, or other polymers; layer-by-layer assembly; templating using carbon nanotube and DNA; and self-assembly at solvent interfaces. Interfacial interactions and surface forces, resulting from surfactant, pH, ionic strength, or surface preparation, for example, are critical in these methods to predictably assemble and stabilize NPs. Alkanethiol self-assembly is the most widely used method of surface-associating NPs. NPs also self-assemble inexpensively at solvent/solvent interfaces via close-packing, but reproducibility is limited by control of system parameters. NPs can be templated

Table 4.2 Fabrication methods for assembling metal NP on 2D surfaces: examples (Adapted from [57])

Techniques			Metals/substrates	Potential applications	Reference
Physical methods	Pulsed laser evaporation		TiO_2/Si	Gas sensors	[58]
	Sputtering		Au/SiO_2	Optical sensors	[59, 60]
	Microwave-assisted deposition		Au/SiO_2	Microelectronic devices	[61]
	Thermal annealing	Colloidal NP	Ag/Si	Microelectronic devices, SERS	[62]
		Evaporated films	Au/SiO_2		[63–65]
Chemical methods	Functionalization	Silanization	Au or Ag/SiO_2	Biosensors, SERS	[66, 67]
	Self-assembly	Alkanethiol	Au/quartz	SERS	[68]
		Dendrimer	Au or Ag/Si		[69]
		Polymer	Au or Ag/Si		[70]
	Layer-by-layer assembly		AuNPs linked with myoglobin/SiO_2	SPR sensors	[71]
	Templating	Carbon nanotube	Ag or Pt/SiO_2	Catalytic applications, biosensors	[72]
		DNA	Au/gold surface		[73, 74]
	Self-assembly at solvent interface		Au/SiO_2	SERS	[75]
Combined	Two-phase solvent interface	Langmuir- Blodgett	Charged Au/SiO_2 or Mica	Optoelectrical/ magnetic applications	[76]
		Substrate pulling	Au, Pt, SiO_2/SiO_2		[77, 78]

Table 4.3 Fabrication methods for assembling metal NP on 2D surfaces: strengths and limitations (Adapted from [57])

Methods		Process	Strengths	Limitations	Reference
Physical methods	Pulsed laser evaporation	• Thermal vaporization of source material to deposit metal	• Deposition rate and final film thickness controlled • High-purity metal films	• High capital and vacuum • Low tolerance to contamination • Weak metal-substrate bond	[79]
	Sputtering	• Physical ejection of surface atoms by momentum transfer to deposit metal	• Material versatility • Easier than vacuum evaporation • Little heat generated	• Low sputter rate • Poor source material use • Contaminated by plasma activation • Weak metal-substrate bond	[72]
	Microwave-assisted deposition	• Microwave irradiation to substrates in colloidal NP suspensions	• Easy and fast deposition	• Substrate pre-silanization needed • Post NP growth process needed	[80]
	Heating metal films	• Thermal heating of ultrathin evaporated metal films	• No stabilizing molecules • No slow diffusion to surface	• High temperatures • NP size and distribution vary	[81]
Chemical methods	Functionalization	• Derivatize molecular anchors on substrate to stabilize NP	• Range of colloidal NP sizes • Ambient conditions used	• Slow diffusion to surface • Primarily sulfur-gold chemistry	[82]
	Templating	• Use secondary molecules to assemble NPs	• Inexpensive • Ambient conditions used	• Labile templates • Size-limited	[83, 84]
	Self-assembly at solvent interface	• Self-assemble close-packed NP arrays at interface of two solvents	• Inexpensive • Environmental-friendly • Large-area array formation	• Noncontrollable system parameters	[75]
Combined	Two-phase solvent interface	Langmuir-Blodgett • Layer forms at interface	• Robust array • Uniform thick • Ambient condition used	• Smooth 2D surface required • Slow deposition • Limited materials	[76]

onto surfaces using biologics like DNA, protein, yeast cells, or bacterial surface layers provide biocompatibility, but these assemblies are limited to ca. nano- to micrometers, and biological templates are labile.

Combined physical/chemical methods such as mechanical Langmuir-Blodgett or substrate pulling are used to assemble hydrophilic NPs such as Au and Pt into ordered monolayers at water/oil or toluene/water interfaces. Langmuir-Blodgett (LB) monolayers robustly assemble NPs, but require exceptionally smooth 2-dimensional substrates and nontrivial capital expense. Weak interactions between NPs and substrates in LB assemblies render them unstable towards various chemical and physical conditions such as heating, aging, and chemical treatment [85].

4.3.1 Electroless Deposition on Substrates

Compared with other physical or chemical methods, redox-driven electroless (EL) deposition of metal has the advantage of its ability to rapidly coat fragile, three-dimensional, and internal surfaces at ambient conditions without the need for conductive substrates or expensive, sophisticated equipment. EL deposition is distinct from electroplating, in which an external current source is applied to reduce cations of a desired metal onto a conductive surface. EL plating autocatalytically reduces diverse metals including Au, Ag, Ni, Cu, and Pt onto various nonconductive substrates based on electrochemical potential [86].

EL-deposited metal films exhibit enhanced uniform and continuity relative to sputtered or evaporated films as shown in Fig. 4.2 [50]. They are also stable to stringent thermal, solvent, and electromagnetic (EM) exposure. In contrast, vacuum-deposited metals bond weakly to substrates due to lattice mismatch. Bonding can be improved by pre-coating, such as with Cr layers or thiol-terminated self-assembled monolayers, but pre-coating limits functionality like the sensitivity of metal film biological sensors [87].

To effect EL deposition, a substrate like silica (Si) or polymer is exposed to a series of solutions of metal ion with increasing reduction potentials. Galvanic substitution of less noble metals by EL deposition avoids use of toxic reagents like cyanide or concentrated hydrofluoric acid used to plate Au(III) from potassium tetrachloroaurate [88]. Hazardous plating solution wastes typically generated after electroplating are eliminated.

EL deposition of nanowires [48], nanocrystals [89], nanodomes [90], thin films on self-assembled AuNP colloid monolayers [82, 91, 92], electrodes [93], and biosensor surfaces have been reported. Deposition rate, bath stability, and resulting crystal structure are influenced by plating solution composition, temperature, pH, agitation, and reducing agent [94–96]. EL plating by galvanic substitution [36, 97–101]. has been used to deposit continuous Au structures like islands and thin films on ceramic and polymer substrates followed by electrical [31, 102] thermal [103], or flame [104] annealing to modify surface properties.

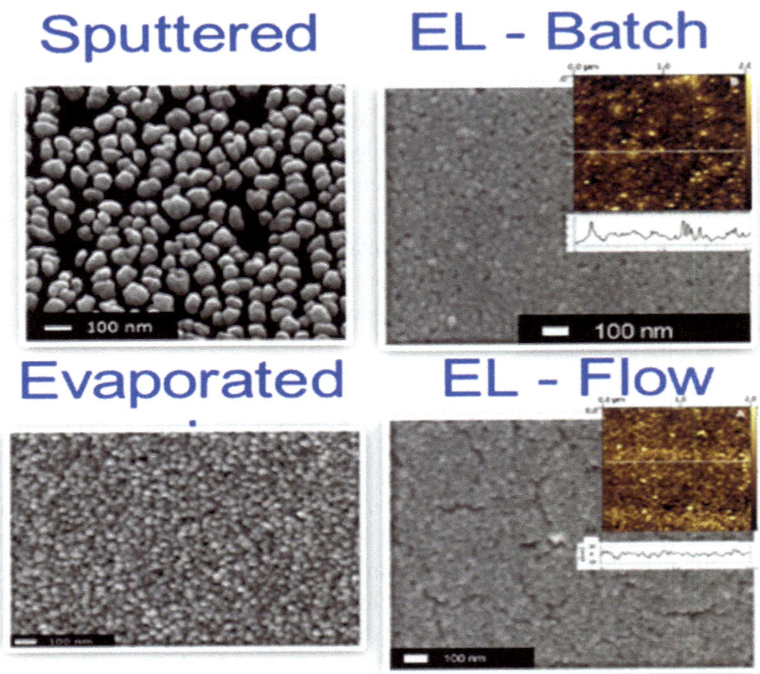

Fig. 4.2 Comparing morphologies for sputtered, evaporated, and electroless deposited (batch and continuous flow) Au gold island thin films

4.3.2 Nanodroplets Self-Assembled from EL-Deposited Films

EL deposition was recently extended to assemble random and ordered NP on silica (Si) substrates by thermal coalescence of deposited metal island film into nanospheres. Thermal reshaping reduces the large radius of curvature at vertices of island films at which high surface tension occurs [105]. This transformation occurs at temperatures lower than bulk melting points. Lower melting points are reported in semiconductor nanocrystals [106] and metal NPs [107] compared to bulk due to higher surface area to volume ratios. In such nanostructures, a quasimolten phase appears that exhibits different structures at temperatures well below melting [108]. Equilibrium surface melting of AuNP in aqueous suspension is reported at several hundreds of degrees K lower than bulk melting temperature [109].

In the reported process, metallic silver was galvanically [97, 110] replaced by Au after Ag had been reduced onto a tin-sensitized [111] Si surface. X-ray diffraction (XRD) showed predominantly (111) orientation in the Au film, which has been associated with increased SERS intensity and resolution. Adjusting the times of EL metal deposition and subsequent heating at 250 °C and 800 °C resulted in monomodal ensembles of AuNPs from (9.5 $\pm$ 4.0) to (266 $\pm$ 22) nm at densities

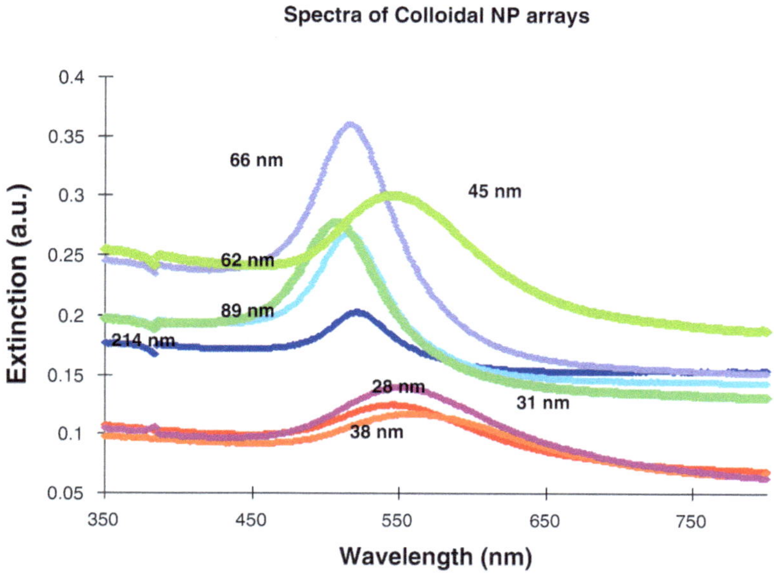

Fig. 4.3 Spectra of various AuNPs formed on Si by heating EL plated thin films

ranging from 2.6×10^{11} to 4.3×10^{8} particles cm^{-2}. These density values were the highest reported at that time for NPs assembled on substrates. Measurements by scanning electron (SEM) and atomic force microscopy (AFM) confirmed these results. Figure 4.3 shows transmission UV–vis spectra of AuNP assembled on Si by heating EL-deposited island thin films.

4.3.3 Electrochemical Mechanism of EL Deposition

Electroless metal deposition consists of a series of four electrochemical steps that can be categorized by the electron donor source: (1) substrate dehydration and sensitization, (2) metal catalyzed reduction, (3) galvanic replacement reduction, and (4) autocatalytic reduction. Description of these steps lacks the attention given to vapor deposition of metal films, which is described as kinetic adsorption and diffusion of adatoms at high gas pressure and temperature [112]. In vapor deposition, nucleation is followed by coalescence with subsequent increase in film thickness via externally donated electrons [113]. Demands to lower process temperature result in polycrystalline thin films. Effects of deposition parameters on crystallinity are described by temperature-dependent structure zone models [114, 115] validated by in situ time-resolved micrographs [116, 117].

An overview of the major steps in electroless deposition of tin, silver, and gold is illustrated in Fig. 4.4 [57].

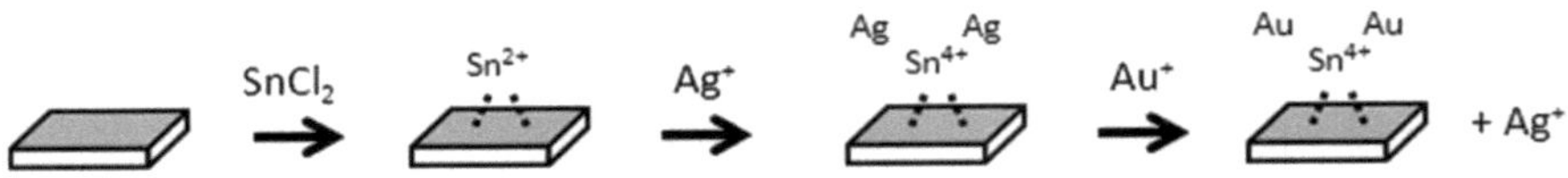

Fig. 4.4 Schematic diagram of the electroless gold-plating procedure: (**a**) substrate pretreatment with 25 % HNO_3, (**b**) tin sensitization, (**c**) silver activation, and (**d**) galvanic displacement of silver by gold

Ceramic substrates are pretreated to remove in/organic contaminant and to promote hydroxylation for subsequent improved adhesion of metal films to the surface [118]. Pretreatments include dilute nitric acid (25–50 %) [55] or piranha solution [118] for glass and concentrated hydrofluoric acid etching for silicon and germanium.

Pretreatment is followed by tin sensitization. Wetting a target surface with an acid solution of $SnCl_2$ generates SnO_2 from the hydrolysis of stannous ion (Sn^{2+} (II)). The stannous ion is adsorbed on the surface where it acts as a molecular anchor for subsequent adsorption of catalytic nuclei [119, 120]. Possible surface reactions that have been suggested for tin sensitization on SiO_2 glass substrate are [121, 122]:

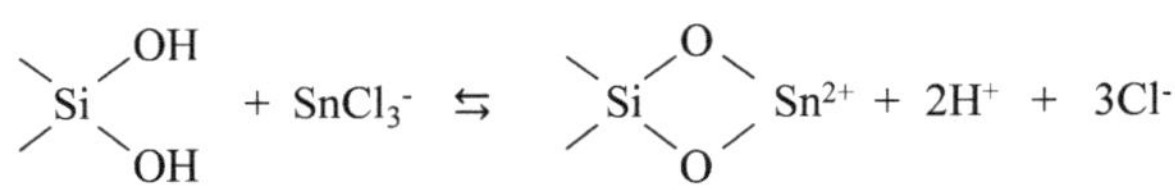

(4.1)

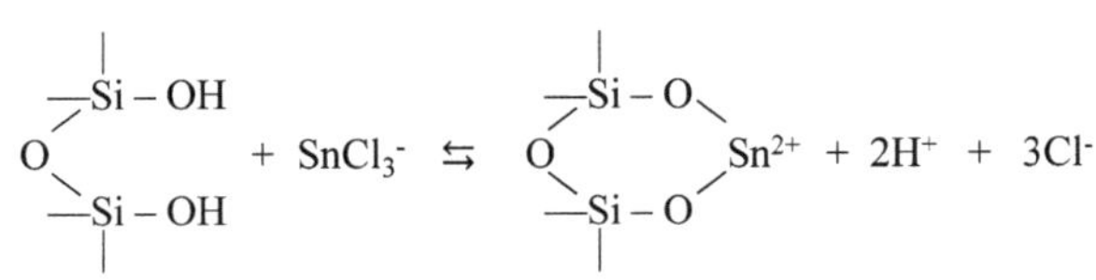

(4.2)

Sn-sensitized silicate is an optically transparent conductor [123] used as a transparent electrode in dye-sensitized solar cells [124]. It has no observable features in transmission UV–vis spectra or scanning electron microscope images, due to electron charging effects in the latter [125].

Exposing Sn-sensitized silicate to ammoniacal silver nitrate solution oxidizes adsorbed Sn(II) to Sn(IV) concurrent with reduction of Ag(I) to elemental Ag(0), which adheres strongly to the surface.

The suggested mechanisms are [126]:

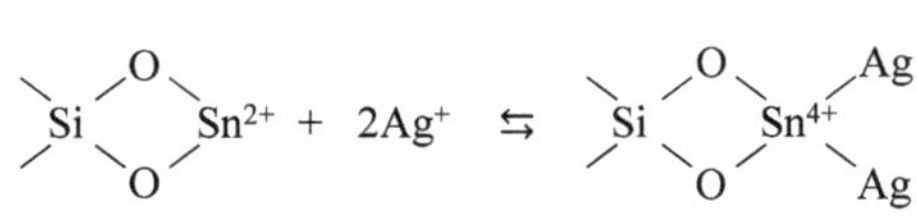

(4.3)

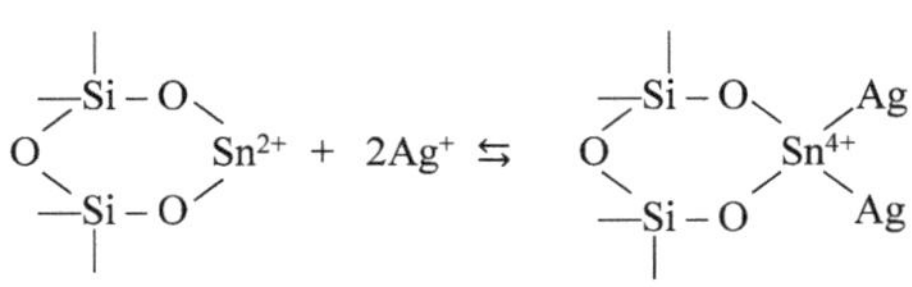

(4.4)

Transmission UV–vis (T-UV) spectra of adherent Ag(0) exhibit a shallow valley centered at ~500 nm, attributable to a Ag island film. Comparable spectra were reported for unannealed Ag island films of both 2-nm [127] and 5-nm [128] thickness.

In situ transmission UV–vis spectroscopy suggests this reaction is limited by available Sn^{2+} and sensitive to light and reductant [126]. Reacting ammoniacal Ag with Sn^{2+} in darkness prevents photochemical (UV) reduction of Ag ions to blackish metallic clusters, ostensibly in the presence of trace dissolved halides. Introducing reductant to adherent Ag(0) induces its self-association into Ag NP and agglomerates; and concomitant light exposure removes Ag(0) from the surface.

Metallic Ag may be galvanically displaced by more noble metal Au(I) ions in the presence of a reducing agent, CH_2O, which plays the role of an electron donor. Galvanically substituted Au particle then acts as a catalytic site to support growth of Au island film which converges and thicken [49]. Proposed mechanisms for galvanic replacement (Eq. 4.5) and autocatalytic Au film growth (Eq. 4.6) are:

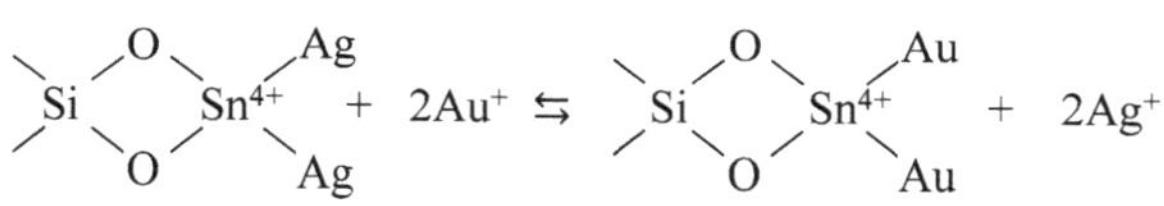

(4.5)

$$2Au^{+} + HCHO + 3OH^{-} \leftrightarrows HCOO^{-} + 2H_2O + 2Au \tag{4.6}$$

EL-deposited Au thin films exhibit a photoluminescence (PL) feature at ~500 nm in T-UV spectra. The PL peak red-shifted from 490 nm to 504 nm with increasing deposition time between 2 min and 6 min [50]. In situ T-UV corroborated by X-ray photoelectron spectroscopy (XPS) indicates transitional formation of Ag/Au alloy during galvanic displacement [129]. Thickened Au island films retain trace elemental Ag at their outermost strata.

4.3.4 Enhanced Nanodroplet Features: Continuous EL Deposition

Bottom-up electroless deposition of metal as it is described and implemented occurs in "batch" mode. Surfaces on which NP are to be deposited are immersed into a series of metal ion solutions. Metal ion and reductant content of the plating solutions varies with time as electroless deposition, galvanic replacement, and reduction occur. For example, concentration of the reduced ion is depleted adjacent to the substrate as metal ions diffuse through a stagnant mass-transport boundary layer towards the surface where deposition occurs. This procedure renders existing batch immersion (I) methods of EL deposition labor and capital intensive as well as time-variant, which limits their scale-ability and reproducibility.

A continuous flow (CF) method for EL deposition was recently introduced, which improved physical and optical properties of deposited metal thin films and NPs [50]. Continuous injection of a series of reducible metal ion solution into a confined rectangular substrate channel produced higher, time-invariant metal deposition rates. This was a result of a higher, steady concentration driving force across a narrower stagnant boundary layer adjacent to the surface. Physical surface morphology of CF and I-EL films was compared using transmission UV–vis spectroscopy (T-UV), scanning electron microscopy (SEM), and atomic force microscopy (AFM). The photoluminescent (PL) features revealed by transmission UV–vis spectra show that CF-EL films are more opaque and less dispersive in spite of shorter plating times.

Nanoparticles (NPs) obtained by thermal transformation of CF-EL Au film were larger and more monodisperse relative to NPs formed by comparable I-EL procedures. Extinction spectra of CF-EL AuNP exhibited higher peak height (PH) due to enhanced localized surface plasmon resonance (LSPR) and narrower full width at half maximum (fwhm) than NP from I-EL films. Comparison with a continuum model for Fickian film diffusion showed optical features CF-EL nanostructures measured by T-UV such as peak height, full width at half max, and resonant extinction maxima each increased in proportion to time-integrated mass-transport coefficient. Au deposition during CF-EL plating was calculated to be 78.3 % faster than I-EL plating. Importantly, these results validate use of a continuum models to predict and control CF-EL Au film deposition at nanometer scales.

4.3.5 Photochemistry and Transport of Droplet Formation

Continuous flow EL deposition also improves understanding of physicochemical mechanisms during self-assembly of noble metal nanodroplets. Equilibrium hydraulic conditions attained during CF-EL allowed direct identification of dynamic, morphological, and physicochemical changes in silver (Ag) during redox-driven self-assembly of metal films and particles on silica surfaces [126]. Spectral changes observed in real time by transmission UV–vis (T-UV) spectroscopy were attributed to morphological and physicochemical transitions by correlation with subsequent time-resolved scanning electron microscopy (SEM) and X-ray photoelectron spectroscopy (XPS) measurements. Specifically, optical features consistent with localized surface plasmon resonance, surface plasmon polaritons, and photoluminescence from Ag and related gold (Au) nanoarchitectures that included clusters, particles, and films were identified in the time-resolved T-UV spectra. These spectra were obtained in situ using a novel continuous electroless metal deposition system outfitted with a transparent flow cell.

For example, silver deposited onto tin-sensitized surfaces nucleated, aggregated, or dispersed nanoparticles upon exposure to reductant and darkness or broadband light, respectively, as shown in Fig. 4.5.

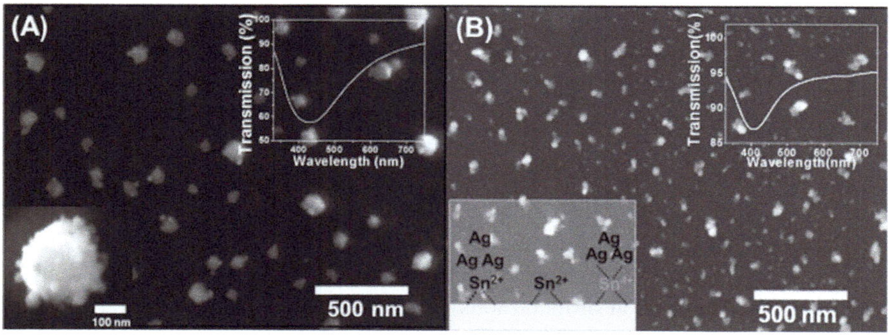

Fig. 4.5 SEM images of wideband light exposure effect to HCHO-treated Ag film on silicate substrate: (**a**) dark condition and (**b**) continuous exposure

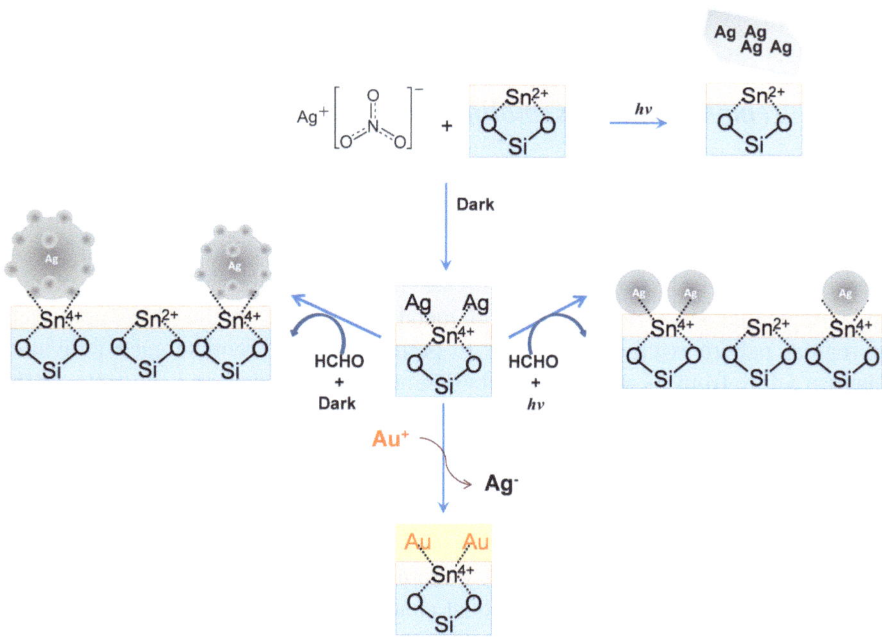

Fig. 4.6 Silver disposition during electroless metal film deposition

Monitoring similar kinetic changes in plasmon features suggested that four previously unrecognized time-dependent physicochemical regimes occur during consecutive EL deposition of Ag and Au onto tin-sensitized silica surfaces. These four regimes are illustrated in Fig. 4.6.

First, addition of ammoniacal silver to tin-sensitized silica substrate results in silver particles and aggregates upon exposure to light (hv) or in self-limited formation of a transparent reduced Ag(0) monolayer in the absence of light. Second, addition of exogenous reductant—a mixture of sodium sulfate and HCHO—transitorily produces Ag NP from Ag film due to axial dispersion of the smaller reductant at a rate faster than predicted for Au(III). These NPs are compound agglomerates which remain associated to the surface in the absence of light or smaller more monodisperse NPs that are stripped from the surface upon exposure to light. Third, surface plasmon resonance features indicate transitional Au/Ag alloy formation, confirmed by XPS. Finally, continued exposure to Au(I) and reductant produces self-catalyzed Au film thickening.

These new methods enhance the mechanistic detail of our understanding of dynamics and disposition of metal ions during EL plating. Their consistency with continuum models of mass transfer also allows design and implementation of improved control over particle nucleation and coalescence as well as morphology and continuity of ultrathin films at the atomic scale in laboratory experiments as well as industrial applications.

4.3.6 Templates for Assembling Ordered Nanodroplets

Ordering nanodroplets into two-dimensional arrays may result in interactions between plasmon polarization and Bragg diffraction [130]. This can enhance performance of the nanodroplets for useful applications like increasing sensor sensitivity [131]. Techniques to fabricate ordered NP arrays consist overall of two sequential steps: templating and metal deposition. In the templating step, a regular pattern is created in order to periodically differentiate each individual nanostructure from its surrounding environment on an underlying substrate. In the deposition step, source material such as Au or Ag metal is preferentially introduced at regularly spaced intervals on the underlying substrate by virtue of the patterned template. Techniques for templating and deposition may be broadly classified into conventional "top-down" and self-assembly "bottom-up" categories.

"Top-down" methods for patterning and deposition rely on precise, directed control of EM waves, electrons, ions, or atoms in extreme environments to produce a uniform, irreversible, one-dimensional change to a target template or substrate. Top-down *electronic* methods for templating use focused beams of electrons (i.e., electron beam lithography, EBL), ions (i.e., focused ion beam, FIB), reactive ion etching (RIE), or interfering light (i.e., laser interference lithography, LIL). Top-down *mechanical* methods for templating employ either nanoscale manipulation of a nanoscale element as in atomic force microscopy (AFM) and dip-pen lithography (DPN) or impression of soft matter with a harder "negative" template as in nanoimprint lithography or microcontact printing. Various thermal, ultraviolet, and mechanical refinements to nanoimprint lithography have been explored to improve conformity, reproducibility, and resolution. Top-down deposition methods consist primarily of the physical methods considered in Tables 4.2 and 4.3.

Top-down patterning and deposition methods provide a direct route to fabricating many ordered nanostructures. But they require significant resources in terms of capital, time, and training. Utility and economic scale-ability of top-down methods is limited by capital and energy requirements, optoelectronic resolutions, serial throughput, and unavailability of desirable shapes or depositable materials.

4.3.7 Bottom-Up Templates for Ordered Self-Assembly

Bottom-up templating and deposition methods rely on manipulation of local surface forces due to phase, electromagnetism, chemical composition, and thermodynamic state to arrange molecular entities into ordered nanostructures. Chemical and combined methods considered in Tables 4.2 and 4.3 comprise a number of widely examined bottom-up approaches.

Nanosphere lithography (NSL) is an example of bottom-up templating. Suspensions of monodisperse micro- to nanoscale silica (Si) or polymer beads naturally coalesce into ordered colloidal crystals that are from millimeters to centimeters in scale as the diluent evaporates [129, 132, 133]. Such colloid crystals provide a hexagonal template on an underlying surface in which interparticle spacing can be tuned using different sized beads. Figure 4.7 compares steps in "top-down" patterning and plating by electron beam lithography and evaporation, respectively, with "bottom-up" patterning and plating by nanosphere lithography and electroless (EL) plating, respectively.

Top-down templating and bottom-up deposition have been combined to produce ordered nanodroplet arrays. Top-down deposition methods deposit nanostructures that have right-angled edges. Spherical AuNPs have been formed by thermally annealing nanostructures that were formed by depositing metal on templates created by EBL [134, 135] and NSL [136, 137] at temperatures ranging from 500 to 900 °C. Mobility and coalescence of the Au in nanostructures induced by heating result in significant changes in morphology. Higher temperatures exaggerates morphological changes like increased average area of individual Au islands and increased gap distance between these islands after thermal annealing Au island films evaporated on SiO_2 or mica substrates at temperatures ranging from 200 to 350 °C [138–141].

Adding an intermediate silanization step, as shown in Fig. 4.7, before metal deposition allows the HCP geometry formed by larger spheres in NSL to be replicated on the substrate [142]. Bare SiO_2 substrate underlying close-packed colloidal SiO_2 beads was masked against EL deposition by silanization using dimethyldichlorosilane (DMDCS). Controlling availability of surface-associated water was critical in steps of the silanization reaction. It allowed tunability of the shape and dimension of resulting nanostructures on the DMDCS-Si bead templated surface.

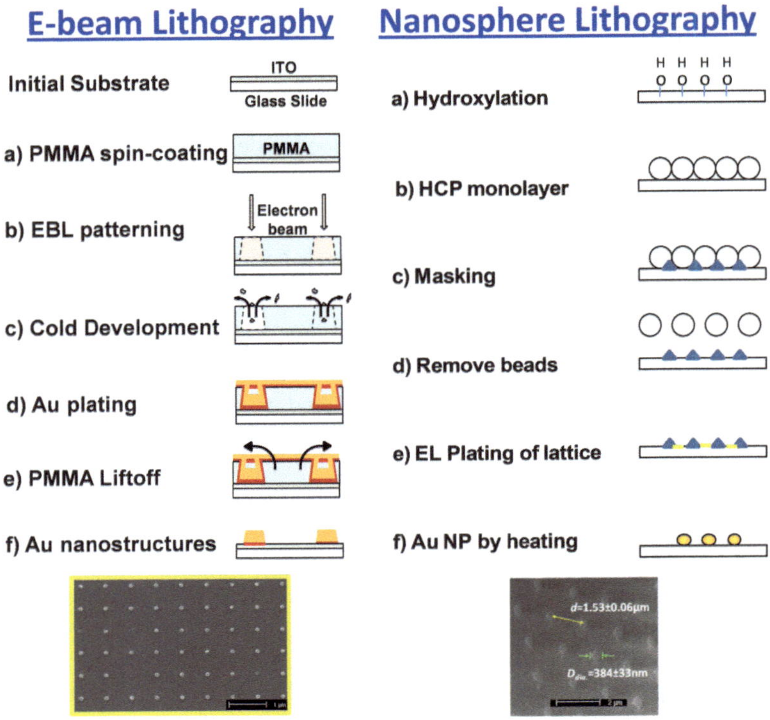

Fig. 4.7 *Top-down* fabrication of square arrays of Au nanocylinders by lithographic electron beam degradation of a poly(methylmethacrylate) resist followed by Au evaporation. *Bottom-up* fabrication of hexagonal closed-packed (HCP) arrays of Au nanospheres by templating nanospheres in a hexagonally close-packed (HCP) monolayer followed by electroless (EL) plating over a hydrophobic mask and thermal annealing. Copyright © Phillip Blake 2008

4.4 Nanodroplets Self-Assembled in 3-Dimensional Matrices

As a consequence of their unique attributes, AuNPs have been incorporated into dense nonporous dielectrics and films in random, three-dimensional assemblies. NPs are reduced from hydrogen tetrachloroaurate ($HAuCl_4$, TCA) Au(III) into polydimethylsiloxane (PDMS) by superficial permeation into cured substrates [143] and by mixing into polymer precursors [144–146]. The balanced redox reaction between Si-H groups and $HAuCl_4$ is reported to be:

$$3\,\overset{|}{\underset{|}{Si}} - H + 2AuCl_4^- + \frac{3}{2}H_2O \rightarrow \frac{3}{2} - \overset{|}{\underset{|}{Si}} - O - \overset{|}{\underset{|}{Si}} - 2Au + 8Cl^- + 6H^+ \qquad (4.7)$$

Spectroscopic and microscopic characterization confirms incorporation of AuNPs into the films. This approach is promising because ease of fabrication,

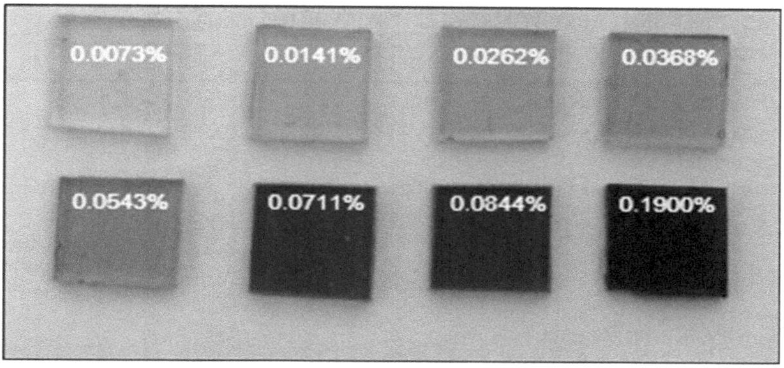

Fig. 4.8 Nanocomposite PDMS films containing increasing gold content as shown

chemical stability, and optical transparency of the films allows facile integration into biomedical [147, 148], sensing [149], and nano- and microelectromechanical systems (N/MEMS) [150–155].

Recently, reduction of Au(III) onto PDMS to form "rAuNP" was compared with direct addition of organic-coated 16-nm nanoparticles (oAuNP) into polymer film precursors as an alternate strategy to AuNP in nanocomposite films [156]. Aqueous solutions of TCA and ethanol suspensions of oAuNPs, respectively, were dispersed into a mixture of Sylgard® 184 PDMS reagents before curing to produce transparent, flexible, nonporous films ca. 680 μm thick. Previously, addition of $KAuCl_4$ to PDMS reagents had yielded brittle gels and foam [157]. Methods were refined to prevent solvent outgassing and uneven NP distribution. Figure 4.8 illustrates plasmonic coloration of PDMS films containing increasing TCA content.

Microscopic and spectroscopic properties of the two types of Au-PDMS composite films were related to thermal behavior induced via laser excitation of LSPR at 532 nm. Previous work with AuNP in aqueous suspensions [158] and in 2D random assemblies on Si [159–161] showed predictable thermalization of incident resonant light using continuum energy balances. Resonant absorption, power consumption, and maximum steady-state temperature of both oAuNP and rAuNP films increased with increasing gold mass percent added. Temperature increased at a faster rate per unit mass of added oAuNP. But higher gold content was achieved by reducing TCA, which resulted in larger overall temperature changes in reduced AuNP films compared with oAuNP films. Table 4.4 compares the thin film characteristics resulting from each method.

The temperature increases recorded in Table 4.4 are comparable to recent simulations that indicate femtosecond [162] or microsecond [163] irradiation of gold(Au)-polymer nanocomposites could produce temperature increases up to 50 K in AuNP themselves or in surrounding media, respectively.

Methods for fabrication and characterizing oAuNP- and rAuNP-PDMS composite films appear useful for applications that use remote, optical induction of local thermalization in order to facilitate heat transport [164], sensing [165],

Table 4.4 Comparing oAuNP-PDMS and rAuNP-PDMS thin film characteristics

	oAuNP-PDMS	rAuNP-PDMS
Embedding AuNP in PDMS films		
Maximum Au content	0.005 % by weight	0.1896 % by weight
Total fraction of Au that is NP	High	Low
Barrier to increase Au content	Variable concentration of commercial AuNP solutions (diluted to 1×10^{13} NP/mL)	Unoptimized trade-off between nucleation vs. growth during reduction
Fabricating AuNP-PDMS films		
Challenges	1. Solvent outgassing 2. Nonuniform NP distribution	1. Film brittleness 2. Solvent outgassing
Processing steps		
Volume of Au added	200 μL suspended NP	10 μL Au^{3+} solution
Time to mix Au into PDMS	10–15 min	4 min
Characterizing AuNP-PDMS films		
Max temp rise at 532 nm	$\Delta T = 21$ °C	$\Delta T = 71$ °C
Temp rise per mass % Au	4,200 °C per %Au	375 °C per %Au
Temp rise per unit absorbance	Lower	Higher

mass transport [166], and spectroscopy [167]. Electromagnetically (EM) active polymer dielectrics in which metallic nanoparticles are embedded have also been suggested for possible use as conductive polymers [168], in DNA ablation [169], in MEMS/NEMS [170], in enhanced solvent flux through microporous cellulose acetate membranes [171], and in drug delivery from PDMS gels and foams [172].

4.5 Future Work

Significant development remains to allow bottom-up templating and self-assembly metal deposition to robustly provide tunable, nanoscale metallic structures desired to improve opto/electronic, theranostic, sensing, spectroscopic, microscopic, and N/MEMS applications. Continued progress in development of models and tools for prediction and real-time monitoring and characterization are needed to allow rational design and optimization of ordered nanoarchitectures with desired features and precise spatiotemporal control. Such improvements could allow implementation of bottom-up templating and deposition methods in large-scale "roll-to-roll" manufacturing. They could also provide avenues to create artificial materials whose significant value arises from their nanoscale structure.

Artificial materials known as metamaterials are of growing interest due to their fascinating electromagnetic behaviors such as cloaking, superresolution, superliminality, and ultratransparency. Metamaterials are assembled from artificial meta-atoms engineered with nanoscale features that confer electromagnetic properties rarely if ever found in nature, primarily negative permittivity and/or negative permeability. Meta-atoms upon which plasmons—collective oscillation of unbound electrons confined to a metal/dielectric interface—may be resonantly induced are particularly interesting for deep-subwavelength light guiding and high-field localization of free electromagnetism. New approaches for templated self-assembly of plasmonic nanodroplets offer scale-able, economic alternatives to traditional top-down fabrication of meta-atoms in metamaterials. Enhanced optical and morphological features of self-assembled plasmonic nanodroplets could reduce anomalous dispersion of meta-atoms in metamaterials to improve performance in a range of optoelectronic devices configured using metamaterials.

References

1. Luo, X., Orlov, A.O., Snider, G.L.J.: Vac. Sci. Technol. B **22**, 3128–3132 (2004)
2. Scheible, D.V., Weiss, C., Kotthaus, J.P., Blick, R.H.: Phys. Rev. Lett. **93**, 186801/1–186801/4 (2004)
3. Fedorovich, R.D., Inosov, D.S., Kiyaev, O.E., Lukyanets, S.P., Marchenko, A.A., Tomchuk, P.M., Bevzenko, D.A., Naumovets, A.G.J.: Mol. Struct. **708**, 67–77 (2004)
4. Passian, A., Lereu, A.L., Farahi, R.H., Ferrell, T.L., Thundat, T.: Thermoplasmonics in thin metal films, chapter 3. In: Jost, A.R. (ed.) Trends in Thin Solid Films Research. Nova, New York (2007)
5. Sockalingum, G.D., Beljebbar, A., Morjani, H., Manfait, M.: Proc. SPIE Int. Soc. Opt. Eng. **3260**, 58–62 (1998)
6. Tolaieb, B., Aroca, R.: Can. J. Anal. Sci. Spectrosc. **48**, 139–145 (2003)
7. Jennings, C.A., Kovacs, G.J., Aroca, R.: Can. J. Phys. Chem. **96**, 1340–1343 (1992)
8. Sockalingum, G.D., Beljebbar, A., Morjani, H., Angiboust, J.F., Manfait, M.: Biospectroscopy **4**, S71–S78 (1998)
9. Vongsvivut, J., Itoh, T., Ikehata, A., Ekgasit, S., Ozaki, Y.: Sci. Asia **32**, 261–269 (2006)
10. Sudo, E., Esaki, Y., Sugiura, M., Murase, A.: Appl. Spectrosc. **61**, 269–275 (2007)
11. Kalyuzhny, G., Vaskevich, A., Schneeweiss, M.A., Rubinstein, I.: Chem. A Eur. J. **8**, 3849–3857 (2002)
12. Ruach-Nir, I., Bendikov, T.A., Doron-Mor, I., Barkay, Z., Vaskevich, A., Rubinstein, I.J.: Am. Chem. Soc. **129**, 84–92 (2007)
13. Lahav, M., Vaskevich, A., Rubinstein, I.: Langmuir **20**, 7365–7367 (2004)
14. Ianoul, A.: Abstracts of the 35th Northeast regional meeting of the American Chemical Society, Binghamton, NY (2006)
15. Zhang, J., Atay, T., Nurmikko, A.V.: Optical detection of brain cell activity using plasmonic gold nanoparticles. Nano Lett. **9**, 519–524 (2009)
16. Roper, D.K., Ahn, W., Taylor, B., Dall'Asén, A.G.: Enhanced spectral sensing by electromagnetic coupling with localized surface plasmons on subwavelength structures. IEEE Sens. **10**, 531–540 (2010)
17. Atwater, H.A., Polman, A.: Plasmonics for improved photovoltaic devices. Nat. Mater. **9**, 205–213 (2010)
18. Palpant, B., Rashidi-Huyeh, M., Gallas, B., Chenot, S., Fisson, S.: Appl. Phys. Lett. **90**, 223105/1–223105/3 (2007)
19. Numata, T., Tatsuta, H., Morita, Y., Otani, Y., Umeda, N.: IEEJ Trans. Electr. Electron. Eng. **2**, 398–401 (2007)

20. Mitsuishi, M., Koishikawa, Y., Tanaka, H., Sato, E., Mikayama, T., Matsui, J., Miyashita, T.: Langmuir **23**, 7472–7474 (2007)
21. Driskell, J.D., Uhlenkamp, J.M., Lipert, R.J., Porter, M.D.: Anal. Chem. **79**, 4141–4148 (2007)
22. Ozawa, H., Kawao, M., Tanaka, H., Ogawa, T.: Langmuir **23**, 6365–6371 (2007)
23. Maier, S.A., Friedman, M.D., Barclay, P.E., Painter, O.: Experimental demonstration of fiber-accessible metal nanoparticle plasmon waveguides for planar energy guiding and sensing. Appl. Phys. Lett. **86**, 071103 (2005)
24. Toderas, F., Baia, M., Baia, L., Astilean, S.: Nanotechnol. **18**, 255702/1–255702/6 (2007)
25. Smythe, E.J., Kickey, M.D., Bao, J., Whitesides, G.M., Capasso, F.: Optical antenna arrays on a fiber facet for *in situ* surface-enhanced Raman scattering detection. Nano Lett. **9**, 1132–1138 (2009)
26. Laurent, G., Félidj, N., Truong, S.L., Aubard, J., Lévi, G., Krenn, J.R., Hohenau, A., Leitner, A., Aussenegg, F.R.: Imaging surface plasmon of gold nanoparticle arrays by far-field Raman scattering. Nano Lett. **5**, 253–258 (2005)
27. Kneipp, K.: Single Mol. **2**, 291–292 (2001)
28. Domingo, C., Resta, V., Sanchez-Cortes, S., Garcia-Ramos, J.V., Gonzalo, J.J.: Phys. Chem. C **111**, 8149–8152 (2007)
29. Kneipp, J., Kneipp, H., Wittig, B., Kneipp, K.: Nano Lett. **7**, 2819–2823 (2007)
30. Turkevich, J., Stevenson, P.C., Hillier, J.: A study of the nucleation and growth processes in the synthesis of colloidal gold. Discuss. Faraday Soc. **11**, 55–75 (1951)
31. Frens, G.: Particle size and sol stability in metal colloids. Colloid Polym. Sci. **250**, 736–741 (1972)
32. Frens, G.: Controlled nucleation for the regulation of the particle size in monodisperse gold suspensions. Nat. (Lond.) Phys. Sci. **241**, 20–22 (1973)
33. Brust, M., Walker, M., Bethell, D., Schiffrin, D.J., Whyman, R.: Synthesis of thiol-derivatised gold nanoparticles in a two-phase liquid-liquid system. Chem. Commun. (7), 801 (1994)
34. Perrault, S.D., Chan, W.C.W.: Synthesis and surface modification of highly monodispersed, spherical gold nanoparticles of 50–200 nm. J. Am. Chem. Soc. **131**(47), 17042–3 (2009)
35. Sakai, T., Alexandridis, P.: Mechanism of gold metal ion reduction, nanoparticle growth and size control in aqueous amphiphilic block copolymer solutions at ambient conditions. J. Phys. Chem. B **109**(16), 7766–7777 (2005)
36. Ray, D., Aswal, V.K., Kohlbrecher, J.: Synthesis and characterization of high concentration block copolymer-mediated gold nanoparticles. Langmuir **2011**(27), 4048–4056 (2011)
37. Barry, J.E., Finkelstein, M., Moore, W.M., Ross, S.D.J.: Org. Chem. **42**, 1292–1298 (1982)
38. Myer, K.: Eshbach's Handbook of Engineering Fundamentals, 5th edn, p. 1234. Wiley, Hoboken, NJ (2009)
39. Myer, K.: Eshbach's Handbook of Engineering Fundamentals, 5th edn, p. 1234. Wiley, Hoboken, NJ (2009)
40. http://www.sparknotes.com/testprep/books/sat2/chemistry/chapter6section7.rhtml
41. Kobasa, I.M., Mazurkevisch, Ya.S., Zozulya, N.I.: Theor. Exp. Chem **39**, 316–321 (2003)
42. Owano, H., Murakami, H., Yamashita, T., Ohigashi, H., Ogata, T.: Synth. Met. **39**, 327–341 (1991)
43. Abdou, H.E., Mohamed, A.A., Fackler Jr., J.P., Burini, A., Galassi, R., Lopez, J.M., Olmos, M.E.: Coord. Chem. Rev. **253**, 1661–1669 (2009)
44. Borensztein, Y., Delannoy, L., Djedidi, A., Barrera, R.G., Louis, C.: Monitoring of the plasmon resonance of gold nanoparticles in Au/TiO2 catalyst under oxidative and reducing atmospheres. J. Phys. Chem. C **114**, 9008–9021 (2010)
45. Ali, H.O., Christie, H.O.: Gold Bull. **17**, 118–127 (1984)
46. Rackham, O., Nichols, S.J., Leedman, P.J., Berners-Price, S.J., Filipovska, A.: Biochem. Pharmacol. **74**, 992–1002 (2007)
47. Kozlov, A.I., Kozlova, A.P., Asakura, K., Matsui, Y., Kogure, T., Shido, T., Iwasawa, Y.: J. Catal. **196**, 56–65 (2000)
48. Menon, V.P., Martin, C.R.: Anal. Chem. **67**, 1920–1928 (1995)

49. Ahn, W., Taylor, B., Dall'Asen, A.G., Roper, D.K.: Langmuir **24**, 4174–4184 (2008)
50. Jang, G.G., Roper, D.K.J.: Phys. Chem. C **113**, 19228–19236 (2009)
51. Blake, P., Ahn, W., Roper, D.K.: Langmuir **26**, 1533–1538 (2010)
52. Lee, S.B., Martin, C.R.: Anal. Chem. **73**, 768–775 (2001)
53. Ahn, W., Taylor, B., Dall'Asén, A.G., Roper, D.K.: Langmuir **24**, 4174–4184 (2008)
54. Takeyasu, N., Tanaka, T., Kawata, S.J.J.: App. Phys. **44**, 1134–1137 (2005)
55. Hrapovic, S., Liu, Y., Enright, G., Bensebaa, F., Luong, J.H.T.: Langmuir **19**, 3958–3965 (2003)
56. Jang, G.-G., Roper, D.K.: Balancing redox activity allows spectrophotometric detection of Au(I) using tetramethylbenzidine dihydrochloride. Anal. Chem. **83**(5), 1836–1842 (2011)
57. Ahn, W.: Novel Electroless gold nano-architectures to enhance photon-plasmon coupling. PhD Thesis, University of Utah (2010)
58. Caricato, A.P., Catalano, M., Ciccarella, G., Martino, M., Rella, R., Romano, F., Spadavecchia, J., Taurino, A., Tunno, T., Valerini, D.: Digest J. Nanomater. Biostruct. **1**, 43–47 (2006)
59. Twardowski, M., Nuzzo, R.G.: Langmuir **18**, 5529–5538 (2002)
60. Hirasawa, M., Shirakawa, H., Hamamura, H., Egashira, Y., Komiyama, H.J.: Appl. Phys. **82**, 1404–1407 (1997)
61. Huang, H., Zhang, S., Qi, L., Yu, X., Chen, Y.: Surf. Coat. Technol. **200**, 4389–4396 (2006)
62. Kim, E., Baeg, K., Noh, Y., Kim, D., Lee, T., Park, I., Jung, G.: Nanotechnology **20**, 355302 (2009)
63. Doron-Mor, I., Cohen, H., Barkay, Z., Shanzer, A., Vaskevich, A., Rubinstein, I.: Chem. Eur. J. **11**, 5555–5562 (2005)
64. Kalyuzhny, G., Vaskevich, A., Schneeweiss, M.A., Rubinstein, I.: Chem. A. Eur. J. **8**, 3849–3857 (2002)
65. Dhawan, A., Muth, J.F.: Nanotechnology **17**, 2504–2511 (2006)
66. Mitsui, K., Handa, Y., Kajikawa, K.: Appl. Phys. Lett. **85**, 4231–4233 (2004)
67. Chumanov, G., Sokolov, K., Gregory, B.W., Cotton, T.M.J.: Phys. Chem. **99**, 9466–9471 (1995)
68. Kim, B., Tripp, S.L., Wei, A.: J. Am. Chem. Soc. **123**, 7955–7956 (2001)
69. Bar, G.: Langmuir **12**, 1172–1179 (1996)
70. Freeman, R.G., Grabar, K.C., Allison, K.J., Bright, R.M., Davis, J.A., Guthrie, A.P., Hommer, M.B., Jackson, M.A., Smith, P.C., Walter, D.G., Natan, M.J.: Science **267**, 1629–1632 (1995)
71. Qi, Z., Honma, I., Ichihara, M., Zhou, H.: Adv. Funct. Mater. **16**, 377–386 (2006)
72. Day, T.M., Unwin, P.R., Wilson, N.R., Macpherson, J.V.: J. Am. Chem. Soc. **127**, 10639–10647 (2005)
73. Hutter, E., Pileni, M.J.: Phys. Chem. B. **107**, 6497–6499 (2003)
74. Sauthier, M.L., Carroll, R.L., Gorman, C.B., Franzen, S.: Langmuir **18**, 1825–1830 (2002)
75. Wang, M., Hu, J., Li, Y., Yeung, E.S.: Nanotechnology **21**, 145608 (2010)
76. Reincke, F., Hickey, S.G., Kegel, W.K., Vanmaekelbergh, D.: Angew. Chem. Int. Ed. **43**, 458–462 (2004)
77. Li, Y., Huang, W., Sun, S.: Angew. Chem. Int. Ed. **45**, 2537–2539 (2006)
78. Wang, Y., Chen, H., Wang, E.: Nanotechnology **19**, 105604 (2008)
79. Kolzer: Vacuum Deposition Processes. http://www.kolzer.com/files/vacuum_deposition_guide.PDF (2004)
80. Huang, H., Zhang, S., Qi, L., Yu, X., Chen, Y.: Surf. Coat. Technol. **200**, 4389–4396 (2006)
81. Kalyuzhny, G., Vaskevich, A., Schneeweiss, M.A., Rubinstein, I.: Chem. A. Eur. J. **8**, 3849–3857 (2002)
82. Jin, Y., Kang, X., Song, Y., Zhang, B., Cheng, G., Dong, S.: Anal. Chem. **73**, 2843–2849 (2001)
83. Hutter, E., Pileni, M.J.: Phys. Chem. B. **107**, 6497–6499 (2003)
84. Day, T.M., Unwin, P.R., Wilson, N.R., Macpherson, J.V.: J. Am. Chem. Soc. **127**, 10639–10647 (2005)
85. Yam, C.M.: Simple acid–base hydrolytic chemistry approach to molecular self-assembly. Ph.D. Dissertation, Department of Chemistry, McGill University (1999)

86. Mallory, G.O., Hajdu, J.B.: Electroless Plating: Fundamentals and Applications, Chapter 1. American Electroplaters and Surface Finishers Society, Orlando, FL (1990)
87. Haes, A.J., Hall, W.P., Chang, L., Klein, W.L., Van Duyne, R.P.: Nano Lett. **4**, 1029–1034 (2004)
88. Ferralis, N., Maboudian, R., Carraro, C.J.: Phys. Chem. C **111**, 7508–7513 (2007)
89. Yasseri, A.A., Sharma, S., Jung, G.Y., Kamins, T.I.: Electrochem. Solid-State Lett. **9**, C185–C188 (2006)
90. Zhao, L., Siu, A.C.-L., Petrus, J.A., He, Z., Leung, K.T.J.: Am. Chem. Soc. **129**, 5730–5734 (2007)
91. Brown, K.R., Natan, M.J.: Langmuir **14**, 726–728 (1998)
92. Menzel, H., Mowery, M.D., Cai, M., Evans, C.E.: Adv. Mater. **11**, 131–134 (1999)
93. Hrapovic, S., Liu, Y., Enright, G., Bensebaa, F., Luong, J.H.T.: Langmuir **19**, 3958–3965 (2003)
94. Ali, H.O., Christie, I.R.A.: Gold Bull. **17**, 118–127 (1984)
95. Okinata, Y.: Chapter 15. In: Mallory, G.O., Hajdu, J.B. (eds.) Electroless Plating: Fundamentals and Applications, pp. 401–420. American Electroplaters and Surface Finishers Society, Orlando, FL (1990)
96. Bhuvana, T., Kulkarni, G.U.: Bull. Mater. Sci. **29**, 505–511 (2006)
97. Nishizawa, M., Menon, V.P., Martin, C.R.: Science **268**, 700–702 (1995)
98. Hou, Z., Abbott, N.L., Stroeve, P.: Langmuir **14**, 3287–3297 (1998)
99. Hou, Z., Dante, S., Abbott, N.L., Stroeve, P.: Langmuir **15**, 3011–3014 (1999)
100. Alvarez-Puebla, R.A., Nazri, G.-A., Aroca, R.F.J.: Mater. Chem. **16**, 2921–2924 (2006)
101. Ferralis, N., Maboudian, R., Carraro, C.J.: Phys. Chem. C **111**, 7508–7513 (2007)
102. Perea-Lopez, N., Rakov, N., Xiao, M.: Rev. Sci. Instrum. **73**, 4399–4401 (2002)
103. Dubrovsky, T.B., Hou, Z., Stroeve, P., Abbott, N.L.: Anal. Chem. **71**, 327–332 (1999)
104. Lauer, M.E., Jungmann, R., Kindt, J.H., Magonov, S., Fuhrhop, J.-H., Oroudjev, E., Hansma, H.G.: Langmuir **23**, 5459–5465 (2007)
105. Habennicht, A., Olapinski, M., Burmeister, F., Leiderer, P., Boneberg, J.: Science **309**, 2043–2045 (2005)
106. Goldstein, A.N., Echer, C.M., Alivisatos, A.P.: Science **256**, 1425–1427 (1992)
107. Buffat, P., Borel, J.-P.: Phys. Rev. A **13**, 2287–2298 (1976)
108. Ajayan, P.M., Marks, L.D.: Phys. Rev. Lett. **60**, 585–587 (1988)
109. Kotaidis, V., Dahmen, C., von Plessen, G., Springer, F., Plech, A.: J. Chem. Phys. **124**, 184702 (2006)
110. Kohli, P., Harrell, C.C., Cao, Z., Gasparac, R., Tan, W., Martin, C.R.: Science **305**, 984–986 (2004)
111. Koura, N.: Chapter 17. In: Mallory, G.O., Hajdu, J.B. (eds.) Electroless Plating: Fundamentals and Applications, pp. 441–462. American Electroplaters and Surface Finishers Society, Orlando, FL (1990)
112. Martin, P.M.: Chapter 12. In: Handbook of Deposition Technologies for Films and Coating-Science, Applications and Technologies, 3nd ed. Elsevier (2010)
113. Kaiser, N.: Appl. Opt. **41**(16), 3053–3060 (2002)
114. Gilmer, G.H., Huang, H., Rubia, T., Torre, J.D., Baumann, F.: Thin Solid Films **365**, 189–200 (2000)
115. Kastner, M., Voigtlander, B.: Phys. Rev. Lett. **82**(13), 2745–2748 (1999)
116. Kodambaka, S., Chopp, D.L., Petrov, I., Greene, J.E.: Surf. Sci. **540**, L611–L616 (2003)
117. Pederson, L.R.: Solar Energy Mater. **6**, 221–232 (1982)
118. Hou, Z., Abbott, N.L., Stroeve, P.: Langmuir **14**, 3287–3297 (1998)
119. Menon, V.P., Martin, C.R.: Anal. Chem. **67**, 1920–1928 (1995)
120. McDermott, J.: Plating of Plastics with Metals, pp. 180–182. Noyes Data Corp., Park Ridge, NJ (1974)
121. Mallory, G.O., Hajdu, J.B.: Electroless plating: fundamentals and applications, Chapter 17. American Electroplaters and Surface Finishers Society, Orlando, FL (1990)
122. Pederson, L.R.: Solar Energy Mater. **6**, 221–232 (1982)

123. Park, H., Park, H., Hill, R.H.: Sens. Actuators A **132**, 429–433 (2006)
124. Tan, B., Toman, E., Li, Y., Wu, Y.J.: Am. Chem. Soc. **129**, 4162–4163 (2007)
125. Jang, G., Hawkridge, M., Roper, D.K.: Silver disposition and dynamics during electroless metal thin film synthesis. J. Mater. Chem. **22**, 21942–21953 (2012)
126. Pederson, L.R.: Solar Energy Mater. **6**, 221–232 (1982)
127. Singer, R.R., Leitner, A., Aussenegg, F.R.J.: Opt. Soc. Am. B **12**, 220–228 (1995)
128. Baba, K., Okuno, T., Miyagi, M.: Appl. Phys. Lett. **62**(5), 437–439 (1993)
129. Jiang, P., McFarland, M.J.: Large-scale fabrication of wafer-size colloidal crystals, macroporous polymers and nanocomposites by spin-coating. J. Am. Chem. Soc. **126**(42), 13778–86 (2004)
130. Dejarnette, D., Roper, D.K., Harbin, B.: Geometric effects on far-field coupling between multipoles of nanoparticles in square arrays. J. Opt. Soc. Am. B **29**(1), 88–100 (2012)
131. Blake, P., Obermann, J., Harbin, B., Roper, D.K.: Enhanced nanoparticle response from coupled dipole excitation for plasmon sensors. IEEE Sens. J. **11**(12), 3332–3340 (2011)
132. Li, H., Low, J., Brown, K.S., Wu, N.: Large-area well-ordered nanodot array pattern fabricated with self-assembled nanosphere template. IEEE Sens. **8**(6), 880–884 (2008)
133. Pan, F., Zhang, J., Cai, C., Wang, T.: Rapid fabrication of large-area colloidal crystal monolayers by a vortical surface method. Langmuir **22**(17), 7101–7104 (2006)
134. Chen, K., Drachev, V.P., Borneman, J.D., Kildishev, A.V., Shalaev, V.M.: Nano Lett. **10**, 916–922 (2010)
135. Ahn, W., Blake, P., Schulz, J., Ware, M.E., Roper, D.K.: Fabrication of regular arrays of Au nanospheres by thermal transformation of electroless-plated films. J. Vac. Sci. Technol. B **28**(3), 638–642 (2010)
136. Zhang, G., Wang, D., Möhwald, H.: Nano Lett. **7**, 127–132 (2007)
137. Goncalves, M.R., Siegel, A., Marti, O.: J. Microsc. **229**, 475–482 (2008)
138. Doron-Mor, I., Barkay, Z., Filip-Granit, N., Vaskevich, A., Rubinstein, I.: Chem. Mater. **16**, 3476–3483 (2004)
139. Doron-Mor, I., Cohen, H., Barkay, Z., Shanzer, A., Vaskevich, A., Rubinstein, I.: Ehcm. Eur. J. **11**, 5555–5562 (2005)
140. Kalyuzhny, G., Vaskevich, A., Schneeweiss, M.A., Rubinstein, I.: Chem. A. Eur. J. **8**, 3849–3857 (2002)
141. Kalyuzhny, G., Vaskevich, A., Ashkenasy, G., Shanzer, A., Rubinstein, I.J.: Phys. Chem. B **104**, 8238–8244 (2000)
142. Ahn, W., Roper, D.K.: Periodic nanotemplating by selective deposition of electroless gold island films on particle-lithographed dimethyldichlorosilane layers. ACS Nano **4**(7), 4181–9 (2010)
143. Zhang, Q., Xu, J., Liu, Y., Chen, H.: *In-situ* synthesis of poly(dimethylsiloxane) – gold nanoparticles composite films and its application in microfluidic systems. Lab Chip **8**, 352–357 (2008)
144. Goyal, A., Kumar, A., Patra, P.K., Mahendra, S., Tabatabaei, S., Alvarez, P.J., John, G., Ajayan, P.M.: In situ synthesis of metal nanoparticle embedded free standing multifunctional PDMS films. Macromol. Rapid Commun. **30**, 1116–1122 (2009)
145. Massaro, A., Spano, F., Cingolani, R., Athanassiou, A.: Experimental optical characterization and polymeric layouts of gold PDMS nanocomposite sensor for liquid detection. IEEE Sens. **11**(9), 1780–1786 (2011)
146. Ryu, D., Loh, K.J., Ireland, R., Karimzada, M., Yagmaie, F., Gusman, A.M.: In situ reduction of gold nanoparticles in PDMS matrices and applications for large strain sensing. Smart Struct. Syst. **8**(5), 471–486 (2011)
147. Mata, A., Fleischman, A.J., Roy, S.: Characterization of polydimethylsiloxane (PDMS) properties for biomedical micro/nanosystems. Biomed. Microdiv. **7**(4), 281–93 (2005)
148. Yu, X., Zhang, D., Li, T., Hao, L., Li, X.: 3-D microarrays biochip for DNA amplification in polydimethylsiloxane (PDMS) elastomer. Sens. Actuators A **108**, 103–107 (2003)

149. Massaro, A., Spano, F., Cingolani, R., Athanassiou, A.: Experimental optical characterization and polymeric layouts of gold PDMS nanocomposite sensor for liquid detection. IEEE Sens. **11**(9), 1780–1786 (2011)
150. Jo, B., Lerberghe, L.M.V., Motsegood, K.M., Beebe, D.J.: Three-dimensional micro-channel fabrication in polydimethylsiloxane (PDMS) elastomer. J. Microelectomech. Syst. **9**(1), 76–81 (2000)
151. Sawano, S., Naka, K., Werber, A., Zappe, H., Konishi, S.: Sealing method of PDMS as elastic material for MEMS. In: Proceedings of the IEEE 21st international conference on MEMS, pp. 419–422, 13–17 Jan 2008
152. Schneider, F., Draheim, J., Kamberger, R., Wallrabe, U.: Process and material properties of polydimethylsiloxane (PDMS) for optical MEMS. Sens. Actuators A **151**, 95–99 (2009)
153. Schneider, F., Fellner, T., Wilde, J., Wallrabe, U.: Mechanical properties of silicones for MEMS. J. Micromech. Microeng. **18**, 065008 (2008)
154. Tong, J., Simmons, C.A., Sun, Y.: Precision patterning of PDMS membranes and applications. J. Micromech. Microeng. **18**, 037004 (2008)
155. Tung, Y., Kurabayashi, K.: A single-layer PDMS-on-silicon hybrid microactuator with multi-axis out-of-plane motion capabilities – part I: design and analysis. J. Microelectromech. Syst. **14**(3), 548–557 (2005)
156. Berry, K.R., Russell, A.G., Blake, P., Roper, D.K.: Gold nanoparticles reduced in situ and dispersed in polymer thin films: optical and thermal properties. Nanotechnology **23**, 375703 (2012)
157. Scott, A., Gupta, R., Kulkarni, G.: A simple water-based synthesis of Au nanoparticle/PDMS composites for water purification and targeted drug release. Macromol. Chem. Phys. **211**, 1640–1647 (2010)
158. Roper, D.K., Ahn, W., Hoepfner, M.: Microscale heat transfer transduced by surface plasmon resonant gold nanoparticles. J. Phys. Chem. C **111**(9), 3636–3641 (2007)
159. Ahn, W., Roper, D.K.: Transformed gold island film improves light-to-heat transduction of nanoparticles on silica capillaries. J. Phys. Chem. C **112**(32), 12214–12218 (2008)
160. Russell, A., McKnight, M., Sharp, A., Hestekin, J., Roper, D.K.: Gold nanoparticles allow opto-plasmonic evaporation from open silica cells. J. Phys. Chem. C **114**(22), 10132–10139 (2010)
161. Russell, A.G., McKnight, M., Hestekin, J., Roper, D.K.: Thermodynamics of optoplasmonic heating in fluid filled gold nanoparticle plated capillaries. Langmuir **27**(12), 7799–7805 (2011)
162. Garwe, F., Bauerschäfer, U., Csaki, A., Steinbrück, A., Ritter, K., Bochmann, A., Bergmann, J., Weise, A., Akimov, D., Maubach, G., König, K., Hüttmann, G., Paa, W., Popp, J., Fritzsche, W.: Optically controlled thermal management on the nanometer length scale. Nanotechnology **19**, 055207 (2008)
163. Zeng, N., Murphy, A.: Heat generation by optically and thermally interacting aggregates of gold nanoparticles under illumination. Nanotechnology **20**, 375702 (2009)
164. Vanherck, K., Hermans, S., Verbiest, T., Vankelecom, I.: Using the photothermal effect to improve membrane separations via localized heating. J. Mater. Chem. **21**, 6079–6087 (2011)
165. Jiang, C., Markutsya, S., Pikus, Y., Tsukruk, V.V.: Freely suspended nanocomposite membranes as highly sensitive sensors. Nat. Mater. **3**, 721–728 (2004)
166. Kang, S.W., Hong, J., Park, J.H., Mun, S.H., Kim, J.H., Cho, J., Char, K., Kang, Y.S.: Nanocomposite membranes containing positively charged polarized gold nanoparticles for facilitated olefin transport. J. Membr. Sci. **321**, 90–93 (2008)
167. Lim, D.K., Jeon, K.S., Hwang, J.H., Kim, H., Kwon, S., Suh, Y.D., Nam, J.M.: Highly uniform and reproducible surface-enhanced Raman scattering from DNA-tailorable nanoparticles with 1-nm interior gap. Nat. Nanotechnol. **6**, 452–460 (2011)
168. Dallas, P., Georgakilas, V., Niarchos, D., Komninou, P., Kehagias, T., Petridis, D.: Synthesis, characterization and thermal properties of polymer/magnetite nanocomposites. Nanotechnology **17**, 2046–2053 (2006)

169. Garwe, F., Bauerschäfer, U., Csaki, A., Steinbrück, A., Ritter, K., Bochmann, A., Bergmann, J., Weise, A., Akimov, D., Maubach, G., König, K., Hüttmann, G., Paa, W., Popp, J., Fritzsche, W.: Optically controlled thermal management on the nanometer length scale. Nanotechnology **19**, 055207 (2008)
170. Schlemmer, C., Betz, W., Berchtold, B., Rüne, J., Santer, S.: The design of thin polymer membranes filled with magnetic particles on a microstructured silicon surface. Nanotechnology **20**, 255301 (2009)
171. Vanherck, K., Hermans, S., Verbiest, T., Vankelecom, I.: Using the photothermal effect to improve membrane separations via localized heating. J. Mater. Chem. **21**, 6079–6087 (2011)
172. Scott, A., Gupta, R., Kulkarni, G.: A simple water-based synthesis of Au nanoparticle/PDMS composites for water purification and targeted drug release. Macromol. Chem. Phys. **211**, 1640–1647 (2010)

Chapter 5
Ordering of Ga Nanodroplets by Low-Energy Ion Sputtering

Sabina Koukourinkova, Zhiming M. Wang, Jiang Wu, Xingliang Xu, Mourad Benamara, Peter Moeck, and Gregory J. Salamo

Abstract Ordered nanostructures have attracted much attention due to their potential in realizing novel device applications. In this chapter, we present a study of ordered nanodroplets fabricated by low-energy ion sputtering on GaAs surfaces. The morphological evolution of a GaAs (001) surface exposed to a Ga^+ focused ion beam was investigated as a function of beam energy, incidence, current, sputter time, and dwell time. The sputter yield of the target, surface roughness, amount of material deposited on the surface, and temperature of the substrate were evaluated. The experimental results show formation of self-assembled Ga metal droplets. Control over the size, density, and ordering of the droplets is possible for various sets of ion beam parameters. The arrays of ordered nanodroplets have potential application as templates as well as a local nanosource in molecular beam epitaxy for fabricating ordered semiconductor structures such as quantum dots.

S. Koukourinkova • M. Benamara • G.J. Salamo
Institute for Nanoscale Materials Science and Engineering, University of Arkansas, Fayetteville, 72701 NC, USA

Z.M. Wang (✉)
Institute for Nanoscale Materials Science and Engineering, University of Arkansas, Fayetteville, 72701 NC, USA

State Key Laboratory of Electronic Thin Film and Integrated Devices, University of Electronic Science and Technology of China, 610054 Chengdu, People's Republic of China
e-mail: zhmwang@gmail.com

J. Wu • X. Xu
State Key Laboratory of Electronic Thin Film and Integrated Devices, University of Electronic Science and Technology of China, 610054 Chengdu, People's Republic of China

P. Moeck
Nano-Crystallography Group, Portland State University, Portland, OR 97207, USA

Oregon Nanoscience and Microtechnologies Institute, Corvallis, OR, USA

Z.M. Wang (ed.), *Nanodroplets*, Lecture Notes in Nanoscale Science and Technology 18,
DOI 10.1007/978-1-4614-9472-0_5,

5.1 Introduction

Within the past decade, ion beam sputtering has been used as a surface nanopatterning technique for fabrication of self-assembled nanostructures such as nanoripples [1–3], nanopits [4, 5], nanoneedles [6], and nanodots [7, 8]. In particular, the ability to achieve ordered self-assembled nanostructures has been of great interest due to the potential applications of ordered nanostructures in next-generation devices [9–12]. Hexagonally ordered nanopits on Ge (001) and $SrTiO_3$ (001) surfaces have been reported for sputtering with Ga^+ ion beam. The nanopits formed on a $SrTiO_3$ (001) surface have been used to induce controlled growth of Cu_2O nanoislands surrounding the nanopits [5]. In addition, sixfold ordered patterns of self-assembled quantum dots have already been achieved by means of low-energy Ar^+ ion sputtering on substrates such as Si [10] and InP [11] with simultaneous stage rotation at off-normal incidence and without stage rotation at normal incidence [13, 14]. Moreover, hexagonal ordering of Ga nanodroplets has been induced on a GaAs (001) surface by Ga^+ ion beam at normal incidence followed by rapid thermal annealing [8]. Later, hexagonal ordering of Ga nanodroplets on GaAs (001) surface independent of crystallographic orientation was reported for off-normal low-energy Ga^+ ion bombardment without simultaneous sample rotation [7]. Wei et al. proposed a theoretical model for droplet ordering based on directional mass loss and transfer.

This chapter demonstrates that ordering of Ga nanodroplets with controlled size and density induced by Ga^+ irradiation on a GaAs surface is reproducible, which is essential for scale-up and practical applications. This chapter also investigates systematically the relationship between the ion beam parameters (energy, current, incident angle, exposure time, and dwell time) and the sputter yield and temperature of the material, surface roughness, volume of Ga deposited on the surface, droplet size, density, and degree of ordering.

5.2 Theory

In order to guide the formation and ordering of nanodroplets during ion beam sputtering, it is essential to understand the dynamics of the kinetic processes involved. Surface roughening and smoothing are regarded as the main competing processes that govern surface modification [15]. These and other kinetic processes depend strongly on the surface morphology as well as the material composition and quality, and the type of ions. Kinetic energy loss from high-energy ions incident on the target material results in generation of various defects such as adatoms and vacancies on the surface as well as vacancies and interstitials in the bulk. Defect generation in materials irradiated with ions has been studied theoretically and experimentally [15–18]. These efforts shed light on understanding the effects of ion beam sputtering on the structural properties of target materials and pave the way for realizing various ordered nanostructures, such as ordered nanoripples and nanodroplets, on different substrates [7, 19–23]. The control over aligned nanostructures can be achieved by engineering the kinetic processes, which are linked to the ion beam energy, ion flux, ion dose, angle of ion incidence, exposure time, as well as the types of target materials and ions.

In particular, the relationship between ion beam energy and sputter yield of the target constitutes the basis of Sigmund theory of surface modulation. Bradley and Harper have further developed a model for surface instability based on Sigmund theory and surface diffusion to explain curvature-dependent sputtering [24]. Following their pioneering work, the Bradley–Harper model has been adopted and modified to explain different patterns induced during sputtering, such as the formation of arrays of hexagonally ordered Ga nanodroplets on a GaAs surface [7]. In this section, these models will be reviewed briefly.

5.2.1 Bradley–Harper Instability Model

In 1973, Sigmund proposed a model describing that a flat surface with microscale variations is not stable under high-dose ion bombardment unless surface smoothing is dominated by atom migration [25]. Bradley and Harper later studied the instability of amorphous targets under ion beam exposure and observed ripple formation. In order to explain the evolution of modulated surfaces during ion sputtering, they developed a model for surface roughening and smoothing on the basis of Sigmund theory while also taking into consideration surface diffusion [24].

The Bradley-Harper instability model assumes the target material has a modulated surface with height variation. The sputtering processes are schematically shown in Fig. 5.1. During sputtering, energy from an ion incident on the surface with radius of curvature R is transferred in the ion impact direction with longitudinal and transverse width of the energy distribution contour α and β at a characteristic penetration depth a [see Fig. 5.1a]. The sputter yield from the incoming ions at point O can be written as

$$Y_0(\varphi) = \frac{n\upsilon(\varphi, R)}{f\left[\cos\varphi - (x/R)\right]\sin\varphi}, \tag{5.1}$$

where n is the number of atoms per unit volume, φ is the incident angle, R is the radius of the curvature at point O, f is the ion flux, and υ is the normal component of the surface erosion velocity as a function of φ and R.

$$\upsilon(\varphi, R) \cong \frac{\Lambda\varepsilon f a}{\sqrt{2\pi B_1}\alpha\beta} \exp\left(-\frac{a^2}{2\alpha^2} + \frac{A^2}{2B_1}\right)\left[\cos\varphi + \Gamma_1(\varphi)\frac{a}{R}\right] \tag{5.2}$$

with

$$\Gamma_1(\varphi) = \frac{A}{B_1}\sin\varphi - \frac{B_2}{2B_1}\left(1 + \frac{A^2}{B_1}\right)\cos\varphi - \frac{AC}{B_1{}^2}\left(3 + \frac{A^2}{B_1}\right)\cos\varphi \tag{5.3}$$

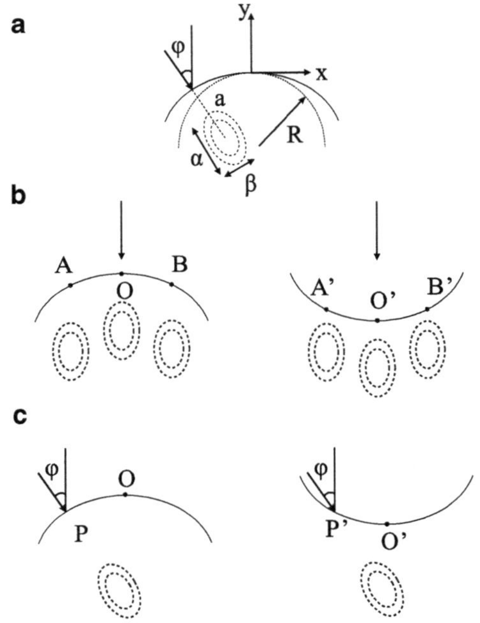

Fig. 5.1 Bradley–Harper instability model [24]: (**a**) an illustration of a surface with curvature radius R bombarded by an ion at an arbitrary angle φ; energy distribution contour profiles during ion bombardment of concave and convex surfaces at (**b**) normal incidence and (**c**) off-normal incidence

Here, ε is the deposited energy, and Λ is a constant relating ε to υ at point O. The coefficients A, B_1, B_2, and C are functions of α, β, a, and φ and are independent of R [24].

From Eq. (5.3), if the angle of incidence $\varphi = 0$, then Γ_1 (φ) is negative, and thus the surface erosion velocity is positive for a negative curvature and negative for a positive curvature. This suggests that ion bombardment at normal incidence removes material faster from a concave surface compared to a convex surface. Figure 5.1b illustrates the energy distribution profiles of ions impacting the surface at normal incidence. The energy deposited at O′ from ion bombardment at A′ and B′ is higher than the energy at O from ions impacting the surface at A and B. Therefore, a concave surface erodes at a faster rate than a convex surface, thereby enhancing surface roughness. In contrast, at off-normal incidence, the convex surface erodes faster in comparison with the concave surface. The energy distribution profiles in Fig. 5.1c show that the ions incident at P can transfer a higher portion of their energy to O than those at P′ can transfer to O′. This is due to the fact that the effective ion flux reaching P is higher than the flux at P′. Therefore, the surface at O erodes faster compared to O′, and the overall surface roughness decreases. Increasing the angle of incidence results in higher sputter yield until a critical angle is achieved.

The Bradley–Harper model describes the ability to control surface nanopatterning by varying the angle of incidence during ion sputtering while simultaneously altering the dynamics of the kinetic processes. Although this model explains nanoripple formation at normal and off-normal incidence and predicts the wavelength and orientation of nanoripples, it does not account for reflected ions at high incidence.

5.2.2 Droplet Formation from a Gallium Layer

While Bradley–Harper model has been used extensively to describe the formation of various patterns achieved by ion sputtering, the model is only applicable to patterns of the same composition as the target material. More specifically, the model provides explanation for ripple formation in amorphous materials. However, changes to the model are continuously being adopted in order to explain complex phenomena in surface modification such as the formation of sixfold patterns of Ga metal nanodroplets on a GaAs surface.

During Ga^+ ion beam sputtering of a GaAs substrate, the incoming ions preferentially remove As atoms and simultaneously sputter Ga atoms. A portion of the incident Ga^+ ions is implanted, and the net gain of Ga atoms from sputtering and implantation results in Ga-rich areas on the surface. Eventually, nucleation of Ga droplets can happen given sufficient sputtering time. Considering the amount of Ga in the droplets as a thin layer over the irradiated surface, the layer thickness can be estimated from the sputter yield of the compound Ga and As from the GaAs substrate, and the sputter yield of pure Ga from the Ga droplet. From simulation of a 5 keV Ga^+ beam normally incident on a GaAs surface using SRIM code [26], the sputter yield of compound Ga and As is $Y(\mathrm{Ga}^*) = 2$ atoms/ion and $Y(\mathrm{As}^*) = 5$ atoms/ion, respectively, and the sputter yield of pure Ga is $Y(\mathrm{Ga}) = Y(\mathrm{As}^*) = 5$ atoms/ion. The Ga layer thickness can then be expressed as

$$L(\mathrm{Ga}) = \frac{3r_{\mathrm{gain}}}{5r_{\mathrm{loss}}} L(\mathrm{GaAs}) \tag{5.4}$$

where r_{gain} is the Ga gain rate from the substrate, r_{loss} is the loss rate of Ga from the droplet, and $L(\mathrm{GaAs})$ is the penetration depth of Ga ions in the GaAs substrate. The amount of Ga deposited on the surface depends on the ion penetration depth and thus is related to the incident angle and energy of the ion beam. Particularly, the penetration depth has a cosine dependence on the angle of ion beam incidence.

5.2.3 Droplet Ordering at Off-Normal Incidence

According to Bradley–Harper model, an amorphous material exposed to ion beam at off-normal incidence evolves into ripples. However, no ripple morphology is observed on a GaAs surface exposed to Ga^+ ion beam. Instead, laterally ordered nanodroplets form. In order to explain the formation of self-aligned Ga nanodroplet arrays on a GaAs surface under the fluence of a Ga^+ ion beam, Wei et al. have developed a model based on the balance between anisotropic supply and loss of Ga atoms [7].

The model proposed by Wei et al. suggests that the profile of energy distribution on the GaAs surface during ion beam irradiation affects nanodroplet ordering.

At normal incidence, the distribution of energy on the surface is isotropic and exhibits a circular shape. As a result, the surface diffusion of Ga atoms is random, and Ostwald ripening drives small droplets to merge into bigger more energetically favorable ones because of their lower surface-to-volume ratio. Therefore, the droplets form randomly with broad size distribution. On the other hand, at off-normal incidence, the anisotropic energy distribution profile has an ellipse contour resulting in directional atom migration on the surface. A consequence of combining the directional atom gain and loss with shadowing and exclusion zone effects is the net supply of Ga atoms between droplets guided by the driving force of the ion beam along the projected beam direction. As a result, droplets adjust their location towards the highest atom supply and form hexagonally ordered droplet arrays given sufficient time of ion bombardment.

5.3 Experiment

The experiments were carried out at room temperature in vacuum pressure of 10^{-7}–10^{-6} Torr on nominally flat, epi-ready semi-insulating GaAs (001) substrates using scanning electron microscope equipped with Ga^+ ion beam source (FEI Nova 200 Dual-Beam SEM/FIB). The GaAs surface was uniformly irradiated using a serpentine scan mode with 50% beam overlap in two dimensions. Ion beam parameters such as incident angle, energy, current, sputter time, and dwell time were varied. Analysis of the surface topography was performed with atomic force microscope (Veeco Dimension 3100 AFM) in standard tapping mode. Silicon cantilevers with radius of curvature $R_c < 10$ nm and average resonant frequency of 170 kHz were used to obtain high-resolution images of the surface.

The response of a GaAs and Ga targets exposed to a Ga^+ beam was evaluated using SRIM software [26, 27]. The simulation was performed using a 5 keV ion beam incident on the target at angles in the range of 0–82°. In addition, the total sputter yield of the GaAs target was determined from experiment and compared to the simulation results. For this purpose, the GaAs surface was uniformly irradiated with an ion beam with energy 5 keV, flux 7.49×10^{14} cm^{-2} s^{-1}, dwell time 1 μs, and sputter time 5 min. Due to limited range of motion of the specimen stage, the incident angle was varied from 0 to 52°. The amount of material removed from the surface was determined using AFM. The experimental data is in good agreement with the simulation results. Figure 5.2a shows the sputter yield for compound Ga and As from the GaAs target and the sputter yield of pure Ga from the Ga target obtained by SRIM. Below a critical incident angle of 62° for compound Ga and 72° for compound As and pure Ga, the sputter yield increases exponentially, and at oblique incidence beyond the critical angle, the sputter yield decreases possible as a result of undeveloped collision cascades or ion reflection. According to the simulation results, the sputter yield of compound Ga is less than the sputter yield of compound As and pure Ga by approximately a factor of 2.

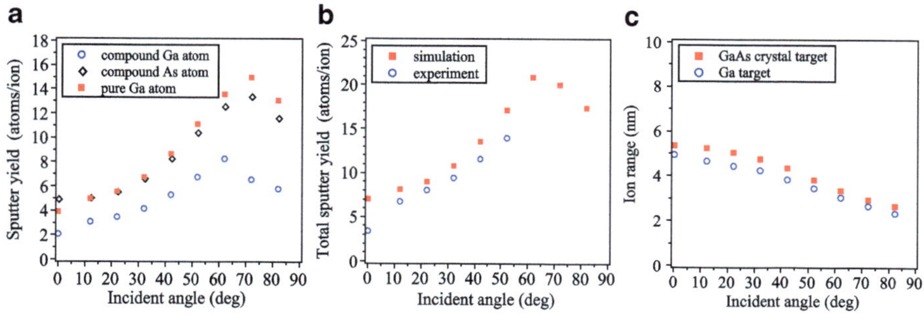

Fig. 5.2 (**a**) Sputter yield of GaAs and Ga targets for various incident angles during irradiation with 5 keV Ga^+ ion beam. Data obtained using SRIM simulation software. (**b**) Total sputter yield of GaAs as a function of incidence during irradiation with 5 keV Ga^+ ion beam: comparison between simulation and experiment. (**c**) Ion range in GaAs and Ga target as a function of incident angle during irradiation with 5 keV Ga^+ ion beam. Data obtained using SRIM simulation software

Figure 5.2b compares the total yield of the GaAs target from simulation and experiment, and similar behavior is observed. The ion range of penetration for each target is also presented in Fig. 5.2c. The range values for incident angles 0–82° vary from 5.4 nm to 2.7 nm, and 4.8 nm to 2.4 nm for the GaAs and Ga targets, respectively.

The effect of ion beam energy on the formation of ordered droplets was also studied for 5 keV ion beam sputtering at an incident angle of 37°, sputter time 5 min, dwell time 1 μs, and flux 1.43×10^{15} cm^{-2} s^{-1} compared to sputtering with 10 keV and 20 keV ion beams with flux 1.31×10^{15} cm^{-2} s^{-1}. The corresponding SEM scans are given in Fig. 5.3a–c. Short-range droplet ordering was observed for sputtering with 5 keV ion beam with periodicity confirmed FFT spectra. The droplets exhibit Gaussian size distribution with average lateral size of 61.0 ± 9.7 nm, droplet height of 14.5 ± 3.1 nm, and density of 8.23×10^{10} cm^{-2} (Fig. 5.3d). Higher-energy beam (10 and 20 keV) produced random positioning of droplets with broad size distribution. Values of the droplet size and height are not included here since the accuracy of the size measurement is hindered by limitations of the image processing software. Figure 5.3e shows droplet density decreases with increasing ion energy, while Fig. 5.3f shows more material is deposited on the surface for higher energies. The total volume of Ga is related to the penetration depth of the ions into the substrate and the angle of incidence.

Uniform sputtering of the GaAs surface was conducted for ion energy 5 keV, constant current (70 pA, 0.12 nA, 0.23 nA, 0.60 nA), sputter time 5 min, dwell time 1 μs, and incidence ranging from 62° to 2° to study droplet ordering, size, and density as a function of incidence. Results show that self-assembled chains of ordered Ga nanodroplets form at incident angles 32° to 62°. SEM scans of droplets formed on a GaAs surface using flux 7.49×10^{14} cm^{-2} s^{-1} are presented in Fig. 5.4a–f. The lateral size and height of the droplets range from 24.6 ± 7.8 nm to 82.0 ± 13.0 nm, and from 8.0 ± 2.1 nm to 18.0 ± 3.6 nm, respectively. The droplet

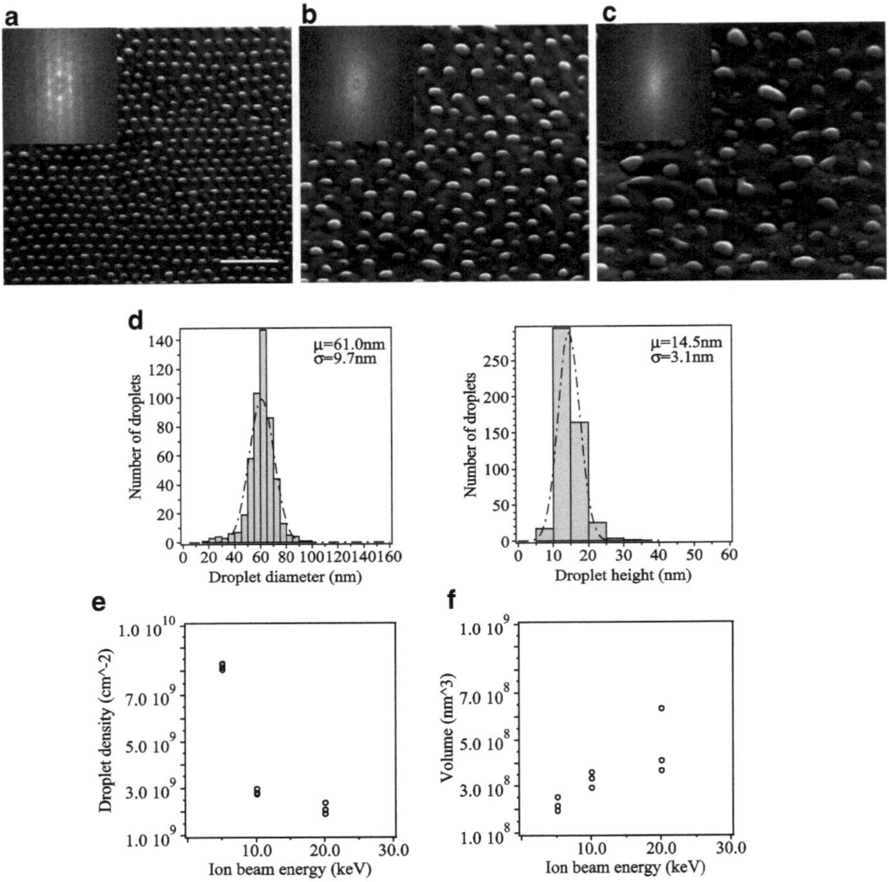

Fig. 5.3 Ordering of Ga metal droplets for sputtering at an incident angle 37°, sputter time 5 min, dwell time 1 μs, 50% beam overlap, and various ion beam energies: (**a**) 5 keV, flux 1.43×10^{15} cm^{-2} s^{-1}, with corresponding frequency histograms for droplet lateral size and height; (**b**) 10 keV, flux 1.31×10^{15} cm^{-2} s^{-1}; (**c**) 20 keV, flux 1.31×10^{15} cm^{-2} s^{-1}, scale bar 500 nm; the *insets* are FFT spectra; (**d**) histograms of the lateral size and height of the droplets; (**e**) dependence of the droplet density on ion beam energy; (**f**) total volume of Ga deposited on the surface as a function of energy

size decreases linearly while the droplet density increases exponentially with increasing incident angle (Fig. 5.4g, h). The decrease in size and the increase in density of the droplets with increasing oblique incidence result from depositing the same amount of material on the surface over a larger area. The volume of Ga supplied to the surface approximates a cosine function of the incident angle as shown in Fig. 5.4i. Consistent with the theoretical model, the surface roughness decreases with increasing incidence, and the local substrate temperature decreases due to broadening of the ion beam over a larger surface area, thus slowing down surface diffusion.

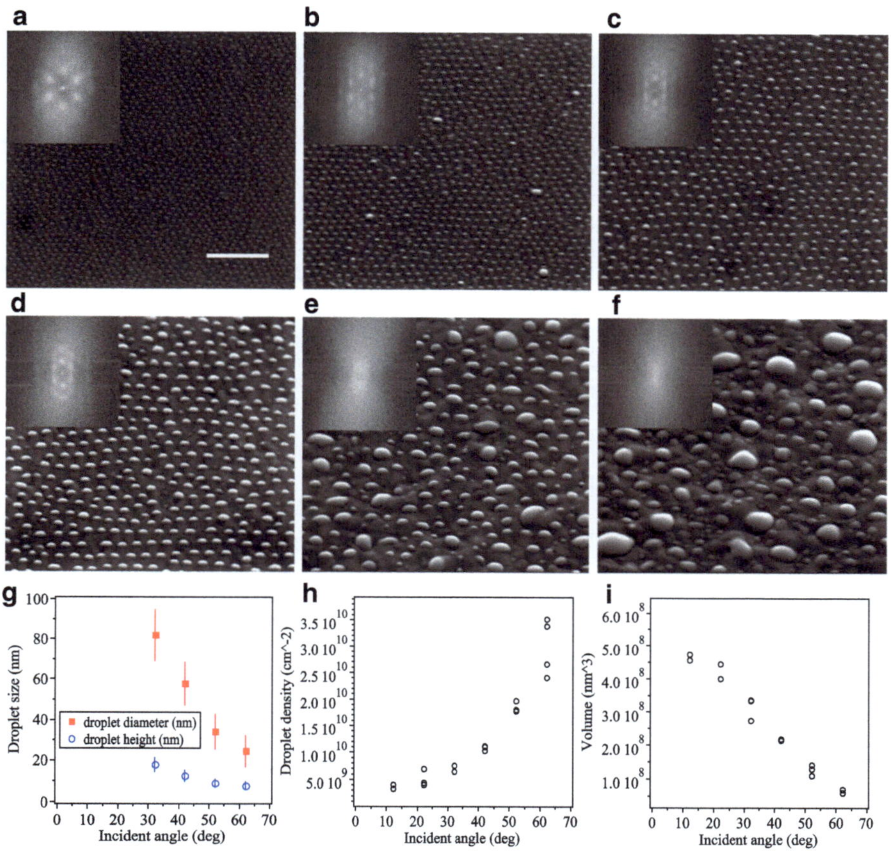

Fig. 5.4 Evolution of the droplet density with incidence: (**a**) 62°, (**b**) 52°, (**c**) 42°, (**d**) 32°, (**e**) 22°, (**f**) 12°. Ion beam parameters: ion beam energy 5 keV, flux 7.49×10^{14} cm^{-2} s^{-1}, sputter time 5 min, dwell time 1 μs, 50% beam overlap, scale bar 500 nm; the insets are FFT spectra; (**g**) dependence of the droplet size on incidence; (**h**) dependence of the droplet density on incidence; (**i**) total volume of Ga deposited on the surface as a function of incidence

Further, we researched the degree of ordering, the size, and density of the droplets as a function of various ion fluences within the determined incidence range for expected droplet ordering. For this purpose, the substrate was sputtered using fluences 1.09×10^{17} cm^{-2}, 2.62×10^{17} cm^{-2}, 4.49×10^{17} cm^{-2}, 8.61×10^{17} cm^{-2}, 2.25×10^{18} cm^{-2}, and 3.75×10^{18} cm^{-2} at high incidence 55°, and fluences 1.69×10^{17} cm^{-2}, 3.23×10^{17} cm^{-2}, and 8.42×10^{17} cm^{-2} at low incidence 37°, beam energy 5 keV, and dwell time 1 μs.

Figure 5.5a–f illustrates the changes in the morphology induced by altering the ion beam current for a surface exposed to a 5 keV ion beam at an incident angle 37°, sputter time 15 min, and dwell time 1 μs. The size and density distribution profiles are presented in Fig. 5.5g, h. The droplets induced by current in the

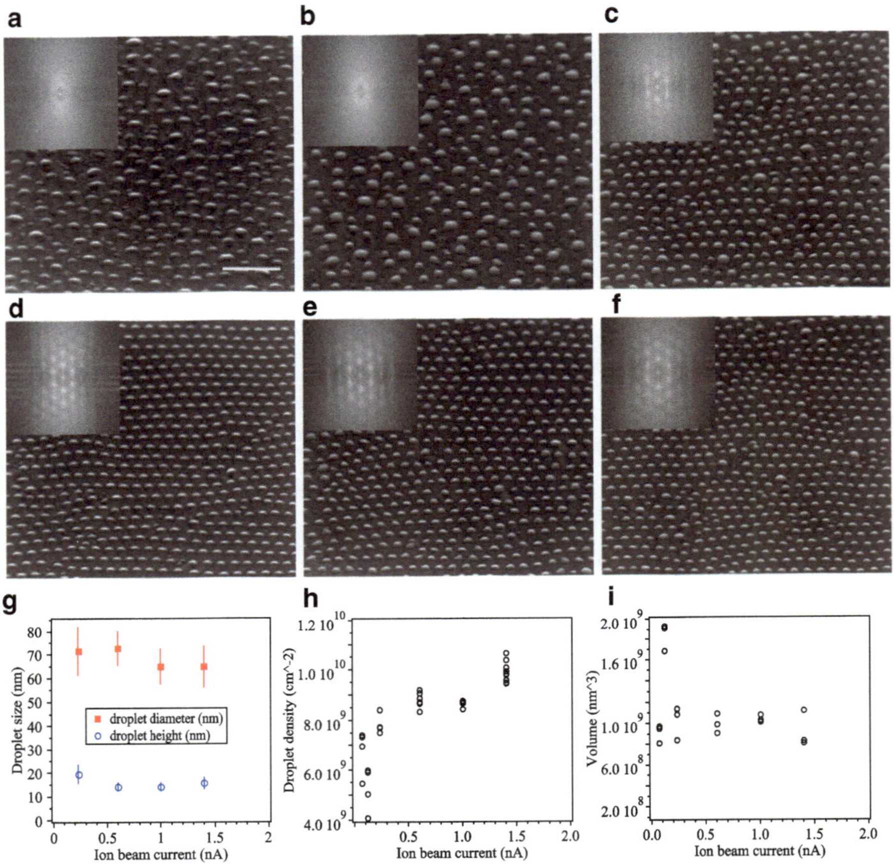

Fig. 5.5 Current-dependent evolution in surface morphology: (**a**) 70 pA, (**b**) 0.12 nA, (**c**) 0.23 nA, (**d**) 0.60 nA, (**e**) 1.0 nA, and (**f**) 1.4 nA. Sputtering was carried out on an area with size 400 μm^2 using ion beam energy 5 keV, incident angle 37°, sputter time 15 min, dwell time 1 μs, and 50% beam overlap, scale bar 500 nm; the *insets* are FFT spectra; (**g**) dependence of the droplet size on ion beam current; (**h**) dependence of the droplet density on ion beam current; (**i**) total volume of Ga deposited on the surface as a function of current

range of 0.23 nA–1.4 nA have average lateral size of 67.1 ± 12.0 nm, height of 15.0 ± 4.0 nm, and density of 9.03×10^9 cm^{-2}. Within this range, the volume of Ga deposited on the surface also remains constant (Fig. 5.5i). The FFT spectra of the scans show that droplet ordering improves with increasing current until saturation occurs. Contrary to expected droplet ordering for incident angles in the range of 32° to 62°, irradiation with low current values (70 pA and 0.12 nA) at angle 37° did not induce ordering. However, irradiating the surface for 129 min using 70 pA current, such that the same fluence is achieved as in the case of 0.60 nA, enhances the self-arrangement of droplets into an ordered pattern. An approximate value of the critical fluence for achieving optimal ordering is suggested to be 2.25×10^{18} cm^{-2}.

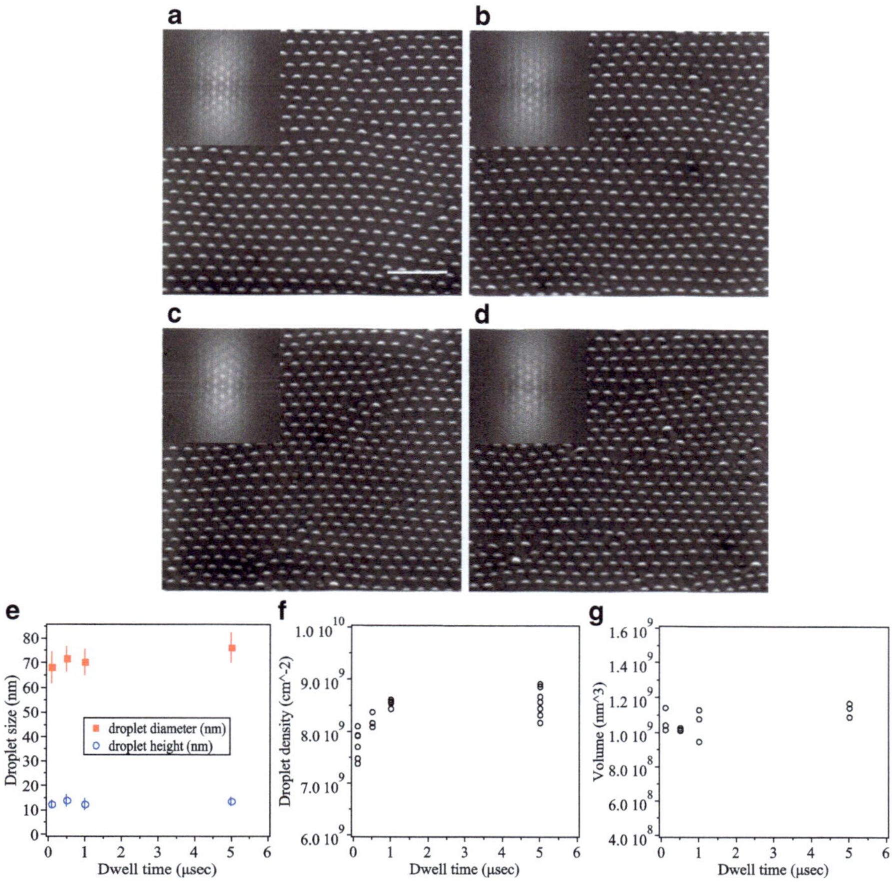

Fig. 5.6 Evolution of Ga droplets with dwell time: (**a**) 0.1 μsec (the *arrow* points to a dislocation in the lattice), (**b**) 0.5 μsec, (**c**) 1.0 μsec, and (**d**) 5.0 μsec. Beam parameters: ion beam energy 5 keV, flux 9.36×10^{14} cm^{-2} s^{-1}, sputter time 40 min, incident angle 37°, 50% beam overlap, scale bar 500 nm; the *insets* are FFT spectra; (**e**) dependence of the droplet size on dwell time; (**f**) dependence of the droplet density on dwell time; (**g**) total volume of Ga deposited on the surface as a function of dwell time

Droplet ordering, size, and density were studied as a function of dwell time. Figure 5.6a–d shows the evolution of a GaAs surface exposed to uniform low-energy irradiation for 40 min at an incident angle 37°, flux 9.36×10^{14} cm^{-2} s^{-1}, and various dwell times (0.1 μs, 0.5 μs, 1.0 μs, and 5.0 μs). The FFT spectral analysis of the images confirms droplet ordering is not influenced by changes in the dwell time. The average values for droplet lateral size, height, and density are 73.1 ± 6.8 nm, 13.9 ± 2.7 nm, and 8.27×10^{10} cm^{-2}, respectively. The droplet size and density and the volume of Ga deposited on the surface remain constant (Fig. 5.6e–g). The substrate temperature, sputter yield, and surface roughness also do not change.

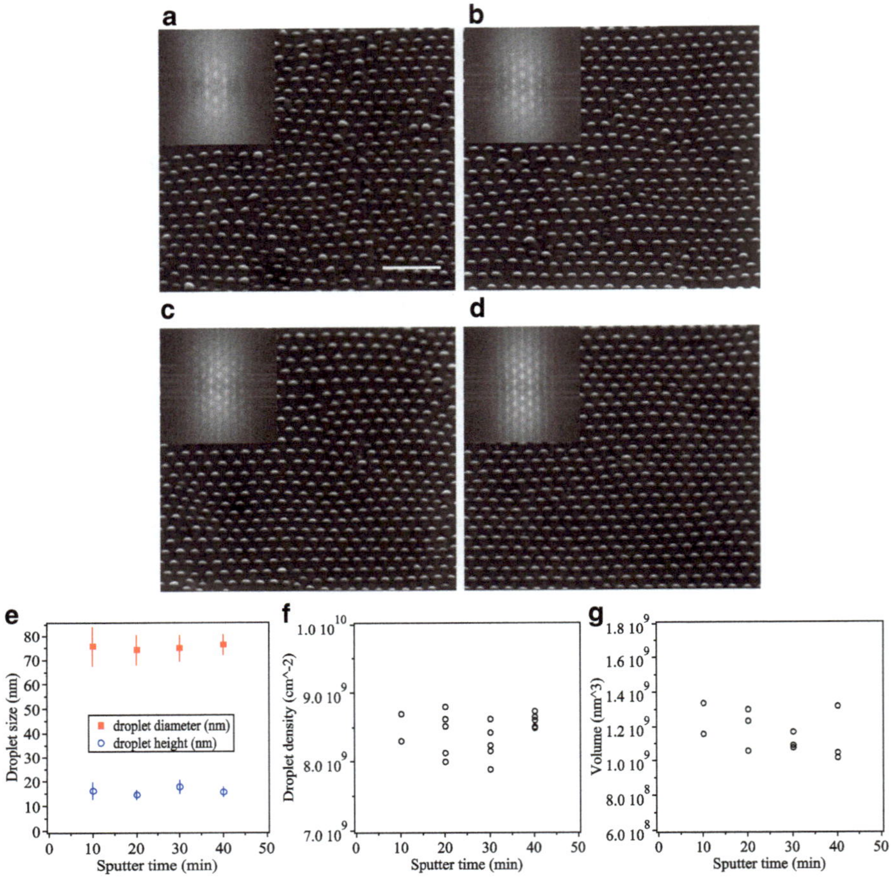

Fig. 5.7 Evolution of Ga droplets with different exposure times: (**a**) 10 min, (**b**) 20 min, (**c**) 30 min, and (**d**) 40 min. Beam parameters: ion beam energy 5 keV, flux 9.36×10^{14} cm^{-2} s^{-1}, incident angle 37°, dwell time 1 μs, 50% beam overlap, scale bar 500 nm; the insets are FFT spectra; (**e**) dependence of the droplet size on sputter time; (**f**) dependence of the droplet density on sputter time; (**g**) total volume of Ga deposited on the surface as a function of sputter time

We also investigated the effect of varying the exposure time on droplet ordering, size, and density. A GaAs surface was exposed to a 5 keV beam at incident angles 32° and 37°, with respective flux values 1.44×10^{15} cm^{-2} s^{-1} and 7.48×10^{14} cm^{-2} s^{-1}, and exposure time varying from 5 to 60 min. Droplet ordering improved until some critical sputter time when saturation in ordering occurred. Figure 5.7a–d shows the evolution of a GaAs surface irradiated uniformly by low energy beam at an incident angle 37° with flux 9.36×10^{14} cm^{-2} s^{-1} for various exposure times. Improvement in droplet ordering with increasing exposure time was observed and confirmed by FFT spectra. The droplets had average density of 8.44×10^9 cm^{-2}, lateral size of 73.9 ± 7.5 nm, and height of 15.2 ± 3.2 nm. The droplet size, density, and volume remain constant with sputter time (Fig. 5.7e–g).

Furthermore, self-organization of the droplets into an ideal ordered pattern was simulated by means of crystallographic image processing (CIP) [28]. Crystallographic image processing is an analytical technique used to determine the plane symmetry group of two-dimensional periodic structures while excluding any image distortion during sample imaging or order disruption introduced by the presence of crystal defects. For this purpose, the deviation of the theoretically accepted (symmetrized) and observed amplitude (F_{res}) and phase (α_{res}) residuals of the Fourier coefficients of the processed image compared to those for each of the seventeen plane symmetry groups was determined from the following relations:

$$F_{\text{res}}(\%) = \frac{\sum_{HK} \left| |F_{\text{obs}}(H,K)| - |F_{\text{sym}}(H,K)| \right|}{\sum_{HK} |F_{\text{obs}}(H,K)|} \tag{5.5}$$

$$\alpha_{\text{res}}(^\circ) = \frac{\sum_{HK} w(H,K) \left| \alpha_{\text{obs}}(H,K) - \alpha_{\text{sym}}(H,K) \right|}{\sum_{HK} w(H,K)} \tag{5.6}$$

where w is relative weight and H and K are Fourier coefficients. The results yield a deviation from plane symmetry group p6mm, the densest hexagonal packing of two-dimensional structures.

The formation of Ga nanodroplets with characteristic size and density and their self-organization into a hexagonal pattern can be explained with changes in the substrate temperature, surface roughness and the sputter rate of the material, the amount of Ga deposited on the surface, and exclusion zone and shadowing effects.

When exposed to a Ga^{+} ion beam, the GaAs surface is in dynamic equilibrium with Ga droplets forming due to deposition and sputtering of GaAs and disappearing as a result of sputtering and evaporation of pure Ga. The Ga droplets nucleate in Ga-rich areas on the surface induced by preferential sputtering of As, and Ga gain from ion implantation and sputtering. The size and density of the droplets are directly related to the rate of surface diffusion governed by the temperature of the substrate due to changes in the ion beam energy, current, and angle of incidence. The substrate temperature is directly proportional to the ion beam energy and current. However, with increasing incidence less energy is deposited on the surface per unit area, and the temperature of the substrate is lower compared to the case of normal incidence. The diffusion length decreases at oblique incidence leading to less droplet ripening. As a result, smaller droplets with high density form. The temperature rise of the substrate during irradiation was estimated to be approximately 9.1 °C. In addition, droplet size is determined by the ratio of Ga atoms gained from implantation and sputtering of the substrate and Ga atoms lost from sputtering of the droplet.

On the other hand, droplet uniformity and ordering are linked to surface smoothness. Typically, smoothing of the surface occurs at high incidence. The surface smoothing and roughening is enhanced by changes in the local sputter yield as a result of presence of surface modulations. The dependence of the sputter yield on the surface curvature constitutes the basis of Bradley–Harper model developed to explain the ripple morphology of planar amorphous surfaces induced by ion beam irradiation. According to the model, an ion incident on the surface penetrates into the bulk material at a characteristic depth determined by the energy of the ion beam and the angle of incidence (Fig. 5.2). The energy carried by the ion beam is deposited beneath the surface, and the energy spread due to collision cascades generated by recoil atoms exhibits Gaussian distribution. The profile of energy spread with relation to the surface varies with incidence. At high oblique incidence, a higher-energy contour is closer to the surface thus inducing higher sputter rate. If the surface is covered by modulations with radius of curvature comparable to the ion range, the energy deposited at a region of interest for both concave and convex surfaces will be from energy profiles of neighboring incident ions; however, the number of ions per unit area is lower for a concave surface compared to a convex surface, thus resulting in an enhanced sputter yield from the convex surface. As a result, irradiation at high incidence induces surface smoothing. On the other hand, at normal incidence, more energy is deposited at a region of interest in a concave surface due to neighboring energy profiles compared to a convex surface, and the sputter yield is enhanced in the concave regions. Therefore, a modulated surface exposed to an ion beam at normal incidence becomes rougher.

Consequently, the sputter yield changes due to local variation of the incidence as a result of shadowing effects resulting in enhanced surface modulations. As a result of sputter yield differences on the surface, a large size distribution of droplets with low density is observed. The sputter yield of the material also provides information about the volume of Ga deposited on the surface. Consistent with the simulated behavior shown in Fig. 5.2, the experimental results in Fig. 5.3 and Fig. 5.4 show an increase in the ion beam energy and incident angle, leading to rougher surface result in volume increase.

Finally, droplet ordering observed at angles greater than 32° is attributed to exclusion zone and shadowing effects. According to the theoretical model for droplet ordering developed by Wei et al., each droplet is surrounded by an exclusion zone within which Ga atoms generated from sputtering diffuse to the droplet [7]. The atoms migrating between the droplets, however, exhibit motion in the direction of the projected beam driven by the force of the beam. Because of the net supply of atoms between the droplets and shadowing effects due to droplets present on the surface, new droplets form in an organized manner. The final droplet array is a sixfold pattern of self-organized droplets.

5.4 Perspectives and Future Work

In conclusion, Ga nanodroplets form by sputtering a nominally flat GaAs surface using focused Ga^+ ion beam. Droplet ordering occurs for low-energy sputtering at off-normal incidence. Control over the ordering, lateral size, height, and density of the droplets is possible by varying ion beam parameters such as ion beam energy, angle of incidence, current, and exposure time. Crystallographic image processing confirms the arrays of ordered droplets ideally belong to p6mm plane symmetry group. The morphological evolution of the GaAs surface exposed to uniform irradiation can be understood on the basis of Bradley–Harper model for curvature-dependent sputtering [24], and a theoretical model for droplet ordering based on shadowing and exclusion zone effects introduced by Wei et al. [7].

Ordered droplet arrays are of interest due to their potential application as templates for patterned growth of strain-free semiconductor nanostructures. To date, epitaxial growth of InAs, GaAs, and InGaAs quantum dot structures has been reported for lithographically patterned GaAs substrates [29, 30]. In particular, Stranski-Krastanov strain-driven growth mode as well as vertical stacking on high-index and patterned surfaces during molecular beam epitaxy are two typical approaches utilized to achieve controlled positioning and improved size homogeneity of quantum dots [31–33]. On the other hand, droplet epitaxy has been studied in recent years as an approach to strain-free growth of quantum dots [34], single and double quantum rings [35, 36], as well as multiple ring nanostructures [37], quantum dot molecules [38–40], quantum dot clusters [41, 42], and nanostructure complexes [43, 44]. Additional control over the size and shape of the epitaxial nanostructures is introduced by means of local droplet etching (LDE) [45], a comparatively new technique for fabricating a nanohole-patterned template during the initial stage of droplet epitaxy [46]. All the above-mentioned approaches to epitaxial growth are characterized by overall random distribution of nanostructures on the surface, lateral positioning dependent on crystallographic orientation, and density fluctuations driven by changes in surface diffusion along different crystallographic directions. To address these issues, we suggest utilizing a self-organized droplet array presented in this paper as a patterned template for engineering nanostructures based on LDE technique.

References

1. Rose, F., Fujita, H., Kawakatsu, H.: Nanotechnology **19**, 035301 (2007)
2. Rusponi, S., Costantini, G., Boragno, C., Valbusa, U.: Phys. Rev. Lett. **81**, 2735 (1998)
3. Kanjilal, A., Prucnal, S., Minniti, M., Skorupa, W., Helm, M., Facsko, S.: J. Appl. Phys. **107**, 073513 (2010)
4. Wei, Q., Zhou, X., Joshi, B., Chen, Y., Li, K.D., Wei, Q., Sun, K., Wang, L.: Adv. Mater. **21**, 2865 (2009)
5. Du, Y., Atha, S., Hull, R., Groves, J., Lyubinetsky, I., Baer, D.: Appl. Phys. Lett. **84**, 5213 (2004)
6. Huang, Y.Z., Cockayne, D.J.H., Ana-Vanessa, J., Cowburn, R.P., Wang, S.G., Ward, R.C.C.: Nanotechnology **19**, 015303 (2007)
7. Wei, Q., Lian, J., Lu, W., Wang, L.: Phys. Rev. Lett. **100**, 076103 (2008)
8. Lugstein, A., Basnar, B., Bertagnolli, E.: J. Vac. Sci. Technol. B Microelectron. Nanometer Struct. **22**, 888 (2004)
9. Zhong, P., Que, W.X.: Nano-Micro. Lett. **2**, 1 (2010)
10. Wu, J., Wang, Z.M., Holmes, K., Marega Jr., E., Zhou, Z., Li, H., Mazur, Y.I., Salamo, G.J.: Appl. Phys. Lett. **100**, 203117 (2012)

11. Naouel, R., Dhaouadi, H., Touati, F., Gharbi, N.: Nano-Micro. Lett. **3**, 242 (2011)
12. Yang, D., Wang, L., Zhang, X., Wang, D., Shen, Z., Li, S.: Nano-Micro. Lett. **3**, 146 (2011)
13. Gago, R., Vázquez, L., Plantevin, O., Sánchez-García, J., Varela, M., Ballesteros, M., Albella, J., Metzger, T.: Phys. Rev. B **73**, 155414 (2006)
14. Frost, F., Schindler, A., Bigl, F.: Phys. Rev. Lett. **85**, 4116 (2000)
15. Chan, W.L., Chason, E.: J. Appl. Phys. **101**, 121301 (2007)
16. Ghaly, M., Nordlund, K., Averback, R.: Philos. Mag. A **79**, 795 (1999)
17. Michely, T., Teichert, C.: Phys. Rev. B **50**, 11156 (1994)
18. Wang, X.S., Pechman, R., Weaver, J.: J. Vac. Sci. Technol. B Microelectron. Nanometer Struct. **13**, 2031 (1995)
19. Avdic, A., Lugstein, A., Schondorfer, C., Bertagnolli, E.: Appl. Phys. Lett. **95**, 223106 (2009)
20. Ran, G., Zhang, J., Wei, Q., Xi, S., Zu, X., Wang, L.: Appl. Phys. Lett. **94**, 073103 (2009)
21. Ziberi, B., Frost, F., Höche, T., Rauschenbach, B.: Phys. Rev. B **72**, 235310 (2005)
22. Zhang, J., Wei, Q., Lian, J., Jiang, W., Weber, W.J., Ewing, R.C.: Appl. Phys. Lett. **92**, 193107 (2008)
23. Lian, J., Zhou, W., Wei, Q., Wang, L., Boatner, L.A., Ewing, R.C.: Appl. Phys. Lett. **88**, 093112 (2006)
24. Bradley, R.M., Harper, J.M.E.: J. Vac. Sci. Technol. A Vac. Surf. Films **6**(2390) (1988)
25. Sigmund, P.: J. Mater. Sci. **8**, 1545 (1973)
26. Ziegler, J. F., Biersack, J.: (1996)
27. Ziegler, J.F., Ziegler, M., Biersack, J.: Nucl. Instruments Meth. Phys. Res. Sect B Beam Interact. Mater. Atoms **268**, 1818 (2010)
28. Moeck, P., Toader, M., Abdel-Hafiez, M., Hietschold, M.: In: AIP Conference Proceedings, p. 294 (2009)
29. Ishikawa, T., Nishimura, T., Kohmoto, S., Asakawa, K.: Appl. Phys. Lett. **76**, 167 (2000)
30. Kiravittaya, S., Rastelli, A., Schmidt, O.: Appl. Phys. Lett. **87**, 243112 (2005)
31. Wang, Z.M., Churchill, H., George, C.E., Salamo, G.J.: J. Appl. Phys. **96**, 6908 (2004)
32. Polimeni, A., Patanè, A., Henini, M., Eaves, L., Main, P.C., Sanguinetti, S., Guzzi, M., Cryst, J.: Growth **201–202**, 276–279 (1999)
33. Kiravittaya, S., Heidemeyer, H., Schmidt, O.: Phys. E Low Dimens. Syst. Nanostruct. **23**, 253 (2004)
34. Keizer, J., Bocquel, J., Koenraad, P., Mano, T., Noda, T., Sakoda, K.: Appl. Phys. Lett. **96**, 062101 (2010)
35. Mano, T., Kuroda, T., Sanguinetti, S., et al.: Nano Lett. **5**, 425 (2005)
36. Lee, J., Wang, Z.M., Abuwaar, Z., Strom, N., Salamo, G.: Nanotechnology **17**, 3973 (2006)
37. Somaschini, C., Bietti, S., Koguchi, N., Sanguinetti, S.: Nano Lett. **9**, 3419 (2009)
38. Lee, J.H., Sablon, K., Wang, Z.M., Salamo, G.J.: J. Appl. Phys. **103**, 054301 (2008)
39. Sablon, K.A., Lee, J.H., Wang, Z.M., Shultz, J.H., Salamo, G.J.: Appl. Phys. Lett. **92**, 203106 (2008)
40. Alonso-González, P., González, L., Martín-Sánchez, J., González, Y., Fuster, D., Sales, D., Hernández-Maldonado, D., Herrera, M., Molina, S.: Growth of Low-Density Vertical Quantum Dot Molecules with Control in Energy Emission, vol. 5, p. 1913. Springer, New York (2010)
41. Creasey, M., Lee, J., Wang, Z., Salamo, G., Li, X.: Nano Lett. (2012)
42. Wang, Z.M., Liang, B., Sablon, K.A., Lee, J., Mazur, Y.I., Strom, N.W., Salamo, G.J.: Small **3**, 235 (2007)
43. Somaschini, C., Bietti, S., Fedorov, A., Koguchi, N., Sanguinetti, S.: J. Cryst. Growth **323**, 279 (2011)
44. Somaschini, C., Bietti, S., Sanguinetti, S., Koguchi, N., Fedorov, A.: Nanotechnology **21**, 125601 (2010)
45. Wang, Z.M., Liang, B.L., Sablon, K.A., Salamo, G.J.: Appl. Phys. Lett. **90**, 113120 (2007)
46. Li, A.Z., Wang, Z.M., Wu, J., Xie, Y., Sablon, K.A., Salamo, G.J.: Cryst. Growth Des. **9**, 2941–2943 (2009)

Chapter 6
Atomistic Mechanisms Underlying the Freezing Behavior of Metal Nanodroplets

Francesco Delogu

Abstract The freezing behavior of nanometer-sized particles of metallic systems is still an open issue, with considerable relevance to a wide spectrum of industrial applications. Understanding the fundamental mechanisms underlying the liquid-to-solid phase transition represents one of the necessary achievements to enable a definite progress for both science and engineering. Aimed at providing a general overview of the molecular dynamics methods that can be used to profitably investigate the response of these systems to a decrease of temperature, this chapter focuses on the phase transition behavior of Au and Ag droplets. In the idea of discussing specific cases as closest as possible to real ones, the control of temperature has been performed by using a collisional thermostat. It is shown that unsupported droplets on the order of a few nanometers in radius can exhibit a relatively complicated dynamics as the freezing point is approached.

6.1 Introduction

It is well known that nanometer-sized systems can exhibit thermodynamic properties quite different from the ones of coarse-grained matter as a consequence of so-called specific and smooth size effects [1–6]. Although no size threshold can be identified on rigorous basis to distinguish between the two effects, it is generally accepted that specific size effects only concern very small clusters, whereas smooth size effects emerge in the size domain between clusters and massive systems [1–6]. In the former case, the deviation of thermodynamic properties from the ones of bulk phases substantially originates from the sensitivity of the potential energy of small clusters to the relative position of the different atoms in the cluster [1–7]. In particular,

F. Delogu (✉)
Dipartimento di Ingegneria Meccanica, Chimica, e dei Materiali, Università degli Studi di Cagliari, via Marengo 2, 09123 Cagliari, Italy
e-mail: francesco.delogu@dimcm.unica.it

Z.M. Wang (ed.), *Nanodroplets*, Lecture Notes in Nanoscale Science and Technology 18, DOI 10.1007/978-1-4614-9472-0_6,

slightly different cluster configurations can exhibit significantly different potential energy, which gives rise to an irregular variation of the thermodynamic properties with size [1–7]. Conversely, the smooth size effects typically induce a regular change of physical and chemical properties with size, which can be described by scaling equations involving relatively simple power laws [1–6, 8].

In principle, the gradual variation of the thermodynamic properties of a given system with its size can be satisfactorily explained by taking into due account the energy contributions to the total Gibbs free energy coming from free surfaces or interfaces [1–6]. Whereas such contributions can be definitely neglected in the case of bulk phases, they can become even dominant in nanometer-sized materials, characterized by an unusually high surface area-to-volume ratio [1–6]. This conceptual framework is fully supported by experimental and theoretical evidences, which clearly suggest that surface, or interface, energy can deeply affect both physical and chemical properties [1–6].

The depression of the melting, or freezing, points of small metallic particles represents one of the most striking examples of thermodynamic behavior deviated by smooth size effects. First hypothesized in 1909 [9], and further investigated in 1953 [10], the depression of the melting, or freezing, points of small metallic particles was experimentally demonstrated starting from 1967 [11, 12].

In systems with reduced dimensionality, such behavior is generally rationalized by invoking a size dependence for the chemical potential [1–6, 9, 13–21]. Along this line, several phenomenological models based on rigorous thermodynamic descriptions have been shown to successfully reproduce the experimental findings, which indicate a depression of the melting, or freezing, points as the system size decreases [1–6, 9, 13–21].

Since these models cannot be satisfactorily validated on a direct experimental basis, an accurate numerical investigation of the heat exchange process during melting, or freezing, is necessary to achieve a comprehensive understanding of the phase transition behavior. In turn, this requires studying the atomic-scale dynamics underlying the thermodynamics and kinetics of the phase transition.

Melting and freezing represent two of the most suitable candidates to perform such analysis. In the following, the attention will be focused precisely on the freezing behavior of nanometer-sized droplets. In particular, the present chapter will be devoted to the discussion of molecular dynamics simulations employing collisional thermostats to simulate the crystallization of a nanometer-sized Au droplet and the coexistence of liquid and solid phases in a nanometer-sized Ag droplet.

Before discussing in detail such cases, the freezing phase transition, and the numerical simulation methods usually employed in its study, will be shortly introduced.

6.2 A General Look to Freezing

Freezing, also termed solidification, is the first-order thermodynamic phase transition in which a liquid phase turns into a solid [22]. In principle, the freezing of a liquid can be induced by changing its thermodynamic state variables pressure P, temperature T, volume V, and chemical composition either independent of each other or in combination with each other [22]. However, the term freezing generally refers to the phase transition induced by a decrease of the temperature T below a threshold value, referred to as the freezing point T_f [22].

According to classical thermodynamics, the freezing point T_f is the temperature at which the liquid and the solid phases are in stable thermodynamic equilibrium [22]. Correspondingly, the Gibbs free energy ΔG_liq of the liquid phase is equal to the one of the solid, ΔG_sol, and the difference in enthalpy identifies the so-called latent heat of freezing ΔH_f [22]. Below the freezing point T_f, the Gibbs free energy ΔG_liq of the liquid phase becomes higher than the one of the solid phase, ΔG_sol [22]. As a consequence, the liquid is no longer the most stable phase and tends to transform into a solid [22]. In the presence of significant contributions from free surfaces, or interfaces, the Gibbs free energies ΔG_sol and ΔG_liq must incorporate additional terms related precisely to the free energy of surfaces, or of interfaces [22].

From a phenomenological point of view, the kinetics of such transformation is relatively well understood. It relies upon a two-stage mechanism involving first the formation of small nuclei of solid phase within the liquid and then their growth [23–25]. The nucleation of the solid phase does not start at the freezing point T_f but rather requires the attainment of a certain degree of undercooling ΔT_f, defined as $T_\mathrm{f} - T$ [23–25]. According to the classical nucleation theory, the degree of undercooling ΔT_f can be related to the minimum Gibbs free energy required to form stable nuclei [23–25]. It can be shown that the Gibbs free energy ΔG needed to form an individual spherical nucleus of radius r is equal to

$$\Delta G = -\frac{4}{3}\,\pi\, r^3\, \Delta G_\mathrm{sol,v} + 4\,\pi\, r^2\, \sigma, \tag{6.1}$$

where $\Delta G_\mathrm{sol,v}$ represents the Gibbs free energy per unit volume of the solid phase and σ is the energy per unit area of the solid–liquid interface [23–25]. Equation (6.1) indicates that the formation of a nucleus is an unfavorable process until its radius r is smaller than a critical value $r^* = -2\ \sigma/\Delta G_\mathrm{sol,v}$ [19–21]. Therefore, the critical Gibbs free energy ΔG^* needed to form the critical nucleus is equal to $16\ \pi\ \sigma^3/(3\ \Delta G^2_{sol,v})$ [23–25]. Now, it is worth noting that the Gibbs free energy per unit volume of the solid phase can be estimated by the expression

$$\Delta G_\mathrm{sol,v} \approx \frac{\Delta H_\mathrm{f,v}}{T_\mathrm{f}} \Delta T_\mathrm{f}, \tag{6.2}$$

where $\Delta H_{f,v}$ is the latent heat of freezing per unit volume and ΔT_f is the degree of undercooling [23–25]. It follows that the critical radius and the critical Gibbs free energy can be expressed as

$$r^* \approx \frac{2\,\sigma\,T_f}{\Delta H_{f,v}} \frac{1}{\Delta T_f} \tag{6.3}$$

and

$$\Delta G^* = \frac{16\,\pi\,\sigma^3\,T_f^2}{3\,\Delta H_{f,v}^2} \frac{1}{\Delta T_f^2}, \tag{6.4}$$

respectively [23–25]. Correspondingly, both the critical radius r^* and the critical Gibbs free energy ΔG^* decrease as the degree of undercooling ΔT_f increases [23–25].

Equations (6.1), (6.2), (6.3), and (6.4) refer to the homogeneous nucleation case, in which nucleation takes place in the bulk of the liquid phase [23–25]. However, freezing often occurs in the presence of solid–liquid or solid–gas interfaces, which result in a heterogeneous nucleation scenario [23–25]. In this latter case, the critical Gibbs free energy for the formation of a nucleus becomes lower due to the thermodynamic contribution coming from the contact of the liquid phase with the solid surfaces [23–25]. In particular, in the presence of a solid,

$$\Delta G^*_{het} = \Delta G^*_{hom} \left(\frac{1}{2} - \frac{3}{4} \cos\theta + \frac{1}{4} \cos^3\theta \right), \tag{6.5}$$

where θ is the angle of contact between the solid surface and the liquid phase [23–25].

The rate at which the nuclei form at a given temperature is equal to

$$J = J_0 \exp\left(-\Delta G^*/R\,T\right), \tag{6.6}$$

where J_0 is a pre-exponential frequency factor and R is the universal gas constant [23–25]. Equation (6.6) and similar ones have been extensively used to analyze the results of experiments focusing on the nucleation of crystalline phases in a liquid [23–25].

The classical nucleation theory and its most recent modifications represent a successful phenomenological model to describe the behavior of thermodynamic phases at the freezing point [23–25]. However, neither the classical thermodynamics nor the classical nucleation theory provides insight into the various processes taking place on the atomic scale when nuclei form and grow. For example, a rigorous theory for the pre-exponential frequency factor J_0 is still lacking [23–25]. Similarly, the energy per unit area of the solid–liquid interface, σ, cannot be predicted on a theoretical basis [25, 26]. As a consequence, the classical nucleation theory is not able to predict absolute nucleation rates. Rather, in most of cases both J_0 and σ are used as adjustable parameters to best fit the available experimental data.

Nevertheless, gaining information on the atomistic processes governing the nucleation of crystalline phases in a liquid matrix is of crucial importance for any advance of knowledge in the field. In this respect, suitably performed numerical simulations employing either Monte Carlo or molecular dynamics methods are playing an increasingly relevant role.

6.3 Some Hints from Numerical Simulations

Computer simulations have been shown to provide valuable help in elucidating the atomistic features of crystal nucleation. For moderate undercooling degrees, both free energy barriers and nucleation rates have been estimated for model systems formed by hard spheres [27, 28] and particles interacting with simple Lennard-Jones potentials [29–32]. For instance, the classical nucleation theory, within the assumptions made for a spherical nucleus in an undercooled liquid, has been shown valid in the critical regime of nucleation for a large range of undercooling degrees and nucleus sizes [33].

Despite their apparent simplicity, these model systems exhibit properties quite close to the ones of real systems, such as colloidal solutions and globular proteins [27–32]. In these cases, the theoretical findings can be compared with experimental results, and a satisfactory agreement has been obtained [34, 35]. However, comparing theoretical predictions and experimental evidences is invariably challenging, being affected by unavoidable experimental uncertainties. As a consequence, developing a comprehensive conceptual framework for both the thermodynamics and the kinetics of crystallization as a function of the degree of undercooling is considerably difficult.

Only for a few model systems, a reliable description of the temperature-dependent kinetics of the crystallization process is available. For example, it has been shown that the crystallization of a Lennard-Jones fluid at fast quenching rates is satisfactorily described by the classical nucleation theory [36]. In particular, the height of the nucleation barrier and the size of the critical nucleus predicted on a theoretical basis are comparable with the ones numerically estimated [36]. In addition, the initial stages of the growth of crystalline embryos have been shown to be characterized by the sudden formation of dense crystalline clusters of small size [36]. This supports the hypothesis that the nucleation of a solid phase in a Lennard-Jones fluid is influenced by the proximity of a pseudo-spinodal singularity [37–39].

On a strictly intuitive basis, it can be expected that the presence of a free surface could significantly affect the above-mentioned freezing behavior. In fact, the free surfaces are the physical *loci* involved in the heat exchange between any given nanometer-sized droplet and its surroundings. Therefore, heat is removed first from surface regions, and their instantaneous temperature can become lower than the average droplet temperature, at least for short time intervals. As a whole, this could suggest that the free surface can act as a preferential nucleation site for the solid phase.

However, it must be noted that the surface region is formed by atoms characterized by the unsaturation of their coordination numbers compared with equilibrium ones [1–6]. It follows that surface regions are less stable than the bulk and exhibit unusual static and dynamic properties that can hardly favor the nucleation of a solid phase [1–6].

This does not mean that surface nucleation must be ruled out. In a few cases, solidification is found to start exactly at the surface of the nanometer-sized droplet, and Au clusters represent one of such cases [40]. Nevertheless, it is much more frequent the situation in which the first stable nuclei of the solid phase appear in the bulk-like region of the droplet. Also in these cases, the free surface affects the nucleation process. For instance, in the case of Al–Ni clusters, nucleation takes place in the bulk-like region of the droplet [41]. Despite this, the nucleation of the solid Al–Ni alloy is affected by surface segregation effects [41].

At the freezing point, the solid and liquid phase can in principle coexist. Depending on the droplet size, metal clusters can exhibit either static or dynamic coexistence [42–48]. In the former case, it is possible to identify a relatively stable interface between the solid and the liquid phases, whereas, in the latter, the whole system oscillates forth and back between solid- and liquid-like dynamics [42–48]. In each case, various effects can be observed superposed to the general features. For example, in the case of Fe–Pt clusters, the dynamic coexistence in a carrier gas has been shown to be the result of a kinetic effect related to the rate of heat dissipation [49].

An expression has been also tentatively proposed to identify the clusters bound to static or dynamic coexistence as a function of their size [50]. Applied to Pb clusters, in agreement with the numerical predictions, such expression predicted that a system of 931 atoms exhibits dynamic coexistence, whereas a system of 1,427 exhibits a static coexistence [50]. However, the detailed mechanism underlying the freezing of metal clusters has not yet been identified.

In this respect, considerable attention has been paid to the structural motifs appearing in clusters during, and as a consequence of, the freezing transition. For example, in the case of Ag clusters, icosahedral structures have been observed for systems with size around the icosahedral magic number 147 and in the range between 245 and 310 [51]. Conversely, decahedral structures appear in the size range between 170 and 240 and face-centered cubic structures around 280 [51]. In all of the cases, molecular dynamics simulations have pointed out a significant dependence of the relative amount of structural motifs on the cooling rate [51]. Entropic effects can be also invoked to explain the formation of icosahedral structures in Ag clusters of 923 atoms [52].

Another series of molecular dynamics simulations on Ag droplets in the size range between 2.5 and 10 nm, cooled at a rate of 100 K ns^{-1}, indicated that the inner structures of the frozen particles can be significantly different from each other, with various combinations of icosahedral and decahedral structures [53]. In addition, several unfamiliar structures with three- and fivefold symmetry and relatively low energy have been found [53].

Body-centered cubic systems in the size range between 2,000 and 31,250 atoms have been also investigated by molecular dynamics methods [54]. It was observed that nucleation occurs on one side of the droplet and then spreads over the other side [54]. Furthermore, the nucleation temperature upon cooling does not vary monotonically with the system size [54].

Due to its importance for a broad spectrum of scientific and technological activities, the study by molecular dynamics simulations of the freezing behavior nanometer-sized metal droplets is still an intriguing, and lively debated, area of fundamental research [55]. Here, it is worth noting that the investigations not only focused on the freezing behavior of isolated metal droplets, but also of more complex systems in which the droplets are embedded within a matrix. This is the case of Au droplets freezing on a metal surface, or inside a carbon nanotube [56–58]. In all of these cases, the freezing behavior exhibits a sensitivity to the nature of the substrate, or of the matrix [56–58].

6.4 Cooling Rates and Cooling Methods

The freezing behavior must be expected to exhibit a dependence on the cooling rate, as well as on the details of the cooling method. The former dependence is quite obvious. The solidification behavior depends on the relative value of the time scales underlying the removal of heat from the metal, on the one hand, and the formation of the crystalline embryos in the liquid, on the other [26]. The experimental evidence concerning the synthesis of bulk metallic glasses clearly indicates that a crystalline phase can nucleate only when the time necessary to form crystalline nuclei is shorter than the one governing the removal of thermal energy from the system [26]. Otherwise, crystallization is kinetically hindered, and the system evolves towards a metastable thermodynamic state characterized by the absence of a crystalline order, and the formation of a glassy structure [26, 59]. As a consequence, crystallization can actually take place if the cooling rate is lower than the crystal nucleation rate [26, 59]. In all of the other cases, kinetic effects can prevent the formation of crystalline embryos, and the crystallization process is replaced by a vitrification one [26, 59].

In the case of numerical simulations employing molecular dynamics methods, this condition assumes a particular importance. In fact, it is quite easy to impose very high cooling rates in simulations, able to induce a system evolution different from the one observed in real experiments, where the cooling rates are typically orders of magnitude lower than the ones employed in calculations [60–62].

The dependence on the cooling method is somehow more subtle. It concerns the numerical method employed to simulate a thermostat [60–62]. In molecular dynamics, the control of temperature, as well as of other physical quantities, depends on the constraints imposed to thermodynamic state variables and state functions to simulate processes under different thermodynamic conditions [60–62].

To this aim, various statistical ensembles have been defined [60–62]. A simple control of temperature can be achieved by directly rescaling the velocity of each particle in the system [63]. The velocity rescaling method adds, or subtracts, energy to, or from, the system quite efficiently. However, it is very coarse and unable to realistically describe the mechanism underlying the dissipation of thermal energy [63]. For this reason, more sophisticated methods have been elaborated to keep temperature constant. The most common ones are the so-called generalized Langevin equation approach [64], the Berendsen method [65], and the Nosè–Hoover thermostat [66–68].

The generalized Langevin equation approach is based on a generalized theoretical description of Brownian motion [64]. Accordingly, the system is regarded as a "solute" embedded in a "solvent" matrix able to impose the desired temperature by both frictional and stochastic effects [64]. In particular, two terms are added to the normal Newtonian equation of motion, namely, frictional and random forces [64]. The frictional force accounts for the frictional drag operated by the "solvent" on the "solute" motion and is usually taken proportional to the particle velocity, but with opposite sign [64]. Instead, the random force is assumed to have no relation to the velocity and position of the particle and is typically sampled from a selected Gaussian distribution with a zero mean [64].

In the case of the Berendsen method, the system is coupled to an imaginary external thermal bath kept at constant temperature [65]. The exchange of thermal energy between the system and the thermal bath is ruled by an algorithm that gradually scales the velocity of the particle in dependence of a time constant for the coupling [65]. As a result, the instantaneous kinetic temperature approaches the desired equilibrium temperature according to a smooth asymptotic trend [65].

In the original Nosè approach, an additional degree of freedom is introduced to describe the heat bath and operate as a time-scaling factor [66–68]. In addition, a fictitious mass is given to the heat bath, to describe the inertia in the heat bath response to instantaneous temperature fluctuations [66–68]. In the Nosè–Hoover thermostat, the time-scaling factor is eliminated, whereas a thermodynamic friction coefficient is introduced [66–68]. Whenever the instantaneous temperature is higher than the desired equilibrium value, the intensity of the frictional force increases in order to induce a lowering of the temperature, and vice versa [66–68]. The value chosen for the fictitious mass of the heat bath is critical to the successful performance of the thermostat [66–68]. A small value of the fictitious mass induces rapid temperature fluctuations, while large values can often result in an unsatisfactory sampling of the phase space [66–68].

In general, the Nosè–Hoover thermostat gives rise to a well-defined canonical distribution of both momentum and coordinates [66–68]. Nevertheless, in the case of small systems or in the presence of high-frequency vibrational modes, the Nosè–Hoover thermostat can fail to generate a canonical distribution [66–68].

This aspect points out another important issue in the control of temperature. In all of the above-mentioned cases, the thermostat is based on the adjustment of the particle velocities according to a given methodology. However, all of the mentioned methodologies are quite far from the real case in which the heat dissipation is mediated by the contact between thermostat and system [69]. The reason is that reproducing the real mechanism of heat flow for massive systems in numerical simulations would be virtually impossible, due to the considerable computational power required [60–62].

The use of realistic methods to control the temperature becomes increasingly viable as the system size is reduced. It follows that reliable calculations based on the explicit use of a material thermostat become possible for nanometer-sized objects immersed in a liquid phase, or embedded in a gaseous environment. In turn, this allows in principle to investigate the response of nanometer-sized systems to a temperature control based on a realistic heat exchange mechanism.

6.5 Gas Thermostats and Realistic Simulations

A gas thermostat can be defined as a collisional thermostat. Typically, it consists of an inert gaseous phase surrounding the system [70, 71]. In principle, the gaseous phase is of infinite extent, and its pressure is low enough to avoid local heating [70, 71]. Its temperature is lower than the system one [70, 71]. The numerical analysis of the energy transfer between the system and the atoms of the inert gas is based on the assumption that individual collisions are followed by a complete microcanonical relaxation of energy [70, 71]. Although this assumption cannot be verified in reality, the limiting law obtained by accepting it strongly correlates with the observed behavior [70, 71]. Furthermore, it allows a systematic prediction of the deviations, thus approaching a satisfactory understanding, and the desired predictive capability [70, 71].

Here, it must be noted that the above-mentioned analysis can be carried out on a perfectly rigorous ground by defining the density of states in the microcanonical ensemble [70, 71]. Nevertheless, the systems are generally much larger in size than the individual atoms of the inert gas [70, 71]. It follows that the systems can be regarded as a thermal reservoir [70, 71]. As a consequence, after the collision, the system and the inert gas atoms can be considered in thermal equilibrium [70, 71]. At each collision, the system loses a certain amount of thermal energy, thus reaching a lower temperature [70, 71]. The temperature reached after any given collision can be estimated by using a simple energy conservation equation [70, 71].

The simple approach described above allows developing reliable rate equations for the change of temperature and energy of the system [70, 71]. Thus, it is possible to compare the results of numerical simulations with both theoretical predictions and experimental findings [70, 71]. In this respect, it is worth noting that realistic simulations can be really performed in the case of nanometer-sized systems, i.e., of aggregates including approximately 10–10^6 atoms [70, 71].

These systems can be synthesized by a variety of methods [72–74]. Among the others, nanometer-sized particles can be obtained by cooling down in a cold inert gas a supersaturated vapor phase, previously prepared by evaporating a bulk material in ultrahigh vacuum [72–75]. In this case, during the initial stages of cluster formation and growth, as well as in the successive heating or cooling processes, the inert gas operates precisely as a collisional thermostat [4, 70, 71].

The frequency f of collisions of the inert gas atoms on the free surface of the nanometer-sized system is given by the expression

$$f \approx \frac{P_g\, A_{cl}}{\left(2\pi\, m_g\, k_B\, T_g\right)^{\frac{1}{2}}}, \tag{6.7}$$

where P_g is the pressure of the inert gas, A_{cl} the area of the system free surface, m_g the gas molar mass, k_B the Boltzmann constant, and T_g the temperature of the inert gas [4, 70, 71]. The average energy gain or loss per collision, δE, determines the heating or cooling rate, respectively, which can be expressed as

$$\frac{dT}{dt} \approx \frac{f\, \delta E}{3N\, k_B}. \tag{6.8}$$

Here, N represents the number of atoms in the nanometer-sized system.

Typical experimental conditions are identified by a pressure P_g of about 1×10^4 Pa, and a temperature T_g around 100 K [76, 77]. Under such conditions, when pure He is employed as the inert gas phase, the frequency f of atomic collisions on the free surface of a cluster of about 3 nm in radius roughly amounts to 1.4×10^{11} Hz. The estimated energy loss per collision δE is on the order of 1×10^{-21} J [70, 71]. As a result, the cooling rate is approximately equal to 0.9 K ns^{-1}.

This latter value indicates that the real cooling processes take place on a time scale that is absolutely out of reach for numerical simulations. At the same time, it must be noted that the thermal relaxation processes after each collision take place at a rate roughly equal to the sound speed [78]. It follows that the temperature equilibration process can be quite fast. For example, in the case of an Au particle with a radius of about 3 nm, where the sound speed is on the order of 3 nm ps^{-1} [79], the temperature equilibrates roughly within 2 ps. This time interval is significantly shorter than the one between two consecutive collisions. Under the above-mentioned experimental conditions, this latter is approximately equal to 7 ps.

The difference of 5 ps suggests that, in principle, time and collisions can be decoupled. More specifically, it is possible to run numerical simulations in which the molecular dynamics of the system is strictly followed only until a complete thermal equilibration after each energy loss is attained. Once the temperature is equilibrated, a new collision event can be simulated without following the system trajectory in the phase space within the time window in which no event takes place. Correspondingly, up to about the 70 % of the whole simulation time can be saved. For instance, molecular dynamics simulation runs of 0.3 ns would be able to give reliable information on the system behavior after 1 ns.

The present chapter discusses the results obtained by molecular dynamics simulations employing the above-mentioned method in the study of the freezing behavior of unsupported nanometer-sized droplets of liquid metals. In particular, attention will be focused on the crystallization of an Au droplet and on the coexistence of the liquid and solid phases in an Ag droplet.

6.6 The Details of the Numerical Simulation Methods

The interaction forces between metal atoms were described by the using the semiempirical many-body tight-binding potential based on the second-moment approximation of the electron band energy [80]. It follows that the cohesive energy U is equal to

$$U = \sum_{i=1}^{N} \left\{ \sum_{j=1}^{N} A e^{-p\left(\frac{r_{ij}}{d}-1\right)} - \left[\sum_{j=1}^{N} \xi^2 e^{-2q\left(\frac{r_{ij}}{d}-1\right)} \right]^{\frac{1}{2}} \right\}, \tag{6.9}$$

where r_{ij} is the distance between atoms i and j, and the parameters A, ξ, p, and q quantify the interatomic potential between any given pair of atoms. The term d represents the nearest-neighbors distance at 0 K and N the total number of atoms. The repulsive part of the potential is expressed in the first member on the right-hand side of Eq. (6.3) as a Born–Mayer pair-wise interaction, whereas the attractive part is expressed in the second member within the framework of the second-moment approximation of tight-binding band energy [80].

The necessary potential parameter values for the different metallic species were taken from literature [80]. Interactions were computed over a distance including the shell of the seventh neighbors. Equations of motion were solved with a fifth-order predictor–corrector algorithm [60] and a time step of 2 fs. A previously tested simulation code, suitably modified to calculate the desired quantities, was used.

The numerical simulations applied molecular dynamics methods. The liquid Au and Ag droplets were created starting from larger crystalline lattices. Calculations were carried out in the isobaric–isothermal ensemble with number of atoms N, pressure P, and temperature T constant [67, 81]. Periodic boundary conditions were applied [60]. Starting from 600 K, the temperature of the crystalline system was raised at high nominal rates, until melting was attained. Afterwards, the temperature was kept constant at a value about 200 K higher than the equilibrium melting point for the time necessary to allow a satisfactory system equilibration [82].

The nanometer-sized Au and Ag droplets of the desired radius R were created by selecting a suitable spherical region at the center of the liquid bulk phase, and removing the atoms outside. Once created, the droplet was relaxed for the necessary time intervals to attain full equilibration.

Subsequent computations aimed at identifying the freezing point of the investigated nanometer-sized droplet were performed in the isochoric–isothermal ensemble with number of atoms N, volume V, and temperature T constant [67, 81].

The temperature was progressively decreased at rates between 0.05 and 10 K ps^{-1} to roughly estimate the freezing point T_f.

The freezing behavior was investigated by monitoring the solid- or liquid-like character of individual atoms. To this aim, a local order parameter $\overline{\psi_i}$ correlating the average positions of pairs of opposite nearest neighbors in the coordination shell was evaluated for any given atom i [83, 84]. The positions of individual atoms were averaged over a time interval of 5 ps. Following previous work [83, 84], the atoms with $\overline{\psi_i}$ values smaller than 0.05, and larger than 0.2, will be hereafter referred to as liquid- and solid-like respectively. Here, it is worth noting that the transition from solid- to liquid-like character, and vice versa, is quite rapid for individual atoms. The same is true for the freezing transition of the nanometer-sized droplet. It follows that, at each instant, only a negligible fraction of the atoms exhibits $\overline{\psi_i}$ values between 0.05 and 0.2 [83, 84].

Once freezing was observed, the stochastic thermal bath was replaced by a He gas collisional thermostat at a pressure P_g of 1×10^4 Pa and temperature T_g of 100 K. Then, calculations were restarted from the atomic configuration recorded 4 K above the freezing point T_f.

The interactions between He atoms were described by using a 12-6 Lennard-Jones pair potential with depth ε of potential well and distance σ at which the potential is zero respectively equal to 10.22 k_B and 2.6 Å [71]. The interactions involving He atoms and metallic ones were also described by a 12–6 Lennard-Jones potential [71]. In the light of the qualitative character of the investigations carried out, the metal species were described as atoms of the rare gas closer to the metal in terms of molar mass [85]. This choice is motivated by the evidence that the exchange of energy at individual collisions is substantially governed by the mass of the atoms come into contact [71]. As a consequence, the details of the interatomic potential curve, in particular of the repulsive branch, are not expected to significantly alter the fundamental physics of the collision process [71]. Although apparently rough, the related approximations appear reasonably acceptable [71]. The interactions between different rare gas atoms were described according to the Lorentz–Berthelot rules [86]. All of the Lennard-Jones interactions were calculated within a cutoff radius extended to the fifth neighbors.

To avoid useless complications, and save computational time, only collisions perpendicular to the free surface of the nanometer-sized droplets, and involving He atoms with an initial kinetic energy equal to 3 $k_B T_g/2$, were simulated. The collision events were randomly distributed over the droplet free surface.

The energy gain, or loss, per collision, δE, was estimated by evaluating the kinetic energy lost, or gained, by the He atoms after the collision. Due to the difference in the mass of He and of the other rare gas atoms used to mimic the collisional behavior of metallic species, the collisions are not able to excite the translational and rotational degrees of freedom of the nanometer-sized droplets.

Successive collisions are separated by 2 ps. Such time interval permits a satisfactory thermal equilibration. Thermal equilibration processes were monitored by evaluating the radial density of kinetic energy, $\rho_E(r)$, at different times t_r after the

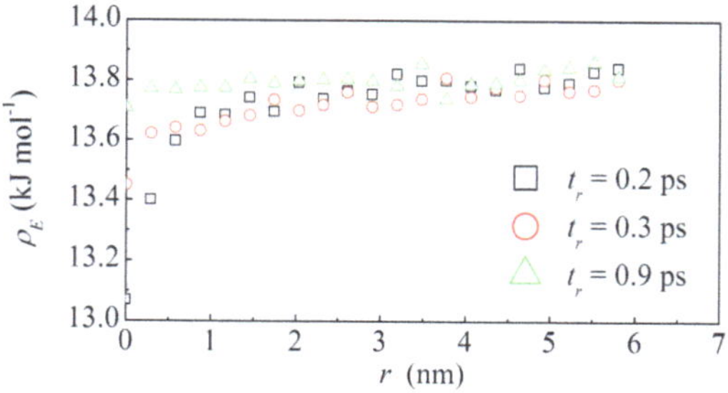

Fig. 6.1 The radial density of the kinetic energy, ρ_E, as a function of the radial distance r from the center of collision. Data refer to the three different relative times indicated

collision has started. Here, the beginning of the collision process is set equal to the instant in which the distance between the impinging He atom and the metallic atom closest to it becomes shorter than the equilibrium distance of the Lennard-Jones potential describing their interaction. Correspondingly, t_r equals zero.

The radial density of kinetic energy, $\rho_E(r)$, represents the average kinetic energy E per atom in the spherical sectors of radius r, centered at the point of collision, in which the droplets can be ideally divided. Therefore, $\rho_E(r)$ can be defined as

$$\rho_E(r) = \frac{1}{n_r} \sum_{i=1}^{n_r} E_i, \tag{6.10}$$

where n_r is the number of atoms in the spherical sector considered and E_i is the kinetic energy of the ith atom.

The typical time behavior of $\rho_E(r)$ is shown in Fig. 6.1, where $\rho_E(r)$ is plotted as a function of the radial distance r of the spherical sectors from the center of collision. Data refer to a droplet of Au with average radius of about 3 nm. The $\rho_E(r)$ values at three different times t_r are shown.

It can be seen that $\rho_E(r)$ undergoes a significant decrease in the neighborhood of the center of collision at short times. This means that a relatively significant energy loss δE takes place at collision, approximately equal to 1.44×10^{-21} J. The energy is almost perfectly re-equilibrated after about 1 ps.

To the above-mentioned energy loss δE corresponds an average temperature decrease per collision, δT, roughly equal to 0.016 K. This value is substantially impossible to directly detect due to the fluctuations in kinetic and potential energy. Nevertheless, such δT value can be profitably used to roughly estimate the number of collisions needed to attain the freezing point T_{f}. For the Au droplet of 3 nm in size, T_{f} is approximately equal to 1,098 K. It follows that about 250 collisions are needed to reach the freezing point T_{f} starting from a temperature of 1,102 K.

The formation of crystalline nuclei was monitored by evaluating a set of local bond-order parameters, hereafter indicated with $q_{6m}(i)$ [27, 40, 87, 88]. These parameters provide a measure of the local structural order within the first coordination shell of any given atom i. For any given atom, nearest neighbors can be identified on the basis of a simple distance criterion. In particular, two atoms can be regarded as nearest neighbors when their distance is shorter than $r_{\min}$, $r_{\min}$ being the position of the first minimum of the pair correlation function of the bulk crystal of the metal considered at the same temperature [60]. Being dependent on the 6th-order spherical harmonics Y_{6m} [27, 40, 87, 88], the parameters $q_{6m}(i)$ permit to distinguish between solid- and liquid-like dynamics based on the correlation degree between the local structural order of neighboring atoms. Following previous work [40], two nearest neighbors can be regarded as connected with each other whenever the correlation function

$$c_{ij} = q_{6m}(i) \cdot q_{6m}^{*}(j) = \sum_{m=-6}^{+6} q_{6m}(i) \cdot q_{6m}^{*}(j) \qquad (6.11)$$

between the vector $\boldsymbol{q}_{6m}(i)$ of components $q_{6m}(i)$ and the complex conjugate vector $\boldsymbol{q}_{6m}^{*}(j)$ of components $q_{6m}^{*}(j)$ takes a value larger than a threshold one of 0.68 [40]. A given atom is assumed to exhibit a solid-like dynamics when at least half of its nearest neighbors are connected with each other in the same cluster [40].

6.7 The Crystallization of a Nanometer-Sized Au Droplet

Simulating a crystallization at low undercooling degrees is a noticeable computational challenge. Under such conditions, the formation of the first crystalline embryos occurs on very long time scales, on the order of 10^9 s for systems including roughly 10^9 particles [24, 27, 88]. Crystallization becomes accessible to molecular dynamics calculations only when the undercooling degree is relatively high, and the activation energy barrier for nucleation is correspondingly relatively low, which enables nucleation to take place at high rates [24, 27, 88].

In the case of a nanometer-sized Au droplet of about 3 nm in radius, the preliminary calculations indicate that the freezing point T_{f} is approximately equal to 1,098 K. This is the temperature at which the Au droplet starts the transition. However, the first atoms exhibiting a behavior precursor of the solid-like dynamics are observed already at about 1,100 K, i.e., 2 K above the estimated freezing point T_{f}. These atoms are connected with each other in the same cluster and appear at the free surface of the Au droplet in the neighborhood of the point at which a He atom has collided. However, they do not yet exhibit a solid-like dynamics.

The first atoms with a transient solid-like character appear at the free surface of the Au droplet only at 1 K below the freezing point T_{f}. Initially, their lifetime is on the order of 0.8 ps, but it becomes longer as the droplet temperature T decreases. The first atoms with a stable solid-like dynamics appear at about 1,096 K, a temperature reached after about 350 He atom collisions.

The number of atoms connected with each other in clusters, n_{c}, and the number of atoms with solid-like dynamics, n_{s}, are shown in Fig. 6.2 as a function of T.

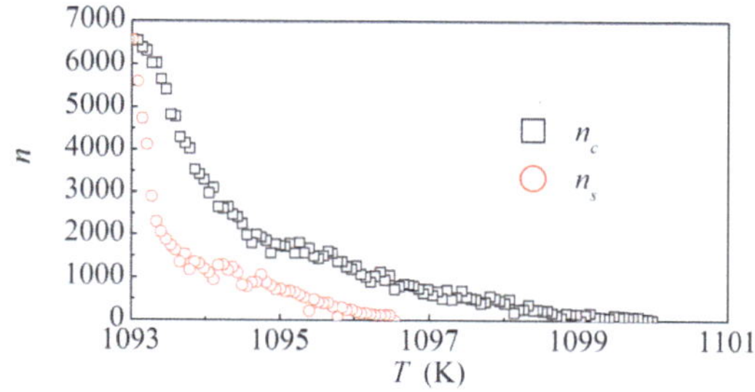

Fig. 6.2 The number of connected, n_c, and solid-like, n_s, atoms as a function of the droplet temperature T

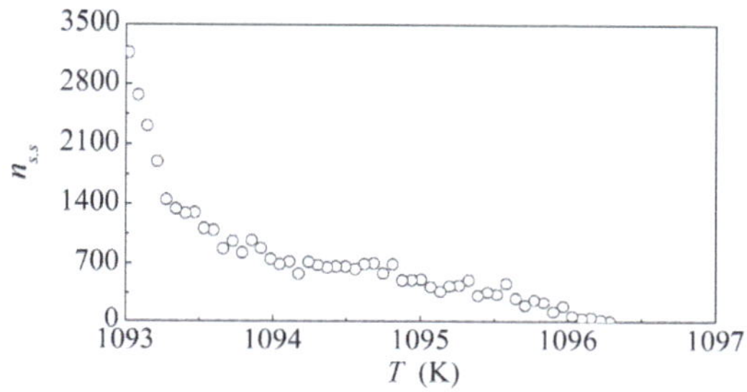

Fig. 6.3 The number of solid-like surface atoms, $n_{s,s}$, as a function of the droplet temperature T

The data indicate that n_c is invariably larger than n_s until the Au droplet completely crystallizes. Freezing exhibits a relatively sharp onset and attains completion within a temperature interval of about 3 K. Initially, the crystallization process mostly involves atoms located at the free surface of the droplet, i.e., atoms located in the spherical shell 2 r_{min} thick at the droplet surface. This can be easily inferred from Fig. 6.3, where the number of surface atoms with solid-like character, $n_{s,s}$, is shown as a function of the temperature T together with the number of atoms with solid-like dynamics, n_s.

It can be seen that $n_{s,s}$ represents up to about 70 % of n_s for a significant fraction of the temperature interval during which the Au droplet crystallizes.

At temperatures below 1,096 K, substantially all of the He atom collisions induce the formation of a cluster of solid-like atoms on the free surface of the droplet, or the size increase of the existing ones. The average number ν of atoms connected in the same cluster increases as the temperature T decreases, amounting to about 4 at 1,096 K and becoming roughly twice at 1,095 K.

In the initial stage of the nucleation process, most of the clusters are not connected with each other. However, the situation changes as the droplet temperature decreases. The general cluster behavior can be reliably inferred from the variation of the number of clusters, N_{cl}, with the droplet temperature T. The data are shown in Fig. 6.4.

A short induction period is initially observed, related to the attainment of the necessary undercooling degree. Shortly after the temperature value of 1,093 K has been reached, N_{cl} undergoes a definite increase up to a maximum value of 35 at about 1,095.5 K. Then, it decreases progressively until N_{cl} becomes equal to 1 at

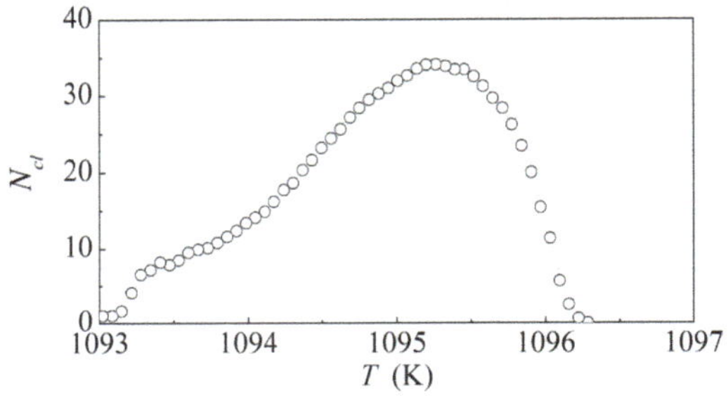

Fig. 6.4 The number N_{cl} of clusters of solid-like atoms as a function of the droplet temperature T

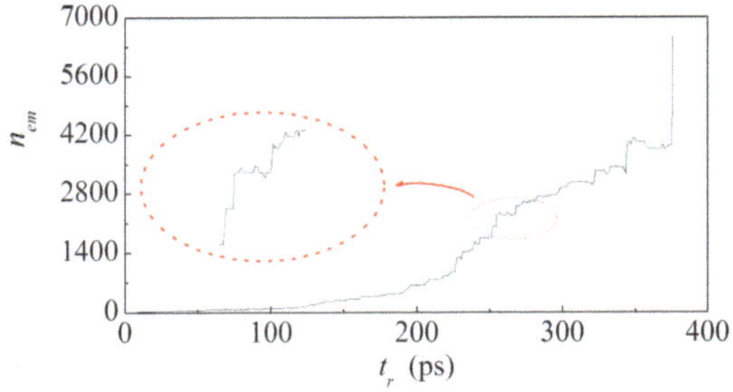

Fig. 6.5 The number n_{em} of solid-like atoms in the embryo considered as a function of the relative time t_{r}

about 1,093 K. This is the temperature value at which the whole nanometer-sized Au droplet has crystallized.

The behavior of the number N_{cl} of clusters can be rationalized by investigating the atomistic processes underlying the formation and growth of the individual clusters. To this aim, a representative case will be discussed in some detail in the following. This case concerns the first cluster able to include 100 atoms with solid-like character.

Such cluster, i.e., the first crystalline embryo including 100 atoms, forms at about 1,095 K, after approximately 450 He atoms have collided. The formation of this cluster is not mediated by the progressive addition of individual atoms with liquid-like dynamics, which subsequently transform into solid-like atoms. Even at 1,096 K, i.e., 1 K above, the addition of individual atoms is relatively slow, taking place on a time interval of about 1 ps. Conversely, the growth of the cluster is essentially governed by the coalescence of neighboring clusters.

The occurrence of coalescence processes involving clusters of different size can be inferred by the data shown in Fig. 6.5, where the number n_{em} of solid-like atoms in the crystalline embryo is plotted as a function of the relative time t_{r}. For clarity, t_{r} has been set equal to zero at the instant in which the collision of a He atom induced the formation of the precursor of the 100-atom cluster, which included only 5 atoms.

It can be seen that n_{em} increases with t_{r} following an irregular trend, characterized by marked fluctuations. A close look at the curve points out that it exhibits a sequence of sudden rises and drops indicating the fast increase and decrease in the number of connected atoms. According to the height of rises and drops, in the initial stages it is much more probable that single atoms detach from the cluster rather than

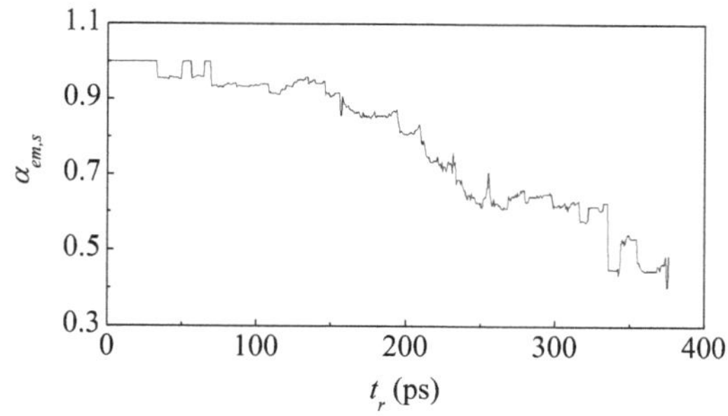

Fig. 6.6 The fraction $\alpha_{em,s}$ of surface solid-like atoms in the embryo as a function of the relative time t_r

become connected with it. Therefore, the cluster growth must be ascribed to another process, which consists of the coalescence of the cluster with neighboring ones. As pointed out by the height of rises and drops, the size of the coalescing clusters increases as the temperature T decreases.

Further information on the cluster growth is obtained by monitoring the fraction $\alpha_{em,s}$ of surface atoms with solid-like dynamics, which can be obtained by calculating the ratio between the number $n_{em,s}$ of solid-like surface atoms and the total number n_{em} of atoms in the cluster. The $\alpha_{em,s}$ values are shown in Fig. 6.6 as a function of t_r. It appears that the cluster consists almost exclusively of surface atoms roughly as far as n_{em} is smaller than 270. When n_{em} increases up to about 2,000, $\alpha_{em,s}$ undergoes a constant decrease. The involvement of the inner regions of the Au droplet in the crystallization process is pointed out by irregular drops at larger n_{em} values.

The above-mentioned behavior is common to all of the observed crystalline embryos. Accordingly, two different stages can be identified in the mechanism of cluster growth.

Initially, the crystalline embryo substantially grows by the coalescence of neighboring clusters including a significant fraction of atoms located at the droplet free surface. In this stage, the cluster growth is mostly related to the formation of a number of small clusters at the Au droplet surface due to the continuous collisions of He atoms. Therefore, the initial increase in the number N_{cl} of clusters must be simply ascribed to the dynamics of the collisional thermostat. Instead, the maximum value of N_{cl} is observed in correspondence of the narrow temperature range in which the rate of cluster coalescence processes is comparable with the frequency of He atom collisions.

Subsequently, as the undercooling degree increases, the addition of individual Au atoms to individual clusters becomes competitive with the cluster coalescence process, and even faster. Here, it is also worth noting that the cluster coalescence processes occur with a probability that increases as the cluster size increases. As a consequence, any atom addition enhances the rate of the cluster coalescence processes. Under such circumstances, the number N_{cl} of cluster decreases at an increasing rate and finally drops to 1.

The possible influence on the above-mentioned behavior of the time interval separating consecutive He atom collisions was also explored. To such aim, additional simulations were performed on Au droplets in the temperature range between 1,094 and 1,095 K by using a time interval between successive He atom collisions equal to 4 ps. The results obtained indicate that no significant change takes place when thermal equilibration after each He atom collision lasts more than 2 ps. The longer time interval allowed to the system to reach the thermal equilibration slightly affects only the average cluster size. In particular, the cluster size slightly decreases as a consequence of the higher rates at which individual atoms detach compared with the ones at which they become connected with the cluster. As a whole, the number of He atom collisions required to start the freezing transition increases about 8 %.

It follows that the numerical methodology and in particular the collisional thermostat employed enable a satisfactory description of the general qualitative behavior of nanometer-sized Au droplets at freezing.

6.8 Coexistence of Solid and Liquid Phases in a Nanometer-Sized Ag Droplet

A further support to the numerical findings regarding the freezing behavior of nanometer-sized Au droplets comes from the numerical study of the mechanism of the freezing phase transition in nanometer-sized Ag droplets.

The numerical investigation has been carried out on a spherical Ag droplet with a radius R of about 5 nm. The isobaric–isothermal molecular dynamics simulations indicate that this system, approximately formed by 31,000 atoms, undergoes the freezing transition at a temperature T_f roughly equal to 1,073 K. As in the case of the nanometer-sized Au droplet, the freezing phase transition is preceded by the appearance of clusters formed by a few connected atoms already at about 1,076 K, i.e., about 3 K above the freezing point T_f. These atoms form small clusters at the free surface of the Ag droplet, invariably in the neighborhood of the points at which He atoms collide.

A transient solid-like dynamics is first exhibited by Ag atoms located at the droplet free surface, at a temperature of about 1.5 K below the freezing point T_f. At this temperature, the lifetime of the transient solid-like dynamics is around 0.7 ps, i.e., very close to the one of the transient solid-like dynamics of atoms in the Au droplet. As the droplet temperature further decreases, the lifetime of solid-like dynamics increases. The first Ag atoms with a stable solid-like character are observed at about 1,071 K, once about 380 He atom collisions have taken place.

The number of atoms connected with each other in clusters, n_c, and the number of atoms with solid-like character, n_s, are shown in Fig. 6.7 as a function of the droplet temperature T.

In analogy with what is observed in the case of the nanometer-sized Au droplet, n_c is larger than n_s until the crystallization of the Ag droplet is complete. The onset of the freezing transition is relatively sharp, and freezing attains completion within a temperature interval of about 4 K.

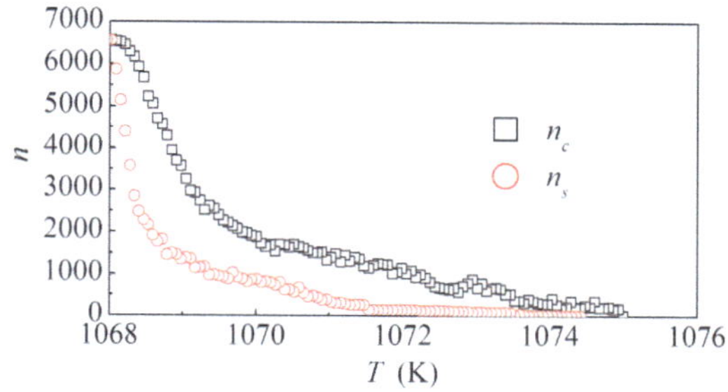

Fig. 6.7 The number of connected, n_C, and solid-like, n_S, atoms as a function of the droplet temperature T

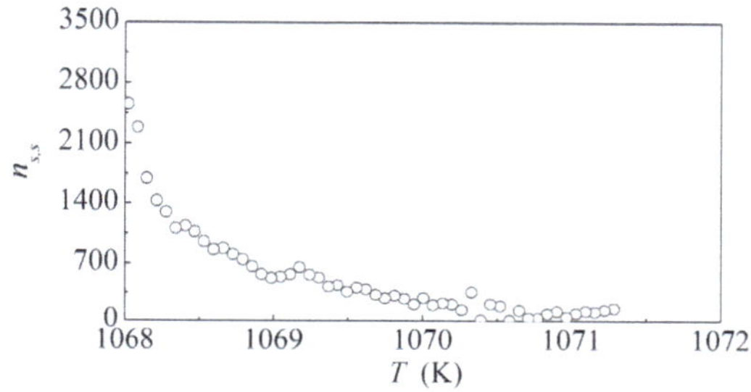

Fig. 6.8 The number of solid-like surface atoms, $n_{S,S}$, as a function of the cluster temperature T

In the initial stages, the solidification process mostly involves atoms located at the droplet free surface. This is clearly pointed out by the data shown in Fig. 6.8, where the total number of atoms with solid-like dynamics, n_S, and of surface atoms with solid-like character, $n_{S,S}$, are plotted as a function of the droplet temperature T. According to the numerical findings, $n_{S,S}$ represents up to about 70 % of n_S for a significant fraction of the temperature interval, or alternatively of the time period, over which the Ag droplet crystallizes.

As the temperature T becomes lower than 1,071 K, the collisions of He atoms with the Ag droplet generate clusters of atoms with solid-like dynamics at the droplet free surface. Similar to what is observed in the case of the Au droplet, the average number ν of atoms connected in clusters increases as the temperature T decreases. More specifically, ν amounts to 5 at 1,071 K and increases up to about 12 at 1,070 K. Initially, most of these small clusters keep independent from each other. In fact, their mobility is quite limited, and they are not able to form large clusters by connecting with each other.

The number of clusters, N_{cl}, is shown in Fig. 6.9 as a function of the droplet temperature T. As in the case of the Au droplet, a short induction period is initially observed. Once the temperature of 1,071 K has been reached, N_{cl} undergoes a relatively smooth increase up to the maximum value of 35 at about 1,070 K. Afterwards, N_{cl} gradually decreases and finally becomes equal to 1 at about 1,067 K. At such temperature, the whole Ag droplet crystallizes.

The change of the number N_{cl} of the clusters formed by solid-like atoms with the droplet temperature T is the result of the atomistic mechanisms underlying the formation and growth of such clusters. These mechanisms can be investigated by molecular dynamics methods. In particular, valuable information can be gained by

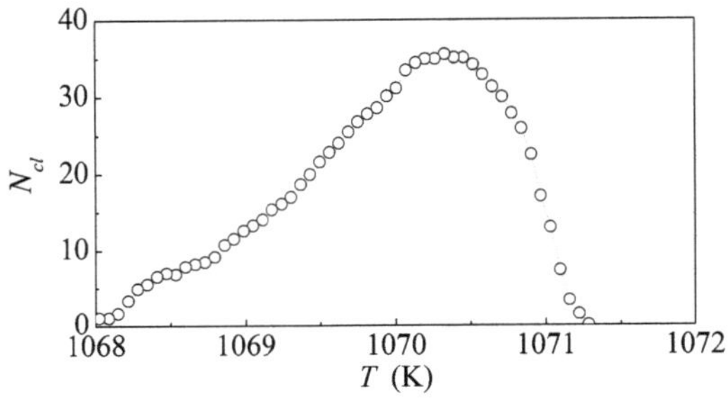

Fig. 6.9 The number N_{cl} of clusters of solid-like atoms as a function of the droplet temperature T

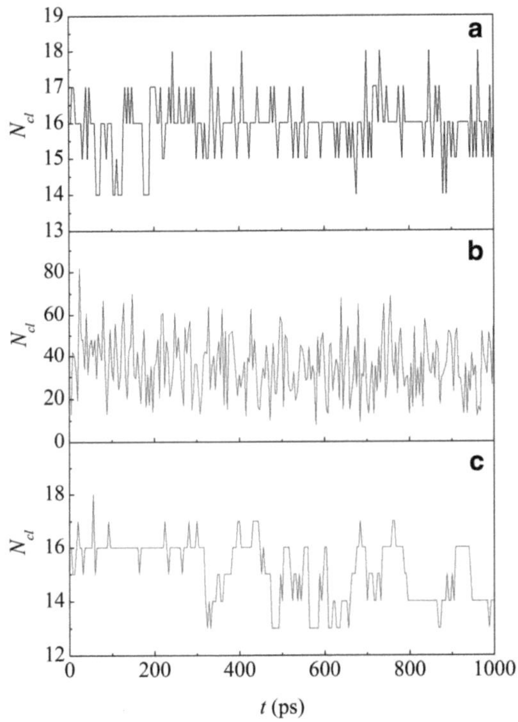

Fig. 6.10 The number N_{cl} of clusters formed by Ag atoms as a function of time t. Data refer to calculations carried out at three different values of the droplet temperature T equal to 1,071.03 K (**a**), 1,070.33 K (**b**), and 1,069.24 K (**c**)

monitoring the total number N_{cl} of clusters, as well as the behavior of individual clusters, at constant temperature.

In this respect, it is worth noting that, within the framework of the calculations carried out to explore the freezing behavior of nanometer-sized droplet, keeping the temperature constant is particularly easy. In fact, it is sufficient to interrupt the sequence of He atom collisions on the free surface of the Ag droplet. Under such conditions, the droplet evolves according to a microcanonical dynamics, which implies constancy for total number of atoms, volume of the system, and total energy [80].

The number N_{cl} of clusters formed by solid-like atoms is shown in Fig. 6.10 as a function of time t for three different temperature values, i.e., for three different degrees of undercooling. The data indicate that N_{cl} is sensitive to the droplet temperature T. At about 1,071.03 K, N_{cl} exhibits fluctuations of small amplitude around an average value of 16. The rises and drops in the N_{cl} value suggest that fluctuations must be substantially ascribed to the clusters' capability of connecting with each other or disconnecting from each other. The clusters are relatively small, and the thermal dynamics result in frequent connection, and disconnection, processes. No definite trend affects the average N_{cl} value.

At about 1,070.33 K, N_{cl} exhibits a similar behavior. The height of rises and drops has significantly increased, indicating that the number of clusters connecting with each other, and disconnecting, has also increased. At the same time, the frequency of fluctuations has also slightly increased. As in the previous case, the average N_{cl} value keeps approximately constant around 35.

At 1,069.24 K, the behavior of N_{cl} is qualitatively different. First, the fluctuations in the N_{cl} value appear significantly less frequent. Second, a relatively definite trend can be observed in the N_{cl} data. In particular, N_{cl} undergoes a decrease roughly from 16 to 14. The height of the drops in N_{cl} values indicates that the size of clusters involved in the connection, and disconnection, processes has decreased. Moreover, the decrease of the N_{cl} value points out that the connection of clusters is favored compared with the disconnection. As a consequence, it must be expected that relatively large clusters tend to form as the droplet temperature T becomes lower than a certain value.

This inference is somehow supported by the numerical evidences obtained by investigating the dynamic behavior of individual clusters at constant temperature. The number ν of atoms connected with each other in the same cluster is shown in Fig. 6.11 for a representative case. The data concern the cluster at the same three values of droplet temperature T previously considered, namely, 1,071.03, 1,070.33, and 1,069.24 K.

As in the case of the number N_{cl} of clusters, the number ν of atoms with solid-like dynamics connected in the same selected cluster is affected by significant fluctuations. The amplitude of these fluctuations increases as the droplet temperature decreases, whereas their frequency decreases. At 1,071.03 and 1,070.33 K, ν exhibits relatively constant average values equal to about 90 and 50 respectively. Instead, at 1,069.24 K, ν undergoes an irregular increase roughly from 100 to 200.

As already inferred from the data regarding the number N_{cl} of clusters, the dynamic behavior of individual clusters is dominated by coalescence processes. Accordingly, a cluster typically becomes connected with neighboring clusters or, oppositely, loses small clusters. However, the analysis of the number ν of atoms in individual clusters also points out that the connection of individual atoms plays a role. In particular, the probability of individual atoms to connect with a preexisting cluster increases as the temperature decreases.

Proceeding further with the analysis of the dynamic behavior of individual clusters, it must be noted that the mobility of clusters does not permit to explain the rapid growth that individual clusters undergo as the droplet temperature decreases, and the consequent decrease of the total number N_{cl} of clusters. Actually, the mobility of even the smallest clusters, formed by 5, 6, or 7 atoms, is extremely limited. The

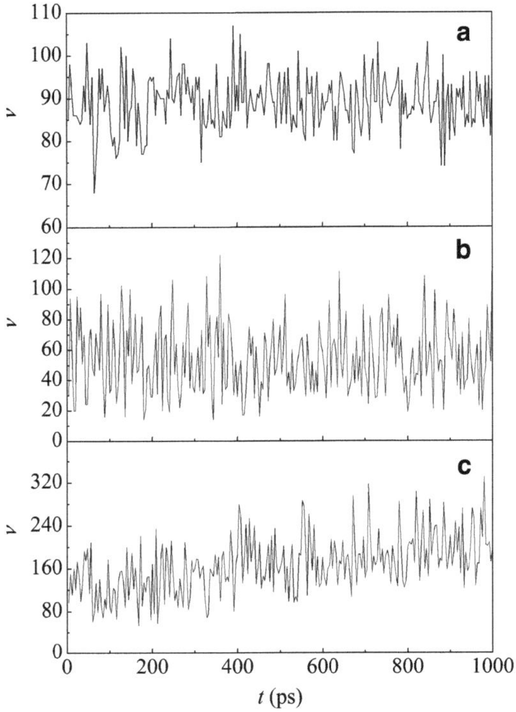

Fig. 6.11 The number ν of atoms connected in the same selected cluster as a function of time t. Data refer to calculations carried out at three different values of the droplet temperature T equal to 1,071.03 K (**a**), 1,070.33 K (**b**), and 1,069.24 K (**c**)

apparent diffusion coefficient of a cluster can be estimated by evaluating the root mean square displacement of its center of mass. For a cluster of 6 atoms, which keeps the same atoms for the whole duration of the observation period, the apparent diffusion coefficient is on the order of 0.8 $nm^2\ ns^{-1}$. This value significantly decreases as the number ν of atoms connected in the cluster increases, and the diffusion can be neglected for clusters of 15 or more atoms.

On this basis, it can be inferred that very long times could be necessary to allow two clusters to connect with each other even if initially distant only 1 nm. Conversely, the molecular dynamics calculations indicate that two clusters can connect with each other on a very shorter time interval, on the order of 50 ps.

The contradiction between the two numerical findings is only apparent, and it can be suitably explained by the dynamics of individual atoms with solid-like character. In fact, it appears that individual atoms are also affected by significant fluctuations in their solid- or liquid-like character. Any given atom undergoes a continuous change in its solid- or liquid-like character. The frequency of such fluctuations and the lifetime of the solid- or liquid-like character depend on the droplet temperature T. Under such circumstances, it can happen that one or more atoms with liquid-like character, located between two neighboring clusters, change their dynamics to the solid-like one. When this takes place, the two neighboring clusters become connected and form a single, larger cluster.

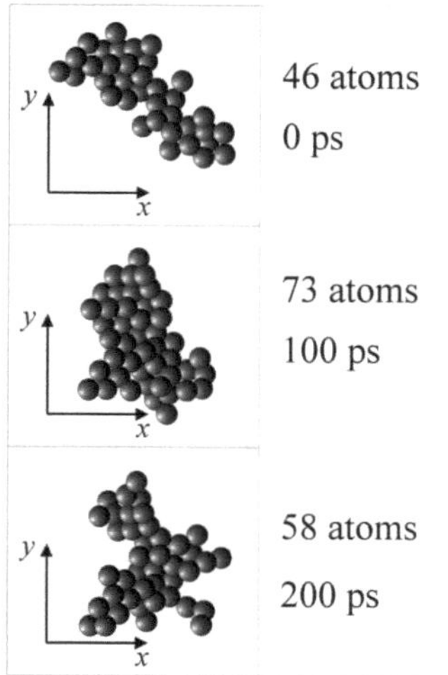

Fig. 6.12 A cluster of solid-like atoms at 1,070.33 K, and at the times indicated. The number ν of atoms connected in the cluster is also indicated

This specific feature suggests for the freezing behavior of the nanometer-sized Ag droplet a percolation scenario, with relatively small and large clusters competing to attain percolation across the whole volume of the droplet. Within this theoretical framework, the amplitude of thermal motion is bound to play a primary role. In fact, the possibility for any given atom to attain a solid-like dynamics starting from a liquid-like dynamics, and vice versa, is intimately related to the translational and vibrational degrees of freedom.

The molecular dynamics simulations also suggest that the dynamic behavior of individual atoms, and of clusters, is strictly dependent on their surroundings. More specifically, the capability of a given solid-like atom to keep its dynamics, or to change it to the liquid-like one, depends on the number of neighboring atoms with solid- and liquid-like character. However, definite thresholds in the number of solid- or liquid-like neighbors necessary for an atom to exhibit a solid- or liquid-like character have not yet been identified.

The study of these intimate aspects of the freezing transition is further complicated by the dynamic character of the transition itself. The solid- or liquid-like character of individual atoms undergoes a continuous change, and the atoms connected in a given cluster change as a consequence. Furthermore, the position of all of the atoms changes with time, and the same is true for the shape, the position, and the size of the clusters. An example is given in Fig. 6.12, where a representative cluster is shown at different times t at the same temperature of 1,070.33 K. For simplicity, relative time is set equal to zero in the instant in which the first cluster configuration was taken. It can be seen that the cluster undergoes significant changes in shape, position, and size in a very short time interval. This aspect is common to all of the different clusters observed.

As the temperature becomes lower than the freezing point T_f and the freezing transition proceeds, the above-mentioned behavior becomes increasingly affected by the finite size of the Ag droplet and by the decreased mobility of individual atoms. Therefore, analyzing the cluster dynamics becomes more and more complicated. Nevertheless, a detailed characterization of the atomic-scale dynamics could still permit a deeper understanding of the mechanisms underlying the freezing transition of nanometer-sized metal droplets embedded in a collisional thermostat.

Acknowledgments A. Ermini, ExtraInformatica s.r.l., is gratefully acknowledged for his helpful assistance. Financial support has been given by the University of Cagliari.

References

1. Moriarty, P.: Nanostructured materials. Rep. Prog. Phys. **64**, 297–381 (2001)
2. Jortner, J., Rao, C.N.R.: Nanostructured advanced materials. Perspectives and advances. Pure Appl. Chem. **74**, 1491–1506 (2002)
3. Hill, T.L.: A different approach to nanothermodynamics. Nano Lett. **1**, 273–275 (2001)
4. Baletto, F., Ferrando, R.: Structural properties of nanoclusters: energetic, thermodynamic, and kinetic effects. Rev. Mod. Phys. **77**, 371–423 (2005)
5. Ozin, G.A., Arsenault, A.C.: Nanochemistry. A Chemical Approach to Nanomaterials. RSC Publishing, Cambridge (2005)
6. Ferrando, R., Jellinek, J., Johnston, R.L.: Nanoalloys: from theory to applications of alloy clusters and nanoparticles. Chem. Rev. **108**, 845–910 (2008)
7. Schmidt, M., Kusche, R., von Issendorf, B., Haberland, H.: Irregular variations in the melting point of size-selected atomic clusters. Nature **393**, 238–240 (1998)
8. Alivisatos, P.: Semiconductor clusters, nanocrystals, and quantum dots. Science **271**, 933–937 (1996)
9. Pawlow, P.Z.: Über die Abhängigkeit des Schmelzpunktes von der Oberflächenenergie eines festen Körpers. Phys. Chem. (Munich) **65**(1–35) (1909)
10. Hollomon, T.H., Turnbull, D.: Nucleation. Prog. Metal Phys. **4**, 333–388 (1953)
11. Wronski, C.R.M.: The size dependence of the melting point of small particles of tin. Br. J. Appl. Phys. **18**, 1731–1737 (1967)
12. Coombes, C.J.: The melting of small particles of lead and indium. J. Phys. F Metal Phys. **2**, 441–448 (1972)
13. Buffat, P., Borel, J.-P.: Size effect on the melting temperature of gold particles. Phys. Rev. A **13**, 2287–2298 (1976)
14. Couchman, P.R., Jesser, W.A.: Thermodynamic theory of size dependence of melting temperature in metals. Nature **269**, 481–483 (1977)
15. Reiss, H., Mirabel, P., Whetten, R.L.: Capillarity theory for the "coexistence" of liquid and solid clusters. J. Phys. Chem. **92**, 7241–7246 (1988)
16. Sakai, H.: Surface induced melting of small particles. Surf. Sci. **351**, 285–291 (1996)
17. Peters, K.F., Cohen, J.B., Chung, Y.-W.: Melting of Pb nanocrystals. Phys. Rev. B **57**, 13430–13438 (1998)
18. Lai, S.L., Guo, J.Y., Petrova, V., Ramanath, G., Allen, L.H.: Size-dependent melting properties of small tin particles: nanocalorimetric measurements. Phys. Rev. Lett. **77**, 99–102 (1996)
19. Efremov, M.Yu., Schiettekatte, F., Zhang, M., Olson, E.A., Kwan, A.T., Berry, L.S., Allen, L.H.: Discrete periodic melting points in nanostructure. Phys. Rev. Lett. **85**, 3560–3563 (2000)
20. Olson, E.A., Efremov, M.Y., Zhang, M., Zhang, Z., Allen, L.H.: Scanning nanocalorimeter for nanoliter scale liquid samples. J. Appl. Phys. **97**, 034304 (2005)
21. Delogu, F.: Thermodynamics on the nanoscale. J. Phys. Chem. B **109**, 21938–21941 (2005)
22. Berry, S.R., Rice, S.A., Ross, J.: Matter in Equilibrium: Statistical Mechanics and Thermodynamics, 2nd edn. Cambridge University Press, Cambridge (2004)

23. Christian, J.: The Theory of Transformations in Metals and Alloys: An Advanced Textbook in Physical Metallurgy. Pergamon Press, Oxford (1965)
24. Kelton, K. F.: Crystal nucleation in liquids and glasses. In: Solid State Physics, Henry Ehrenreich, David Turnbull (Eds.), vol. 45, p. 75. Academic, New York, NY (1991)
25. Porter, D.A., Easterling, K.E.: Phase Transformations in Metals and Alloys, 2nd edn. Chapman and Hall, London (1992)
26. Cahn, R.W., Haasen, P. (eds.): Physical Metallurgy: Fourth, Revised and Enhanced Edition. Elsevier Science BV, Amsterdam (1996)
27. Auer, S., Frenkel, D.: Prediction of absolute crystal-nucleation rate in hard-sphere colloids. Nature **409**, 1020–1023 (2001)
28. Auer, S., Frenkel, D.: Suppression of crystal nucleation in polydisperse colloids due to increase of the surface free energy. Nature **413**, 711–713 (2001)
29. ten Wolde, P.R., Frenkel, D.: Enhancement of protein crystal nucleation by critical density fluctuations. Science **277**, 1975–1977 (1997)
30. Swope, W.C., Andersen, H.C.: 10^6-particles molecular-dynamics study of homogeneous nucleation of crystals in supercooled atomic liquid. Phys. Rev. B **41**, 7042–7054 (1990)
31. ten Wolde, P.R., Ruiz-Montero, M.J., Frenkel, D.: Numerical calculation of the rate of crystal nucleation in a Lennard-Jones system at moderate undercooling. J. Chem. Phys. **104**, 9932–9947 (1996)
32. Moroni, D., ten Wolde, P.R., Bolhuis, P.G.: Interplay between structure and size in a critical crystal nucleus. Phys. Rev. Lett. **94**, 235703 (2005)
33. Bai, X.-M., Li, M.: Calculation of solid–liquid interfacial free energy: a classical nucleation theory based approach. J. Chem. Phys. **122**, 224510 (2005)
34. Wette, P., Schope, H.J., Palberg, T.: Microscopic investigations of homogeneous nucleation in charged sphere suspensions. J. Chem. Phys. **123**, 174902 (2005)
35. Gasser, U., Weeks, E.R., Schofield, A., Pusey, P.N., Weitz, D.A.: Real-space imaging of nucleation and growth in colloidal crystallization. Science **292**, 258–262 (2001)
36. Trudu, F., Donadio, D., Parrinello, M.: Freezing of a Lennard-Jones fluid: from nucleation to spinodal regime. Phys. Rev. Lett. **97**, 105701 (2006)
37. Klein, W., Leyvraz, F.: Crystalline nucleation in deeply quenched liquids. Phys. Rev. Lett. **57**, 2845–2848 (1986)
38. Yang, J.X., Gould, H., Klein, W.: Molecular-dynamics investigation of deeply quenched liquids. Phys. Rev. Lett. **60**, 2665–2668 (1988)
39. Klein, W., Gould, H., Tobochnik, J., Alexander, F.J., Anghel, M., Johnson, G.: Clusters and fluctuations at mean-field critical points and spinodals. Phys. Rev. Lett. **85**, 1270–1273 (2000)
40. Mendez-Villuendas, E., Bowles, R.K.: Surface nucleation in the freezing of gold nanoparticles. Phys. Rev. Lett. **98**, 185503 (2007)
41. Chushak, Y.G., Bartell, L.S.: Freezing of Ni-Al bimetallic nanoclusters in computer simulations. J. Phys. Chem. B **107**, 3747–3751 (2003)
42. Lynden-Bell, R.M., Wales, D.J.: Free-energy barriers to melting in atomic clusters. J. Chem. Phys. **101**, 1460–1476 (1994)
43. Nielsen, O.H., Sethna, J.P., Stoltze, P., Jacobsen, K.W., Norskov, J.K.: Melting a copper cluster: critical-droplet theory. Europhys. Lett. **26**, 51–56 (1994)
44. Schmidt, M., Kusche, R., Hippler, T., Donges, J., Kronmüller, W., von Issendorff, B., Haberland, H.: Negative heat capacity for a cluster of 147 sodium atoms. Phys. Rev. Lett. **86**, 1191–1194 (2001)
45. Lee, J.G., Mori, H.: Solid/liquid two phase structures in isolated nanometer-sized alloy particles. Phys. Rev. B **70**, 144105 (2004)
46. Pochon, S., Mac Donald, K.F., Knize, R.J., Zheludev, N.I.: Phase coexistence in gallium nanoparticles controlled by electron excitation. Phys. Rev. Lett. **92**, 145702 (2004)
47. Schebarchov, D., Hendy, S.C.: Superheating and solid–liquid phase coexistence in nanoparticles with nonmelting surfaces. Phys. Rev. Lett. **96**, 256101 (2006)
48. Delogu, F.: Coupling of structural and energetic fluctuations in Co nanometer-sized particles. Nanotechnology **17**, 2027–2031 (2006)

49. Lümmen, N., Kraska, T.: Solidification kinetics of Fe05Pt05 clusters in an Ar carrier gas: molecular dynamics simulations. Phys. Rev. B **77**, 045425 (2008)
50. Hendy, S.C.: Stability of phase coexistence in atomic clusters. Phys. Rev. B **71**, 115404 (2005)
51. Baletto, F., Mottet, C., Ferrando, R.: Freezing of silver nanodroplets. Chem. Phys. Lett. **354**, 82–87 (2002)
52. Doye, J.P.K., Calvo, F.: Entropic effects on the size dependence of cluster structure. Phys. Rev. Lett. **86**, 3570–3573 (2001)
53. Tian, Z.-A., Liu, R.-S., Peng, P., Hou, Z.-Y., Liu, H.-R., Zheng, C.-X., Dong, K.-J., Yu, A.-B.: Freezing structures of silver nanodroplets: a molecular dynamics simulation study. Phys. Lett. A **373**, 1667–1671 (2009)
54. Shibuta, Y., Suzuki, T.: A molecular dynamics study of the phase transition in bcc metal nanoparticles. J. Chem. Phys. **129**, 144102 (2008)
55. Aguado, A., Jarrold, M.F.: Melting and freezing of metal clusters. Annu. Rev. Phys. Chem. **62**, 151–172 (2011)
56. Ge, R., Clapp, P.C., Rifkin, J.A.: Molecular dynamics of a molten Cu droplet spreading on a cold Cu substrate. Surf. Sci. **426**, L413–L419 (1999)
57. Arcidiacono, S., Walther, J.H., Poulikakos, D., Passerone, D., Koumoutsakos, P.: Solidification of gold nanoparticles in carbon nanotubes. Phys. Rev. Lett. **94**, 105502 (2005)
58. Shi, R., Shao, J., Zhu, X., Lu, X.: On the melting and freezing of Au-Pt nanoparticles confined in single-walled carbon nanotubes. J. Phys. Chem. C **115**, 2961–2968 (2011)
59. Johnson, W.L.: Thermodynamic and kinetic aspects of the crystal to glass transformation in metallic materials. Prog. Mater. Sci. **30**, 81–134 (1986)
60. Allen, M.P., Tildesley, J.: Computer Simulation of Liquids. Oxford Science, Oxford (1989)
61. Rapaport, D.C.: The Art of Molecular Dynamics Simulation. Cambridge University Press, Cambridge (1995)
62. Frenkel, D., Smit, B.: Understanding Molecular Dynamics Simulation: From Algorithms to Application. Academic, London (1996)
63. Dehoff, R.T.: Thermodynamics in Materials Science. McGraw-Hill, New York, NY (1993)
64. Adelman, S.A., Doll, J.D.: Generalized Langevin equation approach for atom/solid-surface scattering: general formulation for classical scattering off harmonic solids. J. Chem. Phys. **64**, 2375–2388 (1976)
65. Berendsen, H.J.C., Postma, J.P.M., van Gunsteren, W.F., Di Nola, A., Haak, J.R.: Molecular dynamics with coupling to an external bath. J. Chem. Phys. **81**, 3684–3690 (1984)
66. Nosè, S.: A molecular dynamics method for simulations in the canonical ensemble. Mol. Phys. **52**, 255–268 (1984)
67. Nosè, S.: A unified formulation of the constant temperature molecular dynamics methods. J. Chem. Phys. **81**, 511–519 (1984)
68. Hoover, W.G.: Canonical dynamics: equilibrium phase-space distributions. Phys. Rev. A **31**, 1695–1697 (1985)
69. Rohsenow, W.M., Hartnett, J.P., Cho, Y.I.: Handbook of Heat Transfer, 3rd edn. McGraw-Hill, New York, NY (1998)
70. Westergren, J., Gronbeck, H., Kim, S.-G., Tomanek, D.: Noble gas temperature control of metal clusters: a molecular dynamics study. J. Chem. Phys. **107**, 3071–3079 (1997)
71. Westergren, J., Gronbeck, H., Rosen, A., Nordholm, S.: Statistical theory of cluster cooling in rare gas. I. Energy transfer analysis for palladium clusters in helium. J. Chem. Phys. **109**, 9848–9858 (1998)
72. Edelstein, A.S., Cammarata, R.C. (eds.): Nanomaterials: Synthesis, Properties and Applications. Taylor and Francis Group, Oxford (1996)
73. Cao, G.: Nanostructures and Nanomaterials. Imperial College Press, London (2004)
74. Rao, C.N.R., Muller, A., Cheetham, A.K. (eds.): The Chemistry of Nanomaterials: Synthesis, Properties and Applications. Wiley-VCH Verlag, Berlin (2005)
75. Cottancin, E., Lermè, J., Gaudry, M., Pellarin, M., Vialle, J.-L., Broyer, M., Prevel, B., Treilleux, M., Melinon, P.: Size effects in the optical properties of Au_nAg_n embedded clusters. Phys. Rev. B **62**, 5179–5185 (2000)

76. Reinhard, D., Hall, B.D., Ugarte, D., Monot, R.: Size-independent fcc-to-icosahedral structural transition in unsupported silver clusters: an electron diffraction study of clusters produced by inert-gas aggregation. Phys. Rev. B **55**, 7868–7881 (1997)
77. Koga, K., Sugawara, K.: Population statistics of gold nanoparticle morphologies: direct determination by HREM observations. Surf. Sci. **529**, 23–35 (2003)
78. Dodson, B.W.: Atomistic simulation of silicon beam deposition. Phys. Rev. B **36**, 1068–1074 (1987)
79. Brandes, E.A., Brook, G.B. (eds.): Smithells Metals Reference Handbook, 7th edn. Butterworth-Heinemann, Oxford (1992)
80. Cleri, F., Rosato, V.: Tight-binding potentials for transition metals and alloys. Phys. Rev. B **48**, 22–33 (1993)
81. Andersen, H.C.: Molecular dynamics simulations at constant pressure and/or temperature. J. Chem. Phys. **72**, 2384–2393 (1980)
82. Delogu, F.: A numerical study of the melting behavior of Au nanotubes. Phys. Rev. B **75**, 235421 (2007)
83. Morris, J.R.: Complete mapping of the anisotropic free energy of the crystal-melt interface in Al. Phys. Rev. B **66**, 144104 (2002)
84. Morris, J.R., Song, X.: The melting lines of model systems calculated from coexistence simulations. J. Chem. Phys. **116**, 9352–9358 (2002)
85. Pollack, G.L.: Extension of the low of corresponding-states to rare-gas solids. Phys. Rev. A **2**, 38–42 (1970)
86. Munster, A.: Statistical Thermodynamics. Springer, Berlin (1974)
87. Steinhardt, P.L., Nelson, D.R., Ronchetti, M.: Bond-orientational order in liquids and glasses. Phys. Rev. B **28**, 784–805 (1983)
88. Auer, S., Frenkel, D.: Numerical prediction of absolute crystallization rates in hard-sphere colloids. J. Chem. Phys. **120**, 3015–3030 (2004)

Chapter 7
Dynamics of Nanodroplets on Structured Surfaces

Markus Rauscher

Abstract Fluids on the nanoscale behave qualitatively different from macroscopic systems. This becomes particularly evident if a free liquid–liquid or liquid–gas interface is close to a solid surface such as in the case of nanodroplets. In contrast to macroscopic drops, hydrodynamic slip, thermal fluctuations, the molecular structure of the liquid, and the range of the intermolecular interactions are important for the structure and the dynamics of such open nanofluidic systems. After a review of the macroscopic modeling and behavior of nonvolatile droplets on structured substrates, we discuss the static and dynamic peculiarities on the nanoscale with special emphasis on theory. In particular we show that nanodroplets experience long-ranged lateral interactions with sharp surface features and that their free energy might be lower on a less wettable part of the substrate surface. A discussion of possible experiments for observing these phenomena is followed by a summary and an outlook.

7.1 Introduction

Droplets are a particularly fascinating manifestation of wetting phenomena which occur as a result of two fluid phases (e.g., a liquid and a gas or two immiscible liquids) coming into contact with a solid surface. And they are not only ubiquitous in our everyday life but also of tremendous technological importance—in many cases because their formation should be avoided or because they should be removed. But they can be also used for patterning surfaces. For this reason we have some intuition how macroscopic droplets on surfaces should behave. In this chapter we show that

M. Rauscher (✉)
Max-Planck-Institut für Intelligente Systeme, Stuttgart, Germany

Institut für Theoretische Physik IV, Stuttgart, Germany
e-mail: rauscher@is.mpg.de

Z.M. Wang (ed.), *Nanodroplets*, Lecture Notes in Nanoscale Science and Technology 18,
DOI 10.1007/978-1-4614-9472-0_7,

our intuition fails on the nanoscale (see also [1–4]) and we discuss theoretical approaches to the dynamics of liquids on the nanoscale, simulation methods, and experimental options.

Speaking of droplets also means speaking about nonvolatile liquids: a drop of a volatile liquid sitting on a chemically and topographically homogeneous surface is unstable: it either evaporates or, if the vapor phase is supersaturated, it grows without bound (or until the vapor reservoir is depleted). As a consequence, wetting phenomena of nonvolatile liquids are much richer. For example, on a homogeneous surfaces in contact with a vapor phase one obtains a homogeneous liquid film of a certain thickness which grows in thickness as one approaches liquid–vapor coexistence [5,6], whereas a nonvolatile liquid can form droplets. And on structured substrates one even observes morphological transitions between different droplet shapes. Strictly speaking, nonvolatility is a question of time scale and in this chapter we assume that the time scale of evaporation is large as compared to the time scale of the motion of the nanodroplets.

In the following we speak about liquid droplets surrounded by a thin gas phase of negligible viscosity on an inert solid substrate. However, apart from the viscous dissipation in the second liquid phase, the results presented below can be generalized to a system of two immiscible liquids in a straightforward way.

The following Sect. 7.2 contains a review of the behavior of macroscopic droplets on chemically (Sect. 7.2.1) and topographically (Sect. 7.2.2) structured substrates, with particular focus on the aspects which are different on the nanoscale. In Sect. 7.3 the dynamics of nanodroplets on structured substrates is discussed. Section 7.3.1 gives a concise overview of the equilibrium dynamic density functional theory (DFT) and the effective interface model which describe the equilibrium properties of nanodroplets. A mesoscopic extension of hydrodynamics is presented in Sect. 7.3.2. In Sect. 7.3.3 several simulation methods applicable to nanofluidics are compared. The peculiar behavior of nanodroplets on topographically and chemically structured substrates is illustrated in Sects. 7.3.4 and 7.3.5, respectively. Section 7.3.6 discusses experiments which might be capable of observing the dynamics of nanodroplets on structured substrates. A summary and outlook can be found in Sect. 7.4.

7.2 Macroscopic Droplets

Throughout this section we assume that gravity has a negligible effect on the shape of the droplets, i.e., that the Bond number $\mathrm{Bo} = \rho g L^2/\gamma$ (with the mass density ρ, the gravitational acceleration g, the surface tension γ, and a characteristic length scale L) is small. For water droplets this would be the case for radii less than a millimeter (the capillary length of water is $L_{\mathrm{cap}} = \sqrt{\gamma/(\rho g)} = 2.6$ mm).

The equilibrium shape of nonvolatile macroscopic droplets on solid substrates is determined by interface energies [7]. Its free energy is given by

$$F = A_{\mathrm{LG}}\,\gamma + A_{\mathrm{LS}}\,\gamma_{\mathrm{LS}} + A_{\mathrm{SG}}\,\gamma_{\mathrm{SG}}, \tag{7.1}$$

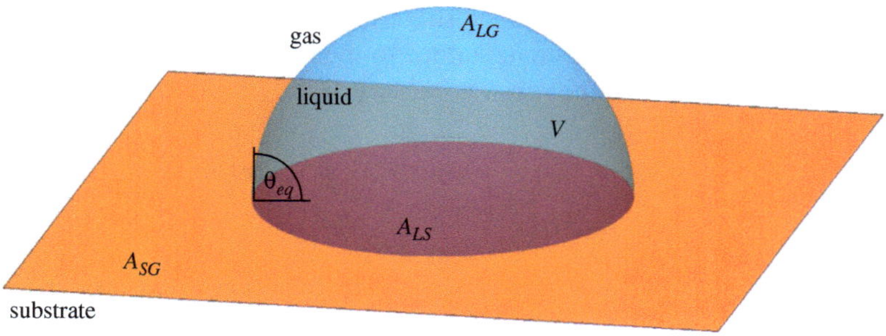

Fig. 7.1 A macroscopic liquid droplet of fixed volume V on a flat and homogeneous substrate forms a hemispherical cap with. The angle θ_{eq} between the liquid–gas interface and the substrate surface is given by Young's law, i.e., it is determined by the interface tensions γ of the liquid–gas interface A_{LG}, γ_{LS} of the liquid–substrate interface A_{LS}, and γ_{SG} of the substrate–gas interface A_{SG}. Here $\gamma_{\text{LG}} = \gamma_{\text{SG}}$ and therefore $\theta_{\text{eq}} = 90°$

with the area of the liquid–substrate interface A_{LS}, the area of the liquid–gas interface A_{LG}, and the area of the substrate–gas interface A_{SG}. The corresponding interface tensions are γ_{LS}, $\gamma = \gamma_{\text{LG}}$, and γ_{SG}, respectively. Fixing the total substrate area $A = A_{\text{SG}} + A_{\text{LS}}$ as well as the liquid volume V and minimizing F with respect to the shape of the liquid–gas interface lead to Young's law for the equilibrium contact angle θ_{eq}

$$\cos\theta_{\text{eq}} = \frac{\gamma_{\text{SG}} - \gamma_{\text{LS}}}{\gamma} \tag{7.2}$$

and to the Euler–Lagrange equation

$$2\gamma H_{\text{LG}} + p = 0, \tag{7.3}$$

which states that the Laplace pressure is constant on the liquid–gas interface. H_{LG} is the mean curvature of the liquid–gas interface and p is the hydrostatic pressure in the droplet. The pressure p is the Lagrange multiplier which fixes the liquid volume V. Therefore the liquid–gas interface has a constant mean curvature and in the case of a flat substrate (7.3) together with Young's equation (7.2) as boundary condition leads to a spherical cap as shown in Fig. 7.1 for the example of $\theta_{\text{eq}} = 90°$.

7.2.1 *Chemically Structured Substrates*

On homogeneous substrates the free energy of the droplet is independent of the position of the droplet. This is also the case for droplet positions on a homogeneous part of a substrate, e.g., on a hydrophilic or on a hydrophobic patch of a chemically structured substrate, or on a flat region of a topographically structured substrate.

On a chemically heterogeneous substrate this is no longer true. In this case, the contributions from the substrate surface to the free energy in (7.1) have to be replaced by an integral over the substrate surface

$$F = A_{\mathrm{LG}}\,\gamma + \int_{A_{\mathrm{LS}}} \gamma_{\mathrm{LS}}(\mathbf{r})\,dA + \int_{A_{\mathrm{SG}}} \gamma_{\mathrm{SG}}(\mathbf{r})\,dA. \qquad (7.4)$$

On a chemically heterogeneous substrate the interface energies $\gamma_{\mathrm{LS}}(\mathbf{r})$ and $\gamma_{\mathrm{SG}}(\mathbf{r})$ depend on the position $\mathbf{r}$ on the surface. If a droplet is positioned on a chemical step between a hydrophilic region and a hydrophobic region, it will move to hydrophilic side and it stops as soon as the complete liquid–substrate interface rests on the hydrophilic part [8] (unless the initial shape of the droplet is far from equilibrium which can result in the droplet being situated completely on the more hydrophobic part of the surface after the initial shape relaxation where it will simply stay [9]). On a surface with a continuous variation of the wetting properties (a so-called chemical gradient), drops can move over distances greater than their diameter [10–12].

Chemically structured substrates have been suggested as an alternative to closed channels for microfluidic applications. This approach is based on the macroscopic observation that nonvolatile liquids on a hydrophobic substrate patterned with hydrophilic stripes will stay on these so-called chemical channels [13, 14] even if driven along the channels [15–17].

7.2.2 *Topographically Structured Substrates*

As mentioned above, the free energy of a droplet is independent of the droplet position only if the substrate is homogeneous and flat. The free energy of a droplet on a topographically structured substrate is also given by (7.1); however, calculating the first variation of F with respect to the shape of the liquid–gas interface and the position of the three-phase contact-line is a nontrivial task. Finite element codes such as the SURFACE EVOLVER can be used to determine the equilibrium shape of droplets numerically [18]. Topographically structured substrates have many technological applications: roughness modifies the wetting properties leading to superhydrophilic (for $\theta_{\mathrm{eq}} < 90^\circ$) or superhydrophobic (surfaces for $\theta_{\mathrm{eq}} > 90^\circ$) surfaces (the first ones are suggested as anti-fogging coatings and the latter ones have self-cleaning properties) [19–22]. As a result gradients in the roughness are expected to induce the motion of droplets in the same way as the chemical gradients mentioned above [23].

Grooves in surfaces can guide liquids in a way similar to chemical channels [24–26]. This is based on the observation that droplets are pinned by edges on surfaces. It has been already observed by Gibbs [27] that at the edge the contact angle is ill defined as shown in Fig. 7.2. It can be measured with respect to either of the sides

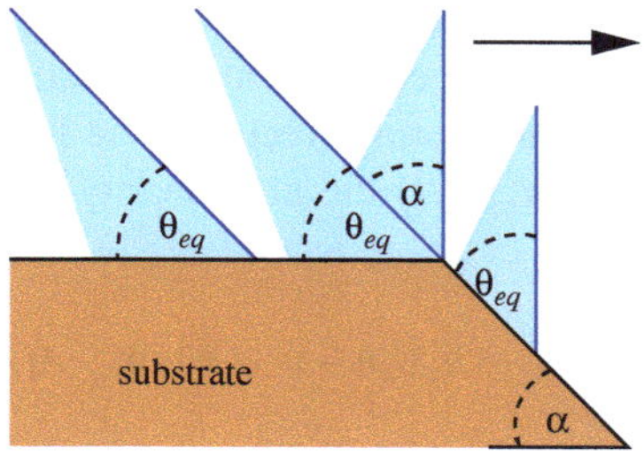

Fig. 7.2 The three-phase contact-line of a droplet approaching an edge in the substrate (here from *left to right*) advances only if the angle of the liquid–gas interface with the substrate is larger than the equilibrium contact angle $\theta_{\rm eq}$. At the edge the contact angle is ambiguous and in order to move down the slope of the edge it has to increase from $\theta_{\rm eq}$ to $\theta_{\rm eq}+\alpha$, with the angle of the slope α

of the edge. In order to overcome the edge the contact angle with the down-hill side of the edge has to be larger than the equilibrium contact angle $\theta_{\rm eq}$ [28]. Actually it has to be even larger than the so-called advancing contact angle, which is larger than $\theta_{\rm eq}$. Even a rounded edge can pin a three phase contact line; however, the pinning strength depends on the edge shape [29]. Pinning at defects is the main reason for the difference of contact angle if the three phase contact line is receding (e.g., in dewetting) or advancing (e.g., for spreading droplets). Only recently it has become possible to directly measure the pinning strength of individual nanoscale defects [30].

Topographically structures can also trigger morphological phase transitions of the shape of droplets, e.g., between drops and filaments in rectangular grooves [24, 26]. And on fibers one observes a transition from a symmetric barrel shape for large volumes and small contact angles to an asymmetric clamshell shape for small volumes and large contact angles [31].

7.2.3 *Dynamics*

For radii larger than a micron the dynamics of nonvolatile liquid droplets is well described by macroscopic hydrodynamic equations. Here we focus on incompressible simple Newtonian liquids surrounded by a vapor or gas of negligible viscosity and density. On the micron-scale the Reynolds number $\mathrm{Re}=\rho UL/\eta$ (with a characteristic velocity U and the liquid viscosity η) is much smaller than one and the liquid velocity $\mathbf{u}(\mathbf{r},t)$ as well as the pressure $p(\mathbf{r},t)$ can be determined by solving the Stokes equation

$$\nabla p = \eta\,\nabla^2\mathbf{u} \tag{7.5}$$

together with the incompressibility condition

$$\nabla\cdot\mathbf{u}=0. \tag{7.6}$$

For incompressible nonvolatile liquids the normal velocity $v = \hat{\mathbf{n}} \cdot \mathbf{u}$ of the moving liquid–gas interface is given by the component of the liquid velocity normal to the interface ($\hat{\mathbf{n}}$ denotes the outward normal vector at the liquid–gas interface). In the Monge parameterization $z = h(x,y,t)$ the nonlinearity of this kinematic condition becomes apparent:

$$\partial_t h = u_z - u_x \, \partial_x h - u_y \, \partial_y h \quad \text{at} \quad A_{\text{LG}}. \tag{7.7}$$

At the surface of impermeable substrates the normal component of the liquid velocity vanishes. Macroscopically it is valid to assume that the tangential velocity is also zero [32]. With the substrate surface at rest this leads to a Dirichlet type boundary condition at the substrate interface

$$\mathbf{u} = 0 \quad \text{at} \quad A_{\text{LS}}. \tag{7.8}$$

Here we assume that the viscosity and pressure of the gas phase are negligible such that the tangential forces on the liquid–vapor interface vanish and the normal component of the stress tensor is balanced by the Laplace pressure from (7.3). In summary we have

$$\sigma \cdot \hat{\mathbf{n}} = 2\,\gamma H_{\text{LG}}\,\hat{\mathbf{n}} \quad \text{at} \quad A_{\text{LG}}, \tag{7.9}$$

with the stress tensor of an incompressible Newtonian fluid

$$\sigma_{ij} = -p\,\delta_{ij} + \eta\,\left(\partial_j u_i + \partial_i u_j\right). \tag{7.10}$$

Up to the motion of the three-phase contact-line between liquid, gas, and substrate, the dynamics of a macroscopic droplet is well described by (7.5) to (7.10). Within these equations, the stress σ and also the dissipation in a moving three-phase contact-line diverges. As a consequence it should not move [5, 33] although everyday experience tells that it does [34]. This is an artifact of this macroscopic model.

7.3 Nanofluidics

On the nanoscale, phenomena which are either irrelevant or summarized in hydrodynamic boundary conditions come into play which lead to qualitative changes in the equilibrium properties as well as in the dynamics of liquids. These are on one hand the finite range of intermolecular forces, thermal fluctuations, and the molecular structure of the liquid which also strongly influence static wetting properties and on the other hand hydrodynamic slip which strongly influences the dynamics [1, 3, 4]. In the following we show how to augment the hydrodynamic equations presented in Sect. 7.2 such that they can be used to describe the dynamics

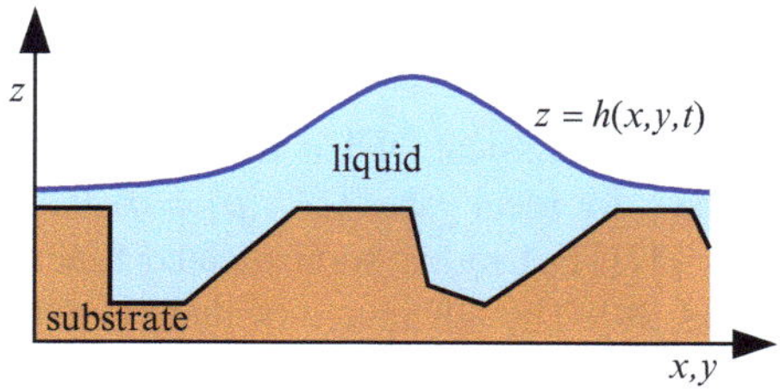

Fig. 7.3 For the clarity of the presentation, we consider a structured substrate aligned with the xy-plane and we assume that the liquid–gas interface can be parameterized by the Monge parameterization $z = h(x, y, t)$

of nanodroplets. Unfortunately it is not yet possible to include the effects of the molecular structure of the liquid into dynamical theories. The static molecular structure of liquids is well understood [35].

For the clarity of the presentation, in the following we consider a (structured) substrate aligned with the xy-plane of our coordinate system and we assume that the liquid–gas interface has no overhangs such that we can use the Monge parameterization $z = h(x, y, t)$ as illustrated in Fig. 7.3.

7.3.1 Density Functional Theory

Great progress has been made in understanding wetting phenomena by using the classical DFT for inhomogeneous systems [36]. One can show that there exists a functional $\Omega[\rho]$ which is minimized by the equilibrium density distribution of a grand canonical ensemble of classical particles (i.e., a system of a given volume V in thermal and chemical equilibrium with a bath at a given temperature T and chemical potential μ). For a one-component system of indistinguishable particle this functional has the form

$$\Omega[\rho] = \mathscr{F}_{\text{id}}[\rho] + \mathscr{F}_{\text{ext}}[\rho] + \mathscr{F}_{\text{ex}}[\rho] - \mu \int_V \rho(\mathbf{r})\, d^3r, \tag{7.11}$$

with the ideal gas part ($\Lambda = h/\sqrt{2\pi m k_{\text{B}} T}$ with the Planck constant h, the Boltzmann constant k_{B}, the molecular mass m is the thermal de Broglie wavelength)

$$\mathscr{F}_{\text{id}}[\rho] = k_{\text{B}} T \int_V \rho(\mathbf{r}) \left(\ln\left(\rho(\mathbf{r})\Lambda^3\right) - 1\right) d^3r, \tag{7.12}$$

with the contribution due to an external potential $\Phi_{\text{ext}}(\mathbf{r})$ which also describes the substrate

$$\mathscr{F}_{\text{ext}}[\rho] = \int_V \rho(\mathbf{r})\, \Phi_{\text{ext}}(\mathbf{r})\, d^3r, \tag{7.13}$$

and with the so-called excess free energy $\mathscr{F}_{\text{ex}}[\rho]$ due to the intermolecular interactions.

The proof of existence of a classical DFT is non-constructive and $\mathscr{F}_{\text{ex}}[\rho]$ is only known exactly in two cases: for ideal gases ($\mathscr{F}_{\text{ex}}[\rho] = 0$) and for a one-dimensional system of hard particles [37]. But rather sophisticated functionals for hard-sphere systems are available [38, 39]. The advantage of these functionals is that they can even capture the molecular structure of the liquid such as layering effects near hard walls. We want to use the DFT results to understand not only the equilibrium properties of droplets but also their dynamics.

Unfortunately, it is not possible up to now to include effects of the molecular structure of the liquid into hydrodynamic theory. However, one can make significant progress by splitting the intermolecular interaction potential into a short-ranged, hard repulsive part and a long-ranged attractive part $\Phi_{\text{att}}(r)$. One effectively gets a system of hard spheres with soft attractions. If one further neglects short distance correlations in the liquid due to packing effects, one can use a local approximation for the hard sphere part of the density functional leading to

$$\mathscr{F}_{\text{ex}}[\rho] = \int_V \left(f_{\text{HS}}(\rho(\mathbf{r})) + \frac{1}{2}\rho(\mathbf{r}) \int \Phi_{\text{att}}(|\mathbf{r}-\mathbf{r}'|)\,\rho(\mathbf{r}')\,d^3r' \right) d^3r, \tag{7.14}$$

with the Carnahan–Starling expression $f_{\text{HS}}(\rho) = k_{\text{B}}T\rho\left(\ln(\rho\Lambda^3) - 1 + \frac{4\eta-3\eta^2}{(1-\eta)^2}\right)$ and the packing fraction $\eta = \frac{4\pi}{3}R^3\rho$ [35]. The (effective) particle radius R depends on the temperature and on details of the repulsive part of the intermolecular interactions.

In a liquid film the liquid density rises from zero right at the substrate surface to the bulk liquid density ρ_{L} within a few molecular diameters. Far from critical points the liquid–gas interface has a width of the order of a few molecular diameters. Within this interface the density changes gradually from the bulk liquid density to the bulk gas density ρ_{G}.

On a mesoscopic length scale the density distribution of liquid molecules within a film of thickness h is reasonably well approximated by step-like profile

$$\rho_{\text{step}}(z,h) = \begin{cases} 0, & \text{for} \quad z < 0 \\ \rho_{\text{L}}, & \text{for} \quad 0 < z < h. \\ \rho_{\text{G}}, & \text{for} \quad h < z \end{cases} \tag{7.15}$$

The density of a film of laterally varying thickness $h(x,y)$ is given by $\rho_{\text{step}}(z,h(x,y))$. Therefore, the density is parameterized by the film thickness $h(x,y)$ and the density functional in (7.11) reduces to a functional of $h(x,y)$, which is called effective interface Hamiltonian [6, 40]. If the curvature of the liquid–gas interface H_{LG} is small (i.e., if the radii of curvature are large as compared to the range of the intermolecular interactions), the surface tension can be written in a local approximation leading to

$$\mathscr{H}[h] = \int \left(\gamma\sqrt{1+(\nabla h)^2} + \omega(x,y,h(x,y)) - ph(x,y) \right) dxdy. \tag{7.16}$$

The so-called effective interface potential $\omega(x,y,z)$ describes the effective interaction between the liquid–substrate and the liquid–gas interface and the pressure p is proportional to the chemical potential μ at liquid–gas coexistence. Far from the substrate surface $\omega(x,y,z)$ goes to zero. The Euler–Lagrange equation corresponding to (7.16)

$$2\gamma H_{\mathrm{LG}} + \Pi(x,y,z) + p = 0 \quad \text{at} \quad z = h(x,y), \tag{7.17}$$

with the disjoining pressure $\Pi(x,y,z) = \partial\omega(x,y,z)/\partial z$, balances the pressure, the Laplace pressure, and the disjoining pressure at the liquid–gas interface.

Although strictly speaking the DFT is only valid for grand canonical systems, it can be used to describe nonvolatile liquids (i.e., canonical ensembles) if one interprets the pressure in (7.17) as the Lagrange multiplier which fixes the liquid volume. In this sense (7.17) generalizes (7.3) to the nanoscale.

Assuming additive and pairwise interactions, the disjoining pressure can be expressed in terms of the attractive part Φ_{att} of the liquid–liquid interaction potential and the liquid–substrate interaction potential Φ_{sub} as an integral over the substrate volume [41]

$$\Pi(\mathbf{r}) = \int\limits_{\text{substrate}} \left(\rho_{\mathrm{L}}^2 \Phi_{\mathrm{att}}(\mathbf{r}-\mathbf{r}') - \rho_{\mathrm{L}}\rho_{\mathrm{S}}(\mathbf{r})\Phi_{\mathrm{sub}}(\mathbf{r})\right) d^3r. \tag{7.18}$$

The external potential in (7.13) is given by $\Phi_{\mathrm{ext}}(\mathbf{r}) = \int_{\mathrm{subs.}} \rho_{\mathrm{S}}(\mathbf{r})\Phi_{\mathrm{sub}}(\mathbf{r}-\mathbf{r}')\,d^3r'$, with the substrate density $\rho_{\mathrm{S}}(\mathbf{r})$. For inhomogeneous substrates the substrate density depends on the position in the substrate. Assuming power law potentials $\sim r^{-\alpha}$ each integration increases the exponent by one. Therefore the contribution from the bulk of the substrate (after integration with respect to x, y, and z) is $\sim r^{-\alpha+3}$ and therefore the one with the largest exponent and the longest interaction range. In experiments the wetting properties of surfaces are often modified by thin coatings. The contribution of such a thin coating to the disjoining pressure is then calculated by effectively integrating with respect to x and y only such that it is $\sim r^{-\alpha+2}$, i.e., much weaker and shorter in range than the contribution from the bulk. Thick liquid films therefore only "see" the bulk of the underlying substrate.

For Lennard-Jones type interaction potentials (the long-ranged attractive part of which is given by (non-retarded) van-der-Waals type dispersion forces) the effective interface potential of a homogeneous and flat substrate have the form

$$\omega(z) = \frac{A_{\mathrm{H}}}{z^2} + \frac{B}{z^3} + \frac{C}{z^8}, \tag{7.19}$$

with the Hamaker constant A_{H} and $C > 0$. $A_{\mathrm{H}} > 0$ and $B < 0$ correspond to the rather generic case of a substrate with a first order wetting transition and $A_{\mathrm{H}} < 0$ and $B = 0$ to a substrate that exhibits critical wetting (see Fig. 7.4). In (7.18) one can generate

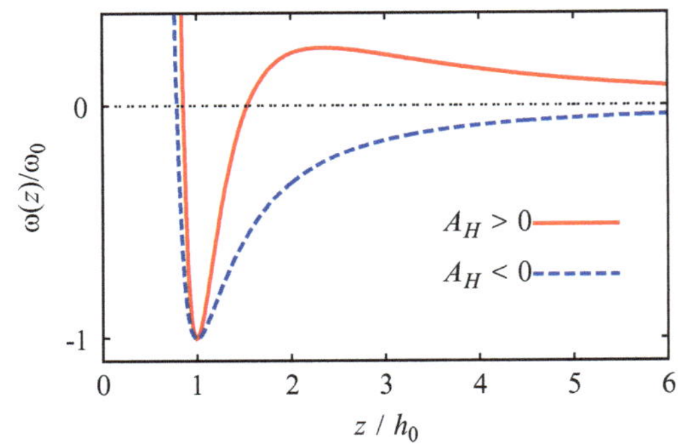

Fig. 7.4 The effective interface potential $\omega(z)$ of a flat and homogeneous partially wetting substrate as a function of the distance z from the substrate surface has a minimum of depth $-\omega_0$ at $z = h_0$. Shown are two examples, one with positive Hamaker constant $A_\mathrm{H} > 0$ (*full red line*) and one with negative Hamaker constant $A_\mathrm{H} < 0$ (*dashed blue line*)

the term proportional to B by assuming a very thin layer of a different material at the substrate surface. In real systems this term is also generated by the atomistic structure of the substrate and therefore it is rather generic [42]. For partially wetting substrates $\omega(z)$ has a minimum at $z = h_0$ of depth $\omega_0 = -\omega(h_0)$. By integrating the Euler–Lagrange equation (7.17) for a drop of infinite size (i.e., for $p = 0$) one can show that $\gamma_\mathrm{SG} - \gamma_\mathrm{LS} = \gamma - \omega_0$ and therefore

$$\cos\theta_\mathrm{eq} = 1 - \frac{\omega_0}{\gamma}. \tag{7.20}$$

The deeper the minimum of $\omega(z)$, the larger the macroscopic equilibrium contact angle θ_eq. $\omega_0 > 2\gamma$ corresponds to complete drying and $\omega_0 = 0$ (i.e., if $\omega(z)$ does not have a minimum at finite distance from the substrate) corresponds to complete wetting $\theta_\mathrm{eq} = 0°$.

Figure 7.5 shows the disjoining pressure for van-der-Waals type dispersion forces in the vicinity of a straight topographic step of height $s = 3h_0$ in an otherwise homogeneous substrate [43]. The material parameters in (a) and (b) are chosen such that the effective interface potential far from the step ($|x| \to \infty$) is given by the solid and dashed line in Fig. 7.4, respectively. The system is translationally invariant in y-direction. For $A_\mathrm{H} > 0$ (Fig. 7.5a) the disjoining pressure is positive for large z. Far from the step, as a function of z there is a local maximum at a distance of about $3h_0$ from the surface and there is a local minimum at a distance of about $1.15h_0$. As one approaches the edge from the left, the minimum becomes deeper and the maximum higher. For $A_\mathrm{H} < 0$ (Fig. 7.5b) the disjoining pressure is negative for large z. Far from the step, as a function of z there is only a minimum at a distance of about $1.2h_0$ from the substrate. As one approaches the step from the left, the minimum becomes more shallow.

Figure 7.5 clearly shows that on the nanoscale the topographic step induces a lateral variation of the disjoining pressure. For van-der-Waals type forces and for large $|x|$ (and finite z) we have $\Pi(x,z) - \Pi_\mathrm{hom}(z) \sim A_\mathrm{h}s/x^4$, with the disjoining

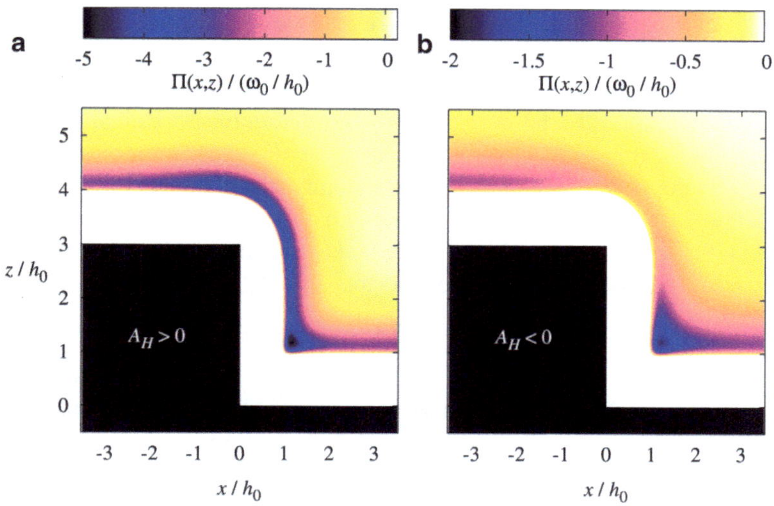

Fig. 7.5 The disjoining pressure $\Pi(x,z)$ at a straight topographic step of height $s = 3\,h_0$ which is aligned with the y-axis for (**a**) $A_\mathrm{H} > 0$ and (**b**) $A_\mathrm{H} < 0$. The parameters are chosen such that the corresponding effective interface potentials far from the step (i.e., for $x \to \infty$) are those shown in Fig. 7.4. The substrate is shown in *black*. In the *white region* next to the substrate the disjoining pressure is extremely large. The *upper boundary* of the *white region* roughly corresponds to the contour line $\Pi(x,z) = 0$. In (**a**) the disjoining pressure is positive for large z and in (**b**) it is negative

pressure of the flat and homogeneous substrate $\Pi_\mathrm{hom}(z) = \lim_{|x|\to\infty}\Pi(x,z)$. As we will see later, a nanodroplet placed in the vicinity of such a step will experience a lateral force and the sign of this force depends on the sign of A_H.

7.3.2 *Mesoscopic Hydrodynamics*

Equation (7.17) not only allows to calculate the equilibrium shape of nanodroplets but it can be also used to generalize the stress boundary condition for the Stokes flow (7.9): the normal component of the stress tensor in the liquid is balanced by the Laplace pressure and the disjoining pressure

$$\sigma \cdot \hat{\mathbf{n}} = (2\,\gamma H_\mathrm{LG} + \Pi)\,\hat{\mathbf{n}} \quad \text{at} \quad A_\mathrm{LG}. \tag{7.21}$$

Equation (7.21) effectively takes into account the long-ranged nature of intermolecular interactions. And it has been shown, that, e.g., dewetting dynamics of thin films can be modeled quantitatively using this approach [44, 45].

Thermal fluctuations can be phenomenologically modeled by a randomly fluctuating stress tensor $\mathbf{S}(\mathbf{r},t)$ of zero mean $\langle \mathbf{S}(\mathbf{r},t)\rangle = 0$ and correlation function [46]

$$\left\langle S_{ij}(\mathbf{r},t)\, S_{lm}(\mathbf{r}',t')\right\rangle = 2\,\eta\, k_{\mathrm{B}}\, T\, \delta(\mathbf{r}-\mathbf{r}')\, \delta(t-t')\, (\delta_{il}\, \delta_{jm} + \delta_{im}\, \delta_{jl}). \tag{7.22}$$

The amplitude of the noise is dictated by the fluctuation dissipation theorem. This form proposed can also be derived from the deterministic Boltzmann equation in a long-wave approximation [47]. Noisy hydrodynamical equations have been discussed in the context of turbulence in randomly stirred liquids [48, 49] as well as for the onset of instabilities in Rayleigh–Bénard convection [50] and Taylor–Couette flow [51]. Since the right-hand side of the Stokes equation (7.5) is the divergence of the deviatoric part of the stress tensor, we get

$$\nabla p = \eta\, \nabla^2 \mathbf{u} + \nabla \cdot \mathbf{S}. \tag{7.23}$$

It has been shown that including thermal fluctuations is crucial for understanding spinodal dewetting of thin films [52], i.e., when a soft mode is present in the system, or in order to overcome energetic barriers. If there is no barrier (or if it is to high to be overcome by thermal fluctuations) and if there is no soft mode, the main effect of thermal fluctuations in a droplet or in a thin film is capillary waves, which can lead to an effective steric repulsion of the liquid–gas interface from the wall, for a review see [53]. This effect can be summarized into the effective interface potential such that in the following we will not explicitly discuss the influence of thermal fluctuations on the dynamics of nanodroplets.

Slip at the liquid–substrate interface is a result of structural changes in a liquid in the direct vicinity of an interface. These structural changes lead to changes in the rheology of the liquid. Another reason for slip is the weak coupling of a liquid to a hydrophobic substrate. If one extrapolates the flow profile in a droplet (e.g., $u_x(z)$ on a flat substrate located in the xy-plane) to the substrate surface, one sometimes finds that the height z at which the velocity goes to zero is negative, i.e., at a point within the substrate. The depth b at which $u_x(b) = 0$ is called the slip length and for simple liquid it is usually on the nanometer scale [32]. Therefore it is irrelevant for macroscopic systems. However, it has been shown that hydrodynamic slip strongly influences dewetting phenomena [54–56].

On the length scale of a few inter-atomic distances and above, slip can be effectively described by replacing (7.8) by a Navier slip boundary condition

$$u_\perp = 0 \quad \text{and} \quad \mathbf{u}_\parallel = b\, \nabla_\perp \mathbf{u}_\parallel \quad \text{at} \quad A_{\mathrm{LS}}. \tag{7.24}$$

Here $u_\perp = \mathbf{u} \cdot \hat{\mathbf{n}}_{\mathrm{S}}$ denotes the component of the flow field perpendicular to the substrate surface (which has to be zero for impermeable substrates) and $\mathbf{u}_\parallel = \mathbf{u} - u_\perp\, \hat{\mathbf{n}}_{\mathrm{S}}$ the components (two in three dimensions) parallel to the substrate surface. $\nabla_\perp = \hat{\mathbf{n}}_{\mathrm{S}} \cdot \nabla$ is the derivative in the direction normal to the substrate surface and $\hat{\mathbf{n}}_{\mathrm{S}}$ denotes the normal vector to the substrate surface pointing into the liquid. The Navier slip boundary states that the slip velocity $\mathbf{u}_\parallel$ is proportional to the shear stress in the direction parallel to the substrate surface $\sigma \cdot \hat{\mathbf{n}}_{\mathrm{S}}$.

7.3.3 Simulation Methods

Equations (7.5), (7.6), (7.7), (7.21), and (7.24) form a nonlinear moving boundary value problem. Solving these equations is further complicated by the fact that the dynamics of nanodroplets is dominated by surface tension, which requires precise calculation of the curvature (i.e., of the second derivative) of the liquid–gas interface. In addition, the disjoining pressure is rather stiff (i.e., it has steep gradients) in the vicinity of the substrate surface.

In a thin film geometry the typical lateral length scale $L_{\|}$ parallel to the substrate surface is much larger than the typical length scale $L_{\perp}$ perpendicular to the substrate surface. In the so-called lubrication approximation for $L_{\perp}/L_{\|} \to 0$ some of these complications can be overcome [33]. This approximation has been used to model thin film flow on topographically [57–62] and chemically [63–65] structured substrates. The disjoining pressure in the vicinity of a substrate surface structure varies on the same small length scale $L_{\perp}$ in normal as well as in lateral direction. Therefore, in the lubrication approximation, the long-ranged lateral variation of the disjoining pressure becomes short-ranged and the lubrication approximation is not suitable for modeling the lateral interaction of a droplet with a substrate structure. However, it has been very successfully used to model the dynamics of liquids on homogeneous substrates [44, 45].

Lattice Boltzmann (LB) simulations have become a very valuable tool for simulating free surface flows [66–70], which has been successfully used to study liquids in contact with structured substrates [71–81]. But although the LB method is often called "mesoscopic" because the free interface between the two fluid phases (between liquid and gas or between two immiscible liquids) is diffuse with a width of a few lattice constants, it is essentially a method to solve macroscopic free interface problems. In most LB simulations of multiphase flows the width of the interface is unphysically large, which, in particular, can influence the dynamics of the three phase contact line. Although it is in principle possible to include arbitrary external potentials, long-ranged intermolecular interactions in the form of an effective interface potential have not been included into the method. One obstacle is the weak but finite compressibility of existing LB methods. In addition, in LB simulations, the material parameters are tightly coupled to the lattice size and to the simulation time step which makes it a nontrivial task to use LB for nanoscale systems.

In phase filed methods the free interface is also diffuse. The motion of the free interface is modeled by coupling a phase field which has a value of, e.g., $+1$ in the liquid phase and -1 in the gas phase (alternatively 0 and $+1$) to the Navier Stokes equations [82]. This method has been used to describe the dynamics of droplets [83] and thin films [84–86]. Within the thin film approximation derived from a phase field model also long-ranged intermolecular interactions have been taken into account [87, 88]. However, all the efforts up to now have been targeting homogeneous substrates only.

The linearity of the Stokes equation (7.5) can be used to turn the moving boundary value problem (7.5), (7.7), (7.21), and (7.24) into an integral equation on the surfaces and interfaces surrounding the liquid. Although this method has been successfully used to describe the motion of three-dimensional droplets [89–92], for clarity we focus on the simpler two-dimensional case in the following. With the stream function $\psi(x,z)$, with $u_x = \partial_z \psi$ and $u_z = -\partial_x \psi$, and the vorticity $\varpi = \partial_z u_x - \partial_x u_z$ one can write the Stokes equation (7.5) in its biharmonic form

$$\nabla^4 \psi = 0 \quad \text{and} \quad \nabla^2 \varpi = 0. \tag{7.25}$$

The vorticity and the stream function are coupled via $\nabla^2 \psi = \varpi$. Using the Green's functions to the equations in (7.25) one can write ψ and ϖ inside the liquid as an integral over the surfaces surrounding the liquid [93]. The boundary conditions can also be written in terms of the ψ and ϖ and one obtains a set of equations for the dynamics of the free surface which only involves the values of ψ and ϖ on the surface [94]. Modeling the dynamics of droplets with this scheme requires explicit front tracking, but it is possible to use this method to simulate the dynamics of two-dimensional nanodroplets on structured substrates [2].

All the results on the dynamics of nanodroplets on structured substrates to be discussed below are based on the equilibrium DFT in (7.11) or on linking DFT to hydrodynamics. However, density functionals are not known exactly and the mesoscopic hydrodynamic (7.5), (7.6), (7.7), (7.21), and (7.24) the shear stress σ (see (7.10)) is considered in a local approximation only. In a more consistent description of the dynamics of a liquid on a length scale at which the finite range of intermolecular interactions matters, the non-locality of the stress tensor should be taken into account.

For over-damped Brownian dynamics (a model for the dynamics of suspended colloidal particles) there is a systematic dynamic extension of the equilibrium DFT. This dynamic DFT is based on an equilibrium approximation for the two-particle correlation function [95–97]. Although several attempts towards a dynamical DFT for simple liquids have been made [98, 99], the main obstacle has not been overcome: in liquids the dominant contribution to the viscosity comes from the distortion of the two point correlation function in a shear flow [100, 101] and the equilibrium approximation used for the dynamic DFT for Brownian particles would set this contribution to exactly zero.

Great insight into the molecular structure and dynamics of liquids can be provided by molecular dynamics (MD) simulations [102], i.e., by solving Newton's equations of motion for all atoms in the system. The reason why quantum effects are negligible in most liquids (except for ultra-cold Helium) is that the thermal wavelength $\Lambda = h/\sqrt{2\pi m k_B T}$ (see (7.12)) is smaller than the average distance of the molecules. The molecular dynamics of both free surface liquid flow on chemically [16, 17, 81, 103, 104] and topographically [105–108] structured surfaces has been studied using MD simulations. Although the length scale considered in MD simulations is necessarily on the nanoscale the effect of the long range of dispersion forces on the dynamics of thin films and droplets is ignored in MD simulations.

This is because the range of intermolecular interactions has to be truncated in order to reduce the computational costs. Without truncation each molecule interacts with each other molecule and for N molecules $N(N-1)$ interaction forces have to be calculated in each time step. If the interaction is truncated (typically at two or three typical molecular distances) the number of force calculations per time step grows only linearly with N and not quadratically.

However, for extremely small systems it might be possible to avoid to truncate the range of interactions. In order to avoid finite size effects one has to keep the gas density low if one wants to keep the number of particles small even in a big simulation box. This can be achieved by considering chain molecules consisting of a few atoms rather than atomic systems [16, 81].

7.3.4 Topographically Structured Substrates

In the vicinity of topographical steps one observes that nanodroplets move laterally [43, 109]. The direction of motion, however, does not depend on the value of the equilibrium contact angle far from the step but on the sign of the Hamaker constant A_H. Near the step shown in Fig. 7.5a nanodroplets move to the left on both sides of the step and in the case shown in Fig. 7.5b to the right, i.e., step up for $A_H > 0$ and step down for $A_H < 0$, respectively. Since the lateral variation of the disjoining pressure is long-ranged, droplets moving away from the step do never stop. However, the droplet velocity decreases rapidly as a function of the distance from the step.

Droplets moving towards the step do not cross it. From the top there is an energy barrier (similar to the Ehrlich–Schwoebel-barrier for an adatom on the terrace of a vicinal crystal surface [110, 111]): to overcome the edge the drop has to deform in such a way that surface area increases which results in an increase of the free energy. This is also true in a macroscopic picture (see Sect. 7.2) or for purely short-ranged interactions. Two-dimensional droplets approaching the step from the bottom stop right in front of the wedge (see Fig. 7.6c) rather than moving into the wedge, although this configuration should represent the global energetic minimum. It is not clear whether this is an artifact of the two-dimensional system or whether there is also a barrier for a three-dimensional droplet. Attempts to directly minimize the droplet free energy in (7.16) using finite element methods [18] (Fig. 7.6a shows a snapshot of an intermediate state of the minimization procedure) have not been conclusive on this respect due to numerical stability problems. However, as shown in Fig. 7.6b, after a rapid initial shape relaxation, during energy minimization the center of mass of the droplet moves in the direction expected for a steepest decent algorithm, i.e., towards the step.

The lateral motion of nanodroplets can be understood as the motion of a droplet on a chemical gradient surface [10, 11] if one defines a "local equilibrium contact angle" by generalizing (7.20) to heterogeneous substrates

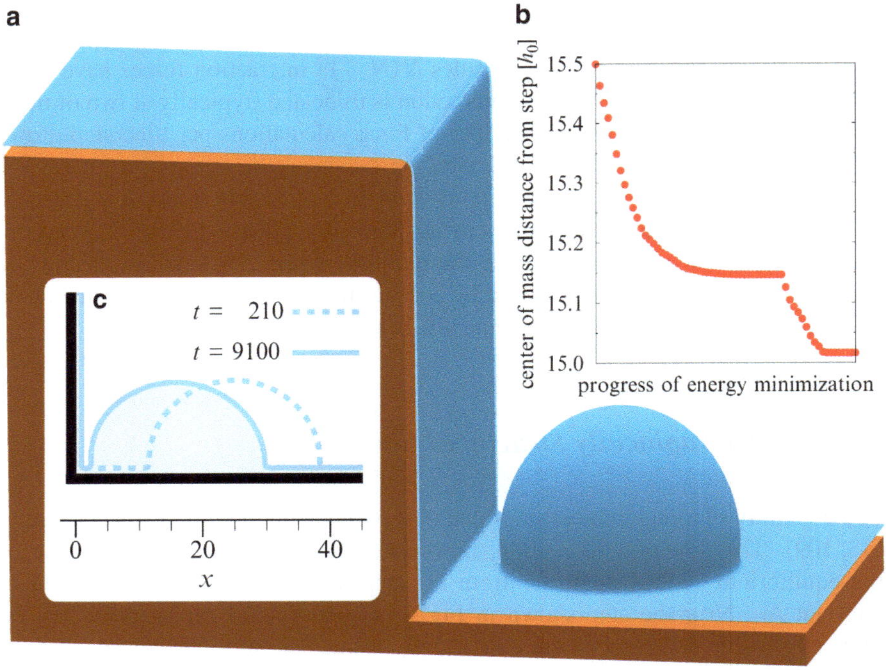

Fig. 7.6 (**a**) A nanodroplet at the base of a topographic step of height $s = 40h_0$ with $A_H > 0$ and an equilibrium contact angle $\theta_{eq} = 90°$. The droplet is connected to a wetting film which covers the whole surface. The shape of the droplet was minimized with the SURFACE EVOLVER [18] which uses a steepest decent algorithm. (**b**) The *inset* shows the horizontal distance of the droplet's center of mass from the step as it evolves in the process of minimization. The position of the droplet has not reached its minimum yet. The existence of an energetic barrier for the droplet to move into the step could not be confirmed or ruled out on the basis of the numerical data. (**c**) Initial (*dashed line*, after shape relaxation) and final (*full line*) state of a two-dimensional boundary element simulation of a droplet at the base of a very high step with $A_H > 0$ and $\theta_{eq} = 90°$ [43]

$$\cos\theta_{eq}(x) = 1 + \int_{h_0(x)}^{\infty} \frac{\Pi(x,z)}{\gamma}\, dz = 1 + \frac{\omega_{min}(x)}{\gamma}, \tag{7.26}$$

with $h_0(x)$ defined via $\Pi(x, h_0(x)) = 0$ and $\partial_z\Pi(x, h(x)) < 0$. In other words, $\omega(x, h_0(x)) = \omega_{min}(x)$ with $\omega_{min}(x) = \min_{0<z<\infty} \omega(x,z)$. Figure 7.7 shows $\omega_{min}(x)$ for the two steps shown in Fig. 7.5. Apart from the immediate vicinity of the step edge, for positive Hamaker constants $\omega_{min}(x)$ increases from right to left and for negative Hamaker constants $\omega_{min}(x)$ increases from left to right. In order to calculate the actual value of $\theta_{eq}(x)$ one further has to specify ω_0/γ. For $\omega_0/\gamma = 1$ one has $\theta_{eq} = 90°$ far from the step. The resulting position-dependent contact angle is indicated on the right abszissa of Fig. 7.7. Since far from critical points h_0 is a microscopic length (between an Ångsrøm and a nanometer) the equilibrium contact angle changes by a few degrees over a few nanometers, i.e., the gradient of $\theta_{eq}(x)$ is much larger than on macroscopic chemical gradient surfaces [10]. Therefore it is not surprising to see the nanodroplets migrate.

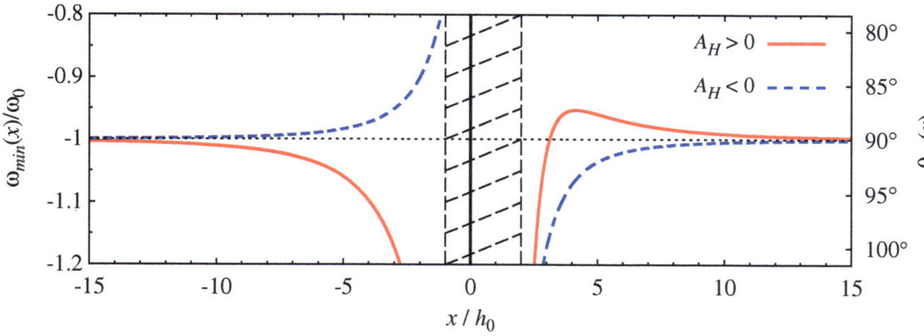

Fig. 7.7 The value of the minimum of $\omega(x,z)$ as a function of z for fixed x plotted vs. x for the cases of $A_H > 0$ (*full red line*, see Fig. 7.5a) and $A_H < 0$ (*dashed blue line*, see Fig. 7.5b). The data points between $x/h_0 = -1$ and $x/h_0 = 2$ (*shaded region*) are not shown since they are to close to the step to be meaningful. The *horizontal line* indicates the value expected for large distances from the step. An increasing value of $\omega_{min}(x)$ corresponds to a decreasing value of $\theta_{eq}(x)$. Therefore $\theta_{eq}(x)$ increases from *left to right* for $A_H > 0$ (apart from the interval $2 < x/h_0 < 4.14$) and it decreases for $A_H < 0$. The corresponding values of $\theta_{eq}(x)$ for a substrate with $\theta_{eq} = 90°$ (i.e., for $\omega_0/\gamma = 1$) far from the step are given at the right abszissa

For $x > 0$ the curve for $A_H > 0$ has a maximum near $x/h_0 \approx 4$. This corresponds to a minimum of $\theta_{eq}(x)$. One might be tempted to interpret this as the signature of an energy barrier for a droplet coming from the right. However, this would be an over-interpretation of the simple picture of a step-induced chemical gradient.

Figure 7.6 shows that within the effective interface model, the step edge is covered by a thin wetting film. On partially wetting substrates the thickness of this film can be a nanometer (1.3 nm for molten polystyrene on silicon [44]), but it is usually on the order of an atomic diameter or even less. This means that it is a layer of adsorbed molecules rather than a liquid film. Nevertheless, this wetting film leads to an effective rounding of the step edge. On the nanoscale the three phase contact line is not pinned at the step edge as shown in Fig. 7.2 but it moves continuously around the edge [112]. However, this has only a small effect on the pinning strength of an edge as does macroscopic rounding of the edge.

The situation is different at step edges. As the step height approaches zero, the disjoining pressure (see for example Fig. 7.5) converges steadily to that of a homogeneous flat substrate (for $z > 0$). In this limit the wetting film on the vertical wall ($\alpha = 90°$) of the step (see Fig. 7.6 for a hight step) smoothly turns into a horizontal line and the slope of the steepest part of the film smoothly becomes zero. However, this means that the step effectively becomes a step with a slanted wall $\alpha < 90°$ which, according to the macroscopic argument illustrated in Fig. 7.2, pins the three phase contact line much weaker. This is also observed in dewetting experiments on substrates with nanometric steps [113]. In this case the critical step height for pinning of the receding contact line was on the order of the radius of gyration of the polymer molecules. The effective interface model described above results in a similar threshold value for the step height.

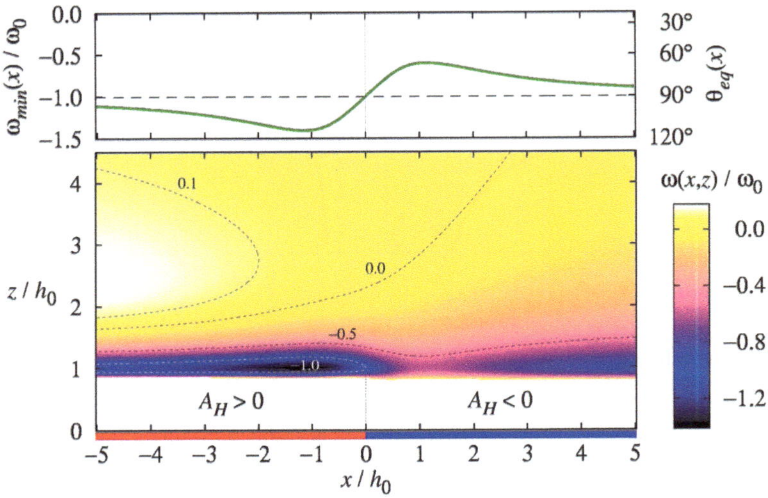

Fig. 7.8 The effective interface potential in the vicinity of a straight chemical step formed by merging two quarter spaces. The *left part* ($x < 0$ and $z < 0$) is made of the material in Fig. 7.5a with $A_H > 0$ and the *right part* ($x > 0$ and $z < 0$) of the material in Fig. 7.5b with $A_H < 0$. ω_0 and therefore θ_{eq} far from the step is equal on both sides. In the *top part* $\omega_{min}(x)$ is shown (*full green line*) as well as the asymptotic value far from the step (*black dashed line*). The corresponding values of $\theta_{eq}(x)$ for a substrate with $\theta_{eq} = 90°$ (i.e., for $\omega_0/\gamma = 1$) far from the step are given at the right abszissa

7.3.5 Chemically Structured Substrates

Using the techniques described above it has been shown that nanodroplets in the vicinity of chemical steps exhibit a similar behavior if the step separates two parts of the substrate which consist of different bulk materials [114, 115]. This is the case, e.g., if one cuts a compound material consisting of inclusions embedded in a matrix, or if one fuses two bulk samples and polishes the surface. In these cases, the Hamaker constants differ on the two sides of the step and one gets the same type of long-ranged lateral interaction of the droplet with the step as in the case of the topographic step.

Figure 7.8 shows the effective interface potential in the vicinity of a straight chemical step formed by merging two materials with the same equilibrium contact angle, i.e., $\omega_{min}(+\infty) = \omega_{min}(-\infty) = \omega_0$. However, the Hamaker constant on the left side is positive and on the right side it is negative. One can clearly see how the left-hand side influences the effective interface potential on the right-hand side and vice versa. For example, close to the step but on its right hand side $\omega(x,z)$ is positive far from the substrate (i.e., for large z) even though the underlying substrate has a negative Hamaker constant. $\omega_{min}(x)$ is shown in the upper part of Fig. 7.8. On both sides of the step $\omega_{min}(x)$ decreases as for increasing x (apart from a narrow region in the direct vicinity of the step. As a consequence, $\theta_{eq}(x)$ increases from left to right and therefore a nanodroplet moves from right to left. However, coming from the right-hand side, it stops before crossing the step.

In general one observes that nanodroplets move towards the side with the larger Hamaker constant if they start at a certain distance from the step (i.e., if they do not span the step), independent of θ_{eq} far from the step [114]. If the more wettable substrate has a larger Hamaker constant, a droplet starting on the less wettable substrate has a chance to cross the chemical step and move away from the step on the more wettable side. However, whether there is a barrier depends on subtle details.

The direction of motion of nanodroplets spanning the chemical step also depends on the droplet size [115]. For the simple picture of heterogeneity-induced chemical gradient in (7.26) to hold, the droplets have to be so high that the effective interface potential at their apex is negligible. In this case, the energy of a droplet on a substrate with a smaller equilibrium contact angle is smaller than on a substrate with a larger equilibrium contact angle. If the droplet is much smaller this is not necessarily the case. A two-dimensional boundary element simulation of a 5 nm high drop of molten polystyrene (a nonvolatile liquid with a high viscosity such that the dynamics of films and droplets can be observed experimentally [44, 52]) on a silicone substrate with a chemical step formed between two regions covered with silicone oxide layers of different thickness shows that the droplet moves to the side with the larger equilibrium contact angle although the Hamaker constants are equal (they are determined by the silicon substrate).

7.3.6 Experimental Perspectives

The dynamics of nanodroplets on structured substrates has not been directly observed yet. One only knows that nanodroplets preferentially condense at topographic steps [116, 117]. One challenge is to prepare suitable substrates, i.e., well-defined topographic and chemical structures. Topographic structures can be produced, e.g., with photolithographic techniques which have been developed for microelectronic devices, and high aspect ratios can be reached. A large number of methods is available for generating a chemical step separating a hydrophilic from a hydrophobic region on a surface (e.g., the boundary between a coated and a non-coated part of the substrate). But in many cases this also leads to a topographic step. However, oxidizing the end groups of self-assembled silane monolayers using the metallic tip of an atomic force microscope allows to prepare well-defined wettability patterns with a 30 nm spatial resolution and with a negligible topographic signature (below 3 Å) [118–120]. The caveat of this method is that only the short-ranged part of the intermolecular interactions is spatially modulated such that the long-ranged lateral interactions between nanodroplets and chemical steps discussed above are negligible in this case.

The main challenge for an experimental verification of the peculiar behavior of nanodroplets at chemical and topographical steps is to prepare and position nanodroplets with nanoscale resolution and then to observe these structures with sufficient spatial and temporal resolution but without influencing them.

Positioning of nanodroplets with nanometer lateral resolution can be achieved with a nano dispenser [121]: the hollow and liquid-loaded tip of an atomic force microscope acts as a pen and allows to deposit droplets with sub-zeptoliter volumes (a cube with 10 nm edge length). Depending of the tip aperture one can reproducibly obtain droplets with diameters down to 70 nm. However, after the transfer of the liquid to the substrate the droplet takes some time to relax its shape [122], and the time scale for this shape relaxation might be comparable to the lateral migration time.

Another option for producing nanodroplets is to pattern a thin film of a material which has a melting point lower than that of the substrate, e.g., by using electron-beam or ion-beam lithography. Regular arrays of islands can be also produced by colloidal monolayer lithography [123]: a densely packed monolayer of colloidal particles is used as a template such that triangular islands form in the gap between three particles. The lateral length scale is controlled by the size of the colloids and islands with an edge length of a few hundred nanometers to a few microns. The volume of the droplets is then controlled by the layer thickness. After annealing the islands turn into nanodroplets. However, as in the case of the nanodispenser the shape relaxation takes time. Melting the islands will speed up shape relaxation but, e.g., for metallic films, the surface tension is so high and the viscosity is so low that inertia comes into play and the droplet actually jump off the substrate [124, 125]. This is an intriguing method for producing nanoparticles but for the purpose of studying the dynamics of nanodroplets this should be avoided.

Nanodroplets are to small to be observed in optical microscopy. X-rays and neutrons are well-established tools for nondestructive structural analysis but scattering methods are hardly suitable for determining the position and shape of an individual droplet and the lack of optical elements limits the resolution of X-ray microscopes. Droplets can be condensed onto surfaces in environmental scanning electron microscopes [126] and they can be observed with high temporal resolution. Spatial resolution of up to a few nanometers is possible but only under optimal conditions and not when imaging soft materials such as droplets.

Scanning probe microscopes have developed into a valuable tool for observing the structure and dynamics of liquid films and droplets (see for example [44, 116, 117]). However, the temporal resolution is fairly low (on the order of minutes, depending on resolution and scan area) and for low viscosity liquids it is difficult to rule out that the tip influences the droplet. However, by quickly freezing the liquid droplet at least the final shape and position of the droplet can be imaged but thermal expansion has to be taken into account.

7.4 Summary and Outlook

Nanodroplets are intriguing objects which behave sometimes in counterintuitive ways—at least this is what theory predicts. These findings are based on static equilibrium DFT and on hydrodynamic equations augmented with features which are relevant on the nanoscale but which are negligible or summarized into boundary conditions on the macroscopic scale (e.g., the equilibrium contact angle, see Fig. 7.1). These are in particular hydrodynamic slip, thermal fluctuations, and the finite range of intermolecular interactions. Neglecting thermal fluctuations, in the stationary limit these mesoscopic hydrodynamic equations lead to the droplet shapes predicted by the DFT. Most functionals are mean field and thus fluctuation effects, in particular capillary waves, are not included.

While hydrodynamic slip plays an important role for the dynamics of the liquid within a moving three phase contact line, it is not to be expected that it changes the qualitative behavior of nanodroplets as this is driven predominantly by the energetics. And it has been shown that the same is true for thermal fluctuations, although they might help to overcome pinning. The main player are the long-ranged dispersion forces. On homogeneous substrates they are summarized into the effective interface potential $\omega(z)$ (see Fig. 7.4) which also determines the equilibrium wetting behavior of the system. On heterogeneous substrates the effective interface potential (see Fig. 7.8 for a chemical step) as well as its derivative with respect to the film thickness, i.e., the disjoining pressure Π (see Fig. 7.5 for a topographic step) depend on the lateral coordinates as well. This leads to a lateral force on a nanodroplet. The sign of this force depends only on the sign of the Hamaker constant A_{H} and at a topographic step nanodroplets are expected to move in the step-up direction for $A_{\mathrm{H}} > 0$ (see Fig. 7.6) and in step-down direction for $A_{\mathrm{H}} < 0$.

At a chemical step the droplets should move towards the side with the larger Hamaker constant. This can be understood if one interprets the lateral variation of the effective interface potential as an effective chemical gradient, i.e., a laterally varying equilibrium contact angle (see Fig. 7.7 for a topographic step and Fig. 7.8 for a chemical step). The behavior at chemical steps is particularly intriguing as it is independent of the equilibrium contact angles far from the step: it can happen that the droplet moves towards the less wettable side. However, in this case it does not cross the step. In particular at chemical steps there is also a strong size dependence. While the free energy of a macroscopic droplet is always smaller on the side with the smaller equilibrium contact angle, for extremely small droplets the situation can be reversed (for an example see Fig. 7.9).

Since this peculiar behavior of nanodroplets is driven by free energy gradients, it is not specific to liquids. Molecular dynamics simulations of gold clusters on graphite with steps show a strong interaction of the step edges with the clusters, but the interactions are purely short-ranged and the steps were only one atom high [127]. On substrates with higher steps and for non-metallic clusters the same phenomena discussed above should be observable.

The development of experimental techniques will determine whether the peculiar dynamics of nanodroplets on structured substrates will ever be observed directly. Maybe clusters are better candidates since they can be studied under high vacuum and in clean conditions on atomically smooth surfaces. However, understanding nanofluidics will be more and more important in order to push the progressive miniaturization of microelectronic, micromechanical, and microfluidic devices even further.

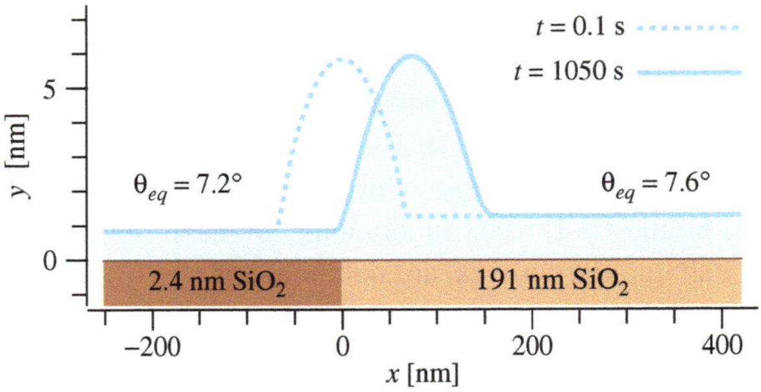

Fig. 7.9 2D boundary element simulation of a small nanodroplet (apex height lower than 7.5 nm) at the chemical step between a silicon substrate with a thin and a thick oxide layer [115]. The Hamaker constant is equal on both sides. Shown are the initial (*dashed line*) and final (*full line*) position of the droplet. Larger droplets move to the *left*

Acknowledgements Markus Rauscher thanks S. Dietrich for supporting this research. The SURFACE EVOLVER calculations were performed by D. Maier and H. Bartsch in their bachelor thesis project.

References

1. Rauscher, M., Dietrich, S.: Ann. Rev. Mater. Res. **38**, 143 (2008)
2. Rauscher, M., Dietrich, S.: Soft Matter **5**(16), 2997 (2009)
3. Rauscher, M., Dietrich, S.: In: Sattler, K.D. (ed.) Handbook of Nanophysics, vol. I: Principles and Methods, chap. 11, pp. 1–23. CRC, Boca Raton (2010)
4. Dietrich, S., Rauscher, M., Napiórkowski, M.: In: Ondarçuhu, T., Aimé, J.P. (eds.) Nanoscale Liquid Interfaces: Wetting, Patterning and Force Microscopy at Molecular Scale. Pan Stanford Publishing, Singapore (2012)
5. de Gennes, P.G.: Rev. Mod. Phys. **57**(3), 827 (1985)
6. Dietrich, S.: In: Domb, C., Lebowitz, J.L. (eds.) Phase Transitions and Critical Phenomena, vol. 12, Chap. 1, pp. 1–218. Academic, London (1988)
7. de Gennes, P.G., Brochard-Wyart, F., Quéré, D.: Capillarity and Wetting Phenomena: Drops, Bubbles, Pearls, Waves. Springer, New York (2004)
8. Ondarçuhu, T., Veyssié, M.: J. Phys. II France **1**(1), 75 (1991)
9. Ondarçuhu, T., Raphaël, E.: C. R. Acad. Sci. Paris II **314**, 453 (1992)
10. Chaudhury, M.K., Whitesides, G.M.: Science **256**(5063), 1539 (1992)
11. Pismen, L.M., Thiele, U.: Phys. Fluids **18**(4), 042104 (2006)
12. Moosavi, A., Mohammadi, A.: J. Phys. Condens. Matter **23**(11), 085004 (2011)
13. Brinkmann, M., Lipowsky, R.: J. Appl. Phys. **92**(8), 4296 (2002)
14. Lipowsky, R., Brinkmann, M., Dimova, R., Franke, T., Kierfeld, J., Zhand, X.: J. Phys. Condens. Matter **17**(9), S537 (2005)
15. Dietrich, S., Popescu, M.N., Rauscher, M.: J. Phys. Condens. Matter **17**(9), S577 (2005)
16. Koplik, J., Lo, T.S., Rauscher, M., Dietrich, S.: Phys. Fluids **18**(3), 032104 (2006)
17. Rauscher, M., Dietrich, S., Koplik, J.: Phys. Rev. Lett. **98**(22), 224504 (2007)
18. Brakke, K.: Exp. Math. **1**(2), 141 (1992)
19. Bico, J., Thiele, U., Quéré, D.: Colloids Surf. A Physicochem. Eng. Aspects **206**(1–3), 41 (2002)

20. Lafuma, A., Quéré, D.: Nat. Mater. **2**(7), 457 (2003)
21. Quéré, D.: Rep. Prog. Phys. **68**, 2495 (2005)
22. Quéré, D.: Ann. Rev. Mater. Res. **38**, 71 (2008)
23. Yao, Z., Bowick, M.J.: Soft Matter **8**(4), 1142 (2012)
24. Seemann, M., Brinkmann, R., Kramer, F.F., Lange, E.J., Lipowsky, R.: Proc. Natl. Acad. Sci. USA **102**(6), 1848 (2005)
25. Khare, K., Herminghaus, S., Baret, J.C., Law, B.M., Brinkmann, M., Seemann, R.: Langmuir **23**(26), 12997 (2007)
26. Herminghaus, S., Brinkmann, M., Seemann, R.: Ann. Rev. Mater. Res. **38**, 101 (2008)
27. Gibbs, J.W.: The Scientific Papers of J. Willard Gibbs, vol. I: Thermodynamics, chap. III, pp. 55–353. Longmans, Green, London (1906)
28. Oliver, J.F., Hu, C., Mason, S.G.: J. Colloid Interface Sci. **59**(3), 568 (1977)
29. Morison, K.R., Sellier, M.: Int. J. Multiphase Flow **39**, 245 (2012)
30. Delmas, M., Monthioux, M., Ondarçuhu, T.: Phys. Rev. Lett. **106**(13), 136102 (2011)
31. Eral, H.B.B., de Ruiter, J., de Ruiter, R., Oh, J.M., Semprebon, C., Brinkman, M., Mugele, F.: Soft Matter **7**(11), 5138 (2011)
32. Lauga, E., Brenner, M.P., Stone, H.A.: In: Tropea, C., Yarin, A.L., Foss, J.F. (eds.) Springer Handbook of Experimental Fluid Mechanics, chap. Part C, pp. 1219–1240. Springer, Berlin, (2007)
33. Oron, A., Davis, S.H., Bankoff, S.G.: Rev. Mod. Phys. **69**(3), 931 (1997)
34. Ralston, J., Popescu, M., Sedev, R.: Ann. Rev. Mater. Res. **38**, 23 (2008)
35. Hansen, J.P., McDonald, I.R.: Theory of Simple Liquids, 2nd edn. Academic, London (1990)
36. Evans, R.: Adv. Phys. **28**(2), 143 (1979)
37. Percus, J.K.: J. Stat. Phys. **15**(6), 505 (1976)
38. Roth, R., Evans, R., Lang, A., Kahl, G.: J. Phys. Condens. Matter **14**(46), 12063 (2002)
39. Hansen-Goos, H., Roth, R.: J. Phys. Condens. Matter **18**(37), 8413 (2006)
40. Napiórkowski, M., Dietrich, S.: Phys. Rev. E **47**(3), 1836 (1993)
41. Robbins, M.O., Andelman, D., Joanny, J.F.: Phys. Rev. A **43**(8), 4344 (1991)
42. Dietrich, S., Napiórkowski, M.: Phys. Rev. A **43**(4), 1861 (1991)
43. Moosavi, A., Rauscher, M., Dietrich, S.: J. Phys. Condens. Matter **21**(46), 464120 (2009)
44. Becker, J., Grün, G., Seemann, R., Mantz, H., Jacobs, K., Mecke, K.R., Blossey, R.: Nat. Mater. **2**(1), 59 (2003)
45. Neto, C., Jacobs, K., Seemann, R., Blossey, R., Becker, J., Grün, G.: J. Phys. Condens. Matter **15**(19), 3355 (2003)
46. Landau, L.D., Lifshitz, E.M.: Fluid Mechanics, Course of Theoretical Physics, vol. 6, 2nd edn. Elsevier, Amsterdam (2005)
47. Mashiyama, K.T., Mori, H.: J. Stat. Phys. **18**(4), 385 (1978)
48. Forster, D., Nelson, D.R., Stephen, M.J.: Phys. Rev. Lett. **36**(15), 867 (1976)
49. Forster, D., Nelson, D.R., Stephen, M.J.: Phys. Rev. A **16**(2), 732 (1977)
50. Hohenberg, P.C., Swift, J.B.: Phys. Rev. A **46**(8), 4773 (1992)
51. Swift, J.B., Babcock, K.L., Hohenberg, P.C.: Physica A **204**(1–4), 625 (1994)
52. Fetzer, R., Rauscher, M., Seemann, R., Jacobs, K., Mecke, K.: Phys. Rev. Lett. **99**(11), 114503 (2007)
53. Mecke, K., Falk, K., Rauscher, M.: In: Radons, G., Rumpf, B., Schuster, H.G. (eds.) Nonlinear Dynamics of Nanosystems, pp. 121–142. Wiley-VCH, Berlin (2010)
54. Blossey, R., Münch, A., Rauscher, M., Wagner, B.: Eur. Phys. J. E **20**(3), 267 (2006)
55. Rauscher, M., Blossey, R., Münch, A., Wagner, B.: Langmuir **24**(21), 12290 (2008)
56. Münch, A., Wagner, B.: J. Phys. Condens. Matter **23**(18), 184101 (2011)
57. Kalliadasis, S., Homsy, G.M.: J. Fluid Mech. **448**, 387 (2001)
58. Bielarz, C., Kalliadasis, S.: Phys. Fluids **15**(9), 2512 (2003)
59. Savva, N., Kalliadasis, S.: Phys. Fluids **21**(9), 092102 (2009)
60. Gaskell, P.H., Jimack, P.K., Sellier, M., Thompson, H.M., Wilson, M.C.T.: J. Fluid Mech. **509**, 253 (2004)
61. Gaskell, P.H., Jimack, P.K., Sellier, M., Thompson, H.M.: Phys. Fluids **18**(1), 013601 (2006)

62. Baxter, S.J., Power, H., Cliffe, K.A., Hibberd, S.: Phys. Fluids **21**(3), 032102 (2009)
63. Kondic, L., Diez, J.A.: Colloids Surf. A Physicochem. Eng. Aspects **214**(1–3), 1 (2003)
64. Kondic, L., Diez, J.: Phys. Fluids **16**(9), 3341 (2004)
65. Mechkov, S., Rauscher, M., Dietrich, S.: Phys. Rev. E **77**(6), 061605 (2008)
66. Shan, X., Chen, H.: Phys. Rev. E **47**(3), 1815 (1993)
67. Shan, X., Chen, H.: Phys. Rev. E **49**(4), 2941 (1994)
68. Swift, M.R., Osborn, W.R., Yeomans, J.M.: Phys. Rev. Lett. **75**(5), 830 (1995)
69. Swift, M.R., Orlandini, E., Osborn, W.R., Yeomans, J.M.: Phys. Rev. E **54**(5), 5041 (1996)
70. Langaas, K., Yeomans, J.M.: Eur. Phys. J. B **15**, 133 (2000)
71. Léopoldès, J., Dupuis, A., Bucknall, D.G., Yeomans, J.M.: Langmuir **19**(23), 9818 (2003)
72. Kuksenok, O., Jasnow, D., Yeomans, J., Balazs, A.C.: Phys. Rev. Lett. **91**(10), 108303 (2003)
73. Dupuis, A., Yeomans, J.M.: In: Bubak, M., van Albada, G.D., Sloot, P.M.A., Dongarra, J.J. (eds.) Computational Science - ICCS 2004. Lecture Notes in Computer Science, vol. 3039, pp. 556–563. Springer, Berlin (2004). dx.doi.org/10.1007/b98005. Cond-mat/0401150
74. Dupuis, A., Léopoldes, J., Bucknall, D.G., Yeomans, J.M.: Appl. Phys. Lett. **87**(2), 024103 (2005)
75. Dupuis, A., Yeomans, J.M.: Langmuir **21**(6), 2624 (2005)
76. Kusumaatmaja, H., Léopoldès, J., Dupuis, A., Yeomans, J.M.: Europhys. Lett. **73**(5), 740 (2006)
77. Kusumaatmaja, H., Yeomans, J: Langmuir **23**(2), 956 (2007)
78. Kusumaatmaja, H., Yeomans, J.M.: Langmuir **23**(11), 6019 (2007)
79. Hyväluoma, J., Harting, J.: Phys. Rev. E **100**(24), 246001 (2008)
80. Harting, J., Kunert, C., Hyväluoma, J.: Microfluid Nanofluid **8**(1), 1 (2009)
81. Dörfler, F., Rauscher, M., Koplik, J., Harting, J., Dietrich, S.: Soft Matter **8**(35), 9221 (2012)
82. Jacqmin, D.: J. Comp. Phys. **155**, 96 (1999)
83. Ding, H., Spelt, P.D.M.: Phys. Rev. E **75**(4), 046708 (2007)
84. Pismen, L.M., Pomeau, Y.: Phys. Rev. E **62**(2), 2480 (2000)
85. Thiele, U., Velarde, M.G., Neuffer, K., Pomeau, Y.: Phys. Rev. E **64**(3), 031602 (2001)
86. Thiele, U., Velarde, M.G., Neuffer, K., Bestehorn, M., Pomeau, Y.: Phys. Rev. E **64**(6), 061601 (2001)
87. Pismen, L.M.: Phys. Rev. E **64**(2), 021603 (2001)
88. Pismen, L.M.: Colloids Surf. A Physicochem. Eng. Aspects **206**(1–3), 11 (2002)
89. Dimitrakopoulos, P., Higdon, J.J.L.: J. Fluid Mech. **336**, 351 (1997)
90. Dimitrakopoulos, P., Higdon, J.J.L.: J. Fluid Mech. **377**, 189 (1998)
91. Dimitrakopoulos, P., Higdon, J.J.L.: J. Fluid Mech. **435**, 327 (2001)
92. Dimitrakopoulos, P.: J. Fluid Mech. **580**, 451 (2007)
93. Kelmanson, M.A.: J. Comp. Phys. **51**, 139 (1983)
94. Mazouchi, A., Gramlich, C.M., Homsy, G.M.: Phys. Fluids **16**(5), 1647 (2004)
95. Marconi, U.M.B., Tarazona, P.: J. Chem. Phys. **110**(16), 8032 (1999)
96. Marconi, U.M.B., Tarazona, P.: J. Phys. Condens. Matter **12**(8A), A413 (2000)
97. Archer, A.J., Rauscher, M.: J. Phys. A Math. Gen. **37**(40), 9325 (2004)
98. Archer, A.J.: J. Phys. Condens. Matter **18**(24), 5617 (2006)
99. Marconi, U.M.B., Melchionna, S.: J. Chem. Phys. **131**(1), 014105 (2009)
100. Green, H.S.: The Molecular Theory of Fluids. North-Holland, Amsterdam (1952)
101. Kreuzer, H.J.: Nonequilibrium Thermodynamics and Its Statistical Foundations. Clarendon, Oxford (1981)
102. Frenkel, D., Smit, B.: Understanding Molecular Simulation, 2nd edn. Academic, San Diego (2002)
103. Yaneva, J., Milchev, A., Binder, K.: J. Chem. Phys. **121**(24), 12632 (2004)
104. Cieplak, M., Koplik, J., Banavar, J.R.: Phys. Rev. Lett. **96**(11), 114502 (2006)
105. Cottin-Bizonne, C., Barrat, J.L., Bocquet, L., Charlaix, E.: Nat. Mater. **2**(4), 237 (2003)
106. Cottin-Bizonne, C., Barentin, C., Charlaix, E., Bocquet, L., Barrat, J.L.: Eur. Phys. J. E **15**, 427 (2004)
107. Cao, B.Y., Chen, M., Guo, Z.Y.: Phys. Rev. E **74**(6), 066311 (2006)

108. Huang, D.M., Cottin-Bizonne, C., Ybert, C., Bocquet, L.: Phys. Rev. Lett. **06**, 064503 (2008)
109. Moosavi, A., Rauscher, M., Dietrich, S.: Phys. Rev. Lett. **97**(23), 236101 (2006)
110. Ehrlich, G., Hudda, F.G.: J. Chem. Phys. **44**(3), 1039 (1966)
111. Schwoebel, R.L., Shipsey, E.J.: J. Appl. Phys. **37**(10), 3682 (1966)
112. Dutka, F., Napiórkowski, M., Dietrich, S.: J. Chem. Phys. **136**(6), 064702 (2012)
113. Ondarçuhu, T., Piednoir, A.: Nano Lett. **5**(9), 1744 (2005)
114. Moosavi, A., Rauscher, M., Dietrich, S.: Langmuir **24**(3), 734 (2008)
115. Moosavi, A., Rauscher, M., Dietrich, S.: J. Chem. Phys. **129**(4), 044706 (2008)
116. Hu, J., Xiao, X.D., Salmeron, M.: Appl. Phys. Lett. **67**(4), 476 (1995)
117. Hu, J., Carpick, R.W., Salmeron, M., Xiao, X.D.: J. Vac. Sci. Tech. B **14**(2), 1341 (1996)
118. Checco, A., Gang, O., Ocko, B.M.: Phys. Rev. Lett. **96**(5), 056104 (2006)
119. Checco, A., Schollmeyer, H., Daillant, J., Guenoun, P., Boukherroub, R.: Langmuir **22**(1), 116 (2006)
120. Checco, A.: Phys. Rev. Lett. **102**(10), 106103 (2009)
121. Fang, A., Dujardin, E., Ondarçuhu, T.: Nano Lett. **6**(10), 2368 (2006)
122. Fabi/'e, L., Ondarçuhu, T.: Soft Matter **8**(18), 4995 (2012)
123. Burmeister, F., Badowsky, W., Braun, T., Wieprich, S., Boneberg, J., Leiderer, P.: Appl. Surf. Sci. **144–145**, 461 (1999)
124. Habenicht, A., Olapinski, M., Burmeister, F., Leiderer, P., Boneberg, J.: Science **309**(5743), 2043 (2005)
125. Boneberg, J., Habenicht, A., Benner, D., Leiderer, P., Trautvetter, M., Pfahler, C., Plettl, A., Ziemann, P.: Appl. Phys. A **93**(2), 415 (2008)
126. Stokes, D.J.: Adv. Eng. Mater. **3**(2), 126 (2001)
127. Yoon, B., Luedtke, W.D., Gao, J., Landman, U.: J. Phys. Chem. B **107**(24), 5882 (2003)

Chapter 8
Atomistic Simulation of Nanodroplet Collisions with a Wall: Fragmentation and Impact Desolvation of Macromolecules

Herbert M. Urbassek and Si Neng Sun

Abstract Impacts of nanodroplets on a hard wall are studied using molecular-dynamics simulation. We focus on water droplets; both pure solvent droplets and droplets filled with a macromolecule are investigated. By choosing a hydrophilic (polyketone) and a hydrophobic (polyethylene) polymer, the effects of the water–polymer interaction can be studied. The process of droplet fragmentation and the ensuing isolation of the embedded macromolecule are investigated. The energy and time dependence of the process is analyzed for various droplet–polymer combinations. By changing droplet size, polymer size, and polymer species separately, we can assess the influence of these factors individually. We demonstrate that the ratio of the impact energy, E, to the cohesive energy, E_{coh}, of the droplet is the key quantity characterizing the droplet fragmentation process. If the impact energy per molecule $E < (0.29-0.4) \cdot E_{\mathrm{coh}}$, the droplet is reflected without fragmenting. Beyond that impact energy fragmentation of the droplet abruptly starts. At $E = E_{\mathrm{coh}}$, the fragmentation process already results in a fine dispersal of the droplet into daughter droplets. The disintegration process continuously increases with collision energy. We find that the polymer is isolated for impact energies E per solvent molecule, which exceed a threshold value of the order of the cohesive energy E_{coh} of the solvent. We find that in this energy regime, the temperature of the isolated polymer increases linearly with E.

8.1 Introduction

Impacts of droplets on a wall constitute a fascinating subject, both from a fundamental point of view [1] and for relevant applications. In the size range of micro- or nanodroplets, these range from inkjet printing to various cluster-surface

H.M. Urbassek (✉) • S.N. Sun
Fachbereich Physik und Forschungszentrum OPTIMAS, Universität Kaiserslautern, Erwin-Schrödinger-Straße, 67663 Kaiserslautern, Germany
e-mail: urbassek@rhrk.uni-kl.de; sunsineng05@gmail.com

Z.M. Wang (ed.), *Nanodroplets*, Lecture Notes in Nanoscale Science and Technology 18,
DOI 10.1007/978-1-4614-9472-0_8,

interaction techniques, including controlled cluster deposition [2], surface cleaning by cluster impact [3], and cluster-impact chemistry [4]. Solvent droplets filled with macromolecules—polymers or proteins—form another interesting research area. A recent study [5] explores the effect of the embedded macromolecule on the collision dynamics of the droplet, in which the bouncing of the droplet off the surface is prevented. Macromolecule-loaded droplets are generated for the purpose of performing mass spectrometry on the embedded macromolecules. It is then necessary to desolvate the molecule. A variety of techniques have been set up for this purpose, such as electrospray ionization (ESI), matrix-assisted laser desorption (MALDI), massive cluster impact (MCI), laser spray, or laser-induced liquid beam ionization (LILBID) [6–11]. A comparatively recent technique is *impact desolvation of electrosprayed microdroplets* (IDEM) [12]. In this method the analyte molecule is solvated in a microdroplet; the droplet is collided with a wall with the aim of desolvating the analyte without destroying it.

At not too small impact energies, the droplet will fragment, but an embedded macromolecule may still bind some water molecules. With increasing impact energy, eventually the macromolecule will be completely isolated. The question on which quantities the optimum energy window—not too small energy in order to fragment the solvent shell, not too large energy in order not to dissociate the macromolecule—is the topic of the present contribution.

Many aspects of droplet- and cluster-wall collisions have been studied theoretically over the years. Macroscopic impacts have been studied theoretically using fluid-dynamical methods; see [13] for a recent study. A recent review [1] assembles the available knowledge on droplet intact reflection (called "bouncing") and fragmentation (subdivided into "splashing," "spreading," and "fingering").

Recently, the desolvation process of polymers [14] and of proteins [15] caused by collisions with a wall has been studied using the technique of molecular-dynamics simulation. The dependence on polymer size and species (hydrophilic or hydrophobic) and on the solvent species and droplet size was systematically investigated. In addition, the fragmentation dynamics and the resulting fragmentation patterns of pure droplets (without embedded macromolecules) have been studied for droplets composed of van-der-Waals bonded atoms [16–19] and molecules [20, 21], and in particular also for water [14, 22, 23].

Here, we employ molecular-dynamics simulations of model systems to obtain a deeper understanding of these questions. We choose a particularly simple macromolecule—a polymer—and water as a common solvent. In other works, we discussed the role of the solvent in detail by studying besides water also atomic and molecular nonpolar solvents (Ar and N_2) [14, 24, 25]. However, we found that it is mainly the cohesive energy of the solvent that determines the fragmentation process and so we shall report here only on water droplets.

In Fig. 8.1 we visualize the collision dynamics of pure and macromolecule-loaded water droplets impinging with identical conditions on a hard wall. As macromolecules a protein (ubiquitin) and a polymer (polyketone) of similar atom number have been chosen. Ubiquitin is a small globular protein found in all eukaryotes; it consists of 76 amino acid residues (1,231 atoms). The impact speed

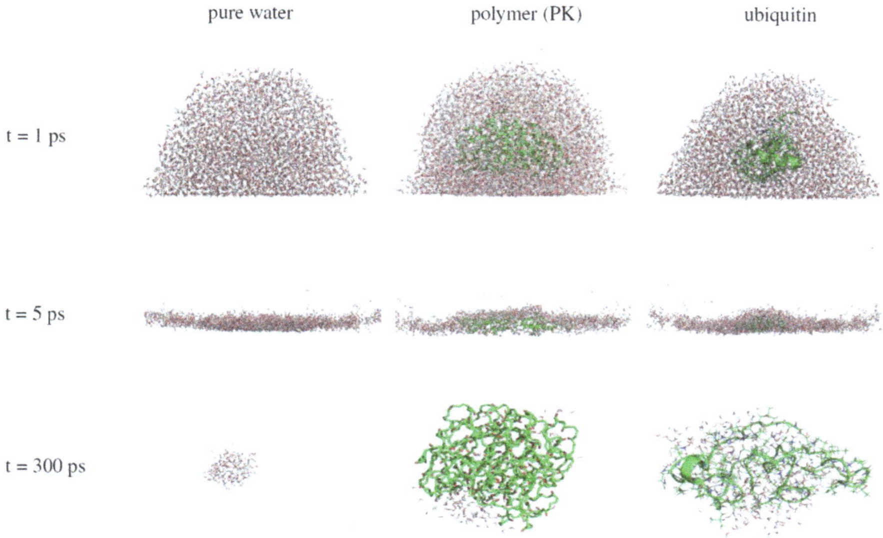

Fig. 8.1 Comparison between pure (*left column*), polymer-loaded (*middle*), and protein-loaded (*right*) droplets, at the times of 1, 5, and 300 ps after collision with a wall (invisible at the bottom of each subfigure) with a velocity of 2.25 km/s. The protein is ubiquitin, the polymer is polyketone consisting of 1,200 atoms; the amount of water molecules is $N = 6{,}000$ in each case

has been selected such that the macromolecules become almost isolated by the collision. The sequence of events seen is identical in all cases: the droplet including its solute molecule becomes strongly distorted in the collision; thereafter the droplet breaks up into many fragments. At 300 ps after the impact, the surviving water cluster of maximum size contains 198 molecules in the case of the pure water droplet; the cluster bound to the solute molecules contains 102 and 222 H_2O molecules in the case of polyketone and ubiquitin, respectively. This comparison demonstrates that it is mainly the impact speed which predicts the degree of isolation of the molecule. This is the main reason why we shall study in this paper mainly polymers as solute molecules, but expect that the results can be transferred to other systems. Of course, the hydrophobicity of the molecule is also relevant and will be studied in Sect. 8.4.3 below.

8.2 Method

We use molecular models to describe the systems and employ molecular-dynamics simulations to investigate the collision processes. Atomistic modeling can provide insight into the phenomena and provide details that are difficult to measure in real experiments. In this section, we give a short introduction to the models we use to study impact-induced processes in nanodroplets.

8.2.1 Water Model

We investigate nanodroplets consisting of water. In a molecular model, each H_2O molecule will be taken into account. There exist many different water models based on empirical interaction potentials, which have been developed to describe the structure of water in its different phases, and its dynamics in various processes. Among these, the most common models are the SPC/E model [26], TIP3P [27], and TIP4P [28]. These models have been developed to describe different properties of water and thus must be chosen only for the desired purpose; none of the H_2O models describes all properties of water fully correctly. For instance, the SPC/E model is suited for describing H-bonds in liquid water; TIP3P models the liquid–gas interface well, and TIP4P is suitable for modeling phase transitions of water [29]. For a comparison of the quality of various water models see [27, 30, 31].

Here we shall employ TIP4P to model the nanodroplets, because the impact process will induce phase transitions in the nanodroplet. This model describes a rigid molecule. In the TIP4P model, O–H bonds have a bond length of 0.9572 Å, the H–O–H angle amounts to 104.52°, the charges on the H atoms are $+0.52e$, and the O atom has a charge of $-1.04e$. However, the partial charge on the O atom is put on a dummy site, which is located on the symmetry axis of the molecule but displaced toward the H atoms by a distance of 0.15 Å; this provides for the correct permanent dipole moment of water. Electrostatic forces are cut off at $r_{\mathrm{cut}} = 8.5$ Å according to the scheme of [32]. In addition, van-der-Waals interactions are included between the O atoms; these are modeled by a Lennard–Jones (LJ) potential:

$$V_{\mathrm{LJ}}(r) = 4\varepsilon\left[\left(\frac{\sigma}{r}\right)^{12} - \left(\frac{\sigma}{r}\right)^{6}\right], \tag{8.1}$$

where r denotes the O–O distance. Here the bond strength is $\varepsilon = 6.72$ meV and the position of zero energy is $\sigma = 3.15365$ Å.

Since TIP4P describes H_2O molecules as rigid, molecular vibrations are not included in this model. In collision processes, molecular vibrations can be excited; due to the high energy of the vibrational quanta of water (0.2 eV for the bending mode and 0.45 eV for the stretching modes), vibrational excitation will only occur at high impact velocities. Then, however, vibrational relaxation is an ultrafast process in water, which takes place in a time scale of 600 fs [33]. During this time, vibrational energy will be relaxed into translational and rotational degrees of freedom. In comparison, the processes of droplet fragmentation require in the order of a few tens of ps. For our purposes it therefore appears justified to use a rigid water model. Note that after the short period of time (a few ps) of the collision itself, the water temperature quickly decreases to below 1,000 K, where vibrational excitation (in thermal equilibrium) can safely be neglected.

8.2.2 Molecular Model of Polymers

We also investigate polymer-loaded droplets, in which a polymer is dissolved in the water droplet. We study two model polymers: polyethylene (PE), CH_3–$(CH_2)_n$–CH_3, and polyketone (PK), CH_3–$(C_2H_4COC_2H_4)_n$–CH_3. For PE, the CH_2 and CH_3 sites in each monomer are considered as united atoms C^* with masses of 14.027 and 15.035 amu, respectively. They are connected with neighboring C^* by sp^3-type bonds. The chemical bonds within the monomers are modeled by Morse potentials both for nearest (1–2) and next-nearest (1–3) neighbors. In addition, the nonbonded intramolecular interactions are modeled by Lennard–Jones potentials. This model as well as the bond parameters are taken from [34].

For PK, we use a part of the *optimized potentials for liquid simulation* (OPLS) force field [35] consisting of bonding and nonbonding interactions. Although the OPLS force field is capable of considering C–H bonds explicitly, we took all –CH_3– and –CH_2– groups as united atoms in order to make the model comparable to that of PE. In this force field, the total potential energy of the polymer takes the form

$$
\begin{aligned}
E_{\mathrm{pot}} = & \sum_{\mathrm{bonds}} D\left[1-e^{-\alpha(r-r_0)}\right]^2 + \sum_{\mathrm{angles}} k_\theta(\theta-\theta_{\mathrm{eq}})^2 \\
& + \frac{1}{2}\sum_{ij}\left\{4\varepsilon_{ij}\left[\left(\frac{\sigma_{ij}}{r_{ij}}\right)^{12}-\left(\frac{\sigma_{ij}}{r_{ij}}\right)^{6}\right]+\frac{q_i q_j e^2}{4\pi\varepsilon_0 r_{ij}}\right\}. \qquad (8.2)
\end{aligned}
$$

In addition, a torsional term is included, which has a similar form as the bond-bending angular term, but is not written out explicitly in Eq. (8.2).

The Morse potential for direct-neighbor C–C bond stretching has parameters $\alpha = 21.93\,\mathrm{nm}^{-1}$, $r_0 = 1.53$ Å, and $D = 3.6192$ eV [36], where α has been determined from the spring constant given in GROMACS. All further parameters were directly taken from the GROMACS-4.0 package [36]. In our PK model, each ketone monomer is followed by two ethylene monomers. Charges are only on the C and O atoms in the polarized ketone monomers and assume a value of $+0.47e$ for C and $-0.47e$ for O. The Lennard–Jones interaction acts only between these atoms in the ketone group as well as between the CH_2 and CH_3 groups.

The polymers consist of N_p atoms. If necessary, we specify the number of atoms in the polymer by a subscript, such as PE_{298}, for instance.

8.2.3 Model of the Surface

The surface is modeled to be a hard repulsive wall; we thus assume that its cohesive energy is far above that of the droplet. This appears to be a good approximation for the impact of water droplets on metal surfaces at not too high impact velocities. Our neglect of any adhesion between the droplet and the wall restricts our analysis to hydrophobic surfaces. The effect of the wall on the droplet molecules is modeled by a purely repulsive rigid external potential [20, 21], which acts on each atom in the nanodroplet; it depends only on the perpendicular distance z of the atom to the wall:

$$V_{\text{wall}}(z) = \begin{cases} 4\varepsilon_w \left[\left(\frac{\sigma_w}{z} \right)^{12} - \left(\frac{\sigma_w}{z} \right)^{6} \right] + \varepsilon_w, & z < 2^{1/6}\sigma_w, \\ 0, & z \geq 2^{1/6}\sigma_w. \end{cases} \tag{8.3}$$

The potential parameters have been chosen as $\varepsilon_w = 3.212\,\text{meV}$ and $\sigma_w = 3.31\,\text{Å}$ [20, 21].

8.2.4 System Preparation and Simulation

We prepare spherical droplets at a temperature of $T_0 = 300\,\text{K}$ containing N H_2O molecules; N is in the range of 2,000–9,000. For the study of polymer-loaded droplets, we dissolve in these either a PK or a PE molecule as follows. The polymer is created as a linear chain and then forced to coil up using a spherical repulsive potential [37]. Then, it is set in a cubic simulation box, which is then filled up with water. After energy minimization, a spherical droplet containing the polymer and the specified number of water molecules is cut out, heated to 300 K, and equilibrated for a few hundred ps.

The droplets are initialized to an impact velocity v with a direction toward the wall. We shall denote the initial energy per water molecule as E; it is composed of the kinetic energy of translation toward the wall and the thermal energy as

$$E = \frac{6}{2} kT_0 + \frac{1}{2} mv^2, \tag{8.4}$$

where m is the molecular mass and k is the Boltzmann's constant. In the following we shall set the initial energy E in relation to the cohesive energy $E_{\text{coh}} = 520\,\text{meV}$ [38] of water.

The impact velocity v is related to the energy per molecule as

$$v = \sqrt{\frac{E}{E_{\text{coh}}} - 0.15} \cdot 2.36\,\text{km/s}. \tag{8.5}$$

The droplet has sonic speed $v_{\text{sound}} = 1.50\,\text{km/s}$ (for liquid water at 300 K [39]) at a collision energy of $E/E_{\text{coh}} = 0.55$.

The GROMACS package [36] is used to calculate the dynamics of the water/polymer system, while our own code is used to analyze the data. The leapfrog algorithm with a time step of 1 fs is used to integrate numerically the equations of motion. The SETTLE algorithm [40] is employed to include the constraints posed by the rigid water molecules in the dynamics.

8.2.5 Data Analysis

Fragments

After colliding with the wall, the nanodroplet may fragment. We identify the water molecules which are still bound to the macromolecule as the *central fragment.* Fragments are identified using the cluster detector of Stoddard [41]. It works such that all water molecules which are within their respective interaction radius ($r_{\mathrm{cut}} = 8.5$ Å)—or for which an uninterrupted chain of interacting water molecules can be found—are assigned to the same cluster.

The relative number of atoms in the central fragment, N_{fr}, will be denoted as

$$f = \frac{N_{\mathrm{fr}}}{N}. \tag{8.6}$$

Temperatures

We determine the temperature of water by distinguishing between the translational T_{trans} and rotational T_{rot} degrees of freedoms of H_2O molecules. These temperatures will be evaluated as the average temperature in the nanodroplet or in its fragments.

We proceed as follows. Around each H_2O molecule i, an ensemble is defined to which all molecules within the cutoff radius, r_{cut}, belong. The rotational temperature T^i_{rot} of molecule i is defined as

$$\frac{3}{2} N_i k T^i_{\mathrm{rot}} = \sum_{j=1}^{N_i} \sum_{\alpha=1}^{3} \frac{m_\alpha}{2} (\boldsymbol{v}^{\mathrm{rot}}_{j\alpha})^2. \tag{8.7}$$

Here, N_i is the number of molecules in the ensemble around molecule i, and $\alpha = 1, 2, 3$ counts the atoms (mass m_α) in a water molecule. The rotational velocity of atom α in molecule j is given by $\boldsymbol{v}^{\mathrm{rot}}_{j\alpha} = \boldsymbol{v}_{j\alpha} - \boldsymbol{v}_j$, where $\boldsymbol{v}_{j\alpha}$ is the velocity of the atom and $\boldsymbol{v}_j$ is the center-of-mass velocity of molecule j.

Analogously, the translational temperature, T^i_{trans}, of molecule i is defined as

$$\frac{3}{2} N_i k T^i_{\mathrm{trans}} = \frac{m}{2} \sum_{j=1}^{N_i} \boldsymbol{v}'^2_j. \tag{8.8}$$

Here, m is the mass of a water molecule and $\boldsymbol{v}'_j$ is the relative velocity of molecule j in the ensemble,

$$\boldsymbol{v}'_j = \boldsymbol{v}_j - \boldsymbol{V}_i, \tag{8.9}$$

and $\boldsymbol{V}_i$ is the center-of-mass velocity of the whole ensemble.

T_{trans} and T_{rot} of the entire *central fragment* are then given as averages over T^i_{trans} and T^i_{rot} of all molecules in the central fragment.

Finally, the polymer temperature, T_{polymer}, is determined via

$$\frac{3}{2}N_{\text{p}}kT_{\text{polymer}} = \sum_{i=1}^{N_{\text{p}}} \frac{m_i}{2}(\boldsymbol{v}_i - \boldsymbol{V}_{\text{p}})^2. \tag{8.10}$$

Here, N_{p} is the number of (united) atoms in the polymer, m_i the mass of (united) atom i, and $\boldsymbol{v}_i$ its velocity. Furthermore, $\boldsymbol{V}_{\text{p}}$ is the polymer center-of-mass velocity.

8.3 Fragmentation of Pure Droplets

In this section we consider collision processes of pure nanodroplets with a wall. As the droplet meets the wall, the translational kinetic energy is transferred into internal energy; hence the temperature and pressure in the droplet strongly increase. If the impact energy E exceeds a threshold value, the droplet will be disintegrated and fragments of different sizes are formed. The fragmentation process of a nanodroplet is visualized in Fig. 8.2 and will be discussed in detail below. In this case, a nanodroplet with 6,000 H_2O molecules impinges on a repulsive (hydrophobic) wall with velocity $v_0 = 2.25$ km/s. In order to understand the behavior of the droplet under different impacts, we define the total impact energy (per molecule) as the sum of the translational energy of motion and the thermal energy in the droplet, Eq. (8.4). We scale the impact energy E with the cohesive energy E_{coh} of the material; the ratio E/E_{coh} is a measure of the total impact energy with respect to the binding energy present in the substance.

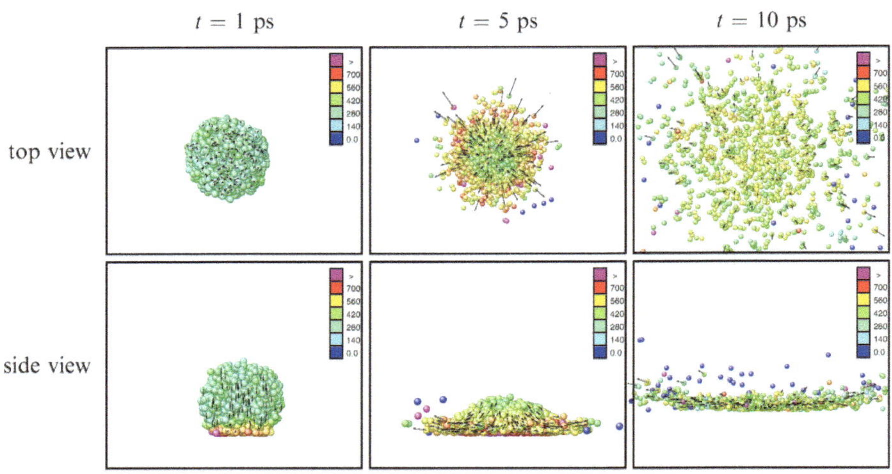

Fig. 8.2 Impact-induced fragmentation of a pure water nanodroplet; same system as in Fig. 8.1 (*left*). The droplet impinges on the wall with a velocity of 2.25 km/s corresponding to $E/E_{\text{coh}} = 1.06$. *Color* codes the local temperatures, the *black arrows* indicate the velocity and direction of the flow

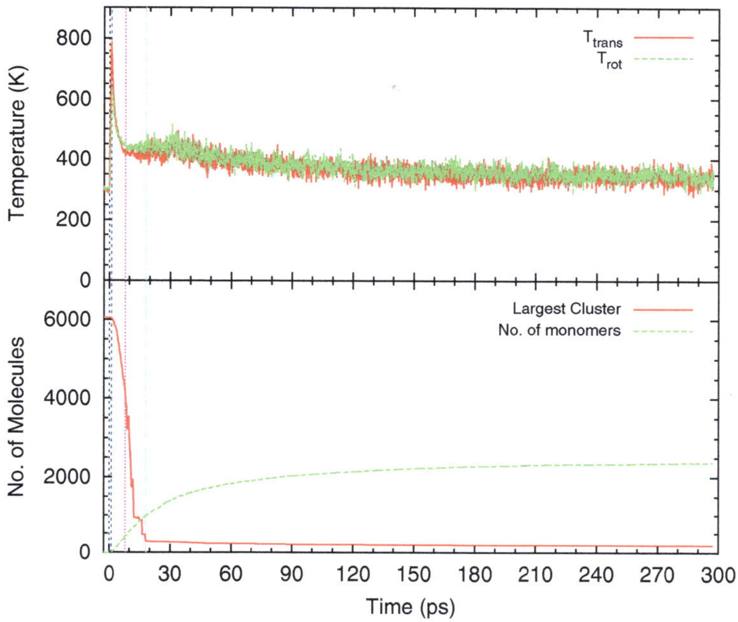

Fig. 8.3 Temporal evolution of the fragmentation process of a pure nanodroplet containing $N = 6{,}000$ H_2O molecules. The droplet impinges on the wall with a velocity of 2.25 km/s, corresponding to $E/E_{coh} = 1.06$. *Top*: evolution of translational (T_{trans}) and rotational (T_{rot}) temperature. *Bottom*: evolution of the number of molecules in the largest fragment, m_{max}, and of the number of free monomers, N_{mono}. The *dashed black line* indicates the time when the droplet collides with the wall. The other *vertical lines* divide the fragmentation process into four stages: compression, lateral jetting, void formation, and evaporation

8.3.1 Temporal Evolution

The time evolution of the temperature and of the fragmentation process of the collision visualized in Fig. 8.2 is shown in Fig. 8.3. Water temperature is measured using two quantities: T_{trans} is the temperature in translational degrees of freedom of H_2O molecules in the largest fragment and T_{rot} is the temperature in rotational degrees of freedom.

We can identify four different development stages in Fig. 8.3:

1. The *compression stage* takes place from 0 to 1.2 ps. In this stage, the kinetic energy of translational motion is converted to internal energy, and hence, the temperature dramatically increases.
2. From 1.2 to 8.1 ps, in the *lateral-jetting stage*, the high compressive pressure present in the bottom part of the droplet drives the reexpansion of the droplet. This expansion occurs mainly sideways, i.e., along the wall, while the remainder

of the droplet still continues moving toward the wall, until it is spread out as a thin layer on top of the wall. Simultaneously, some H_2O monomers may be sprayed out of the droplet with high velocities. The snapshot at 5 ps in Fig. 8.2 visualizes the lateral-jetting process.

3. After the compressive pressure has been relaxed sideways, tensile pressure builds up and starts tearing the droplet. Voids appear in the thin disk, to which the droplet has been transformed, and tear it into fragments of different sizes. This *void-formation* or multi-fragmentation stage takes place from 8.1 to 18.2 ps. In this stage, the size of the remaining droplet decreases abruptly. The void-formation process is shown in the 10-ps snapshot of Fig. 8.2.
4. After 18.2 ps, the droplet has been disintegrated to small clusters and monomers. The ensuing dynamics is governed by the evaporation of H_2O molecules from the fragments and produces a large number of monomers (*evaporation stage*).

After a systematic study of different impact energies [24], we find that the molecular-dynamics results are quite compatible with what is known about the fate of macroscopic droplets after collisions with a wall [42]. With increasing impact energy, the fate of nanodroplets can be described as (intact) bouncing off the surface, onset of fragmentation (wetting of the surface) and full fragmentation. These three regimes start at impact energies of $E/E_{coh} = 0.23$, $E/E_{coh} = 0.51$ and $E/E_{coh} = 1.03$ for the water droplets studied here.

The droplet dynamics occurring for different impact energies can be made comparable by normalizing time to the period of time needed to cover the droplet radius R when moving with a velocity $\sqrt{2E/m}$:

$$\tau = \frac{R}{\sqrt{2E/m}}. \tag{8.11}$$

Figure 8.4 assembles the fragmentation dynamics for three impacts; the size of the largest fragment, m_{max}, has been chosen to illustrate the fragmentation process. In the case of bouncing (bottom part of Fig. 8.4, $E/E_{coh} = 0.3$), the droplet survives the collision and no significant molecule loss can be observed. As the nanodroplet collides with the wall, it is deformed by the momentum pressing it onto the wall; then, however, it is reflected from the wall and resumes its original spherical form due to surface tension. If the impact energy is increased to $E \cong E_{coh}$ (middle part of Fig. 8.4), the fragmentation will disintegrate the droplet. As the impact energy is further increased (top part of Fig. 8.4, E/E_{coh}=3.0), the dynamics does not change much when plotted in normalized time and is similar with the case of $E/E_{coh} = 1.0$.

Additionally, we find that independently of the droplet material, the dynamics in normalized time is similar [24], and that the threshold for onset of droplet fragmentation is in the order of $E/E_{coh} \cong 0.29$–0.40 [24]. For full disintegration of the droplet into monomers, on the other hand, an energy of $E \cong 100E_{coh}$ is needed [20].

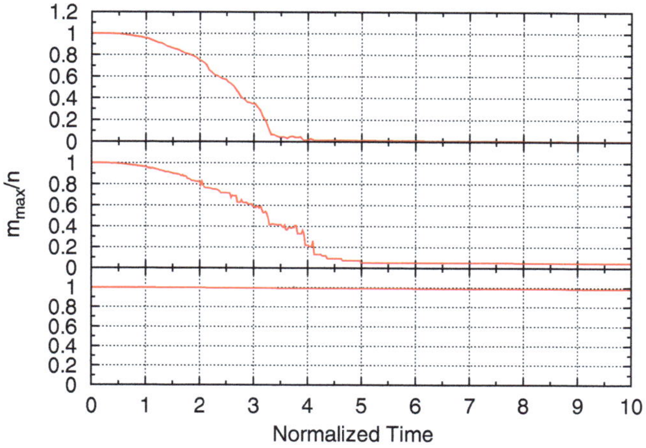

Fig. 8.4 Comparison of the evolution of the size of the largest fragment m_{max} (normalized to the initial droplet size N) for different impact energies. The droplet contains 9,000 H_2O molecules. The scaled impact energies amount to $E/E_{coh} = 3.00$, 1.00, and 0.30 *from top to bottom*. Time t is scaled to the characteristic time τ, Eq. (8.11), the droplet needs to fly through a distance equal to the droplet radius. Data taken from [24]

8.3.2 Dependence on Impact Energy

In this section we investigate the dependence of droplet fragmentation on impact energy. To describe the fragmentation process in a quantitative way, we introduce four measures: the total number of fragments, N_{fr}, the average fragment size, $\langle m \rangle$, the size of the largest fragment, m_{max}, and the number of free monomers created, N_{mono}.

In Fig. 8.5 we changed the impact energy systematically between $0.2E_{coh}$ and $2.1E_{coh}$, and determined these quantities at the end of the simulations. From the results we observe:

1. The onset of the fragmentation regime is seen in particular in the increase in the number of fragments and in the number of free monomers. Taking as a threshold the energy when $N_{fr}/N > 10\,\%$, fragmentation sets in at impact energies E/E_{coh} around 0.35–0.4. At smaller impact energies, the tensile pressures produced in the lateral-jetting phase are not strong enough to tear the droplet.
2. The onset of the fragmentation regime is accompanied by a sharp falloff of the average fragment size $\langle m \rangle$, which occurs at $E/E_{coh} \cong 0.29$–0.4. These strong changes in the fragment distributions are reminiscent of a phase transition and have been characterized as such [17, 43–47].
3. A regime of fully developed fragmentation might be tentatively characterized by requiring that no large fragment of the initial droplet survives; we have set $m_{max}/N < 10\,\%$ as the limit. Fully developed fragmentation then starts at $E/E_{coh} = 0.63$–0.72.

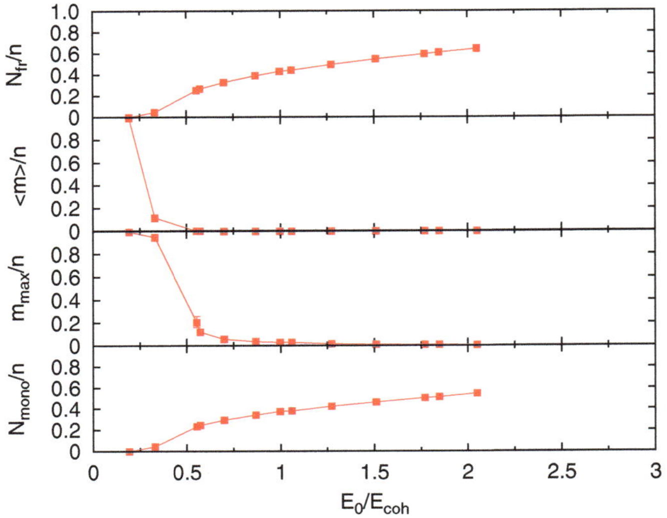

Fig. 8.5 Dependence of fragmentation parameters (including error bars) of a pure H_2O droplet containing $N = 9{,}000$ molecules at the end of the simulation time on the scaled impact energy E/E_{coh}. *From top to bottom*: number of molecules in the central fragment, N_{fr}, average fragment size, $\langle m \rangle$, the size of the largest fragment, m_{max}, and the number of free monomers created, N_{mono}. Data taken from [24]

4. While the complete disintegration of the droplet requires considerably higher impact energies, we observe that already more than 50 % of the droplet has been fully disintegrated into its molecular constituents at $E/E_{coh} = 1.7$.

The fragmentation behavior has also been studied for droplets consisting of other materials; the results—when scaling the impact energy to the cohesive energy of the droplet—are astonishingly similar [24]. It may be concluded that the cohesive energy of the material is the essential parameter describing droplet fragmentation.

8.3.3 Understanding the Fragment Formation

Theories of fragmentation usually predict exponential distributions $f(m) \propto m^{-\alpha} \exp(-bm^c)$, where $f(m)$ denotes the distribution of fragment size m. Such theories originate either by assuming fragmentation to occur in local thermodynamic equilibrium such that b depends on the local temperature achieved during fragmentation [48–50] or by interpreting fragmentation via percolation theory such that b depends on the percolation probability at the time of fragmentation [1, 51]. Both these theories predict a pure power-law decay at their respective critical points, where cluster production at all sizes is most abundant: the thermodynamic model at the critical point of the liquid–gas phase transition, where b vanishes and the power

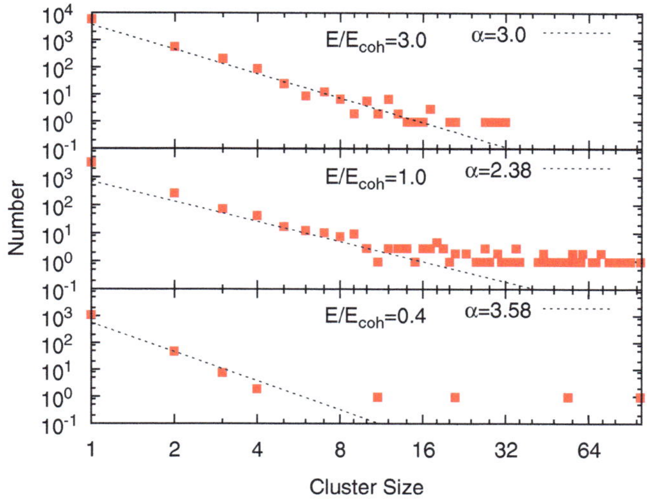

Fig. 8.6 Distribution of fragment sizes of a $(H_2O)_{9000}$ nanodroplet for three normalized impact energies E/E_{coh}. The *dashed line* is a fit to the power-law distribution, Eq. (8.12). The values of the exponent α are indicated. Data taken from [24]

exponent α obeys $2 < \alpha < 2.5$ mildly dependent on the materials properties, and the percolation model at the critical point of percolation $\alpha = 2.18$ [52].

In Fig. 8.6 we display the distribution function of fragment size $f(m)$ for three impact energies $E/E_{coh} = 0.4$, 1.0, and 3.0 for a nanodroplet with 9,000 H_2O molecules. In order to compare with theory, we fit their low-m behavior to a power-law distribution,

$$f(m) \propto m^{-\alpha}. \tag{8.12}$$

We note that the monomers were not used in this fit since their contribution is determined by evaporation rather than by the fragmentation process. It is observed that at the smallest impact energy, $E/E_{coh} = 0.4$, the distribution is steepest, while the slowest decay is observed for $E/E_{coh} = 1.0$. For larger impact energies, the distribution steepens again. The reason for this dependence is intuitively clear: At large impact energies, droplet energization is sufficient not only to tear the droplet into many pieces but also to destroy any larger fragments. Thus, both at small and at large impact energies, predominantly small clusters and a steep fragment distribution result. Hence, at intermediate energies, the maximum cluster abundance can be expected. Comparing with the theoretical value, we observe that the minimum exponent observed in our simulation, $\alpha = 2.38$ at E/E_{coh}, is in fair agreement with those of the two models just described at their respective critical points. We note that in a previous investigation of cluster formation during fragmentation of a solid under intense short-pulse laser irradiation, it was observed that cluster formation is most abundant when $E \cong E_{coh}$ [53] and can be described by a power-law distribution in the small-fragment regime [54].

In summary, our simulation results corroborate the finding of [55] that exponential distributions may be too steep to describe fragment-size distributions in droplet-wall collisions.

8.4 Collisions of Polymer-Loaded Droplets

Because the nanodroplet-wall collisions lead to droplet fragmentation, such processes can be used to disintegrate the solvent and isolate the embedded macromolecules, which are dissolved in these droplets. This is the principle of a recently developed experimental method in mass-spectrometry [12] known as IDEM. In experiment, the macromolecules are dissolved in water or methanol; droplets of this solution can be produced using the electrospray technique. Based on the knowledge discussed in the last section, we shall investigate the isolation of macromolecules from nanodroplets, induced by droplet-wall collisions.

In experiment, this method is usually employed for the isolation of biological molecules, such as proteins. Here, in order to make a more systematic investigation of the isolation process, we will consider polymers (polyketone and polyethylene), whose sizes can be easily changed in our model system. This allows us to obtain insight into the dependence of the isolation process on the size of the macromolecules. Furthermore, proteins are composed of amino acids, which have either polar (hydrophilic) or nonpolar (hydrophobic) properties. Thus, proteins offer both hydrophilic and hydrophobic aspects. We use polyketone as a polymer with purely hydrophilic character and polyethylene as an example of a purely hydrophobic molecule.

8.4.1 Desolvation

Figure 8.7 displays a series of snapshots which show the essential stages of the collision, droplet-fragmentation, and polymer-desolvation processes. Time is set to $t = 0$, when the droplet starts to feel the interaction with the wall. The collision leads to a strong vertical droplet compression, in the course of which the translational center-of-mass energy of the droplet is converted to internal energy. Similarly as for the pure droplets discussed in Sect. 8.3 above, the droplet starts to spread out laterally on the surface. This lateral-jetting stage starts here at 5 ps and is fully developed at 8 ps. In agreement with the behavior of pure droplets, this jetting process produces small solvent clusters and monomers moving sideways along the surface with high velocity, while in the central region the system is still dense and continuous. After the lateral-jetting phase, the high energy density inside the central region leads to the fragmentation of the central region, during which a wide spectrum of small water clusters and free H_2O molecules is formed. This fragmentation stage has ended at the snapshot shown at 49 ps, where the polymer

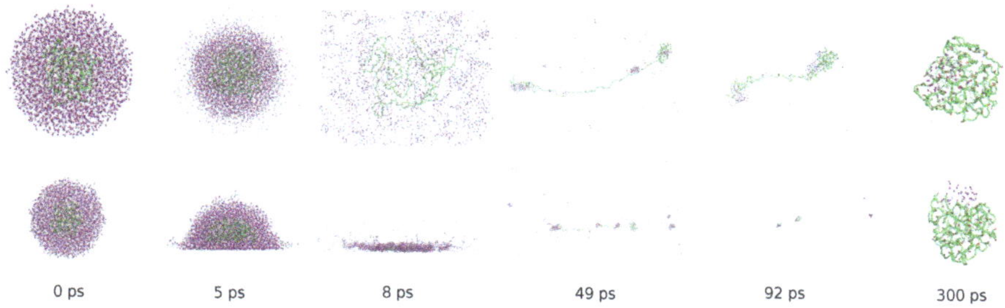

Fig. 8.7 Snapshots of the isolation process of a PK_{604} molecule embedded in a $(H_2O)_{6000}$ droplet, which hits at time $t = 0$ a hard wall with energy $E/E_{coh} = 1.275$. For these parameters, the polymer does not become completely desolvated. *Upper row*: top view, *lower row*: side view. *Purple*: water molecules. *Green*: polymer (PK). From [14]

is seen to be almost completely desolvated, while a few water clusters fly away from it. The polymer itself has undergone a considerable conformation change due to the violent collision process; its coiled structure has become unwound, while it slowly moves away from the surface. A similar dynamics has recently been observed in experiment [5]. In the next snapshot (300 ps), the polymer resumes its original coiled form, while the water on it assembles to a remaining droplet; this constitutes the *central fragment*, on which our attention is focused, because its time evolution describes the desolvation process. The changes in the number of water molecules in the central fragment occurring after 49 ps are due to water evaporation.

The impact-induced heating may become problematic for the mass-spectrometric identification, because macromolecules may be decomposed via unimolecular reactions by high temperature. Since in mass spectrometry analyte molecules are recorded typically at a few µs after the collision with the wall, the analyte molecule must survive at least a corresponding time scale (µs) in order to provide correct measurements. We shall take up this issue in Sect. 8.5.

8.4.2 Temporal Evolution

In order to understand quantitatively how the system is heated in the compression stage, we follow the temporal evolution of the temperature of polymer and solvent in Fig. 8.8a, while the evolution of the fraction f, Eq. (8.6), of water molecules still present in the central fragment bound to the polymer is displayed in Fig. 8.8b.

The droplets move with a velocity of $v = 2.50$ km/s and thus cover a distance equal to the droplet diameter (5–7 nm) in 2–3 ps. The maximum temperatures of both solvent and polymer occur at 3 ps, see Fig. 8.8a; this marks the compression stage, during which the droplet is heated by conversion of center-of-mass energy to thermal energy.

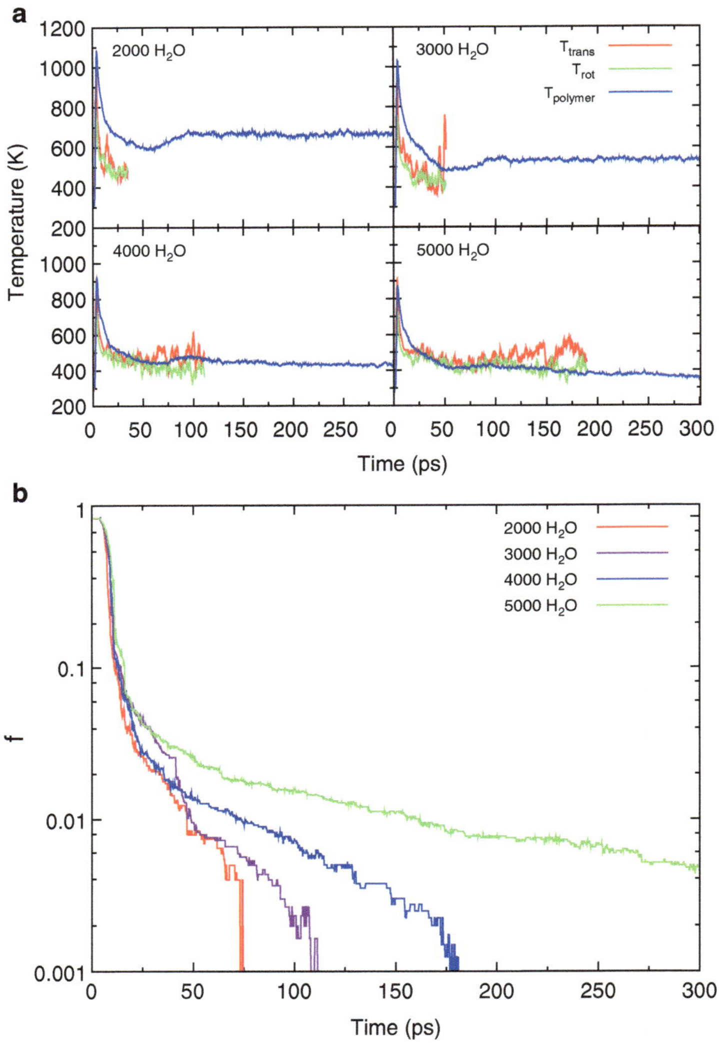

Fig. 8.8 (*Top*) Time evolution of the rotational (T_{rot}) and translational (T_{trans}) temperatures of water molecules in the central fragment, and of the polymer (PK_{604}) temperature, T_{polymer}, after collision with a wall with energy $E/E_{\mathrm{coh}} = 1.275$. This corresponds to a supersonic impact, compare Eq. (8.5). The initial size of the water droplet is indicated in the subfigures. Water temperatures are shown only as long as at least 30 water molecules are in the central fragment. (*Bottom*) Time evolution of the relative number, f, of water molecules in the central fragment. From [14]

The heating process can be described in more detail on a molecular level. The sudden stopping of the front molecules by the wall lets the succeeding molecules run into them; this not only leads to a compression of the liquid but also leads to an increase in the pressure, since molecules are at short intermolecular distances and hence within the range of the repulsive part of the intermolecular potential. The relaxation of this pressure leads to both hydrodynamic flow and heating; the flow occurs sideways (parallel to the wall) since the direction normal to the surface is blocked by the succeeding parts of the droplet. Pressure relaxation is accompanied by energy dissipation, which leads to heating. As Fig. 8.8a demonstrates, both the rotational motion of the water molecules is excited and the translational temperature, that is, the random thermal motion of the molecular centers-of-mass increases. The simultaneous rise of solvent and polymer temperature indicates the simultaneous heating of both solvent and polymer. Figure 8.8a also indicates that the maximum value of T_{polymer} decreases with increasing droplet size, assuming values of 1,100, 1,020, 925, and 892 K, respectively, for increasing droplet size. The reason for this trend is as follows: The lower part of the energy dampens the collision with the wall and transports part of the energy away by the lateral jetting process. Thus, a larger droplet provides for a softer "cushion" to dampen polymer impact onto the hard wall.

After the compression, the temperatures quickly decrease; this occurs during the lateral-jetting and multifragmentation stages described above. The initial quick decrease lasts for a time of around 20 ps. Note that the solvent temperature is *below* that of the polymer, indicating an energy flow from the polymer to the solvent; the compression-heated polymer acts as a heat source, which is cooled by the fragmenting and evaporating solvent. During the fragmentation process, bond breaking in the solvent reduces its temperature efficiently; in an equilibrium process this would be characterized as "evaporational cooling," and in this highly nonequilibrium process, the physics is similar. Note that in this regime, rotational and translational temperature of the solvent are equal, indicating that the solvent itself is thermally equilibrated.

The later fate of the system is strongly influenced by the droplet size. The smaller droplets studied ($N = 2{,}000$ and 3,000) quickly desolvate the polymer. Here, after the compression stage (at $t = 20$ ps) only few solvent molecules remain attached to the polymer—between 60 and 270, depending on the initial droplet size, compare Fig. 8.8b. This means that polymer and solvent are only in poor thermal contact; the polymer temperature is 100–200 K above that of the solvent. The strong temperature fluctuations of the solvent immediately before isolation are due to the small ensemble size. Note that the polymer temperature increases again in the time window of 50–100 ps, after contact with the solvent has been lost. This temperature increase is due to conversion of polymer configurational energy to internal kinetic energy (temperature), while the polymer resumes its coiled structure after it has been strongly stretched immediately after the collision.

For the larger droplets ($N = 4{,}000$ and 5,000), however, good thermal equilibrium is established also in the evaporation phase. Here the polymer temperature continuously decreases, while the polymer is in contact with the solvent. The polymer reheating effect described above has been strongly dampened ($N = 4{,}000$) or is even entirely missing ($N = 5{,}000$) due to the thermal contact with the solvent.

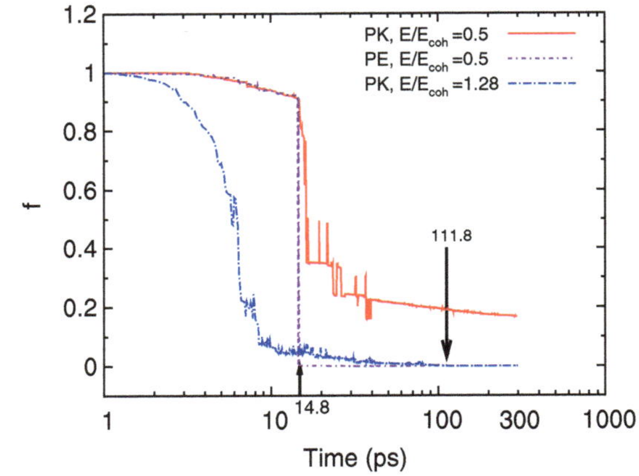

Fig. 8.9 Time evolution of the relative number, f, of water molecules in the central fragment for PE and PK (each with 298 atoms) embedded in a $(H_2O)_{3000}$ droplet at various impact energies. The *arrows* denote times of isolation and are discussed in the text. From [14]

Complete polymer isolation occurs at 81.6, 113.7, 183.4 ps for droplet sizes of 2,000, 3,000, 4,000 molecules; the $N = 5{,}000$ droplet does not completely desolvate the polymer during the 300 ps simulation time.

8.4.3 *Comparison Between Hydrophilic and Hydrophobic Macromolecules*

The time evolution of the relative size, f, of the central fragment also allows to characterize the influence of the nature of the polymer–solvent interaction on the desolvation process. For this purpose, we compare in Fig. 8.9 the evolution of the desolvation of a hydrophobic PE with that of a hydrophilic PK polymer. All other parameters (droplet size $N = 3{,}000$, energy $E/E_{\text{coh}} = 0.5$, size of polymer 298 atoms) were kept unchanged.

The time evolution is identical until a time of 14.8 ps, where the droplet has lost only 8 % of its molecules by lateral jetting. Then the fragmentation process sets in, and instantaneously completely desolvates PE at 14.8 ps, while still around 30 % of the solvent is bound to PK. At this time, the droplet has been spread out onto the wall (see the side view shown in Fig. 8.7 at time $t = 8$ ps) and is torn apart laterally. The fragmentation leads to desolvation since the polymer–water binding forces are weaker than the water–water bonds; then the fragmenting water cloud prefers to keep bonds among themselves and loosens its bonds with the polymer. The desolvation process occurs so suddenly at this low impact energy because the expansion of the water shell is radially symmetric such that the water decouples more or less simultaneously from the polymer. In the course of time, PK sheds off more water molecules, partly in the form of larger fragments, and eventually by continuous evaporation. At the end of the simulation 16 % of the original molecules remained bound to PK, while PE already has desolvated at 14.8 ps. The "spikes" seen are due to short-lived droplet–droplet interactions in the still quite dense vapor

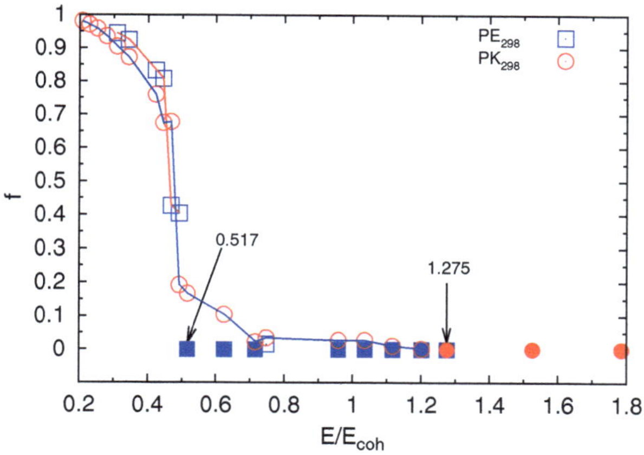

Fig. 8.10 Dependence of the relative number, f, of water molecules in the central fragment at the end of the simulations ($t = 300$ ps) in dependence of the scaled impact energy $E/E_{\rm coh}$. Initially, the droplet contained 3,000 water molecules and a PK or PE polymer with 298 atoms. *Filled symbols* indicate an isolated polymer, $f = 0$. From [14]

cloud. This comparison shows clearly the decisive influence of the solvent–polymer interaction on the desolvation process.

We quantify the solvent–polymer interaction by an interaction energy $E_{\rm ss}$. We calculate it by embedding the polymer in bulk solvent; after an energy minimization we determine the total solvent–polymer interaction energy by summing the respective van-der-Waals and electrostatic terms; $E_{\rm ss}$ is then obtained by dividing the total interaction energy by the number of solvent molecules within the cutoff radius (8.5 Å) of the polymer. We obtain $E_{\rm ss} = 33.45$ and 15.65 meV for PK and PE, respectively, in water.

We find that PK only becomes isolated at higher impact energies, $E/E_{\rm coh} = 1.28$, see Fig. 8.10. We include also the time evolution of the central fragment for this high impact energy in Fig. 8.9. Here complete isolation is achieved after 111.8 ps. Of course, due to the higher droplet impact speed, water loss occurs earlier, starting at 3 ps, with large losses due to fragmentation at 7 ps. Already at 10 ps, the central fragment only contains 6 % of its initial size, and this size decreases continuously by evaporation until complete isolation.

In Fig. 8.10, we display the dependence of the relative number, f, of water molecules in the central fragment on the impact energy E and compare droplets containing a PE and a PK polymer. While collisions with energies $\leq 0.4 E_{\rm coh}$ disperse the droplet only a little ($\leq$20 % of its water molecules are lost), an increase of the impact energy to $0.5 E_{\rm coh}$ strongly shatters the droplet, such that less than 20 % of the initial number of molecules remain bound to the polymer. As discussed above, the PE is then completely isolated, while the hydrophilic PK still binds water to it. From the change in slope of the $f(E)$ data, we observe that at $E/E_{\rm coh} \cong 0.5$ the PK isolation enters a new regime, namely, the evaporation of the remaining water

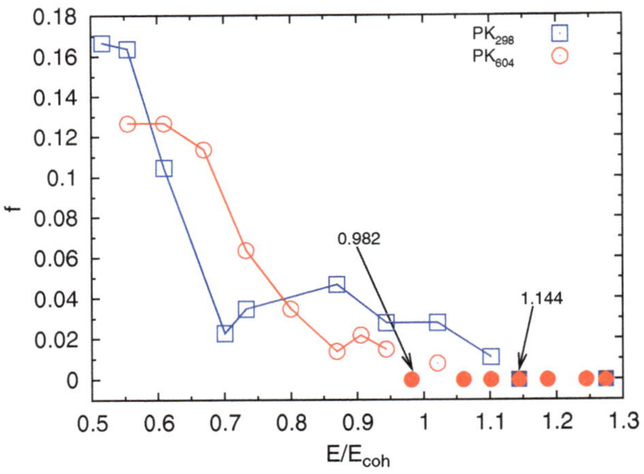

Fig. 8.11 Dependence of the relative number, f, of water molecules in the central fragment at the end of the simulation ($t = 300\,\text{ps}$) on the scaled impact energy E/E_{coh}. Initially the droplet contained 3,000 water molecules and a PK polymer with 298 or 604 atoms. *Filled symbols* indicate $f = 0$. From [14]

molecules from the PK molecule. This regime is missing for the PE molecules, because here all water molecules have been shed off the polymer during the fragmentation stage.

8.4.4 *Effect of Polymer Size*

In order to explore the effect of polymer size on the critical isolation energy E_{isol}, we compare the desolvation of PK with 298 and 604 atoms, respectively, from a water droplet. The dependence of the relative number of solvent atoms in the central fragment on impact energy is displayed in Fig. 8.11. We concentrate on the energy range above $E/E_{\text{coh}} = 0.5$, where the interesting physics of polymer isolation happens.

We find that, on average, the *smaller* polymer binds *more* solvent molecules and, consequently, has a higher isolation energy, $E_{\text{isol}}/E_{\text{coh}} = 1.14$ as compared to 0.98 for the larger polymer. However, the effect is not pronounced; it only amounts to 16 % for a threefold change in polymer mass. The main reason for the easier isolation of the large polymer is the extra amount of translational energy it carries with it, and which can be used to heat and fragment the droplet and desolvate the polymer. In addition, a larger polymer can receive more thermal energy in the collision and store it so as to reduce the evaporational cooling effect of the solvent. Furthermore a larger polymer will release more solvent molecules during the collision-induced unfolding process, compare Fig. 8.7.

8.5 Collision-Induced Heating of the Macromolecule

In this section we focus on the heating of polymers during the impact-induced processes, and its dependence on the sizes of polymer and of the droplet as well as the impact energy [25]. As discussed in the last section, the translational kinetic energy is converted to thermal energy in the polymer during the compression stage, and the polymer can exchange energy with the surrounding solvent molecules before it becomes isolated. If the thermal energy of the polymer exceeds the activation energy for unimolecular decay, it will be decomposed to fragments and reach a mass spectrometer in the form of fragments. This motivates us to investigate the dependence of the polymer temperature at the end of the isolation process on the above-mentioned quantities.

8.5.1 Factors Affecting the Polymer Temperature

We summarize the dependence of the equilibrated polymer temperature at the end of the simulation, T_{end}, and its peak value during the compression phase, T_{max}, on polymer size N_p in Fig. 8.12. Two droplet sizes with $N = 3{,}000$ and $N = 6{,}000$ are compared. We find a roughly linear dependence of both T_{max} and T_{end} on N_p. Thus, for the same impact velocity v and droplet size N, the peak temperature of the polymer, T_{max}, is only affected by N_p. The final temperature, T_{end}, results from the cooling of T_{max} by the polymer–solvent coupling and the internal energy redistribution described above. Figure 8.12 also demonstrates that larger droplets,

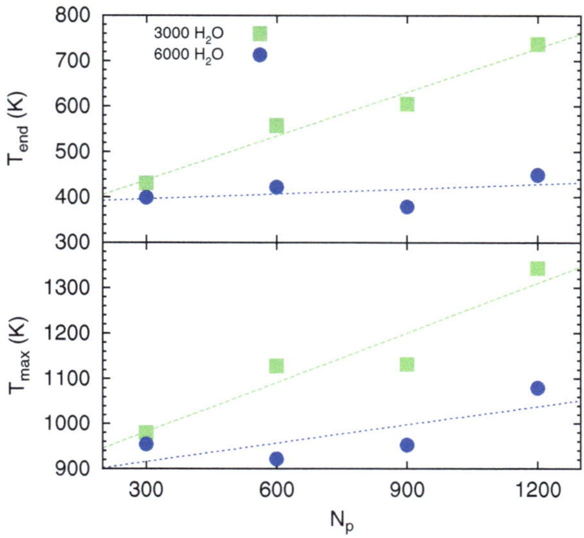

Fig. 8.12 Final temperature T_{end} (*top*) and maximum temperature T_{max} (*bottom*) of polyketone molecule in dependence of their number of atoms N_p. Nanodroplets containing $N = 3{,}000$ and $N = 6{,}000$ H_2O molecules are compared. The collision speed with the wall was $v = 2.5\,\mathrm{km\,s^{-1}}$, the *lines* are linear fits to the data. From [25]—Reproduced by permission of the Royal Society of Chemistry

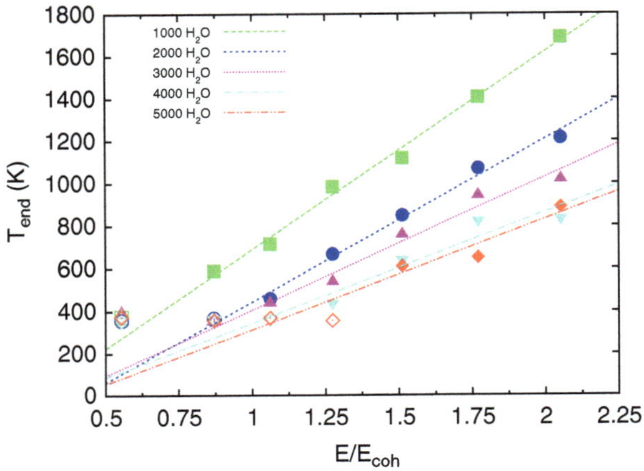

Fig. 8.13 The polymer temperature $T_{\rm end}$ at the end of the simulations in dependence of the scaled impact energy per solvent molecule, $E/E_{\rm coh}$. The simulations were performed with $N_{\rm p} = 600$. The *lines* are linear fits to the data. The *filled symbols* indicate successful isolation, *open symbols* indicate cases where the polymer is still bound to H_2O molecules. From [25]—Reproduced by permission of the Royal Society of Chemistry

N, reduce both $T_{\rm max}$ and $T_{\rm end}$. This appears plausible since a larger number of H_2O molecules can dampen more efficiently the droplet-wall collision, such that in the compression phase less energy is conveyed to the polymer; additionally, a larger amount of water molecules will take more energy from the polymer in the post-compression phase.

The dependence of the final polymer temperature, $T_{\rm end}$, on the scaled impact energy, $E/E_{\rm coh}$, is shown in Fig. 8.13; also here a linear dependence can be observed. At small impact energies and for large droplets, the polymer is not fully isolated (open symbols in Fig. 8.13); here $T_{\rm end}$ assumes a rather constant value of around 400 K. This value is determined by the boiling temperature of water, since the polymer is still coupled to water and water can efficiently cool by evaporation. After the polymer becomes isolated, $T_{\rm end}$ increases linearly with the impact energy E. This implies that for the same droplet size N, the solvent always receives *the same fraction* of the initial energy from the polymer.

However, the increase of $T_{\rm end}$ with E strongly depends on the droplet size N. In large droplets, $T_{\rm end}$ will not increase much with impact energy because more energy is transferred from the polymer to the solvent. In view of the complexity of the polymer isolation process, it is astonishing to find a simple linear dependence of $T_{\rm end}$ on $E/E_{\rm coh}$.

8.5.2 Polymer Stability After the Impact Process

The rate of unimolecular reactions of the isolated hot polymer after the end of the simulation time can be calculated using the RRKM theory [56] as

$$k_{\rm RRKM} = \frac{1}{2\pi} \frac{\Sigma(E_{\rm kin} - E_0)}{\rho(\omega)}. \tag{8.13}$$

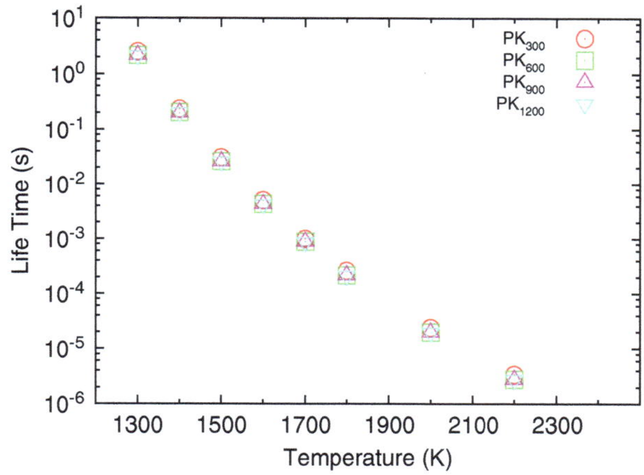

Fig. 8.14 The lifetime of polyketone molecules with various sizes, N_p, as a function of their temperature. From [25]—Reproduced by permission of the Royal Society of Chemistry

Here E_0 denotes the activation energy for the unimolecular decay (we set it equal to the dissociation energy D of the C–C bonds in the polymer) and $\sum(E_{kin} - E_0)$ denotes the number of transition states between energy 0 and $(E_{kin} - E_0)$. The density of states $\rho(\omega)$ can be determined using the method described in [25]; $\rho(\omega)$ is calculated at 300 K.

The polymer lifetime is shown as a function of $T_{polymer}$ in Fig. 8.14. We find that the lifetime does not depend sensitively on N_p. For temperatures above 2,200 K, it approaches the µs time scale. Since mass-spectrometric experiments are performed on this time scale, this result shows that polymer temperatures above 2,200 K need to be avoided. This value sets an upper limit to the impact energies that can be used for molecule desolvation.

8.6 Conclusions

We have investigated the fragmentation behavior of pure nanodroplets and of polymer-loaded droplets. In both cases, the processes occurring after impact on a repulsive wall are similar and one can identify four development stages: compression, lateral jetting, cluster formation, and evaporation. Among these, the third (cluster formation) stage is essential for disintegrating the droplet. Quantitatively, we find:

1. The droplet fragmentation starts already at low impact energies per molecule, $E = (0.29–0.4)E_{coh}$, where E_{coh} is the cohesive energy of the solvent.
2. For an impact energy of $E = E_{coh}$, the droplet fragmentation is already fully developed, resulting in a fine dispersal of the droplet into fragments. If a polymer is dissolved in the droplet, it will not yet be fully isolated.
3. The distribution of cluster sizes produced by droplet fragmentation obeys a power law, in good agreement with theoretical predictions.

4. The hydrophilic properties of macromolecules have significant influence on the isolation process. Hydrophilic polymers attract H_2O molecules to stay bound to them during the droplet fragmentation. While they boil off some solvent molecules after the fragmentation stage ended, evaporation is a slow process and the polymer may not be isolated successfully. Hydrophobic polymers will be isolated quite promptly after the fragmentation stage started, on a time scale of only a few 10 ps.
5. Smaller polymers are more difficult to isolate since they receive less energy during the impact, and thus have less chances to boil off remaining solvent molecules after the droplet fragmentation.
6. While polymers may become heated dramatically during the impact, their temperature decreases thereafter due to evaporation cooling by the solvent.
7. The temperature of an isolated polymer increases linearly with the impact energy. It also increases with the size of the polymer.
8. Larger nanodroplets will more efficiently cool the polymer after the impact, and hence the polymer isolation process will last longer, and also the final polymer temperature, T_{end}, will be lowered.
9. For mass-spectrometric applications, there exists a large window in impact energy to use, since the lifetime of the polymer molecule after isolation lies in the order of μs. Suitable impact energies are above $E/E_{coh} > 1$.

Furthermore, we argue that the isolation processes for embedded proteins are similar to those of polymers, since proteins contain both hydrophobic and hydrophilic groups and thus are intermediate between our two polymer models.

References

1. Yarin, A.L.: Annu. Rev. Fluid Mech. **38**, 159 (2006)
2. Harbich, W.: In: Metal Clusters at Surfaces: Structure, Quantum Properties, Physical Chemistry. Springer Series in Cluster Physics, p. 107. Springer, Berlin (2000)
3. Yamada, I., Matsuo, J., Insepov, Z., Aoki, T., Seki, T., Toyoda, N.: Nucl. Instrum. Methods B **164–165**, 944 (2000)
4. Christen, W., Even, U.: J. Phys. Chem. A **102**, 9420 (1998)
5. Smith, M.I., Bertola, V.: Phys. Rev. Lett. **104**, 154502 (2010)
6. Jarrold, M.F.: Annu. Rev. Phys. Chem. **51**, 179 (2000)
7. Melchionna, S., Briganti, G., Londei, P., Cammarano, P.: Phys. Rev. Lett. **92**, 158101 (2004)
8. Sogbein, O.O., Simmons, D.A., Konermann, L.: J. Am. Soc. Mass Spectrom. **11**, 312 (2000)
9. Sadeghi, M., Wu, X., Vertes, A.: J. Phys. Chem. B **105**, 2578 (2001)
10. Patriksson, A., Marklund, E., van der Spoel, D.: Biochemistry **46**, 933 (2007)
11. Mahoney, J.F., Perel, J., Ruatta, S.A., Martino, P.A., Husain, S., Cook, K., Lee, T.D.: Rapid Commun. Mass Spectrom. **5**, 441 (1991)
12. Aksyonov, S.A., Williams, P.: Rapid Commun. Mass Spectrom. **15**, 2001 (2001)
13. Eggers, J., Fontelos, M.A., Josserand, C., Zaleski, S.: Phys. Fluids **22**, 062101 (2010)
14. Sun, S.N., Urbassek, H.M.: J. Phys. Chem. B **115**, 13280 (2011)
15. Thirumuruganandham, S.P., Urbassek, H.M.: Int. J. Mass Spectrom. **289**, 119 (2010)
16. Cleveland, C.L., Landman, U.: Science **257**, 355 (1992)

17. Raz, T., Even, U., Levine, R.D.: J. Chem. Phys. **103**, 5394 (1995)
18. Christen, W., Even, U., Raz, T., Levine, R.D.: Int. J. Mass Spectrom. Ion Proc. **174**, 35 (1998)
19. Christen, W., Even, U., Raz, T., Levine, R.D.: J. Chem. Phys. **108**, 10262 (1998)
20. Zimmermann, S., Urbassek, H.M.: Eur. Phys. J. D **39**, 423 (2006)
21. Zimmermann, S., Urbassek, H.M.: Phys. Rev. A **74**, 063203 (2006)
22. Tomsic, A., Schröder, H., Kompa, K.L., Gebhardt, C.R.: J. Chem. Phys. **119**, 6314 (2003)
23. Tomsic, A., Andersson, P.U., Markovic, N., Pettersson, J.B.C.: J. Chem. Phys. **119**, 4916 (2003)
24. Sun, S.N., Urbassek, H.M.: Phys. Rev. E **84**, 056315 (2011)
25. Sun, S.N., Urbassek, H.M.: Soft Matter **8**, 4708 (2012)
26. Berendsen, H.J.C., Grigera, J.R., Straatsma, T.P.: J. Phys. Chem. **91**, 6269 (1987)
27. Jorgensen, W.L., Chandrasekhar, J., Madura, J.D., Impey, R.W., Klein, M.L.: J. Chem. Phys. **79**, 926 (1983)
28. Jorgensen, W.L., Madura, J.D.: Mol. Phys. **56**, 1381 (1985)
29. Vega, C., Abascal, J.L.F., Sanz, E., MacDowell, L.G., McBride, C.: J. Phys.: Condens. Matter **17**, S3283 (2005)
30. Robinson, G.W., Zhu, S.B., Singh, S., Evans, M.W.: Water in Biology, Chemistry and Physics: Experimental Overviews and Computational Methodologies. World Scientific, Singapore (1996)
31. Guillot, B.: J. Mol. Liq. **101**, 219 (2002)
32. Fennell, C.J., Gezelter, J.D.: J. Chem. Phys. **124**, 234104 (2006)
33. Dokter, A.M., Woutersen, S., Bakker, H.J.: Phys. Rev. Lett. **94**, 178301 (2005)
34. Delcorte, A., Garrison, B.J.: J. Phys. Chem. C **111**, 15312 (2007)
35. Jorgensen, W.L., Tirado-Rives, J.: Proc. Natl. Acad. Sci. USA **102**, 6665 (2005)
36. van der Spoel, D., Lindahl, E., Hess, B., Groenhof, G., Mark, A.E., Berendsen, H.J.C.: J. Comp. Chem. **26**, 1701 (2005)
37. Sun, S.N., Urbassek, H.M.: Appl. Phys. A **101**, 71 (2010)
38. Johnson, R.E.: Energetic Charged-Particle Interactions with Atmospheres and Surfaces. Springer, Berlin (1990)
39. Lemmon, E.W., McLinden, M.O., Friend, D.G.: In: Linstrom, P.J., Mallard, W.G. (eds.) NIST Chemistry WebBook, NIST Standard Reference Database Number 69. National Institute of Standards and Technology, Gaithersburg (2011). Retrieved from 2 October 2011: http://webbook.nist.gov
40. Miyamoto, S., Kollman, P.A.: J. Comput. Chem. **13**, 952 (1992)
41. Stoddard, S.D.: J. Comput. Phys. **27**, 291 (1978)
42. Bussmann, M., Chandra, S., Mostaghimi, J.: Phys. Fluids **12**, 3121 (2000)
43. Sotolongo-Costa, O., Moreno-Vega, Y., Lloveras-González, J.J., Antoranz, J.C.: Phys. Rev. Lett. **76**, 42 (1996)
44. Strachan, A., Dorso, C.O.: Phys. Rev. C **55**, 775 (1997)
45. Strachan, A., Dorso, C.O.: Phys. Rev. C **58**, R632 (1998)
46. Kun, F., Herrmann, H.J.: Phys. Rev. E **59**, 2623 (1999)
47. Katsuragi, H., Sugino, D., Honjo, H.: Phys. Rev. E **68**, 046105 (2003)
48. Fisher, M.E.: Physics **3**, 255 (1967)
49. Fisher, M.E.: Rep. Prog. Phys. **30**, 615 (1967)
50. Urbassek, H.M.: Nucl. Instrum. Methods B **31**, 541 (1988)
51. Yarin, A.L.: Free Liquid Jets and Films: Hydrodynamics and Rheology. Longman Scientific & Technical, Harlow (1993)
52. Stauffer, D., Aharony, A.: Introduction to Percolation Theory, 2nd edn. Taylor and Francis, London (1994)
53. Colla, T.J., Urbassek, H.M.: Comput. Mater. Sci. **6**, 7 (1996)
54. Upadhyay, A.K., Urbassek, H.M.: Phys. Rev. B **73**, 035421 (2006)
55. Tomsic, A., Gebhardt, C.R.: J. Chem. Phys. **123**, 064704 (2005)
56. Baer, T., Mayer, P.M.: J. Am. Soc. Mass Spectrom. **8**, 103 (1997)

Chapter 9
Polymer Films with Nanosized Liquid-Crystal Droplets: Extinction, Polarization, Phase, and Light Focusing

Valery A. Loiko

Abstract Extinction and polarization state of light transmitted through a polymer-dispersed liquid-crystal (PDLC) film with nanosized, spherical, nonabsorbing nematic droplets is investigated theoretically. Scattering properties of a single droplet are described by the Rayleigh–Gans approximation. Propagation of coherent light field is described in the frame of the Foldy–Twersky theory. The concept of multilevel order parameters is used. Equations for extinction coefficients, phase shift, and polarization of transmitted light for layers with random and oriented droplets are written and discussed. Conditions for circular and linear polarization of light are determined and investigated. Polarization-independent focusing of light by films made of PDLC is considered. The results are in good coincidence with the known experimental data.

9.1 Introduction

Matter that has properties of a liquid and of a solid crystal is named liquid crystal (LC). They consist of anisotropic molecules, typically rodlike and disklike shapes. Liquid crystals flow like a liquid and its molecules can be oriented in a crystal-like way. Molecule orientation order is sensitive to mechanical strength; thermal, magnetic, and electrical fields; and other external factors. Owing to that these materials are of very considerable promise for science and technology. Liquid crystals are also known as anisotropic fluids and mesomorphic materials [1–3].

There is a set of types of LC materials, which are distinguished by their different properties. There are the ones where liquid crystal is dispersed in a polymer matrix. They are known as polymer-dispersed liquid crystals (PDLCs). This is a polymer

V.A. Loiko (✉)
B.I. Stepanov Institute of Physics of the National Academy of Sciences of Belarus,
68 Nezalezhnasti Avenue, Minsk 220072, Belarus
e-mail: loiko@dragon.bas-net.by

Z.M. Wang (ed.), *Nanodroplets*, Lecture Notes in Nanoscale Science and Technology 18,
DOI 10.1007/978-1-4614-9472-0_9,

film with embedded droplets sandwiched between two transparent plates with transparent electrodes. Such sandwiched structure is named as an LC cell. Under an external field applied molecules of LC change their orientation. As a result, its optical properties are changed. Thus, characteristics of light transmitted through the cell (or light reflected by the cell) vary with the applied field. By selecting proper materials and methods of the PDLC cell preparation, a lot of electro-optical devices based on this effect are designed [4–7]. Nevertheless these materials are still under investigation.

While bulk liquid-crystal cells typically need crossed polarizers to get electro-optical effect [1–3] (but there are some exceptions [8–11]), PDLC films typically do not require polaroids for producing that effect. They are characterized by higher light stability, mechanical strength, flexibility, and resistance to external actions compared with the devices based on bulk LCs. Their thickness is in the range from a few micrometers to a few tens of micrometers. The size of droplets is in the range from ten nanometers to tens of micrometers.

Films with droplets compared with the wavelength of the incident radiation exhibit strong light scattering. Their transmission depends on the applied field rotating optical axes (directors) of the LC droplets and/or changing the internal structure of these particles [12, 13]. A lot of installations for modulating the intensity of visible [6, 7] and infrared [4, 5, 7, 14, 15] ranges are constructed.

Films with nanosized liquid-crystal droplets exhibit small scattering. They are usually characterized by high transmittance, which is not strongly influenced by changes of optical axes of droplets. As a result the description of light propagation reduces in many cases to the analysis of coherent (directed) [16] component of the transmitted radiation. Here, the main field effect consists in a change in the effective refractive index (similar to the Kerr effect) under the action of an applied electric field [17]. They enable one to modulate phase [18] and polarization [19, 20] of transmitted radiation. Thanks to this, the cells possess a wide range of applications, from lenses [21, 22], lens arrays [23] with tunable focal length, and other optical elements to diffusive random lasers [24] and vertical cavity surface-emitting lasers [25].

In any liquid-crystal elementary volume, there is a preferred orientation of molecules. A unit vector indicating this direction is called "director." There is a set of stable director configurations in liquid-crystal droplets. They depend on the boundary conditions on the surface of droplet. Two groups are separated: tangential (parallel) and homeotropic (perpendicular) orientation of director to the droplet surface.

A bipolar configuration is shown schematically in Fig. 9.1. There are two singular points at opposite poles of a droplet. In this configuration, LC molecules are aligned along the drawn lines from one pole to another and form a cylindrically symmetric structure. The symmetry axis is defined by two point defects of the LC director located on the droplet surface. The unit vector along this axis is called the director of the droplet. It indicates the direction of its optical axis. Just this kind of droplets is under consideration in here. In the applied electric field, the droplet director tends to change its orientation, while the elastic forces oppose to this alignment.

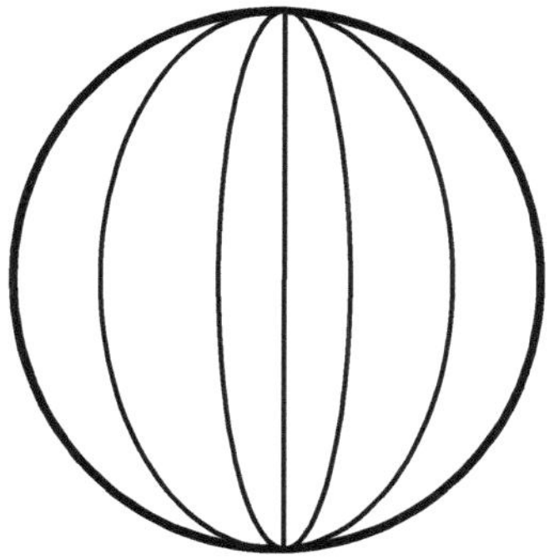

Fig. 9.1 Bipolar director configuration in a droplet. Directors are aligned along the drawn lines

Rigorous theory establishing connection between the morphological properties of a PDLC film and the characteristics of transmitted light is still under development, although the main principles of their operation are rather clear. Rayleigh and Rayleigh–Gans approaches, anomalous diffraction approximation (ADA), Wentzel–Kramers–Brillouin (WKB) approach, discrete dipole approximation (DDA), and the Mie theory [4, 5, 26–33] are used to describe scattering by a separate droplet.

In this chapter we consider the passage of light through a layer with nematic nanosized bipolar droplets. The model [34, 35] is based on the Rayleigh–Gans approximations [36, 37] to describe scattering by a single droplet and the Foldy–Twersky theory [38, 39] to describe coherent field propagation in the layer. The concept of multilevel order parameters [5, 26] is employed to characterize orientation of LC molecules inside droplets and droplets in a sample.

The general equations for light propagation in the cells with nanosized liquid-crystal droplets with random and partially ordered droplet directors are written and analyzed. Intensity, extinction, phase, and polarization of transmitted light are investigated. The attention is paid to the conditions, where circular and linear polarization of transmitted light is implemented. Characteristics of polarization-insensitive lenses are investigated.

9.2 Amplitude Scattering Matrix of Droplet

At first consider a single liquid-crystal droplet. Let the droplet be illuminated by a linearly polarized monochromatic plane wave (Fig. 9.2). Every small elementary volume dV of the droplet reemits an incident electromagnetic wave. In the far zone, these waves add up with due regard for the phase difference arising as a result of different arrangements of scattering elementary volumes in the droplet. Therefore, the radiation propagates at angles that are absent in the incident wave (i.e., there occurs light scattering) with a change in the polarization state.

The unit vector **d** indicates the direction of the orientation of the optic axes of liquid-crystal molecules in this elementary volume. This vector is referred to as the

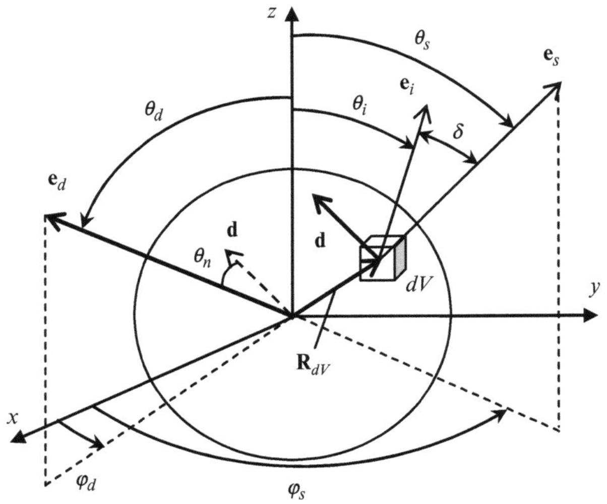

Fig. 9.2 Schematic drawing of the liquid-crystal droplet and the elementary scattering volume dV illumination. Vector $\mathbf{R}_{dV}$ specifies the location of this volume with respect to the droplet center; $\mathbf{e}_\mathrm{i}$ and $\mathbf{e}_\mathrm{s}$ are the unit vectors in the directions of propagation of the incident and scattered waves, respectively; $\mathbf{e}_\mathrm{d}$ is the droplet director; $\mathbf{d}$ is the director of the elementary volume of the droplet; angle θ_n specifies the deviation of the director of the elementary volume of the droplet from the director of the droplet $\mathbf{e}_\mathrm{d}$; angles θ_s and θ_d define orientations of the vectors $\mathbf{e}_\mathrm{s}$ and $\mathbf{e}_\mathrm{d}$ with respect to the axis z; angles φ_s and φ_d define orientations of the projections of vectors $\mathbf{e}_\mathrm{s}$ and $\mathbf{e}_\mathrm{d}$ onto the plane (xy) with respect to axis x, respectively

director of the elementary volume in the liquid-crystal droplet. Because there is no physical polarity along the director axis, vectors $\mathbf{d}$ and $-\mathbf{d}$ are equivalent. The unit vector $\mathbf{e}_\mathrm{d}$ is the director of the liquid-crystal droplet.

It specifies the average orientation of the liquid-crystal director in the droplet. The direction of propagation of the incident wave is designated by the unit vector $\mathbf{e}_\mathrm{i}$. In every elementary volume dV, the incident light wave induces the dipole moment:

$$\delta\mathbf{P} = \varepsilon_\mathrm{p}\widehat{\alpha}_{dV}\mathbf{E}_\mathrm{i}\exp\left(ik\mathbf{R}_{dV}\cdot\mathbf{e}_\mathrm{i}\right), \tag{9.1}$$

where ε_p is the permittivity of the polymer, $\widehat{\alpha}_{dV}$ is the polarizability tensor of the elementary volume of the liquid crystal in the coordinate system related to the incident wave, $\mathbf{E}_\mathrm{i}$ is the electric vector of the incident wave, $k = 2\pi/\lambda_\mathrm{p}$ is the module of the wave vector in the polymer matrix, and λ_p is the wavelength in polymer (polymer binder).

In the far zone the electric vector of the field scattered by the elementary volume of the liquid-crystal droplet in the direction $\mathbf{e}_\mathrm{s}$ can be written as follows:

$$\delta\mathbf{E}_\mathrm{s} = -\frac{k^2}{4\pi r}\exp\left(ik\left(\mathbf{r}+\mathbf{R}_{dV}\left(\mathbf{e}_\mathrm{i}-\mathbf{e}_\mathrm{s}\right)\right)\right)\mathbf{e}_\mathrm{s}\times\left[\mathbf{e}_\mathrm{s}\times\left[\mathbf{e}_\mathrm{s}\times\widehat{\alpha}_{dV}\mathbf{E}_\mathrm{i}\right]\right]. \tag{9.2}$$

Here r is the distance from the droplet center to the observation point.

The amplitude scattering matrix of the elementary volume of the droplet is determined by the relation [36, 37]:

$$\begin{pmatrix} \delta \mathbf{E}_s \cdot \mathbf{e}_1^s \\ \delta \mathbf{E}_s \cdot \mathbf{e}_2^s \end{pmatrix} = -\frac{\exp(ikr)}{ikr} \begin{pmatrix} \delta S_{11} & \delta S_{12} \\ \delta S_{21} & \delta S_{22} \end{pmatrix} \begin{pmatrix} \mathbf{E}_i \cdot \mathbf{e}_1^i \\ \mathbf{E}_i \cdot \mathbf{e}_2^i \end{pmatrix}. \tag{9.3}$$

Here the unit vectors $\mathbf{e}_1^s$, $\mathbf{e}_2^s$ specify any two mutually perpendicular directions of the polarization of the scattered wave; the three vectors $\mathbf{e}_1{}^s$, $\mathbf{e}_2{}^s$, $\mathbf{e}_s$ form a right-hand triad. The unit vectors $\mathbf{e}_1^s$, $\mathbf{e}_2^s$ specify the mutually perpendicular directions of the polarization of the incident wave, and the three vectors $\mathbf{e}_1^i$, $\mathbf{e}_2^i$, $\mathbf{e}_i$ also form a right-hand triad.

By using relationships (9.1) and (9.2), we can express the amplitude scattering matrix elements as follows:

$$\delta S_{\alpha\beta} = -\frac{ik^3}{4\pi} \exp(i\psi) \left(\mathbf{e}_\alpha^s \widehat{\alpha}_{dV} \mathbf{e}_\beta^i \right), \tag{9.4}$$

where $\psi = k\mathbf{R}_{dV}(\mathbf{e}_i - \mathbf{e}_s)$; $\alpha, \beta = 1, 2$.

In order to determine the elements of the amplitude scattering matrix for the entire liquid-crystal droplet, it is necessary to integrate expression (9.4) over the droplet volume. The quantities dependent on the location of the elementary volume in the droplet are the phase factor $\exp(i\psi)$ and the polarizability tensor $\widehat{\alpha}_{dV}$. Determination of the polarizability tensor $\widehat{\alpha}_{dV}$ in each individual elementary volume is a complex and tedious problem. To simplify the problem, we use the concept of the orientational order parameters [26]. On the basis of this concept, we introduce the average tensor $\widehat{\alpha}$, so that it is a constant over the entire droplet volume. Tensor components depend on the configuration of the liquid-crystal molecules inside the droplet.

9.3 Average Polarizability Tensor of a Liquid-Crystal Droplet

The permittivity tensor and the polarizability tensor for the elementary volume of the liquid-crystal droplet are related by the equation:

$$\widehat{\alpha}_{dV} = \left(\frac{1}{\varepsilon_p} \widehat{\varepsilon}_{dV} - \widehat{\mathrm{I}} \right) dV, \tag{9.5}$$

where $\widehat{\mathrm{I}}$ is the 3-by-3 unit matrix.

In an arbitrary coordinate system, the permittivity tensor components can be written as

$$\varepsilon_{ij} = \varepsilon_\perp \delta_{ij} + \left(\varepsilon_\parallel - \varepsilon_\perp \right) d_i d_j, \tag{9.6}$$

where d_i and d_j are the components of the director of the elementary volume in this coordinate system; δ_{ij} is the Kronecker symbol; $i, j = 1, 2, 3$; and $\varepsilon_{\parallel}$ and $\varepsilon_{\perp}$ are the principal values of the permittivity tensor of the liquid crystal (the elements of the dielectric permittivity parallel and perpendicular to the elementary volume director **d**).

To study the response of PDLCs, we use the following order parameters [5, 26]: the molecular order parameter S, which takes into account the orientation of molecular directors due to the thermal fluctuations, and the order parameter of the liquid-crystal droplet S_{d}, which determines the degree of orientation of droplet molecules with respect to the droplet director $\mathbf{e}_{\mathrm{d}}$ (this order parameter is related to the director configuration within the droplet).

The product SS_{d} is a measure of the orientational order of the LC molecules in the droplet:

$$SS_{\mathrm{d}} = \frac{1}{2}\left\langle 3\cos^2\theta_{\mathrm{n}} - 1\right\rangle, \tag{9.7}$$

where the angle θ_{n} specifies the deviation of the director of the elementary volume of the liquid-crystal droplet from the droplet director (Fig. 9.2) and the angle brackets indicate averaging over the orientations of the liquid-crystal molecules in the droplet. The value of $SS_{\mathrm{d}} = 1$ corresponds to the complete molecules ordering, and the order parameter $SS_{\mathrm{d}} = 0$ corresponds to the random orientation of molecules.

Note that we consider droplets with the bipolar configuration. Liquid-crystal molecules are oriented in a cylindrically symmetric manner with respect to the droplet director $\mathbf{e}_{\mathrm{d}}$ (Fig. 9.1). The average permittivity tensor elements of the droplet in the coordinate system related to the droplet director as follows from relationships (9.6) and (9.7) are

$$\varepsilon_{11} = \varepsilon_{22} = \frac{1}{3}\left(2\varepsilon_{\perp} + \varepsilon_{\parallel} - \Delta\varepsilon_{\mathrm{d}}\right) \equiv \varepsilon_{\mathrm{do}}, \tag{9.8}$$

$$\varepsilon_{33} = \frac{1}{3}\left(2\varepsilon_{\perp} + \varepsilon_{\parallel} + 2\Delta\varepsilon_{\mathrm{d}}\right) \equiv \varepsilon_{\mathrm{de}}. \tag{9.9}$$

Here by $\varepsilon_{\mathrm{do}}$ and $\varepsilon_{\mathrm{de}}$, we denote average permittivity elements of the droplet. They are the droplet effective refractive indexes, which depend on the director orientation inside the droplet:

$$\Delta\varepsilon_{\mathrm{d}} = \left(\varepsilon_{\parallel} - \varepsilon_{\perp}\right) SS_{\mathrm{d}} = \Delta\varepsilon SS_{\mathrm{d}}. \tag{9.10}$$

$$\Delta\varepsilon = \varepsilon_{\parallel} - \varepsilon_{\perp} \tag{9.11}$$

is the permittivity anisotropy of the liquid crystal. The off-diagonal tensor elements are equal to zero. For nematic liquid crystals permittivities ε_{o} and ε_{e} for the ordinary and extraordinary waves are [1, 4, 5]

$$\varepsilon_{\mathrm{o}} \equiv \varepsilon_{\perp}, \varepsilon_{\mathrm{e}} \equiv \varepsilon_{\parallel} \tag{9.12}$$

In some cases it is convenient to write Eqs. (9.8) and (9.9) in another form [40]:

$$\varepsilon_{\mathrm{do}} = \varepsilon_{\mathrm{iso}} - \frac{1}{3}\Delta\varepsilon \cdot SS_{\mathrm{d}}, \tag{9.13}$$

$$\varepsilon_{\mathrm{de}} = \varepsilon_{\mathrm{iso}} + \frac{2}{3}\Delta\varepsilon \cdot SS_{\mathrm{d}}. \tag{9.14}$$

Here

$$\varepsilon_{\mathrm{iso}} = \frac{2\varepsilon_{\mathrm{o}} - \varepsilon_{\mathrm{e}}}{3}. \tag{9.15}$$

There are linear dependencies of the droplet effective refractive indexes $\varepsilon_{\mathrm{do}}$ and $\varepsilon_{\mathrm{de}}$ on the parameter SS_{d}.

Using relationship (9.5) we write the polarizability tensor of the elementary volume of the liquid-crystal droplet in the coordinate system related to the incident wave:

$$\widehat{\alpha} = A^T \widehat{\alpha}_{dV} A. \tag{9.16}$$

Here A is the transition matrix from coordinate system $\mathbf{e}_1{}^i$, $\mathbf{e}_2{}^i$, $\mathbf{e}_{\mathrm{i}}$ to coordinate system associated with the director of the droplet. Since the choice of the axes in the last frame in a plane perpendicular $\mathbf{e}_{\mathrm{d}}$ is arbitrary, without loss of generality, we choose one axis in the plane ($\mathbf{e}_{\mathrm{i}}$, $\mathbf{e}_{\mathrm{d}}$) and the other one—in the perpendicular plane [41].

It is a conveniently presented polarizability tensor as a sum of isotropic and anisotropic components. The last depends on the orientation of the liquid-crystal droplet:

$$\widehat{\alpha} = \left(\left(\frac{\varepsilon_{\mathrm{do}}}{\varepsilon_{\mathrm{p}}} - 1\right)\widehat{\mathrm{I}} + \frac{\Delta\varepsilon_{\mathrm{d}}}{\varepsilon_{\mathrm{p}}}\widehat{\mathrm{T}}\right) dV, \tag{9.17}$$

$$\widehat{T} = \begin{pmatrix} \left(\mathbf{e}_1^i \cdot \mathbf{e}_{\mathrm{d}}\right)^2 & \left(\mathbf{e}_1^i \cdot \mathbf{e}_{\mathrm{d}}\right)\left(\mathbf{e}_2^i \cdot \mathbf{e}_{\mathrm{d}}\right) & \left(\mathbf{e}_1^i \cdot \mathbf{e}_{\mathrm{d}}\right)\left(\mathbf{e}_{\mathrm{i}} \cdot \mathbf{e}_{\mathrm{d}}\right) \\ \left(\mathbf{e}_1^i \cdot \mathbf{e}_d\right)\left(\mathbf{e}_2^i \cdot \mathbf{e}_d\right) & \left(\mathbf{e}_2^i \cdot \mathbf{e}_d\right)^2 & \left(\mathbf{e}_2^i \cdot \mathbf{e}_{\mathrm{d}}\right)\left(\mathbf{e}_1^i \cdot \mathbf{e}_{\mathrm{d}}\right) \\ \left(\mathbf{e}_1^i \cdot \mathbf{e}_{\mathrm{d}}\right)\left(\mathbf{e}_{\mathrm{i}} \cdot \mathbf{e}_{\mathrm{d}}\right) & \left(\mathbf{e}_2^i \cdot \mathbf{e}_{\mathrm{d}}\right)\left(\mathbf{e}_{\mathrm{i}} \cdot \mathbf{e}_{\mathrm{d}}\right) & \left(\mathbf{e}_{\mathrm{i}} \cdot \mathbf{e}_{\mathrm{d}}\right)^2 \end{pmatrix}. \tag{9.18}$$

After integration of Eq. (9.4) over the droplet volume, we obtain the formula for the elements of the amplitude scattering matrix of the liquid-crystal droplet:

$$S_{mn} = -\frac{ik^3}{4\pi} F \mathbf{e}_m^s \widehat{\alpha} \mathbf{e}_n^i. \tag{9.19}$$

Here m, $n = 1, 2$.

Pay attention that the average polarizability tensor is a constant. It means that we have to integrate only multiplier $\exp(i\psi)$. This function depends on the size and the shape of the liquid-crystal droplet and on the scattering angle. For droplets of spherical shape [42],

$$F = \int\limits_V \exp(i\psi)\,dv = \int\limits_V \exp\left(ik\mathbf{R}_{dV}\left(\mathbf{e}_\mathrm{s} - \mathbf{e}_\mathrm{i}\right)\right)dV$$
$$= \frac{4\pi}{a^3}\left(\sin aR - aR\cos aR\right), \tag{9.20}$$

where $a = k\sqrt{2\left(1-\left(\mathbf{e}_\mathrm{i}\cdot\mathbf{e}_\mathrm{s}\right)\right)}$, R is the droplet radius.

It should be noted that for forward scattering ($\mathbf{e}_\mathrm{i} = \mathbf{e}_\mathrm{s}$) function F is equal to the droplet volume V. Substituting the polarizability tensor (Eq. 9.17) and performing the necessary transformations, we find

$$S_{mn} = -\frac{ik^3}{4\pi}F\left[\left(\frac{\varepsilon_\mathrm{do}}{\varepsilon_\mathrm{p}} - 1\right)\left(\mathbf{e}_m^s\cdot\mathbf{e}_n^i\right) + \frac{\Delta\varepsilon_\mathrm{d}}{\varepsilon_\mathrm{p}}\left(\mathbf{e}_m^s\cdot\mathbf{e}_\mathrm{d}\right)\left(\mathbf{e}_n^i\cdot\mathbf{e}_\mathrm{d}\right)\right]. \tag{9.21}$$

Relationship (9.21) permits one to calculate the elements of the amplitude scattering matrix of droplets with a cylindrically symmetric configuration. These matrix elements make it possible to determine light extinction coefficients of the polymer film. Owing to the anisotropy of the liquid-crystal droplets, the extinction coefficients depend on the incidence angle.

Now, we decompose the incident and scattered waves into two components: parallel and perpendicular to the scattering plane. The scattering plane is the plane determined by vectors $\mathbf{e}_\mathrm{i}$ and $\mathbf{e}_\mathrm{s}$ (Fig. 9.2; see also Fig. 9.11 below).

The two components of the wave scattered by a single droplet are determined in terms of the amplitude scattering matrix [36] as follows:

$$\begin{pmatrix} \mathbf{E}_\mathrm{s}\cdot\ \mathbf{e}_\parallel^s \\ \mathbf{E}_\mathrm{s}\cdot\ \mathbf{e}_\perp \end{pmatrix} = -\frac{\exp(ikr)}{ikr}\begin{pmatrix} S_2 & S_3 \\ S_4 & S_1 \end{pmatrix}\begin{pmatrix} \mathbf{E}_\mathrm{i}\cdot\mathbf{e}_\parallel^i \\ \mathbf{E}_\mathrm{i}\cdot\mathbf{e}_\perp \end{pmatrix}. \tag{9.22}$$

Here $\mathbf{E}_\mathrm{s}$ and $\mathbf{E}_\mathrm{i}$ are the electric vectors of the scattered and incident waves, respectively; S_j ($j = 1, 2, 3, 4$) are the elements of the amplitude scattering matrix.

Under the Rayleigh–Gans approximation, the elements are [40–42]

$$S_1 = -\frac{ik^3}{4\pi}F\left[\left(\frac{\varepsilon_\mathrm{do}}{\varepsilon_\mathrm{p}} - 1\right) + \frac{\Delta\varepsilon_\mathrm{d}}{\varepsilon_\mathrm{p}}\left(\mathbf{e}_\perp\cdot\mathbf{e}_\mathrm{d}\right)^2\right], \tag{9.23}$$

$$S_2 = -\frac{ik^3}{4\pi}F\left[\left(\frac{\varepsilon_\mathrm{do}}{\varepsilon_\mathrm{p}} - 1\right)\left(\mathbf{e}_\mathrm{i}\cdot\mathbf{e}_\mathrm{s}\right) + \frac{\Delta\varepsilon_\mathrm{d}}{\varepsilon_\mathrm{p}}\left(\mathbf{e}_\parallel^s\cdot\mathbf{e}_\mathrm{d}\right)\left(\mathbf{e}_\parallel^i\cdot\mathbf{e}_\mathrm{d}\right)\right], \tag{9.24}$$

$$S_3 = -\frac{ik^3}{4\pi}F\frac{\Delta\varepsilon_\mathrm{d}}{\varepsilon_\mathrm{p}}\left(\mathbf{e}_\parallel^s\cdot\mathbf{e}_\mathrm{d}\right)\left(\mathbf{e}_\parallel^i\cdot\mathbf{e}_\mathrm{d}\right), \tag{9.25}$$

$$S_4 = -\frac{ik^3}{4\pi} F \frac{\Delta\varepsilon_{\mathrm{d}}}{\varepsilon_{\mathrm{p}}} \left(\mathbf{e}_{\|}^{i} \cdot \mathbf{e}_{\mathrm{d}}\right) \left(\mathbf{e}_{\perp} \cdot \mathbf{e}_{\mathrm{d}}\right). \tag{9.26}$$

Let the droplet be illuminated by a wave propagating parallel to the axis z ($\theta_{\mathrm{i}} = 0$). At this case for the forward scattering ($\theta_{\mathrm{s}} = 0$) at a fixed orientation of the director $\boldsymbol{e}_{\mathrm{d}}$, the elements of the amplitude scattering matrix S of the droplet in the laboratory system of coordinates xyz are [40]

$$S_1 = -\frac{ik^3}{4\pi} V \left[\left(\frac{\varepsilon_{\mathrm{do}}}{\varepsilon_{\mathrm{p}}} - 1\right) + \frac{\Delta\varepsilon_{\mathrm{d}}}{\varepsilon_{\mathrm{p}}} \sin^2\theta_{\mathrm{d}} \sin^2\left(\varphi_{\mathrm{d}} - \varphi_{\mathrm{s}}\right)\right], \tag{9.27}$$

$$S_2 = -\frac{ik^3 V}{4\pi} \left[\begin{array}{l} \left(\frac{\varepsilon_{\mathrm{do}}}{\varepsilon_{\mathrm{p}}} - 1\right)\cos\delta + \frac{\Delta\varepsilon_{\mathrm{d}}}{\varepsilon_{\mathrm{p}}} \cos\delta \sin^2\theta_{\mathrm{d}} \cos^2\left(\varphi_{\mathrm{d}} - \varphi_{\mathrm{s}}\right) \\ - \frac{\Delta\varepsilon_{\mathrm{d}}}{2\varepsilon_{\mathrm{p}}} \sin\delta \sin^2\theta_{\mathrm{d}} \cos\left(\varphi_{\mathrm{d}} - \varphi_{\mathrm{s}}\right) \end{array}\right], \tag{9.28}$$

$$S_3 = -\frac{ik^3}{4\pi} V \frac{\Delta\varepsilon_{\mathrm{d}}}{2\varepsilon_{\mathrm{p}}} \left[\begin{array}{l} \cos\delta \sin^2\theta_{\mathrm{d}} \sin 2\left(\varphi_{\mathrm{d}} - \varphi_{\mathrm{s}}\right) - \\ - \sin\delta \sin 2\theta_{\mathrm{d}} \sin\left(\varphi_{\mathrm{d}} - \varphi_{\mathrm{s}}\right) \end{array}\right], \tag{9.29}$$

$$S_4 = -\frac{ik^3}{4\pi} V \frac{\Delta\varepsilon_{\mathrm{d}}}{2\varepsilon_{\mathrm{p}}} \sin^2\theta_{\mathrm{d}} \sin 2\left(\varphi_{\mathrm{d}} - \varphi_{\mathrm{s}}\right). \tag{9.30}$$

Here, δ is the scattering angle determined by the vectors $\boldsymbol{e}_{\mathrm{s}}$ and $\boldsymbol{e}_{\mathrm{i}}$; angle φ_{s} defines the orientation of the projections of vector $\boldsymbol{e}_{\mathrm{s}}$ (Fig. 9.2) onto the plane (xy) with respect to the axis x (it is the angle between the scattering plane ($\boldsymbol{e}_{\mathrm{i}}\boldsymbol{e}_{\mathrm{s}}$) and the plane ($zx$)); angle φ_{d} defines the orientations of the projections of vector $\boldsymbol{e}_{\mathrm{d}}$ onto the plane (xy) with respect to the axis x; and angle θ_{d} defines the orientation of Eq. (9.30) the vector $\boldsymbol{e}_{\mathrm{d}}$ with respect to the axis z.

As a rule, films consist of polydisperse ensembles of droplets. Therefore, analysis of a coherent field transmitted through the film requires the knowledge of averaged elements of the scattering matrix. Averaged over the sizes and orientations of droplets scattering matrix elements are

$$\langle S_1 \rangle = -\frac{ik^3}{4\pi} \langle V \rangle \left[\left(\frac{\varepsilon_{\mathrm{do}}}{\varepsilon_{\mathrm{p}}} - 1\right) + \frac{\Delta\varepsilon_{\mathrm{d}}}{3\varepsilon_{\mathrm{p}}} \left(1 + 2S_x \sin^2\varphi_{\mathrm{s}} + 2S_y \cos^2\varphi_{\mathrm{s}}\right)\right], \tag{9.31}$$

$$\langle S_2 \rangle = -\frac{ik^3}{4\pi} \langle V \rangle \left[\begin{array}{l} \left(\frac{\varepsilon_{\mathrm{do}}}{\varepsilon_{\mathrm{p}}} - 1\right)\cos\delta + \frac{\Delta\varepsilon_{\mathrm{d}}}{3\varepsilon_{\mathrm{p}}} \left(1 + 2S_x \sin^2\varphi_{\mathrm{s}} \right. \\ \left. + 2S_y \cos^2\varphi_{\mathrm{s}}\right) \cos\delta \end{array}\right], \tag{9.32}$$

$$\langle S_3 \rangle = \langle S_4 \rangle = 0. \tag{9.33}$$

Here, the angular brackets $\langle\rangle$ denote averaging over the sizes of droplets and orientation of their directors; $\langle V\rangle$ is the average volume of droplets; S_x, S_y, and S_z are the x, y, and z components of the film order parameter tensor. They characterize the orientation ordering of the directors of LC droplets in the coordinate system xyz:

$$S_x = \frac{1}{2}\left(3\left\langle \sin^2\theta_{\mathrm{d}}\cos^2\varphi_d\right\rangle - 1\right), \tag{9.34}$$

$$S_y = \frac{1}{2}\left(3\left\langle \sin^2\theta_{\mathrm{d}}\cos^2\varphi_{\mathrm{d}}\right\rangle - 1\right), \tag{9.35}$$

$$S_z = \frac{1}{2}\left(3\left\langle \cos^2\theta_{\mathrm{d}}\right\rangle - 1\right). \tag{9.36}$$

These components are related by the equation:

$$S_x + S_y + S_z = 0. \tag{9.37}$$

Pay attention that at random orientation, S_x and S_y components are expressed via S_z component as follows:

$$S_x = S_y = -\frac{1}{2}S_z. \tag{9.38}$$

9.4 Transmittance of the PDLC Film

Consider PDLC film (Fig. 9.3) of thickness l with a large number of heterogeneous in size and orientation droplets. Every droplet is characterized by its director $\boldsymbol{e}_{\mathrm{d}}$, indicating the direction of preferred orientation of the liquid-crystal molecules within it. Suppose further that at the absence of an applied electric field, droplets have a preferred orientation direction $\langle \boldsymbol{e}_{\mathrm{d}}\rangle$ (note that directions $+\langle \boldsymbol{e}_{\mathrm{d}}\rangle$ and $-\langle \boldsymbol{e}_{\mathrm{d}}\rangle$ are physically equivalent).

We choose a laboratory coordinate system as follows: the z-axis is perpendicular to the film plane and the x- and y-axes are in the plane of the film. We choose the direction of the x-axis so that the vector $\langle e_{\mathrm{d}}\rangle$ is in the plane (xz). The angle θ_d^{av} sets the deviation of the mean director $\langle e_{\mathrm{d}}\rangle$ on the z-axis. Unit vectors $\boldsymbol{e}_{\mathrm{i}}$ and $\boldsymbol{e}_{\mathrm{s}}$, as before, indicate the direction of propagation of the incident and scattered waves, and angles θ_{i}, θ_{s}, φ_{i}, φ_{s} are their polar and azimuthal angles, respectively.

So when we deal with the object consisting of a large number of LC droplets, we can use statistical methods to describe characteristics of transmitted light. The transmitted radiation can be considered as a sum of coherent (average) and incoherent (fluctuating) components. The coherent component is determined by means of the Foldy–Twersky equation [38]. In the problem under consideration,

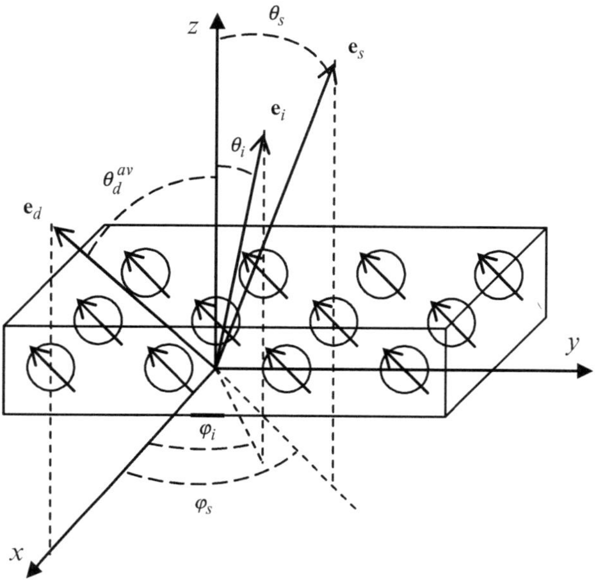

Fig. 9.3 Schematic presentation of the PDLC film and the illumination conditions. Notations are in the text

we deal with polarized light; therefore, we have to generalize it to the vector case. We write this equation for the vector case as

$$\begin{pmatrix} \langle \mathbf{E}_{\mathrm{e}} \rangle \\ \langle \mathbf{E}_{\mathrm{o}} \rangle \end{pmatrix} = \begin{pmatrix} \psi_{\mathrm{e}}(z) & 0 \\ 0 & \psi_{\mathrm{o}}(z) \end{pmatrix} \Bigg|_{z=l} \begin{pmatrix} \cos\alpha \\ \sin\alpha \end{pmatrix} \mathbf{E}_{\mathrm{i}}. \tag{9.39}$$

Here, $\langle \mathbf{E}_{\mathrm{o}} \rangle$ and $\langle \mathbf{E}_{\mathrm{e}} \rangle$ are the ordinary and extraordinary components of the coherent field transmitted through a PDLC film. Functions $\psi_{\mathrm{e,o}}(z)$ are the solutions to the integral equations:

$$\Psi_{\mathrm{e;o}}(z) = \exp(ikz) \left(1 - q \langle S_{\mathrm{e;o}}(0) \rangle \int_0^z \exp(-ikz_s)\, \psi_{\mathrm{e;o}}(z_s)\, dz_s \right), \tag{9.40}$$

where $q = 2\pi k^{-2} N$, N is the number of LC droplets per unit volume, and $\langle S_{\mathrm{e,o}}(0) \rangle$ are the amplitude scattering functions, averaged over the droplet sizes and the orientations of their directors, for the extraordinary and ordinary waves at zero scattering angle.

A solution to Eq. (9.40) is given by

$$\Psi_{\mathrm{e;o}}(z) = \exp(ik_{\mathrm{e,o}} z). \tag{9.41}$$

Here, k_e and k_o are the wave propagation constants of the extraordinary and ordinary waves:

$$k_{e,o} = k + iq\exp\left(ik_{e,o}z\right)\left\langle S_{e;o}(0)\right\rangle. \tag{9.42}$$

Amplitude scattering functions $S_e(0)$ and $S_o(0)$ are calculated with the use of formula (9.32):

$$S_e(0) = \left\langle S_2\right\rangle_{\varphi_s=0,\delta=0} = -\frac{ik^3}{4\pi}\left\langle V\right\rangle\left[\frac{\varepsilon_{do}}{\varepsilon_p} - 1 + \frac{\Delta\varepsilon_d}{3\varepsilon_p}\left(1+2S_x\right)\right], \tag{9.43}$$

$$S_e(0) = \left\langle S_2\right\rangle_{\varphi_s=\pi/2,\delta=0} = -\frac{ik^3}{4\pi}\left\langle V\right\rangle\left[\frac{\varepsilon_{do}}{\varepsilon_p} - 1 + \frac{\Delta\varepsilon_d}{3\varepsilon_p}\left(1+2S_y\right)\right], \tag{9.44}$$

Using Eqs. (9.39), (9.40), (9.41), (9.42), (9.43), and (9.44), we find the extraordinary and ordinary components of the transmitted light [40, 42]:

$$\mathbf{E}_s^e = \mathrm{Re}\frac{\mathrm{E}_s\cdot\mathbf{e}_e^i}{E_i\exp(ikl)} = \frac{1}{E}\exp\left(-\frac{1}{2}\gamma_e l\right)\cos\Phi_e\left(\mathbf{E}_i.\mathbf{e}_e^i\right), \tag{9.45}$$

$$\mathbf{E}_s^0 = \mathrm{Re}\frac{\mathbf{E}_s\cdot\mathbf{e}_0^i}{E_i\exp(ikl)} = \frac{1}{E_i}\exp\left(-\frac{1}{2}\gamma_0 l\right)\cos\Phi_0\left(\mathbf{E}_i\cdot\mathbf{e}_0^i\right). \tag{9.46}$$

Here $\gamma_{e,o}$ are extinction indices and $\Phi_{e,o}$ are phases of the extraordinary and ordinary waves, respectively. They are determined as follows:

$$\gamma_{e,o} = \frac{4\pi N}{k^2}\mathrm{Re}\left\langle S_{2,1}(0)\right\rangle, \tag{9.47}$$

$$\Phi_{e,o} = \frac{4\pi Nl}{k^2}\mathrm{Im}\left\langle S_{2,1}(0)\right\rangle. \tag{9.48}$$

The transmission coefficients for extraordinary and ordinary waves as obvious from Eqs. (9.45) and (9.46) are

$$t_{e,o} = \exp\left(-\frac{1}{2}\gamma_{e,o}l\right). \tag{9.49}$$

Pay attention that at random orientation of droplets $\gamma_e = \gamma_o$ and $t_e = t_o$.

Characteristics of light transmitted through a PDLC film are investigated in details below.

9.5 Phase Shift

The phase shift between extraordinary and ordinary components of the transmitted light can be determined by Eq. (9.48):

$$\Delta\Phi = \Phi_0 - \Phi_e = \frac{lkC_V}{2\cos\theta_i}\frac{\Delta\varepsilon_d}{\varepsilon_p}\left(\left\langle(\mathbf{e}_e\cdot\mathbf{e}_d)^2\right\rangle - \left\langle(\mathbf{e}_o\cdot\mathbf{e}_d)^2\right\rangle\right). \tag{9.50}$$

Here θ_i is an incident angle of light; l is the film thickness; $C_V = N\langle V\rangle$ is a volume concentration of droplets [42]. Note that the volume fraction of droplets is less than the volume fraction of LC in the sample, because some portion of the liquid crystal is dissolved in the polymer matrix.

In the case of normal incidence, Eq. (9.50) is reduced to

$$\Delta\Phi_n = \frac{lkC_V}{2}\frac{\Delta\varepsilon_d}{\varepsilon_p}\left(\langle\sin^2\theta_d\cos^2\phi_d\rangle - \langle\sin^2\theta_d\sin^2\phi_d\rangle\right), \tag{9.51}$$

where $\Delta\Phi_n$ is the phase shift at normal incidence of radiation and θ_d and φ_d are the tilt and azimuth angles of a droplet director $\boldsymbol{e}_d$, respectively.

To describe orientation distribution of LC droplet directors in the film, a tensor order parameter $\widehat{S}$ of a PDLC film with components

$$S_{ij} = \frac{1}{2}\left\langle 3(\mathbf{e}_d)_i(\mathbf{e}_d)_j - \delta_{ij}\right\rangle, \tag{9.52}$$

is used [5]. Here $(\mathbf{e}_d)_i, (\mathbf{e}_d)_j$, $i,j = 1,2,3$ are the components of the droplet director [see Eqs. (9.34), (9.35), and (9.36)]: $S_{11} = S_x$, $S_{22} = S_y$, $S_{33} = S_z$.

The components of the order parameter tensor numerically characterize the orientational ordering of the directors of the droplets with respect to the corresponding axes of the laboratory coordinate system. These parameters are determined by the orientation of LC droplet directors and depend on an applied external field. They allow us to describe different types and transformations of the distribution of directors of liquid-crystal droplets without knowledge of the explicit expressions for the dependence on the controlling field applied to the film.

We assume, for simplicity, that droplet director azimuths are uniformly distributed over the interval $[-\varphi_d^{max}, +\varphi_d^{max}]$ (note that the vector $\langle e_d\rangle$ is in the plane (xz)) and polar angles are uniformly distributed over the interval $[\theta_d^{av} - \theta_d^{max}, \theta_d^{av} + \theta_d^{max}]$ at zero applied to the film electric field. θ_d^{av} is the average polar angle for droplet director orientation (Fig. 9.3.).

In this case we can write the components S_x and S_y in terms of component S_z as follows:

$$S_x = \frac{1}{2}\left((1 - S_z)\frac{\sin 2\phi_d^{max}}{2\phi_d^{max}} - S_z\right), \tag{9.53}$$

$$S_y = \frac{1}{2}\left((S_z - 1)\frac{\sin 2\phi_{\mathrm{d}}^{\max}}{2\phi_{\mathrm{d}}^{\max}} - S_z\right). \tag{9.54}$$

Z component of the tensor can be written as follows [43]:

$$S_z = \frac{1}{4} + \frac{3}{8}\left(\frac{1}{3}\frac{\sin 3\theta_{\mathrm{d}}^{\mathrm{av}}}{\sin \theta_{\mathrm{d}}^{\mathrm{av}}}\frac{\sin 3\theta_{\mathrm{d}}^{\max}}{\sin \theta_{\mathrm{d}}^{\max}} - 1\right), \tag{9.55}$$

Below the tensor order parameters for some asymptotic orientational structures of the layer are presented as the examples:

1. Planar-oriented structure of optical axes of liquid-crystal droplets along the x-axis, $\theta_d^{av} = \pi/2$, $\theta_d^{max} = 0$, $\varphi_d^{max} = 0$ ($S_x = 1$, $S_y = -1/2$, $S_z = -1/2$)
2. Homeotropic structure with directors of droplets oriented along the z-axis, $\theta_d^{av} = 0$, $\theta_d^{max} = 0$, $\varphi_d^{max} = 0$ ($S_x = -1/2$, $S_y = -1/2$, $S_z = 1$)
3. Planar random orientation in the (x, y) plane of the layer (x,y), $\theta_d^{av} = \pi/2$, $\theta_d^{max} = 0$, $\varphi_d^{max} = \pi$ ($S_x = 1/4$, $S_y = 1/4$, $S_z = -1/2$)
4. Random orientation in the volume, $\theta_d^{max} = \pi/2$, $\theta_d^{max} = \pi$ ($S_x = S_y = S_z = 0$)

Under an applied field, droplets tend to be aligned along the direction of the field (we consider LC with positive birefringence $\Delta n = n_{\mathrm{e}} - n_{\mathrm{o}} > 0$, $n_{\mathrm{e}} = \sqrt{\varepsilon_{\mathrm{e}}}$, $n_{\mathrm{o}} = \sqrt{\varepsilon_{\mathrm{o}}}$ are refractive indexes of LC for extraordinary and ordinary waves, respectively). So long as the field is directed normally to the sample, the change in the orientation order degree of LC droplets can be described by means of the order parameter tensor component S_z.

We suppose that the angular distribution of the droplet directors is uniform within a solid angle $\Delta\Omega$. At $E = 0$ angle$\theta_d^{av} = \pi/2$ $\theta_d^{av} = \pi/2$ and angle $\theta_d^{av} = 0$. It means that $[-\varphi_d^{max} \le \varphi_{\mathrm{d}} \le +\varphi_d^{max}]$ and $[\pi/2 - \theta_d^{max} \le \theta_{\mathrm{d}} \le \pi/2 + \theta_d^{max}]$. Here φ_d^{max} and θ_d^{max} are the maximal deviation angles of the droplet directors from axis x in the (xy) and (xz) planes, respectively. Then [40],

$$S_z = \frac{1}{4} + \frac{3}{4}\left(\frac{u}{4E^2} + \frac{3E^4 - 2E^2 - 1}{16E^3 \sin \theta_{\mathrm{d}}^{\max}} \ln\left|\frac{u + 2E\sin\theta_{\mathrm{d}}^{\max}}{u - 2E\sin\theta_{\mathrm{d}}^{\max}}\right|\right), \tag{9.56}$$

$$u = \sqrt{(E^2 - 1)^2 + 4E^2 \sin^2\theta_d^{max}}. \tag{9.57}$$

Here E is the reduced (normalized) strength of an applied electric field. $E = E_{\mathrm{a}}/E_{\mathrm{th}}$, where E_{a} is the applied (control) electric field and E_{th} is the field magnitude to reorient droplet [5, 26].

At $\theta_d^{max} = \pi/2$, Eq. (9.22) describes transition from a random structure of droplet directors to the homeotropic one [5, 26]:

$$S_z = \frac{1}{4} + \frac{3\left(E^2 + 1\right)}{16E^2} + \frac{3\left(3E^2 + 1\right)\left(E^2 - 1\right)}{32E^3}\ln\left|\frac{E + 1}{E - 1}\right|. \tag{9.58}$$

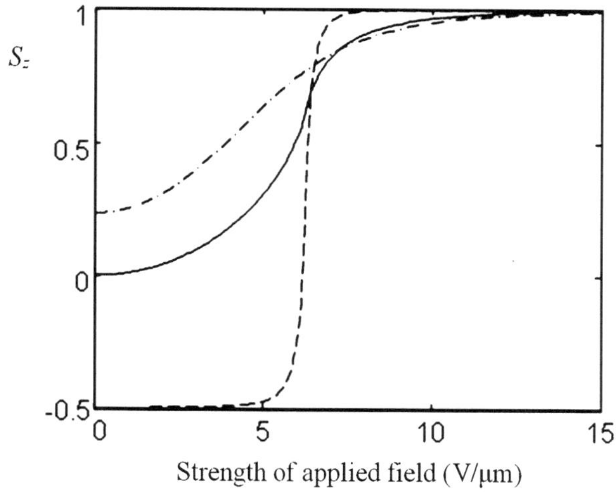

Fig. 9.4 Order parameter S_z as function of field E_a applied to the film along the z-axis. Transition from a planar (($\theta_d^{av} = \pi/2$, $\theta_d^{max} = 5°$) *dashed line*), random (($\theta_d^{av} = \theta_d^{max} = \pi/2$) *solid line*), and partially ordered (($\theta_d^{av} = 45°$, $\theta_d^{max} = 10°$)) *chain line*) structures to the homeotropic one. $E_{th} = 6.25$ V/μm

The dependence of order parameter S_z as function of the applied field E_a is shown in Fig. 9.4 for different initial angles θ_d^{max} (realized at zero applied field). One can see that the transition from a partially ordered structure of droplet directors to the homeotropic structure is characterized by stronger dependence of the order parameter on the applied field and offers more possibilities for modulating optical radiation than transition from the random to the homeotropic one.

At the oblique incidence, the phase shift between extraordinary and ordinary waves can be written as follows:

$$\Delta\Phi_{obl} = \Phi_o - \Phi_e = \frac{lkc_v}{2\cos\theta_i}\frac{\Delta\varepsilon_d}{\varepsilon_p} \times \left((1 - S_z)\left(2\cos^2\theta_i - \sin^2\theta_i\cos 2\phi_i\right)\frac{\sin 2\phi_d^{max}}{6\phi_d^{max}} + S_z\sin^2\theta_i\right). \quad (9.59)$$

Dependence of the phase shift $\Delta\Phi_{obl}$ on the angle of incidence θ_i and the electric field strength for partial orientation of droplets in the film is displayed in Fig. 9.5.

For strongly oriented droplets, the first term in Eq. (9.59) is dominant. Maximal phase shift changes are implemented for small angles of incidence. Phase shift has maximal values at small angles of incidence and low fields when droplet directors are mainly oriented along the x-axis, while achieving the greatest difference in the refractive indices of the film for the extraordinary and ordinary waves.

Note that phase shift very weakly depends on the angle of incidence when the wave vector of the incident wave is in the plane yz (angle φ_i is close to $\pm\pi/2$).

Dependence of the phase at oblique incidence of light as function of the angle of incidence θ_i and the applied electric field E_a is displayed in Fig. 9.6.

At random orientation of droplet directors, maximum phase shift is achieved at high fields and high angles of incidence. This is explained by the fact that at low fields the film is similar to the isotropic one. In a strong field, it has a preferred

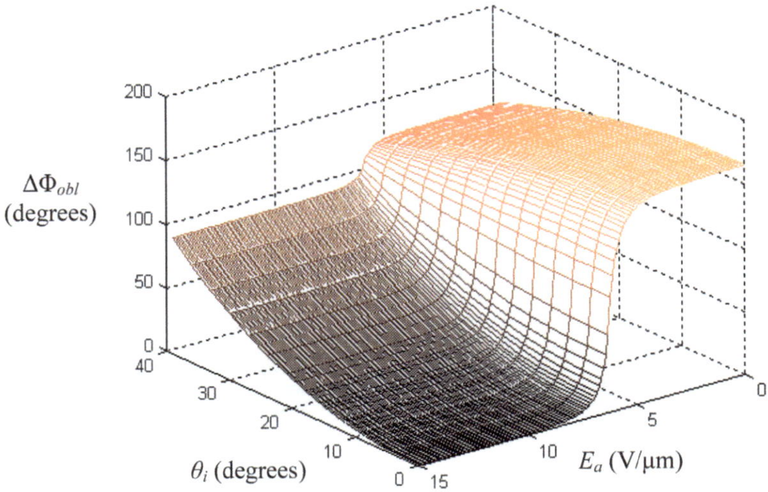

Fig. 9.5 Phase shift $\Delta\Phi_{\text{obl}}$ at oblique incidence of light for the partially orientation droplets in the film as function of incidence angle θ_{i} and the electric field strength E_{a}. In the absence of an external field (E_{a} =0) $\varphi_d^{max} = \theta_d^{max} = 10°$, $\theta_d^{av} = 90°$. The wavelength of the incident radiation $\lambda = 0.6328$ μm. The refractive index of the polymer $n_{\text{p}} = 1.524$ ($\varepsilon_{\text{p}} = n_p^2$), the refractive indices of the liquid crystal $n_\perp = 1.511$ и $n_\| = 1.74$ ($\varepsilon_\| = (n_\|)^2$); the layer thickness $l = 41.3$ μm, the volume concentration of the liquid crystal $C_{\text{V}} = 0.075$, the molecular order parameter $S = 0.6$, the order parameter of the liquid-crystal droplet $S_{\text{d}} = 0.7$, the average droplet radius $\langle R \rangle = 0.05$ μm, and the parameter of the gamma distribution [see Eq. (9.73)] $\mu = 16$

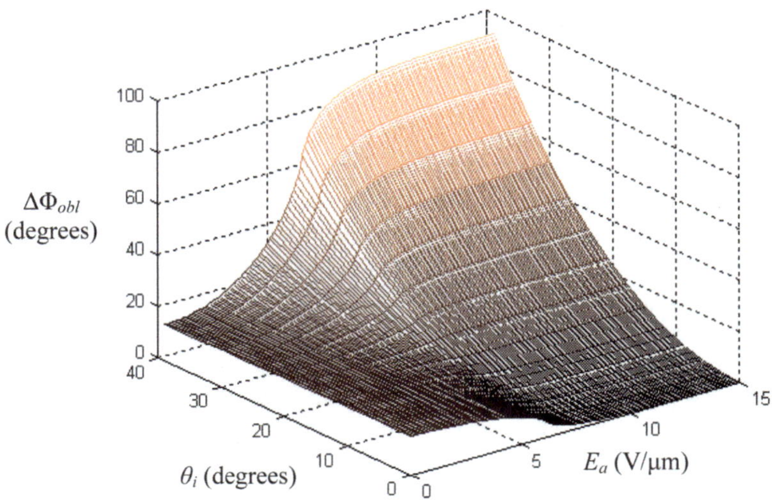

Fig. 9.6 Phase shift $\Delta\Phi_{\text{obl}}$ at oblique incidence of light as function of angle of incidence θ_{i} and electric field E_{a}. Initial droplet director distribution (in the absence of an external field, $E_{\text{a}} = 0$) is close to the random ($\varphi_d^{max} = \theta_d^{max} = 80°$, $\theta_d^{av} = 90°$). The other parameters are the same as in Fig. 9.5

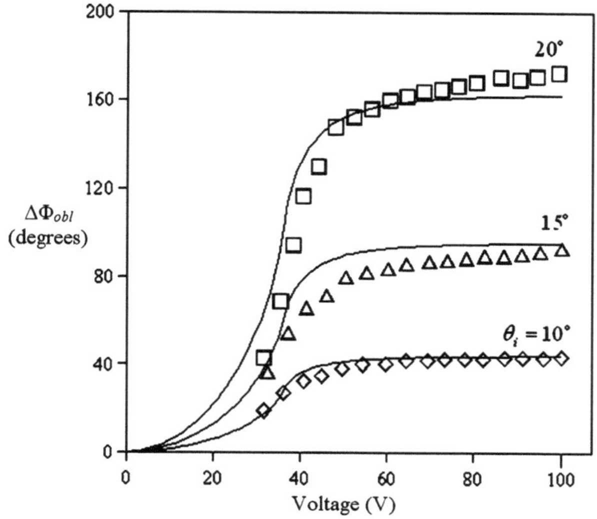

Fig. 9.7 Phase shift between extraordinary and ordinary components of transmitted light at oblique illumination by a laser with $\lambda = 0.6328$ μm versus applied voltage at different incident angle θ_i. Theoretical results (*solid curves*), experimental data [46] (*symbols*). Thickness of film $l = 36$ μm, volume concentration of droplets $C_v = 0.4$, molecular order parameter $S = 0.92$, order parameter of droplet $S_d = 0.818$. $n_o = 1.511$, $n_e = 1.74$, $n_p = 1.55$. The samples were prepared by the phase-separation method using fluid prepolymer EPON 815 (Shell Chemical Co.) and MK 107 (Wilmington Chemical Corporation) and as curing agent Capcure-3-800 (Shell Chemical Co.) and B component BOSTIC (Boston s.p.A.). The LC E7 (British Drug House) was used

direction (optical axis). Therefore, the larger the angle of incidence of light (the angle between the optical axis and the direction of propagation), the greater is the difference in refractive indexes of the film and the larger phase difference is achieved.

Below we consider two limiting cases that allow one to get the highest phase difference: (i) oblique incidence of light on the film with random director distribution and (ii) normal incidence of light on a layer of with partially oriented director droplets.

(i) At random droplet distribution $\varphi_d^{max} = \pi/2$. In this case the phase difference $\Delta\Phi_{obl}$ between the extraordinary and ordinary components is

$$\Delta\Phi_{obl} = \frac{lkc_v}{2\cos\theta_i}\frac{\Delta\varepsilon_d}{\varepsilon_p}S_z\sin^2\theta_i. \tag{9.60}$$

The theoretical results as shown in [44, 45] are in good agreement with the experimental data [46]. Dependence of the phase shift $\Delta\Phi_{obl}$ on the applied voltage is displayed in Fig. 9.7.

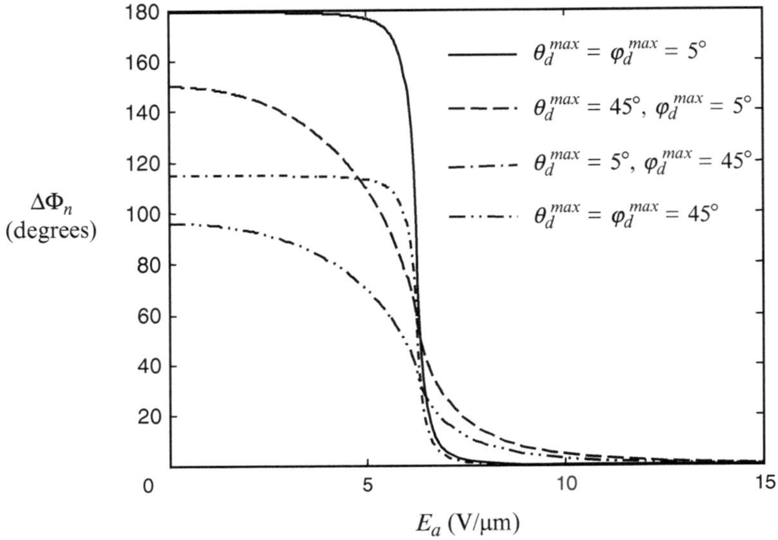

Fig. 9.8 Phase shift $\Delta\Phi_n$ between extraordinary and ordinary waves versus the applied electric field strength E_a at normal illumination. The other parameters are the same as in Fig. 9.5

(ii) At partially oriented droplets and normal illumination ($\theta_i = 0°$), we have

$$\Delta\Phi_n = \frac{lkc_v}{3}\frac{\Delta\varepsilon_d}{\varepsilon_p}\frac{\sin 2\phi_d^{max}}{2\phi_d^{max}}(1 - S_z). \tag{9.61}$$

Maximal possible value of the phase shift at normal illumination is achieved when all directors are oriented along the x-axis ($S_z = -1/2$):

$$\Delta\Phi_n^{max} = \frac{lkc_v}{2}\frac{\Delta\varepsilon_d}{\varepsilon_p}\frac{\sin 2\phi_d^{max}}{2\phi_d^{max}}. \tag{9.62}$$

The dependence of the phase shift on the strength of applied electric field is displayed in Fig. 9.8. It shows the results of calculations of phase $\Delta\Phi_n$ at different initial orientation of the LC droplets.

Analysis shows that the maximal value of the phase shift is determined by both angles θ_d^{max} and φ_d^{max}. Meanwhile, the slope of the curve depends only on the angle θ_d^{max}. This implies that a PDLC sample with small partially oriented droplets can be utilized as a phase plate working either in the "sharp" regime (the phase shift decreases abruptly from maximal value to zero with applied electric field) or in the "smooth" one (the phase shift is changed gradually with applied electric field).

Dependence of the phase shift $\Delta\Phi_n$ versus strength of the applied electric field E_a and angle φ_d^{max} at normal illumination is shown in Fig. 9.9.

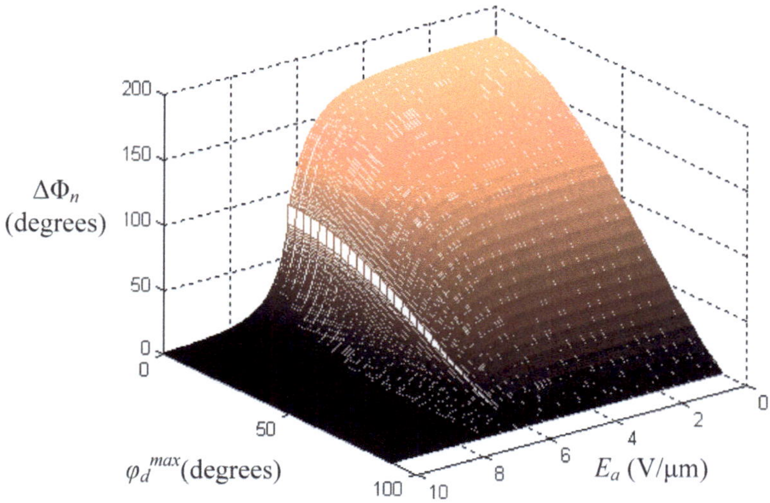

Fig. 9.9 Phase shift $\Delta\Phi_{\mathrm{n}}$ at normal incidence of light versus strength of the applied electric field E_{a} and angle φ_d^{max} at $\theta_d^{max} = 5°$. The other parameters are the same as in Fig. 9.5

9.6 Extinction Indices

Since we consider nonabsorbing LC, the elements of the amplitude scattering matrix are pure imaginary values. It means that we cannot use Eq. (9.47) to calculate the extinction indices. Thus, we faced the problem of the extinction indices determination and have to use another way [36]:

$$\gamma_{\mathrm{e,o}} = N \int\limits_{4\pi} \left\langle \frac{d\sigma}{d\Omega} (\mathbf{e} = \mathbf{e}_{\mathrm{e,o}}) \right\rangle \sin\theta_{\mathrm{s}} d\theta_{\mathrm{s}} d\phi_{\mathrm{s}}. \tag{9.63}$$

Here N is concentration of the LC droplets; θ_{s} and φ_{s} are the tilt and azimuth angles of the vector $\mathbf{e}_{\mathrm{s}}$; $\mathbf{e}$ is unit vector in the direction of the incident light polarization; $\mathbf{e}_{\mathrm{o}}$ and $\mathbf{e}_{\mathrm{e}}$ are unit vectors which point the direction of the extraordinary and ordinary waves polarization, respectively; and $d\sigma/d\Omega$ is the differential scattering cross section for a single LC droplet [29, 36]:

$$\frac{d\sigma}{d\Omega} = \frac{1}{k^2}\left(\left|S_2\mathbf{e}\cdot\mathbf{e}_{||}^{i} + S_3\mathbf{e}\cdot\mathbf{e}_{\perp}^{i}\right|^2 + \left|S_4\mathbf{e}\cdot\mathbf{e}_{||}^{i} + S_1\mathbf{e}\cdot\mathbf{e}_{\perp}^{i}\right|^2\right). \tag{9.64}$$

By means of Eqs. (9.63) and (9.64), the extinction indices can be calculated only numerically [41].

At small azimuthal angles φ_{i}, the extinction coefficient for extraordinary wave γ_{e} strongly depends on the angle of incidence θ_i, whereas the extinction coefficient for the ordinary wave γ_{o} depends on this angle very weakly. By contrast, at the

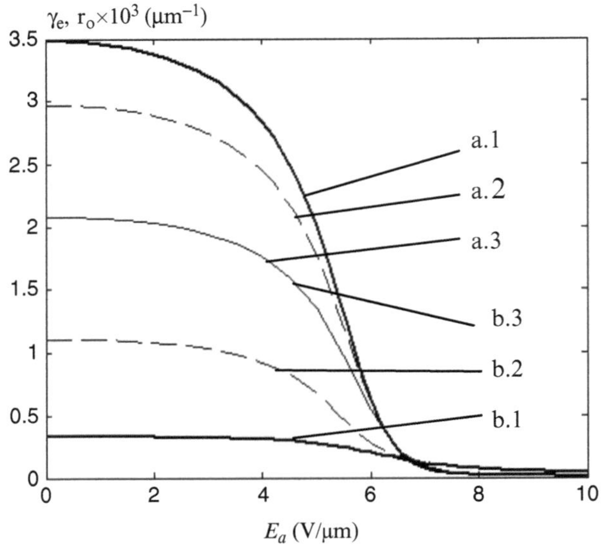

Fig. 9.10 Dependences of the light extinction coefficient for the extraordinary wave γ_e (curves a.1, a.2, a.3) and for the ordinary wave γ_o (curves b.1, b.2, b.3) on the strength of the electric field applied to the layer at the angles φ_d^{max}: 1 – $\varphi_d^{max} = 5°$; 2 – $\varphi_d^{max} = 45°$; 3 – $\varphi_d^{max} = 90°$. $\theta_i = 0°$. The other parameters are the same as in Fig. 9.5

azimuthal angles φ_i close to $\pi/2$, the extinction coefficient for the ordinary wave γ_o substantially depends on the angle of incidence, whereas the extinction coefficient γ_e depends on this angle only slightly. However, both extinction coefficients (γ_e and γ_o) at any angles of incidence are on the order of 10^{-3} μm^{-1}. This, for example, means that the transmittance of the film with a thickness of 40 μm is no less than 85 %. Owing to high transmittance, the films containing nanosized dispersed liquid-crystal droplets can be used as optical modulators with low-energy losses for many applications, for example, in telecommunication systems. The influence of droplet director ordering in the film plane on the light extinction coefficients is illustrated in Fig. 9.10.

An increase in the angle φ_d^{max} (it characterizes the degree of orientational ordering) leads to a decrease in the extinction coefficient for the extraordinary wave and, contrastingly, to an increase in the extinction coefficient for the ordinary wave. In the limiting case of a random orientation of droplets in the film plane ($\varphi_d^{max} = \pi/2$), extinction coefficients coincide.

The calculations from relationship (9.63) are complicated by the fact that the distribution of the orientations of liquid-crystal droplets in the film is known only in the rare cases.

In order to obtain convenient analytical expressions for the extinction coefficients, we use the approximation of the effective amplitude scattering matrix. The real layer containing liquid-crystal droplets with different sizes and different

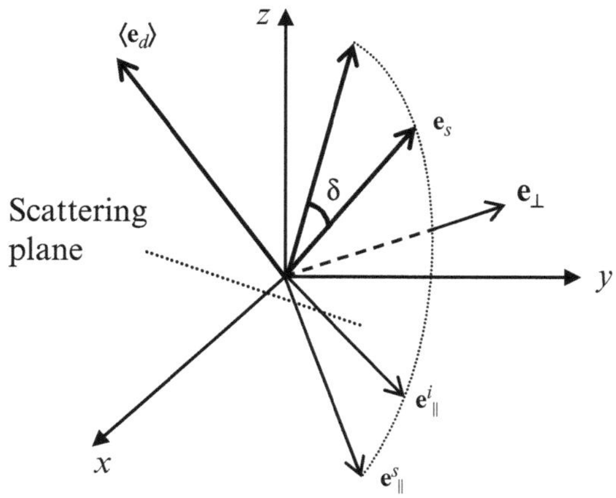

Fig. 9.11 Schematic representation of the laboratory frame, the scattering plane, and the unit vectors. The vector $\langle e_\mathrm{d} \rangle$ is the preferred orientation direction of the droplet directors. The z-axis is perpendicular to the sample plane and the vector $\langle e_\mathrm{d} \rangle$ is in the x–z plane. The scattering plane is the plane which the wave vectors of incident $\boldsymbol{e}_\mathrm{i}$ and scattered $\boldsymbol{e}_\mathrm{s}$ waves belong to; δ is the angle between $\boldsymbol{e}_\mathrm{i}$ and $\boldsymbol{e}_\mathrm{s}$; $e_\perp$, $e_\parallel{}^\mathrm{i}$, and $e_\parallel{}^\mathrm{s}$ are components of the polarization vectors of the incident and scattered waves

orientations is replaced by an effective layer containing identical droplets with the average optical characteristics. In order to determine the amplitude matrix of light scattering by these droplets, the matrix elements defined by relationships (9.23), (9.24), (9.25), and (9.26) are averaged over the sizes of droplets and over the orientations of their directors.

The effective extinction coefficients of the layer are determined by integrating the equations for the differential scattering cross sections over the full solid angle. In order to simplify the problem, we calculate the differential scattering cross sections as the half sum of the differential scattering cross sections for the two limiting positions of the vectors $\mathbf{e}_\mathrm{i}$, $\boldsymbol{e}_\mathrm{s}$, and $\langle e_\mathrm{d} \rangle$ at which these cross sections take on minimum and maximum values. Vectors $\mathbf{e}_\mathrm{i}$, $\boldsymbol{e}_\mathrm{s}$, and $\langle e_\mathrm{d} \rangle$ determine unit vectors of the incident and scattering waves and the average direction of the droplet directors (Fig. 9.11).

In the first case the scattering plane ($\mathbf{e}_\mathrm{i}$, $\boldsymbol{e}_\mathrm{s}$) coincides with the principal plane ($\mathbf{e}_\mathrm{i}$,$\langle e_\mathrm{d} \rangle$). Then the differential scattering cross sections are

$$\left(\frac{d\sigma}{d\Omega}\right)_\mathrm{e}^{(1)} = \left(\frac{d\sigma}{d\Omega}\left(\mathbf{e}=\mathbf{e}_\mathrm{e}\right)\right)^{(1)} = \frac{1}{k^2}\left(\left|\left\langle S_2^{(1)}\right\rangle\right|^2 + \left|\left\langle S_4^{(1)}\right\rangle\right|^2\right), \qquad (9.65)$$

$$\left(\frac{d\sigma}{d\Omega}\right)_\mathrm{o}^{(1)} = \left(\frac{d\sigma}{d\Omega}\left(\mathbf{e}=\mathbf{e}_\mathrm{o}\right)\right)^{(1)} = \frac{1}{k^2}\left(\left|\left\langle S_3^{(1)}\right\rangle\right|^2 + \left|\left\langle S_1^{(1)}\right\rangle\right|^2\right). \qquad (9.66)$$

In the second case the scattering plane is perpendicular to the principal plane. The differential scattering cross sections are

$$\left(\frac{d\sigma}{d\Omega}\right)_{\mathrm{e}}^{(2)} = \left(\frac{d\sigma}{d\Omega}\left(\mathbf{e}=\mathbf{e}_{\mathrm{e}}\right)\right)^{(2)} = \frac{1}{k^2}\left(\left|\left\langle S_3^{(2)}\right\rangle\right|^2 + \left|\left\langle S_1^{(2)}\right\rangle\right|^2\right), \tag{9.67}$$

$$\left(\frac{d\sigma}{d\Omega}\right)_{\mathrm{o}}^{(2)} = \left(\frac{d\sigma}{d\Omega}\left(\mathbf{e}=\mathbf{e}_{\mathrm{o}}\right)\right)^{(2)} = \frac{1}{k^2}\left(\left|\left\langle S_3^{(2)}\right\rangle\right|^2 + \left|\left\langle S_1^{(2)}\right\rangle\right|^2\right). \tag{9.68}$$

We determine "effective" extinction indices of the cell as follows:

$$\gamma_{\mathrm{e,o}}^{\mathrm{eff}} = \pi N \int_0^{\pi} \left[\left(\frac{d\sigma}{d\Omega}\right)_{\mathrm{e,o}}^{(1)} + \left(\frac{d\sigma}{d\Omega}\right)_{\mathrm{e,o}}^{(2)}\right] \sin\theta_{\mathrm{s}} d\theta_{\mathrm{s}}. \tag{9.69}$$

At normal light incidence Eq. (9.69) is transformed into [41]

$$\gamma_{\mathrm{e,o}}^{\mathrm{eff}} = \frac{Nk^4 g}{16\pi}\left(\frac{\varepsilon_{\mathrm{do}}}{\varepsilon_{\mathrm{p}}} - 1 + \frac{\Delta\varepsilon_{\mathrm{d}}}{3\varepsilon_{\mathrm{p}}}\left(1 + 2S_{x,y}\right)\right)^2, \tag{9.70}$$

where

$$g = \int_0^{\pi} \langle F\rangle^2 \left(\cos^2\theta_{\mathrm{s}} + 1\right) \sin\theta_{\mathrm{s}} d\theta_{\mathrm{s}}, \tag{9.71}$$

$\langle F\rangle$ is the function given by Eq. (9.20) averaged over the LC droplet size.

Assume that LC droplet radii R are distributed according to the gamma distribution with the probability density:

$$\rho_{\mu}(R) = \frac{\mu^{\mu} R^{\mu-1}}{\langle R\rangle^{\mu}\Gamma(\mu)} \exp\left(-\mu\frac{R}{\langle R\rangle}\right), \tag{9.72}$$

where $\langle R\rangle$is the mean radius of LC droplets; μ is the distribution parameter. Then

$$\langle F\rangle = \frac{4\pi}{a^3}\left(t^2 - 1\right)^{-\frac{\mu+1}{2}} \left(\begin{array}{l} \sin\left((\mu+1)\arctan t\right) \\ -\frac{t(\mu+1)}{\sqrt{t^2+1}} \cos\left((\mu+2)\arctan t\right) \end{array}\right), \tag{9.73}$$

$$t = \frac{a\langle R\rangle}{\mu+1}. \tag{9.74}$$

Comparison of the described simplified approach for the extinction index calculation with the results obtained by Eq. (9.63) has shown their reasonable coincidence [41].

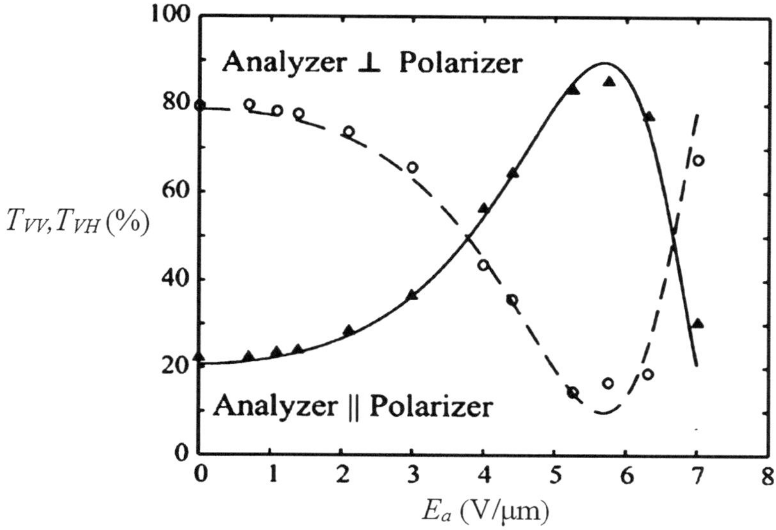

Fig. 9.12 Transmittance at parallel T_{VV} and crossed T_{VH} polarizer and analyzer versus applied electric field strength. Theoretical results obtained by the described model (curves) ($SS_d = 079$). Experimental data [19] (*symbols*)

Equation (9.70) allows one to determine effective extinction indices by the order parameter tensor components. Note that the function g depends on the polydispersity of droplets. It does not change with the external field. It was used to calculate the ellipticity and azimuth of the polarization state of transmitted light. Calculation error for these values as follows from numerical analysis is about 1 %.

The results of calculations for transmittance are displayed in Fig. 9.12 (lines) in comparison with the experimental data [19] (symbols). The described theoretical results were compared [42] with the experimental data of [19]. The sample was prepared using LC BL24 from Merck Industrial Chemicals ($n_e = 1.7174$, $n_o = 1.5132$ at 589 nm) and polymer NOA81 from Norland Products Inc. ($n_p = 1.5662$ at 589 nm). The volume concentration of LC droplets C_v was about 1 % and the mean radius of droplets was measured to be 50 nm. The sample was illuminated with infrared radiation ($\lambda = 1.3$ μm) along its plane. Electric field was applied normally to the sample plane. The incident light was polarized at angle $\alpha = 45°$ to the z-axis by polarizer. Analyzer was placed behind the sample and was oriented in two ways: parallel and perpendicular to the polarizer. Transmittance at parallel T_{VV} and crossed T_{VH} polarizer and analyzer was measured.

It is obvious that the theoretical model is in good agreement with the experiment. Moreover, since the only fitting parameter is the product SS_d, it might expect that the method proposed can be employed to determine the order parameters by the transmittance measurements.

9.7 Polarization of Transmitted Light

Let the incident beam be parallel to the direction of applied electric field and normal to the film surface. At this case [see Eqs. (9.45) and (9.46)] amplitudes of extraordinary α_e and ordinary α_o waves are

$$a_{e,o} = t_{e,o} \begin{pmatrix} \cos\alpha \\ \sin\alpha \end{pmatrix}. \tag{9.75}$$

Here α is the polarization angle of the incident wave relative to the x-axis of the laboratory frame; $t_{e,o}$ are the transmission coefficients:

$$t_{e,o} = \exp\left(-\frac{1}{2}\gamma_{e,o} l\right). \tag{9.76}$$

Using the above equations, one can write the expression for polarization ellipse:

$$\left(\frac{E_x}{a_e}\right)^2 + \left(\frac{E_y}{a_o}\right)^2 - 2\frac{E_x E_y}{a_e a_o}\cos\Delta\Phi_n = \sin^2\Delta\Phi_n. \tag{9.77}$$

The semiaxes of the polarization ellipse are

$$A^2 = (a_e\cos\beta)^2 + (a_o\sin\beta)^2 + a_e a_o \sin 2\beta\cos\Delta\Phi_n, \tag{9.78}$$

$$B^2 = (a_e\sin\beta)^2 + (a_o\cos\beta)^2 - a_e a_o \sin 2\beta\cos\Delta\Phi_n, \tag{9.79}$$

where

$$tg2\beta = 2\frac{a_e a_o}{a_e^2 - a_o^2}\cos\Delta\Phi_n. \tag{9.80}$$

Consider ellipticity η and azimuth ξ of the transmitted wave. Ellipticity η is a ratio of the semiminor axis of the ellipse of polarization to the semimajor one:

$$\eta = \sqrt{\frac{(a_e\sin\xi)^2 + (a_o\cos\xi)^2 - a_e a_o \sin 2\xi\cos\Delta\Phi_n}{(a_e\cos\xi)^2 + (a_o\sin\xi)^2 + a_e a_o \sin 2\xi\cos\Delta\Phi_n}}. \tag{9.81}$$

Azimuth ξ is an angle between the major axis of the polarization ellipse and the x-axis of the laboratory frame counted counterclockwise with respect to the positive direction of axis z. If $A > B$, the angle $\xi = \beta$. Otherwise, if $\beta < \pi/2$, the angle $\xi = \beta + \pi/2$, and if $\beta > \pi/2$, then $\xi = \beta - \pi/2$.

Transition from a partially ordered structure of droplet directors to a homeotropic structure is displayed schematically in Fig. 9.13. The phase difference $\Delta\Phi$ (which depends on the order parameter S_z) changes from 0 to π [40].

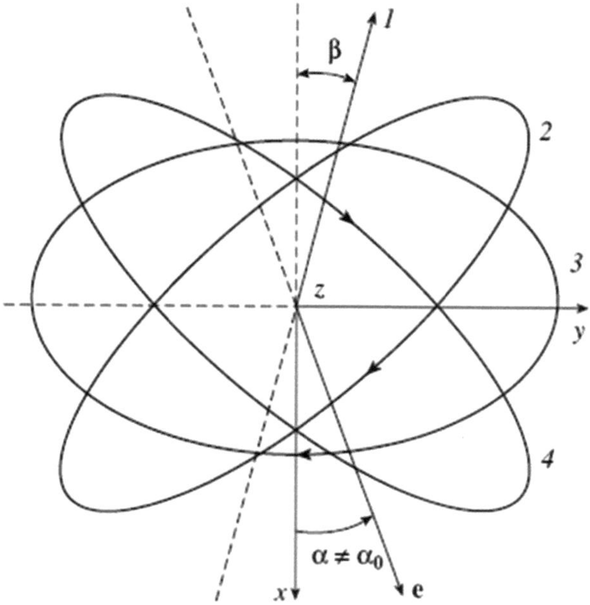

Fig. 9.13 Schematic view of the shape and orientation of the polarization ellipse. The *arrows* indicate the sign of rotation of the electric vector. Notations are in the text

It follows from Eqs. (9.78), (9.79), and (9.80) that, for $\Delta\Phi = \pi$, a wave transmitted through a film at an angle of β ($\tan \beta = (t_e/t_o)\tan \alpha$) is linearly polarized (line *1* in Fig. 9.13). For $\Delta\Phi = 0$, the transmitted wave holds the linear polarization of the incident wave (line *5* in Fig. 9.13). When $\pi > \Delta\Phi > 0$, the transmitted wave is elliptically polarized. The polarization ellipse rotates clockwise for the optical anisotropy $\Delta\varepsilon > 0$ (ellipses *2*, *3*, and *4* in Fig. 9.13). Maximal ellipticity is obtained for $\Delta\Phi = \pi/2$ (ellipse *3* in Fig. 9.13). When $\Delta\Phi = \pi/2$ and a wave is incident at an angle of $\alpha_o(\tan \alpha_o = t_e/t_o)$, then the transmitted wave is circularly polarized ($\eta = 1$). Figure 9.13 illustrates the rotation of the polarization ellipse and of the plane of polarization of the transmitted wave for polarization angles $\alpha \neq \alpha_o$ of the incident wave. The electric vector rotates clockwise from the positive direction of axis z (right elliptic polarization).

The calculated dependences of the ellipticity and the azimuth on the strength of the external electric field applied to the PDLC sample are shown in Fig. 9.14. The phase shift between extraordinary and ordinary waves (see dashed line) is displayed in this figure as well.

In a general case the transmitted light is elliptically polarized. Under certain conditions linear or circular polarizations can be implemented. For the case under consideration when the strength of external field is equal to 4.5 V/μm (corresponding to the phase shift between extraordinary and ordinary waves $\Delta\Phi_n = 180°$), the ellipticity $\eta = 0$ (see Fig. 9.14). Consequently, the transmitted light is linearly polarized. At this strength the polarization angle of the transmitted light is characterized by the azimuth $\xi = 127°$. This means that the polarization plane of the transmitted light is rotated (polarization angle α of incident light is 50°).

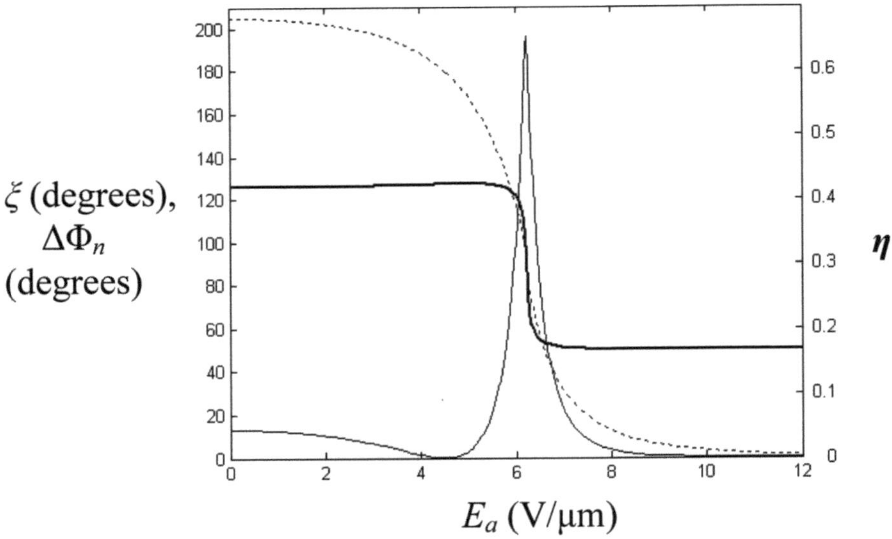

Fig. 9.14 Ellipticity η (*thin solid line*), azimuth ξ (*bold solid line*), and phase shift $\Delta\Phi_{\mathrm{n}}$ (*dashed line*) of the transmitted light at normal illumination versus the electric field strength E_{a}. Polarization angle of the incident light $\alpha = 50°$; the angles $\theta_d^{max} = 25°$ and $\theta_d^{max} = 5$. The other parameters are the same as in Fig. 9.5

One more case of linear polarization of the transmitted light is displayed in Fig. 9.14. For a strong applied field, the ellipticity $\eta = 0$ and the azimuth equals to the polarization angle of the incident light. This implies that the transmitted light keeps its initial polarization state. If the phase shift $\Delta\Phi_{\mathrm{n}} = \pi/2$ (for the sample under consideration it occurs at 6.25 V/μm), the ellipticity has a peak, and the azimuth is equal to $\pi/2$.

The transmitted light is linearly polarized at angle

$$\tan \xi_{\mathrm{lin}} = (-1)^n \left\{ \exp\left[\frac{k^3 gn}{8\langle V \rangle} \left(\frac{\varepsilon_{\mathrm{do}}}{\varepsilon_{\mathrm{p}}} - 1 + \frac{2n\pi\varphi_{\mathrm{d}}^{\mathrm{max}}}{Lkc_{\mathrm{v}} \sin\left(2\varphi_{\mathrm{d}}^{\mathrm{max}}\right)} \right) \right] \tan\alpha \right\}, \tag{9.82}$$

if the phase shift $\Delta\Phi_{\mathrm{n}} = \pi n$ (n is an integer). For $\langle R \rangle$ tending to zero the transmitted light polarization angle $\xi_{\mathrm{lin}} = \pi - \alpha$ for $\Delta\Phi_{\mathrm{n}} = (2n+1)\pi$ and $\xi_{\mathrm{lin}} = \alpha$ for $\Delta\Phi_{\mathrm{n}} = 2n\pi$.

Dependences of polarization angle ξ_{lin} of the transmitted linearly polarized light versus the angle φ_d^{max} and the average LC droplet radius $\langle R \rangle$ at different polarization angles of the incident light are displayed in Fig. 9.15.

It is evident from Fig. 9.15 that at small $\langle R \rangle$ the angle $\xi_{\mathrm{lin}} = \pi - \alpha$. Thus the polarization plane can be rotated by more than 90° by means of a PDLC cell with nanosized LC droplets. The other parameters entering into Eq. (9.82) have rather weak influence on the polarization angle ξ_{lin}.

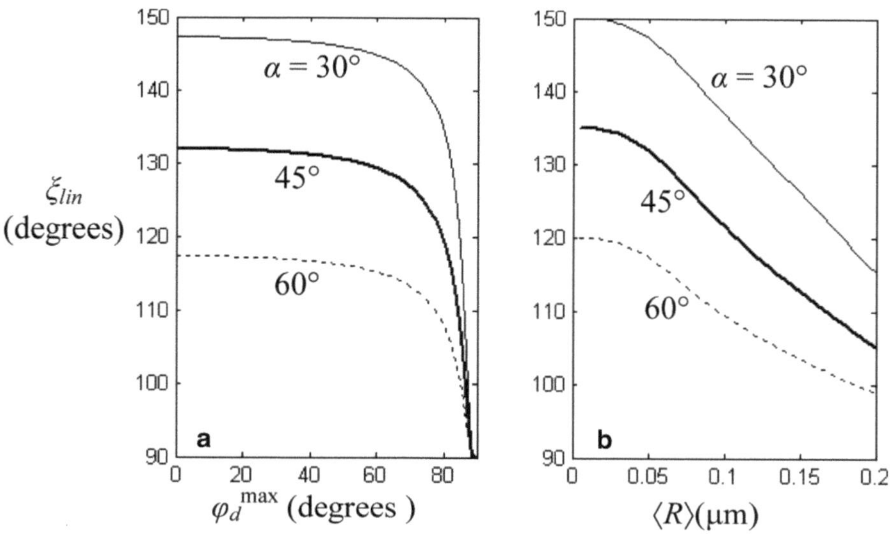

Fig. 9.15 Polarization angle ξ_{lin} of the transmitted linearly polarized light versus the angle φ_d^{max} (**a**) and the average LC droplet radius $\langle R \rangle$; (**b**) at different polarization angles of the incident light. The other parameters are the same as in Fig. 9.5

Rotation angle Φ_{rot} of the polarization plane, obviously, is determined by the relation:

$$\Phi_{\text{rot}} = \xi_{\text{lin}} - \alpha. \tag{9.83}$$

Dependence of the rotation angle on the average radius of LC droplets $\langle R \rangle$ and angle φ_d^{max} is displayed in Fig. 9.16.

The rotation angle depends linearly on the angle α:

$$\Phi_{\text{rot}} = \pi - 2\alpha. \tag{9.84}$$

Dependence of the ellipticity on the polarization angle of incident light and the strength of external electric field is displayed in Fig. 9.17. There are two polarization angles of the incident light such that the circular polarization of transmitted light (the ellipticity $\eta = 1$) is implemented.

Dependence of ellipticity η versus the applied field and the angle φ_d^{max} is shown in Fig. 9.18.

We now consider the cases in which it is possible to get circular polarization of the transmitted wave. From Eq. (9.81) it follows that ellipticity $\eta = 1$ if the incident light is polarized at angle

$$\alpha_{\text{circ}} = \arctan\left(\exp\left(\frac{-k^3}{8\langle V\rangle}\left(\frac{\varepsilon_{do}}{\varepsilon_{\text{p}}} - 1 + \frac{\pi\,(n+1/2)\,\phi_d^{\max}}{lkc_{\text{V}}\sin\left(2\phi_{\text{d}}^{\max}\right)}\right)\right)\right) \tag{9.85}$$

or at angle $\alpha'_{\text{circ}} = \pi - \alpha_{\text{circ}}$, and phase shift $\Delta\Phi_{\text{n}} = (n + 1/2)\pi$. Dependence of the angle α_{circ} on the average LC droplet radius $\langle R \rangle$ is displayed in Fig. 9.19.

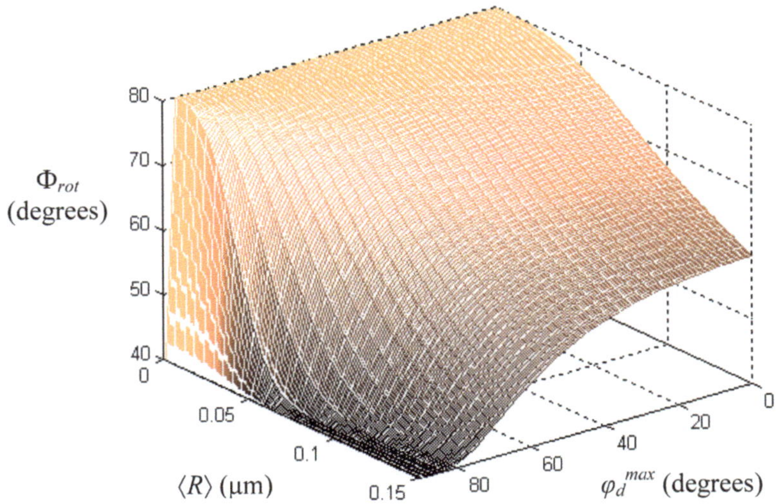

Fig. 9.16 Angle Φ_{rot} versus average radius of LC droplets $\langle R \rangle$ and angle φ_d^{max}. The other parameters are the same as in Fig. 9.5

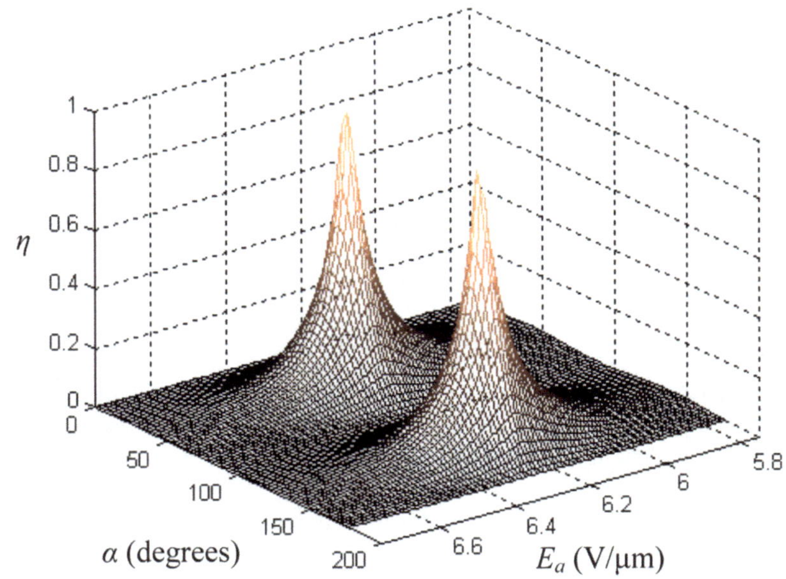

Fig. 9.17 Ellipticity η versus polarization angle α of the incident light and the applied electric field strength. Normal illumination. The other parameters are the same as in Fig. 9.5

The angle α_{circ} tends to $\pi/4$ at small size of LC droplets and does not exceed this magnitude for LCs with positive birefringence. By means of Eqs. (9.82) and (9.85), droplet size can be estimated through the optical measurements.

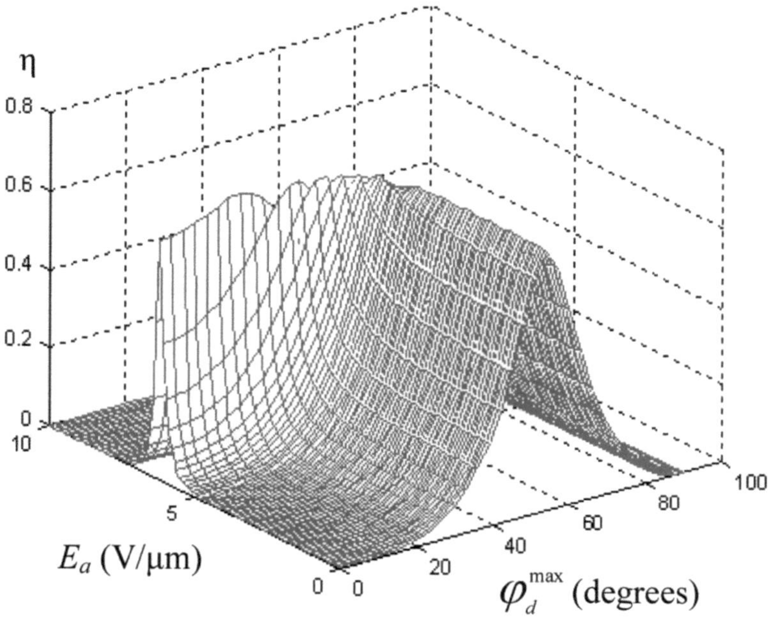

Fig. 9.18 Ellipticity η versus the applied field E_a and the angle φ_d^{max}. Polarization angle $\alpha = 50^0$; the average LC droplet diameter $\langle d \rangle = 75$ nm, $\theta^{\max} = 5^0$. The other parameters are the same as in Fig. 9.5

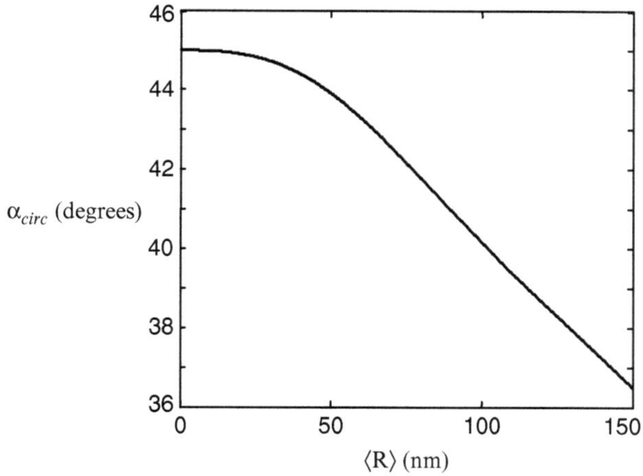

Fig. 9.19 Polarization angle of the incident light α_{circ} at which the transmitted light is circularly polarized versus the average droplet radius $\langle R \rangle$. The results are calculated at $\varphi_d^{max} = 5°$ at the parameters specified in Fig. 9.5

Note that not always it is possible to produce light circular polarization at the output of the film. There are initial orientations of the LC droplets when it is unavailable. Maximum ellipticity depends on the size of droplets and the degree of orientational order in the plane (*xy*). The example is shown in Fig. 9.20. For the considered case angle φ_d^{max} should not exceed 35°, to obtain circular polarization of the transmitted light.

9.8 Liquid-Crystal Lens with Electrically Variable Focus Length

LC lenses usually have polarization-dependent properties. Using the PDLCs allows one to make polarization-insensitive lenses. Below, based on the described results, we analyze electrically tunable polarization-independent light focusing by two types of PDLC lenses with nematic LC droplets with random (at zero field) orientation of directors. One of these called for brevity "flat lens" is a plane parallel PDLC layer with axially symmetric distribution of liquid-crystal droplet concentration. The other is a combination of a PDLC layer with a glass lens. We call it for brevity "combined lens" [47].

9.8.1 *Flat Lens*

Consider a plane parallel PDLC film of thickness l with axially symmetric distribution of liquid-crystal droplet concentration confined between plane parallel transparent electrodes, with an aperture of diameter D. The (*xy*) plane of the laboratory coordinate system x, y, z is the front surface of the PDLC film, and the z-axis is normal to the film (see Fig. 9.21).

At zero field, the LC directors are randomly oriented (the order parameter $S_z = 0$). When a uniform electric field is applied along the normal to the film, the droplet directors align with the field. The characteristics of light transmitted through the PDLC film with droplet directors reoriented in this manner are independent of the light polarization at normal incidence.

We use the Foldy–Twersky equation [38, 40] to determine the coherent transmitted field. The mean coherent field amplitude in the plane $z = l$ is expressed as

$$\langle E\,(x,y,l)\rangle = E_{\mathrm{i}}\exp(iKl), \tag{9.86}$$

where E_{i} is the amplitude of the incident wave and K is the propagation constant of the film:

$$K = k + iq\,\langle S(0)\rangle\,. \tag{9.87}$$

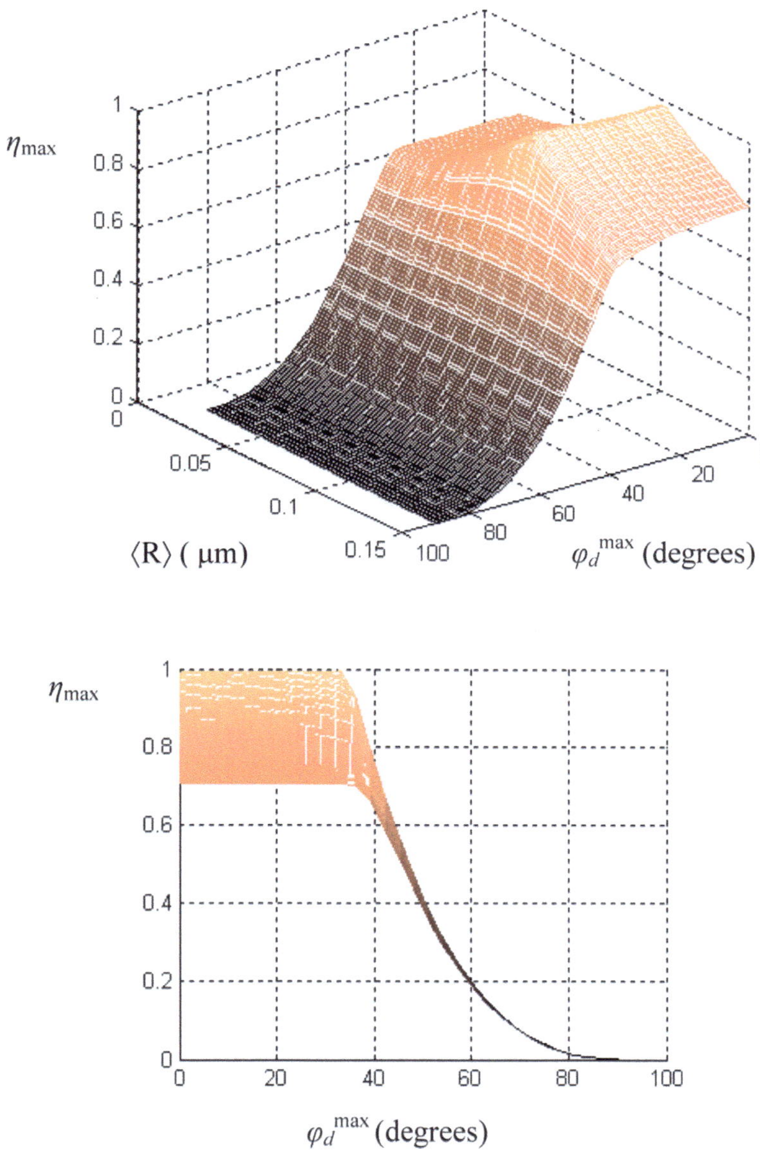

Fig. 9.20 Maximum ellipticity as function of the average radius of droplets $\langle R \rangle$ and angle φ_d^{max}*(top)*, the projection of a three-dimensional graph in the plane of maximum ellipticity angle φ_d^{max}*(bottom)*

Here, $k = 2\pi/\lambda_p$ (λ_p is the wavelength of the incident light in the polymer), $q = 2\pi N_v/k^2$ (N_v is the concentration of droplets), and $\langle S(0) \rangle$ is the forward amplitude scattering averaged over LC droplet size and director orientation.

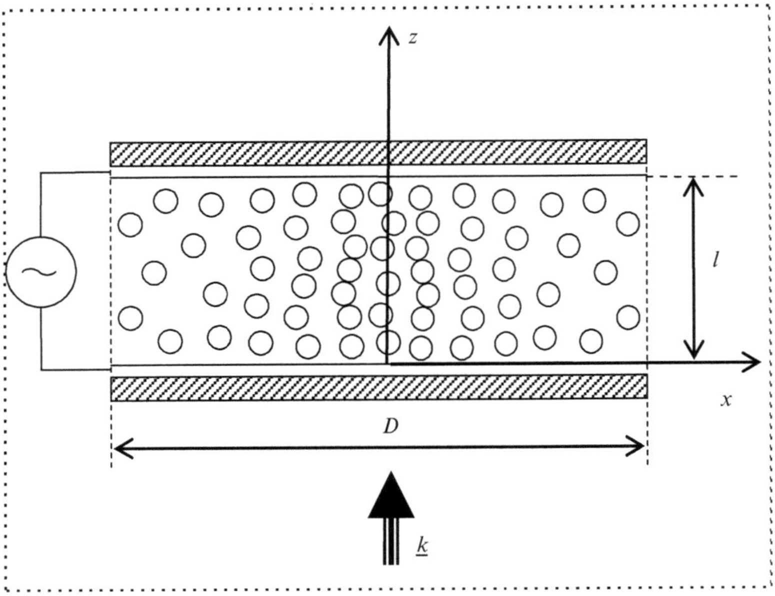

Fig. 9.21 Schematic representation of a flat lens with axially symmetric droplet distribution

It follows from Eqs. (9.86) and (9.87) that the PDLC film can be treated as a uniform retarder with the complex refractive index [45, 47, 48]:

$$\tilde{n} = \frac{K}{k_{\mathrm{i}}} = n_{\mathrm{p}} \left(1 + \frac{2\pi}{k_{\mathrm{p}}^3} N_{\mathrm{v}} \left\langle S(0) \right\rangle \right), \tag{9.88}$$

where $k_i = 2\pi/\lambda_{\mathrm{i}}$ is the wave vector of the incident on the film light, λ_{i} is the incident wavelength, and n_{p} is the refractive index of the polymer.

The optical characteristics and director orientations of LC droplets are changed by the applied field, and the averaged amplitude scattering $\langle S(0) \rangle$ changes accordingly. Using the Rayleigh–Gans approximation, we can write an expression for the averaged amplitude scattering $\langle S(0) \rangle$ of nanosized droplets as follows [40, 45]:

$$\langle S(0) \rangle = -\frac{ik_{\mathrm{p}}^3}{4\pi} \langle V \rangle \left[\frac{\varepsilon_{\mathrm{do}}}{\varepsilon_{\mathrm{p}}} - 1 + \frac{\varepsilon_{\mathrm{de}} - \varepsilon_{\mathrm{do}}}{\varepsilon_{\mathrm{p}}} \left(1 - S_f \right) \right], \tag{9.89}$$

where $\langle V \rangle$ is the mean LC droplet volume, ε_{p} is the dielectric constant of the polymer, $S_f \equiv S_z$ is the field-dependent scalar order parameter of the PDLC film, and $\varepsilon_{\mathrm{do}}$ and $\varepsilon_{\mathrm{de}}$ denote the effective ordinary and extraordinary dielectric constants of droplet, respectively. They are determined by Eqs. (9.13) and (9.14).

Using relations (9.86), (9.87), (9.88), and (9.89), one can find expressions for the absolute value t_a of amplitude transmittance and the real part of the complex refractive index of the PDLC film:

$$t_{\mathrm{a}} = \exp\left(-\frac{1}{2}\gamma l\right), \tag{9.90}$$

$$\mathrm{Re}\tilde{n} = n_{\mathrm{f}} = n_{\mathrm{p}}\left\{1 + C_{\mathrm{V}}\frac{1}{2}\left[\frac{\varepsilon_{\mathrm{o}}}{\varepsilon_{\mathrm{p}}} - 1 + \frac{\Delta\varepsilon}{3\varepsilon_{\mathrm{p}}}\left(1 - SS_d S_f\right)\right]\right\}, \tag{9.91}$$

where γ is the attenuation coefficient, $C_{\mathrm{v}} = N\langle V\rangle$ is the LC volume fraction in the film, and n_f is the effective refractive index of the PDLC film.

The film with axially symmetric distribution of refractive index can be used as a polarization-independent lens. According to Eq. (9.91), the spatial distribution of the real part of refractive index can be varied by creating an axially symmetric distribution of LC droplet concentration in the film and/or applying an axially symmetric field.

Suppose that the LC droplet concentration is described by the Gaussian axially symmetric distribution:

$$C_{\mathrm{v}} = C_{\mathrm{v}}(r) = C_{\mathrm{V}}^{c}\exp\left(-\alpha r^2/R^2\right), \tag{9.92}$$

$$\alpha = \ln\left(C_{\mathrm{V}}^{c}/C_{\mathrm{V}}^{b}\right), \tag{9.93}$$

where C_V^c and C_V^b are the LC droplet concentrations at the center and periphery of the lens, respectively; $r^2 = x^2 + y^2$ determines the distance from the lens center; and $R = D/2$ is the lens radius.

At the paraxial approximation, the refractive power of the lens is

$$\frac{1}{f} = -\frac{1}{n}\int_0^1\left(\frac{d^2\mathrm{Re}\tilde{n}(r)}{d^2 r}\right)_{r=0} dz. \tag{9.94}$$

Here n is the ambient refractive index and f is the focal length of the lens.

According to expression (9.91), when the LC droplet concentration decreases toward the periphery of the lens, the second derivative of $\mathrm{Re}\tilde{n}(r)$ is negative at $r = 0$; i.e., $f > 0$ (the lens is converging). When the LC droplet concentration increases toward edge, we have $f < 0$ (the lens is diverging).

Combining distribution (9.92) with (9.91) and (9.94), we obtain

$$f = \frac{R^2}{l}\frac{1}{2\alpha}\frac{n}{n_{\mathrm{p}}C_{\mathrm{V}}^{c}A}. \tag{9.95}$$

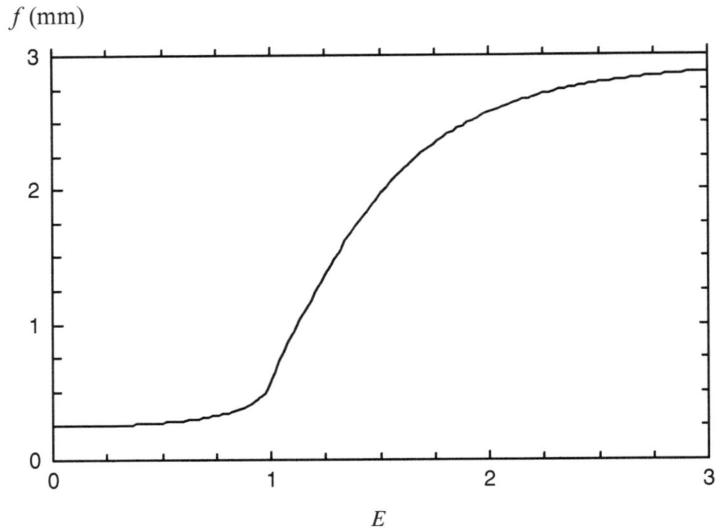

Fig. 9.22 Focal length f versus normalized applied field E for flat lens. $D = 108$ μm, $l = 36$ μm, $C_V^c = 0.8$, $C_V^b / C_V^c = l_b / l_c = 0.05$. The LC refractive indices $n_o = 1.511$ and $n_e = 1.457$, $n_p = 1.524$, $n_g = 1.74$, $n = 1$, $\lambda = 0.6328$ μm, $S = 0.94$, and $S_d = 0.82$

Here

$$A = \frac{1}{2}\left[\frac{\varepsilon_o}{\varepsilon_p} - 1 + \frac{\Delta\varepsilon}{3\varepsilon_p}\left(1 - SS_dS_f\right)\right]. \tag{9.96}$$

In general case the focal length corresponding to an axially symmetric LC droplet concentration can be written as

$$f = \frac{R^2}{l} G \frac{n}{n_p C_V^c A}, \tag{9.97}$$

where G is a function depending on the distribution of LC droplet concentration. For the converging lens, $G > 0$. For the diverging lens, $G < 0$.

Dependence of the focal length versus normalized applied field (E) for the flat lens is shown in Fig. 9.22.

9.8.2 Combined Lens

A combined lens consisting of a glass lens, a substrate, transparent electrodes, and a PDLC is schematized in Fig. 9.23.

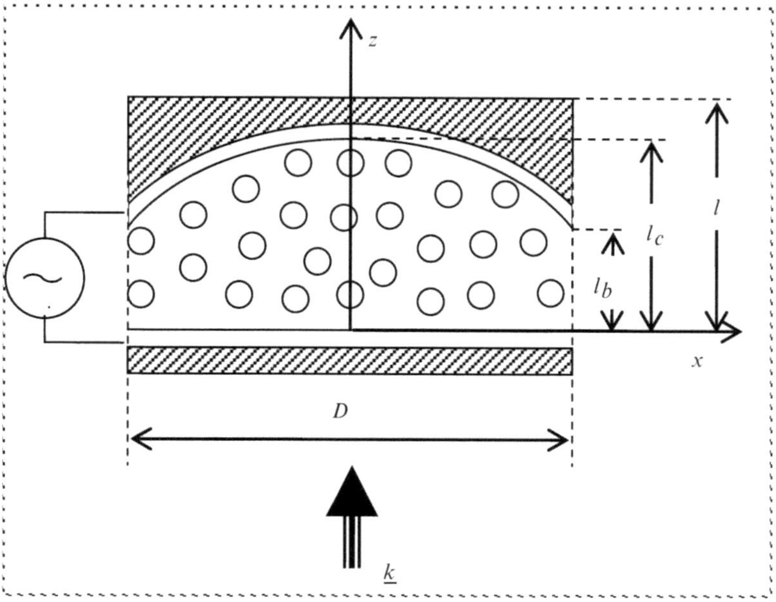

Fig. 9.23 Schematic representation of a combined lens comprising a PDLC with uniform LC droplet distribution and a glass lens

Using the Foldy–Twersky equation [38], we obtain the following expression for the amplitude of the coherent field component at $z = l$ (here l is the overall thickness of the combined PDLC–glass lens):

$$\langle E \rangle = E_i t_a \exp\left(i\Phi(r)\right). \tag{9.98}$$

Here $\Phi(r)$ is the axially symmetric phase distribution in a beam transmitted through the PDLC film and glass lens, E_i is the incident field amplitude, and t_a is the absolute value of the amplitude transmittance.

Expression (9.98) is a generalized solution to the Foldy–Twersky equation for an axially symmetric optical system. It is obtained by taking into account the axial symmetry of the field-induced reorientation of droplet directors. For an axially asymmetric combined lens, the phase of the field at $z = l$ is

$$\Phi(r) = kz(r) + k_g\left(l - z(r)\right). \tag{9.99}$$

Here k is the module of the wave vector in the film and k_g is the module of the wave vector in the glass; the function $z(r)$ describes an axially symmetric PDLC–glass interface.

Equation (9.98) can be rewritten as

$$\langle E \rangle = E_i t_a \exp\left(ik_i l \mathrm{Re}\tilde{n}(r)\right). \tag{9.100}$$

The real part of the complex refractive index of the combined lens is

$$\mathrm{Re}\tilde{n}(r) = n_g + \left(n_f - n_g\right)\frac{z(r)}{l}, \tag{9.101}$$

where n_g is the refractive index of the glass. The value n_f is the effective refractive index of the film (Eq. 9.91).

Note that the nonuniformity of the applied field due to the curvature of an electrode in the system depicted in Fig. 9.23 is neglected.

If the PDLC–glass interface (see Fig. 9.23) is a convex spherical segment of radius R_0, then

$$z(r) = h\left[1 + C_1\left(\sqrt{1 - C_2\frac{r^2}{R^2}} - 1\right)\right], \tag{9.102}$$

where $h = l_c - l_b$ (l_c and l_b denote the PDLC film thickness at the center and periphery of the lens, respectively),

$$C_1 = \frac{R_0}{h} = \frac{1}{2}\left(\frac{R^2}{h^2} + 1\right), \tag{9.103}$$

$$C_2 = \frac{R^2}{R_0^2} = \frac{4(R/h)^2}{\left((R/h)^2 + 1\right)^2}, \tag{9.104}$$

and R is the lens radius. For function $z(r)$ (Eq. 9.102), there are relations:

$$z(r) = \begin{cases} h, r = 0 \\ 0, r = R \end{cases}, \tag{9.105}$$

$$\left(\frac{d^2 z(r)}{d^2 r}\right)_{r=0} = -\frac{h}{R^2}C_1 C_2 = -\frac{1}{R_0}. \tag{9.106}$$

The focal length f of a lens with convex PDLC–glass interface is

$$f = \frac{R_0 n}{n_f - n_g} = \frac{DBn}{n_p\left(1 + C_V A\right) - n_p}. \tag{9.107}$$

Here

$$B = \frac{1}{4}\left(\frac{D}{2l_c\alpha} + \frac{2l_c\alpha}{D}\right), \tag{9.108}$$

$$\alpha = 1 - \frac{l_b}{l_c}. \tag{9.109}$$

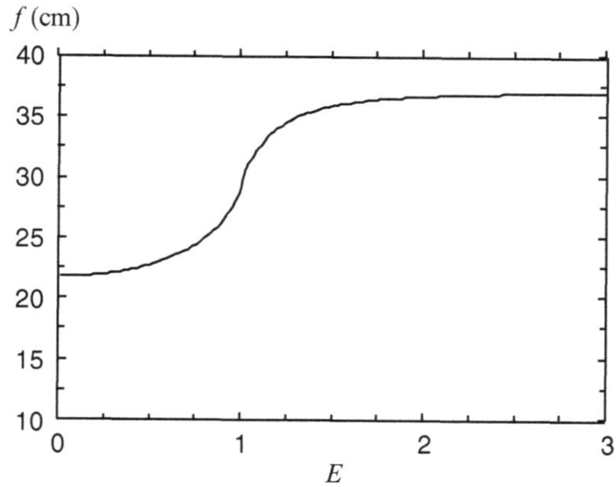

Fig. 9.24 Focal length f versus normalized applied field E for combined lens with $D = 10$ μm, $l_c = 500$ μm, and $l_b/l_c = 0.05$. The other parameters are the same as in Fig. 9.22

It is clear from Eqs. (9.108) and (9.109) that the focal length of the combined lens is determined by the PDLC film thickness at the center and periphery of the cell, being independent of the glass thickness. If $n_f < n_g$, then $f > 0$, and vice versa (Fig. 9.24).

This implies that the glass and liquid-crystal refractive indices can be adjusted by varying the applied voltage so that the lens is switched between converging and diverging. The following expression for absolute value of the amplitude transmittance of the lens depicted in Fig. 9.23 takes place:

$$t_a = \exp\left(-\frac{1}{2}\gamma(lb + z(r))\right), \tag{9.110}$$

where the attenuation coefficient γ is determined using Eq. (9.63).

Finally, we note that the focal length of a lens with an aspherical PDLC–glass interface is

$$f = \pm\frac{R_c n}{n_f - n_g}, \tag{9.111}$$

where R_c is the radius of curvature of the interface at the center of the lens. The plus and minus signs correspond to convex and concave interfaces, respectively. The dependence of focal length f on normalized applied field E is shown in Fig. 9.25.

As noted above, a lens with a proper chosen refractive indexes of glass and liquid crystal may be either converging ($f > 0$) or diverging ($f < 0$). Figure 9.25 shows the focal length plotted versus E for a converging–diverging lens.

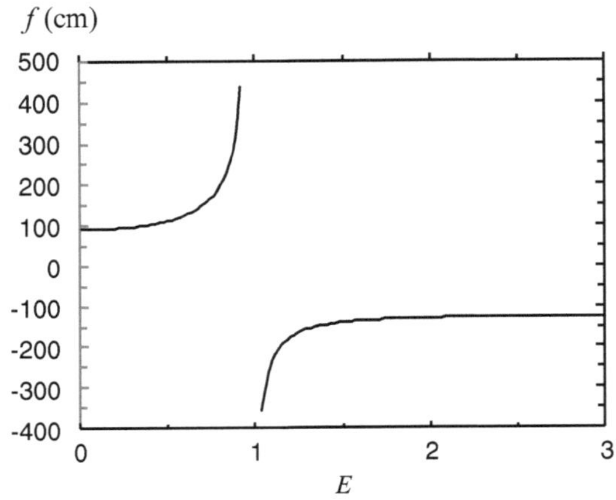

Fig. 9.25 Focal length f versus normalized applied field E for the converging–diverging combined lens with $D = 10$ μm, $l_c = 500$ μm, $l_b/l_c = 0.05$. $n_g = 1.55$. The other parameters are the same as in Fig. 9.21

9.9 Conclusions

In this chapter, we considered methods for analyzing the light extinction, polarization, and phase by polymer film containing nanosized liquid-crystal droplets at normal and oblique incidence of light. The methods are convenient for use in practical applications because they do not require knowledge of the exact distribution of the droplet director in the layer. They operate with the order parameters. The results enable one to analyze characteristics of transmitted radiation as a function of the morphology and structure of the film. The approach is valid for samples with spherical droplets and can be extended for films with elliptical droplets. The model is in good agreement with the experimental data.

Numerical results for extinction of ordinary and extraordinary components of transmitted light are presented. Phase shift between extraordinary and ordinary components and the polarization state of light transmitted through PDLC film with nanosized spherical partially oriented nematic droplets are investigated. Dependence of the polarization state of the transmitted light on the properties of a sample is established. The conditions to get circular and linear polarization of transmitted light are found.

Equations to calculate the focal lengths of the polarization-independent lenses as functions of their geometric parameters, film morphology, and applied field are derived. Analysis of the focal length of lenses as a function of lens geometry is made. It is shown that the combined lens can be used as the converging–diverging one.

The results can be used to design various devices for light modulation based on polymer-dispersed liquid-crystal materials: optical filters, tunable polarizers, phase plates, polarization plane rotators, devices for mapping of optical information, and lenses with tunable focal lengths. The last can be used in fiber optical switch design, laser collimation, phase-conjugate optics, ophthalmic optics, and photographic optics.

The described methods make it possible to solve the inverse scattering problem, i.e., to determine the parameters of polymer films with small liquid-crystal droplets (to evaluate the size and shape of the liquid-crystal droplets and the degree of their orientational ordering) from the characteristics of transmitted radiation.

References

1. Collings, P.J.: Liquid Crystals. Nature's Delicate Phase of Matter. Princeton University Press, Princeton and Oxford (2002)
2. Blinov, L.M.: Structure and Properties of Liquid Crystals. Springer, Dordrecht (2011)
3. Pasechnik, S.V., Chigrinov, V.G., Shmeliova, D.V.: Liquid Crystals, Viscous and Elastic Properties. Wiley-VCH, Berlin (2009)
4. Drzaic, P.S.: Liquid Crystals Dispersions. World Scientific, Singapore (1995)
5. Simoni, F.: Nonlinear Optical Properties of Liquid Crystals. World Scientific, Singapore (1997)
6. Drzaic, P.S., Gonzales, A.M.: Refractive index gradients and light scattering in polymer-dispersed liquid crystal films. Appl. Phys. Lett. **62**, 1332–1334 (1993)
7. McKenna, L., Miller, L.S., Peterson, I.R.: Polymer dispersed liquid crystal films for modulating infra-red radiation. Polymer **45**, 6977–6984 (2004)
8. Andreev, A.L., Bobylev, Yu.P., Gubasaryan, N.A., Kompanets, I.N., Pozhidaev, E.P., Fedosenkova, T.B., Shoshin, V.M., Shumkina, Yu.P.: Electric-field-controlled light scattering in ferroelectric liquid crystals. J. Opt. Technol. **72**, 701–707 (2005)
9. Andreev, A.L., Bobylev, Yu.P., Fedosenkova, T.B., Yambaev, I.B., Kompanets, I.N., Pozhidaev, E.P., Shoshin, V.M., Shumkina, Y.P.: Dynamic parameters of electrically controlled light scattering in helix FLC cells. J. SID **14**, 643–648 (2006)
10. Loiko, V.A., Konkolovich, A.V., Miskevich, A.A.: Transient light scattering in helix ferroelectric liquid crystal cells. Liq. Cryst. **36**, 365–370 (2009)
11. Loiko, V.A., Konkolovich, A.V., Miskevich, A.A.: Optical model of transient light scattering in ferroelectric liquid crystals. J. Exp. Theor. Phys. **108**, 535–545 (2009)
12. Zyryanov, V.Ya., Krakhalev, M.N., Prishchepa, O.O, Shabanov., A.V.: Orientational structure transformations caused by the electric-field-induced ionic modification of the interface in nematic droplets. JETP Lett. **86**, 383–388 (2007)
13. Loiko, V.A., Zyryanov, V.Ya., Maschke, U., Konkolovich, A.V., Miskevich, A.A.: Small-angle light scattering and transmittance of polymer films, containing liquid crystals droplets with inhomogeneous boundary conditions. J. Quant. Spectrosc. Radiat. Transf. **113**, 2585–2592 (2012)
14. Dick, V.P., Loiko, V.A.: Electrically controlled dispersion filters for the visible and mid-infrared regions. J. Opt. Technol. **79**, 403–406 (2012)
15. Ibragimov, T.D.: Christiansen effect in alumina particles – liquid crystal system. J. Appl. Spectrosc. **76**, 752–755 (2009)
16. Ivanov, A.P., Loiko, V.A., Dick, V.P.: Light Propagation in Densely Packed Disperse Media. Science and Technics, Minsk (1988)
17. Matsumoto, S., Sugiyama, Y., Sakata, S., Hayashi, T.: Electro-optic effect, propagation loss, and switching speed in polymers containing nanosized droplets of liquid crystal. Liq. Cryst. **27**, 649–655 (2000)
18. Hands, P.J.W., Kirby, A.K., Love, G.D.: Phase modulation with polymer-dispersed liquid crystals. Proc. SPIE **5894**, 193–200 (2005)
19. Matsumoto, S., Houlbert, M., Hayashi, T., Kubodera, K.: Fine droplets of liquid crystals in a transparent polymer and their response to an electric field. Appl. Phys. Lett. **69**, 1044–1046 (1996)

20. Amimori, I., Priezjev, N.V., Pelcovits, R.A., Crawford, G.P.: Optomechanical properties of stretched polymer dispersed liquid crystal films for scattering polarizer applications. J. Appl. Phys. **93**, 3248–3252 (2003)
21. Ren, H., Fan, Y.-H., Wu, S.-T.: Tunable Fresnel lens using nanoscale polymer-dispersed liquid crystals. Appl. Phys. Lett. **83**, 1515–1517 (2003)
22. Bloisi, F., Ruocchio, C., Terrecuso, P., Vicari, L.: Optoelectronic polarizer by PDLC. Liq. Cryst. **20**, 377–379 (1996)
23. Ren, H., Fan, Y.-H., Lin, Y.-H., Wu, S.-T.: Tunable-focus microlens arrays using nanosized polymer-dispersed liquid crystal droplets. Opt. Commun. **247**, 101–106 (2005)
24. Gottardo, S., Cavalieri, S., Yaroshchuk, O.V., Wiersma, D.S.: Quasi-two-dimensional diffusive random laser action. Phys. Rev. Lett. **93**, 263901 (2004)
25. Levallois, C., Verbrugge, V., Dupont, L., de Bougrenet de la Tocnaye, J.-L., Caillaud, B., Le Corre, A., Dehaese O., Folliot H., Loualiche S.: 1:55-μm optically pumped tunable VCSEL based on a nano-polymer dispersive liquid crystal phase modulator. Proc. SPIE **61850W** (2006)
26. Kelly, J.R., Palffy-Muhoray, P.: The optical response of polymer dispersed liquid crystals. Mol. Cryst. Liq. Cryst. **243**, 11–29 (1994)
27. Kiselev, A.D., Yaroshchuk, O.V., Dolgov, L.O.: Ordering of droplets and light scattering in polymer dispersed liquid crystal films. J. Phys. Condens. Matter **16**, 7183–7197 (2004)
28. Mertelj, A., Copic, M.: Anisotropic diffusion of light in polymer dispersed liquid crystals. Phys. Rev. E **75**, 011705 (2007)
29. Zumer, S., Doane, J.W.: Light scattering from a small nematic droplet. Phys. Rev. A **34**, 3373–3386 (1986)
30. Loiko, V.A., Konkolovich, A.V., Miskevich, A.A.: Order parameter of elongated liquid crystal droplets: the method of retrieval by the coherent transmittance data. Phys. Rev. E **74**, 031704–031707 (2006)
31. Loiko, V.A., Molochko, V.I.: Polymer dispersed liquid crystal droplets: methods of calculation of optical characteristics. Liq. Cryst. **25**, 603–612 (1998)
32. Loiko, V.A., Molochko, V.I.: Influence of the director field structure on extinction and scattering by a nematic liquid-crystal droplet. Appl. Opt. **38**, 2857–2861 (1998)
33. Loiko, V.A., Konkolovich, A.V., Miskevich, A.A.: Using the Wentzel–Kramers–Brillouin method to analyze the transmission and reflection of light by a layer of liquid-crystal drops dispersed in a polymeric matrix. J. Opt. Technol. **78**, 455–459 (2011)
34. Aphonin, O.A., Panina, Y.V., Pravdin, A.B., Yakovlev, D.A.: Optical properties of stretched polymer dispersed liquid crystal films. Liq. Cryst. **15**, 395–407 (1993)
35. Loiko, V.A., Konkolovich, A.V., Miskevich, A.A., Maksimenko, P.G.: Propagation of light in the polymer dispersed liquid-crystal film with nanosized nematic droplets. Proc. SPIE **6023**, 49–58 (2005)
36. van de Hulst, G.: Light Scattering by Small Particles. Wiley, NewYork (1957)
37. Bohren, C., Huffman, D.: Absorption and Scattering of Light by Small Particles. Wiley, New York (1983)
38. Ishimaru, A.: Wave Propagation and Scattering in Random Media. Academic, New York (1978)
39. Mishchenko, M.I., Travis, L.D., Lacis, A.A.: Multiple Scattering of Light by Particles: Radiative Transfer and Coherent Backscattering. Cambridge University Press, Cambridge (2006)
40. Loiko, V.A., Konkolovich, A.V.: Polarization of light transmitted through a polymer film with nanosized droplets of liquid crystal. J. Exp. Theor. Phys. **99**, 343–351 (2004)
41. Loiko, V.A., Maksimenko, P.G., Konkolovich, A.V.: A method for calculating the light extinction coefficients of a polymer layer with small nematic liquid-crystal droplets. Opt. Spectrosc. **105**, 791–797 (2008)
42. Lisinetskaya, P.G., Konkolovich, A.A., Loiko, V.A.: Polarization properties of polymer-dispersed liquid-crystal film with small nematic droplets. Appl. Opt. **48**, 3144–3315 (2009)

43. Loiko, V.A., Maschke, U., Zyryanov, V.Ya., Konkolovich A.V., Miskevich A.A.: Angular structure of radiation scattered by monolayer of polydisperse droplets of nematic liquid crystal. Opt. Spectrosc. **110**, 110–118 (2011)
44. Loiko, V.A., Konkolovich, A.V., Maksimenko, P.G.: Polarization and phase of light transmitted through polymer dispersed liquid crystal film. J. Soc. Inf. Disp. **14**, 595–601 (2006)
45. Loiko, V.A., Konkolovich, A.V.: Phase change of a plane wave propagating through a polymer film with encapsulated nanodimensional nematic liquid crystal particles. J. Exp. Theor. Phys. **96**, 489–495 (2003)
46. Basile, F., Bloisi, F., Vicari, L., Simoni, F.: Optical phase shift of polymer-dispersed liquid crystals. Phys. Rev. E. **48**, 432–438 (1993)
47. Loiko, V.A., Konkolovich, A.V.: Focusing of light by polymer-dispersed liquid-crystal films with nanosized droplets. J. Exp. Theor. Phys. **103**, 935–943 (2006)
48. Dick, V.P., Loiko, V.A.: Optical phase shift by polymer dispersed liquid crystal films with fine droplets. J. Phys. D Appl. Phys. **37**, 1834–1840 (2004)

Chapter 10
Clusters and Nanoparticles in Superfluid Helium Droplets: Fundamentals, Challenges and Perspectives

Shengfu Yang and Andrew M. Ellis

Abstract Helium nanodroplets provide a cold and confined environment that offers many possibilities for the formation and investigation of clusters and nanoparticles. Here we present a review describing the fundamental properties of helium droplets and address in particular their application to and importance in the study of clusters and nanoparticles. We highlight several key experiments on atomic and molecular clusters and then turn our attention to very recent work on using helium droplets for nanoparticle synthesis. Finally, we look to the future and consider some areas where the growth of new nanoparticles via this route may be beneficial.

10.1 Introduction

Atomic/molecular clusters are collections of atoms/molecules that are bound together, often by intermolecular forces. Such clusters have received intensive experimental and theoretical investigations in recent decades from chemists and physicists, who have been endeavouring to understand the interactions between atoms and molecules and how they scale with size.

Clusters can vary from a collection of a few atoms/molecules through to objects on the nanometre size scale. The latter, with dimensions up to a few hundred nanometres, are called nanoclusters or, more usually, nanoparticles. At even larger particle sizes, the material has essentially the same properties as the bulk substance except the large surface to volume ratio. Compared with bulk materials, nanoparticles have distinct and fascinating properties that can depend strongly on their size, shape, structure and chemical composition; hence, nanoparticles can in principle be 'tuned' for specific applications, and many exciting possibilities have already been identified in medicine, electronic and optoelectronic devices, catalysis

S. Yang (✉) • A.M. Ellis
Department of Chemistry, University of Leicester, Leicester, LE1 7RH, UK
e-mail: sfy1@le.ac.uk; ame2@le.ac.uk

Z.M. Wang (ed.), *Nanodroplets*, Lecture Notes in Nanoscale Science and Technology 18, DOI 10.1007/978-1-4614-9472-0_10,

and even consumer products such as clothing and sunscreen creams. Unsurprisingly, nanoparticles have now become the focus of a huge scientific and technological research effort, and major challenges include the high degree of control of the growth of nanoparticles in order to achieve the desired properties and performance, as well as a good understanding of fundamental science at the nanoscale.

The most commonly used method to produce atomic and molecular clusters is the molecular beam technique [1], in which a gaseous sample at high pressure expands through a small orifice into a vacuum chamber to form a supersonically expanding jet. If this passes through a skimmer it forms a collimated beam of atoms/molecules moving at approximately equal velocities. In a typical molecular beam, the translational temperature is generally a few K, the cooling being produced by adiabatic expansion of the gas. To form clusters, either a seeded supersonic beam technique or a pickup technique can be applied. The former employs the expansion of a gas mixture, usually containing an inert carrier gas in large excess, such as Ar or He, along with a small quantity of the desired atomic or molecular dopants. This seeded supersonic molecular beam is a celebrated tool in the arsenal of the physical chemist and has been widely used to produce molecular complexes, as well as being used to study individual molecules at very low temperatures [2, 3]. The pickup technique generally employs the supersonic expansion of an inert gas, such as argon or helium, which then gathers gaseous molecules and/or atoms in their path [4, 5], allowing clusters to be formed within this low temperature gas environment.

Another means to stabilise atomic and molecular clusters is to embed them in a cold and solid environment, which is the so-called matrix isolation technique [6]. Matrix isolation generally involves molecules and atoms being trapped in an inert solid matrix, such as one formed by molecular hydrogen, nitrogen or rare gases at cryogenic temperatures. Due to the chemical inertness of the matrix materials, particularly for the rare gases, guest molecules are only slightly perturbed, allowing them to qualitatively retain the structures and electronic properties of the free molecules. In addition, the cold solid matrix confines the guest molecules so that even chemically reactive species that are difficult to maintain in the gas phase can be isolated and stabilised for long periods of time [7]. Since its discovery in 1954 [8], cryogenic matrix isolation has become a powerful experimental technique for the investigation of molecules and small molecular clusters, and is particularly useful in studies of transient species, such as atoms, molecular radicals and ions [9–13]. Other examples include the synthesis of rare gas complexes, which are very challenging to make using conventional techniques because of the intrinsic chemical inertness of rare gases, especially, for the lighter rare gases Ar [14] and Ne [15].

Helium nanodroplets [16–20] are derived from a combination of matrix isolation and molecular beam techniques and form the subject matter of this chapter. Unlike other review articles and book chapters [16, 20–22], where the emphasis has been on the broad utility of helium droplets in physical chemistry, here we focus specifically on clusters and nanoparticles. We will begin with a description of some of the basic properties of helium droplets that are particularly important in a cluster context and then proceed to illustrate the key issues through a review of several previous studies.

Finally, we turn to the future and ponder some remaining challenges and suggest some new experiments using helium droplets that may lead to exciting developments in nanoscience and nanotechnology.

10.2 Fundamentals of Helium Droplets

10.2.1 What Are Helium Droplets?

Helium is the only substance that remains liquid at the absolute zero of temperature, 0 K. This property derives from a combination of the low mass of atomic helium and the exceptionally weak interactions between two helium atoms owing to the fully occupied 1*s* orbital. The zero-point vibrational energies are comparable to the very weak interatomic binding energies, thus preventing individual atoms from being restricted to a particular lattice site (except at high pressure). The critical temperature of ^{4}He is 5.2 K, and helium cannot be liquefied at temperatures above this even with a large excess pressure [23].

^{4}He is the most abundant isotope of helium, and our emphasis here is on this isotope. Unlike other elements in our universe, helium is the only element that does not possess a triple point. Instead, it has a so-called λ-point, which separates the normal liquid phase (He-I), the superfluid phase (He-II) and the solid phase (see Fig. 10.1). The superfluid phase kicks in at 2.17 K for bulk helium [23], while above this temperature the helium behaves in many respects like most other liquids. Below this temperature the liquid helium becomes a superfluid, with vanishingly small viscosity and an ultra-high thermal conductivity. In addition, superfluid helium shows bizarre behaviour such as the fountain effect, film flow and creep, and quantized vortices [24, 25]. The ability to form the superfluid phase derives from the bosonic nature of ^{4}He atoms: at a sufficiently low temperature, the system can form a single macroscopic wave function in which single particle identities are lost. The superfluidity is due to this partial Bose–Einstein condensation of the liquid helium.

Helium droplets are large helium clusters typically composed of 10^3–10^6 helium atoms [20], although even larger droplets are accessible [26]. Unlike bulk liquid helium, whose temperature can be continuously adjusted up or down using an appropriate cryostat, the temperature of superfluid helium droplets is nonadjustable. After being formed, they undergo a rapid evaporative loss of helium atoms until a sufficiently low temperature is reached to prevent any further evaporation. This steady state is known to be close to 0.38 K, as confirmed by determining the rotational temperatures of molecules inside helium droplets using spectroscopy [27–29]. Given that this temperature is well below the λ-point, the resulting droplet is expected to consist of superfluid helium.

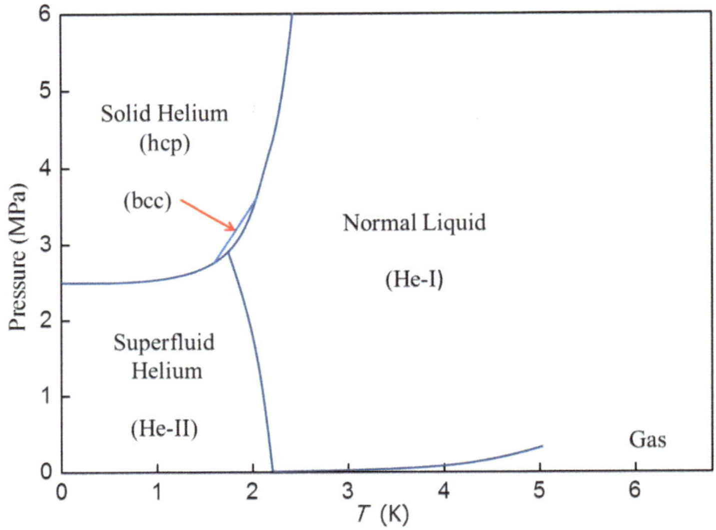

Fig. 10.1 The phase diagram of ^{4}He

Direct evidence for the superfluidity of ^{4}He droplets was first obtained by comparing the effect of ^{3}He and ^{4}He on the spectra of molecular dopants contained within these droplets. In contrast to ^{4}He droplets, ^{3}He droplets are not expected to be superfluid. This is because ^{3}He only becomes superfluid at temperatures below 0.003 K [29], whereas ^{3}He droplets have a steady-state temperature of 0.15 K after evaporative loss of helium atoms. When a probe molecule such as OCS is added to a reasonably large ^{4}He droplet, the infrared vibration–rotation spectrum shows well-resolved rotational structure similar to that in the gas-phase spectrum produced when OCS is seeded into a supersonic argon beam (see Fig. 10.2). On the other hand, in ^{3}He the rotational structure for OCS is completely lost and this is behaviour typical of a normal liquid, where collisional quenching is rife [30]. The stark difference between ^{4}He and ^{3}He droplets provides strong evidence that ^{4}He droplets are superfluid. From here onwards we concentrate on ^{4}He droplets.

10.2.2 Formation of Helium Droplets

As indicated in the phase diagram in Fig. 10.1, a very low temperature is needed to liquefy helium. Helium droplets can be generated by agitating bulk liquid helium, for example, by a piezoelectric transducer placed a few millimetres below the surface of the liquid [31]. Other means, such as laser levitation [32] and magnetic field-assisted levitation [33], can also be applied to form helium droplets from bulk liquid helium. However, producing helium droplets by this means is expensive, and the droplets produced are often extremely large, for example, up to a diameter of 2 cm [34].

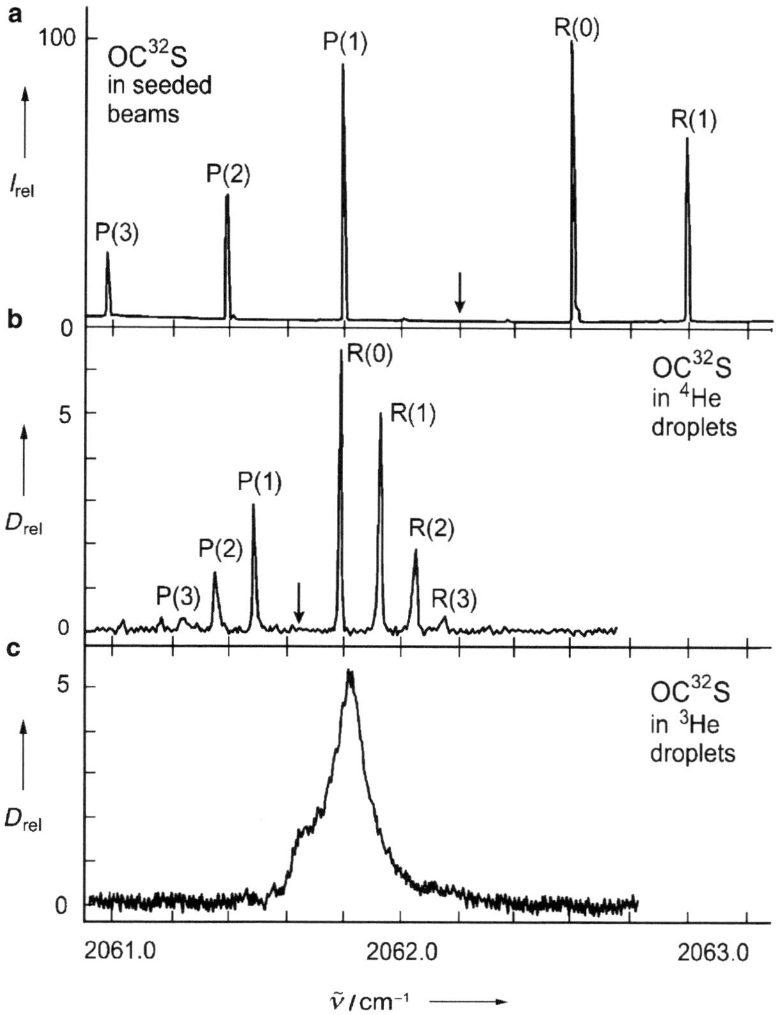

Fig. 10.2 Comparison of the rotationally-resolved vibrational spectra of $OC^{32}S$ in different environments: (**a**) the free molecule in an argon supersonic beam, (**b**) in pure ^4He droplets with an average size of 6,000 atoms, (**c**) in pure ^3He droplets containing 1.2×10^4 atoms [30] (reproduced with permission from [20])

The most commonly used and most cost-effective way to produce nanoscale helium droplets is via the supersonic expansion of pre-cooled helium through a pinhole nozzle [16, 17], typically with a 2–5 μm diameter aperture. At present most existing helium droplet sources apply continuous expansion of helium gas [16, 19] or liquid [26, 35]. The typical helium droplet experiment consists of a cryostat for initial cooling of the helium, a skimmer to create a collimated helium droplet beam, pickup cells where gaseous dopants are added and differential pumping to avoid

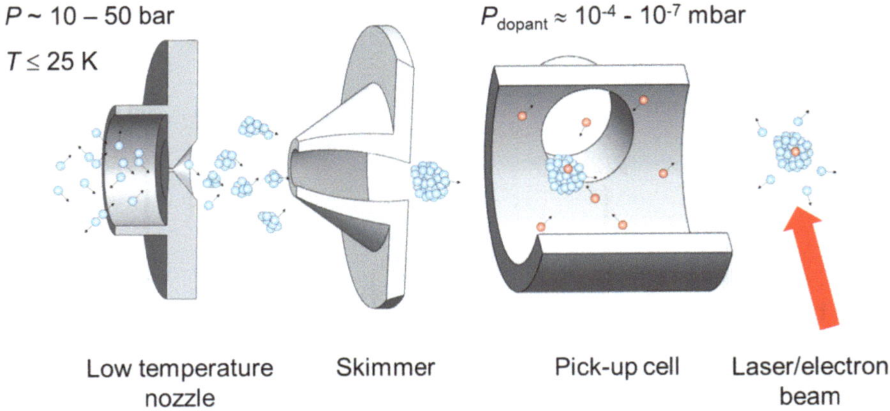

Fig. 10.3 Typical scheme for the formation of helium droplets and their subsequent doping and probing

excess collisions that can destroy the helium droplets (see Fig. 10.3). In addition, diagnostic equipment, such as mass spectrometers, is often employed to characterise the helium droplet source and the pickup conditions.

Continuous helium droplet sources are now well characterised and empirical scaling laws have been applied to describe the size distribution of helium droplets. At a typical pressure at 20 bar, a beam of helium droplets is formed by an isentropic supersonic expansion in the subcritical regime for $T_0 > 9.2$ K [36] where the resulting helium droplets show a size distribution that can be adequately described by a lognormal function [20], i.e.:

$$P(N) = \frac{1}{\sqrt{2\pi}N\sigma} \exp\left(-\frac{(\ln N - \mu)^2}{2\sigma^2}\right) \tag{10.1}$$

Here σ is the width of the distribution, μ is the maximum, and $\langle N \rangle = \exp(\mu + (\sigma^2/2))$ is the average size of helium droplets expressed as the number of helium atoms [37]. Below ~9.2 K, the droplets are formed by fragmentation of the liquid state [38], for which the size of helium droplets has been measured to follow an exponential distribution [39], i.e.:

$$P(N) = \frac{1}{\langle N \rangle} \exp\left(-\frac{N}{\langle N \rangle}\right) \tag{10.2}$$

In this regime the size of helium droplets rapidly increases when the source temperature drops, and the average size can increase by several orders of magnitude within a few Kelvin.

Equations (10.1) and (10.2) describe the distribution of helium droplet sizes under certain experimental conditions. The average size, $<N>$, is determined by

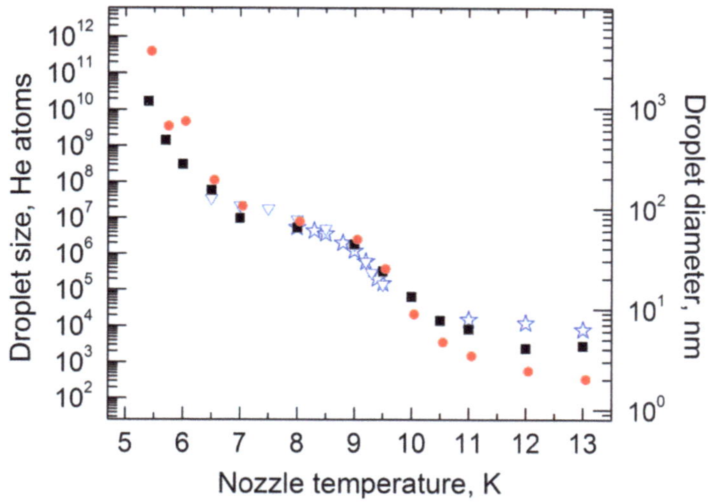

Fig. 10.4 Average helium droplet size versus nozzle temperatures for a continuous helium droplet source operating with a 5 μm nozzle and $P_0 = 20$ bar. Results obtained via measurements of the attenuation of helium monomer in collisions with helium and argon gases are shown by *squares* and *circles*, respectively. *Open triangles* [43] and *stars* [36] were obtained from previous deflection measurements of helium droplet sizes (reproduced with permission from [42])

the source conditions, i.e. the aperture size of the nozzle along with the helium stagnation pressure and temperature. Based on these parameters empirical scaling laws have been suggested to predict the average size of helium droplets as a function of temperature, pressure and the aperture diameter [40, 41]. However, scaling laws can only be applied in a certain range of source temperatures and can be unreliable outside of this range, especially, for large helium droplets produced at temperatures below 10 K. Recently, Gomez et al. have carried out comprehensive measurements of the average size of helium droplets composed of 10^3–10^{10} helium atoms by attenuation of the droplet beam through collisions with argon and helium gases at room temperature [42]. With similar experimental conditions, i.e. a nozzle diameter of 5 μm and a stagnation pressure at 20 bar, it is possible to estimate the size of helium droplets at each temperature by interpolation using the results generated by Gomez et al. (see Fig. 10.4).

A special case is the expansion of liquid helium through a small aperture at very low temperature. Grisenti and Toennies have expanded liquid ^{4}He at pressures $P_0 = 0.5$–30 bar and temperatures $T_0 = 1.5$–4.2 K into a vacuum through a 2 μm nozzle [26]. Under these conditions the liquid helium forms a microjet, which eventually fragments into large droplets due to Rayleigh oscillations. The droplet beam produced under such conditions consists of micron-sized droplets ($<N> \geq 10^9$), which have been found to be highly monodisperse by direct microscopy observations.

Pulsed helium droplet sources have also been developed [44]. The nozzle diameter in a pulsed system is usually much larger than for continuous nozzles and is typically around 0.5 mm. Compared with continuous helium droplet sources, pulsed sources consume much less helium owing to the short duty cycles. In addition, pulsed helium droplet sources can produce helium droplet beams with much larger fluxes than continuous beams [45] and can be tuned by the source conditions in a manner similar to that of continuous sources [46]. Another feature of pulsed helium droplet beams is that a wide range of helium droplet sizes can be selected within a single pulse by probing the appropriate portion of the droplet pulse [47]. More recently, a high-repetition-rate pulsed helium droplet source has been developed using the Even–Lavie pulsed valve, which can be operated at a repetition rate up to 500 Hz [48].

10.2.3 Key Properties of Helium Droplets Relevant to the Formation of Clusters

Pickup and Location of the Dopants

Doping helium droplets with atoms and molecules is straightforward using the so-called pickup technique, which was originally conceived by Scholes and his co-workers for doping impurities in argon clusters [4, 5]. In the case of helium droplets, the droplets pass through a so-called pickup region (see Fig. 10.3) containing the atoms or molecules that the user wishes to add to the droplets. Helium droplets are highly sticky and when the dopant species collides with the helium droplets, they will be captured with a near unity pickup probability [49]; the pickup cross-section is close to that of their geometric cross-sections. Once picked up, the dopant can penetrate the surface and energetically will prefer to locate itself somewhere near the centre of the droplet. The only known exceptions to this are some metal atoms and their clusters with diffuse valence electron distributions, such as the alkalis and some alkaline earth metals [20, 50]. Theory indicates that the alkali atoms locate themselves in a dimple on the surface of a helium droplet, as illustrated in Fig. 10.5 [51].

Ancilotto and co-workers have provided a criterion for the location of impurities in liquid helium in terms of the dimensionless parameter, λ [52], where λ is given by

$$\lambda = \frac{\rho \varepsilon r_e}{2^{1/6} \sigma} \tag{10.3}$$

The quantities ρ and σ are the density and the surface tension of liquid He, respectively, and ε and r_e are the well depth and the equilibrium distance of the helium–impurity interaction, respectively. Essentially, this parameter measures the balance between the energy of the impurity when located inside the helium versus the surface energy of the liquid helium required to create a cavity for the dopant.

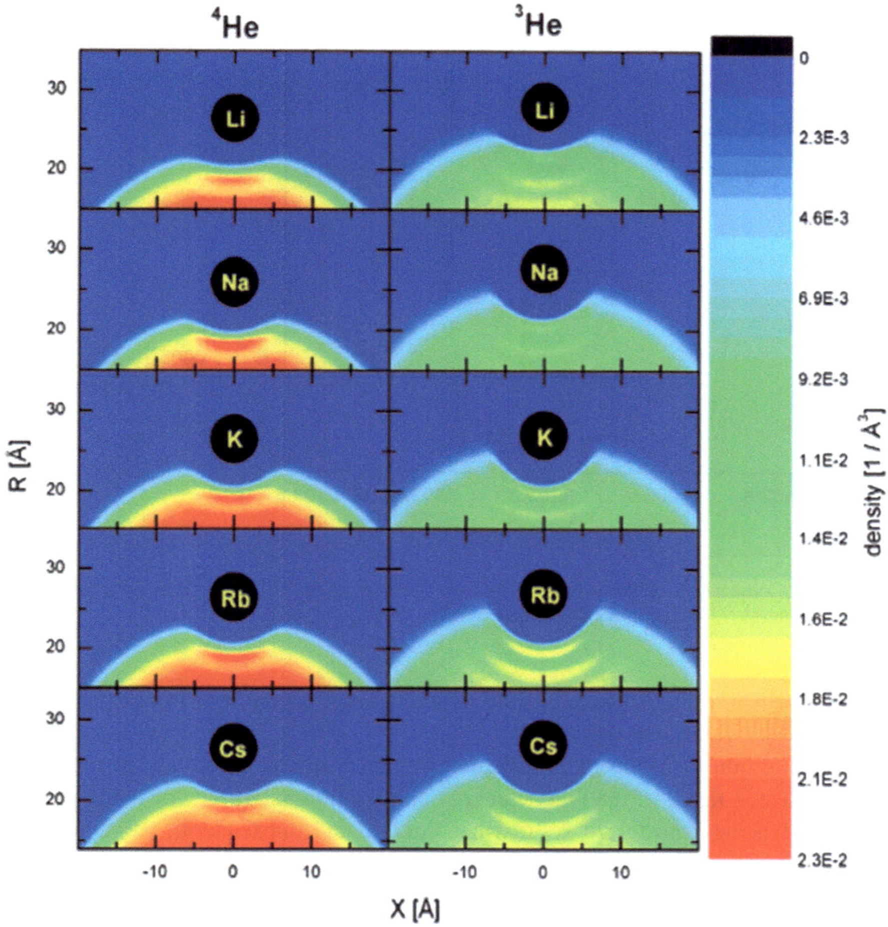

Fig. 10.5 Density profiles of ^{3}He and ^{4}He nanodroplets with $N = 1{,}000$ atoms doped with alkali atoms (reproduced with permission from [51])

According to this model if $\lambda < 1.9$, the impurity sits on the surface of helium droplets and no solvation occurs, while for $\lambda > 1.9$ the impurity will reside inside the droplet. Since this criterion is necessarily an approximate description, one must be cautious when the value of λ is fairly close to 1.9: in this case only a detailed calculation may unveil the solvation properties of a given impurity in He droplets.

For Mg, $\lambda = 2.6$ for ^{4}He and $\lambda = 4.6$ for ^{3}He, so on these grounds it is expected to have an internal location. A comparison of infrared spectra of HCN–Na and HCN–Mg is given in Fig. 10.6, which was originally carried out by Douberly and Miller [53]. The HCN–Mg complex is fully embedded in helium droplets and therefore its spectrum resembles that in the gas phase with clearly identified rotational structure. In contrast the HCN in the HCN–Na complex is submerged within the droplet, but

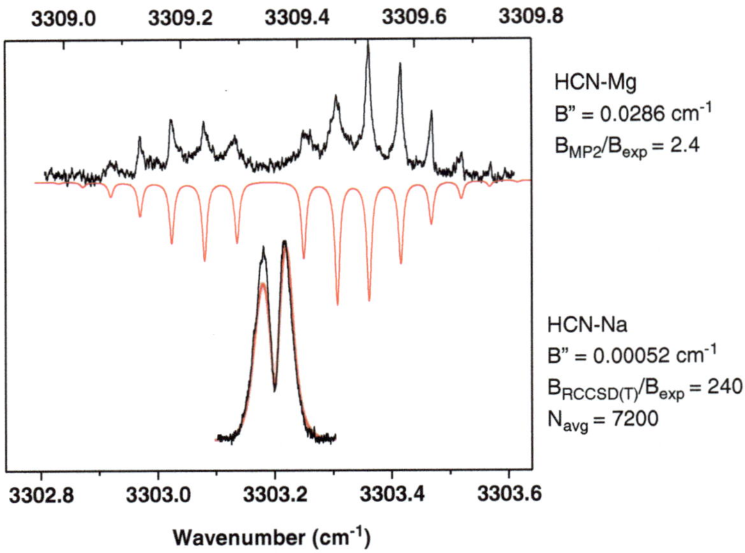

Fig. 10.6 Comparison of the infrared vibration–rotation spectra of HCN–Mg and HCN–Na in helium droplets (reproduced with permission from [53])

the Na resides at the surface, leading to a very large moment of inertia and thus a much smaller rotational constant *B*.

The addition of a dopant to a helium droplet leads to a decrease in the initial droplet size, a result of the kinetic energy and the internal energy of the impacting molecule being transferred to thermal energy and subsequently dissipated by evaporative loss of helium, leading to droplet shrinkage. Thus for a second pickup event, the pickup probability will decline because of the reduced geometric cross-section of the droplet, although the effect will be small if the loss of helium is small relative to the overall helium content of the droplet. The loss of helium atoms and the reduction of the droplet size will tend to be most significant when the dopant species interact strongly: good examples are those metal clusters that form strong metal–metal bonds. For example, the addition of one Ag atom to a pre-existing Ag cluster liberates approximately 3 eV energy [54], resulting in the loss of 4,800 helium atoms. Thus, if the starting droplet size is small, it will take the addition of only a few Ag atoms before complete evaporation of the helium.

If the change in the pickup cross-section as dopants are added is neglected, the pickup probability, P_k, of helium droplets can be described using Poisson statistics [16]:

$$P_k(z) = \frac{z^k}{k!}\exp(-z) \tag{10.4}$$

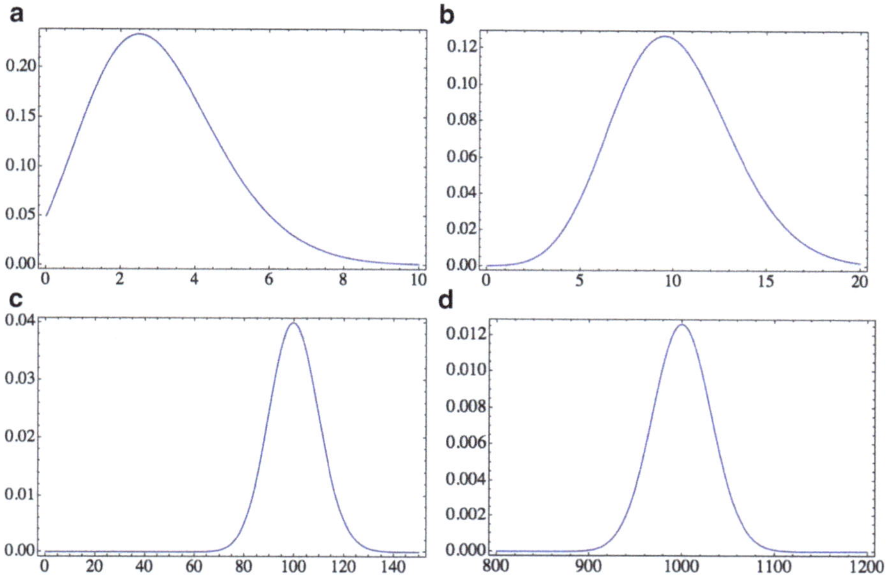

Fig. 10.7 The pickup probability of dopants (Poisson distribution) for (**a**) $<n> = 3$, (**b**) $<n> = 10$, (**c**) $<n> = 100$ and (**n**) $<n> = 1{,}000$. For further details see main text

Here k is the number of atoms or molecules picked up by a droplet and $z = \sigma n l$ is the so-called pickup parameter, with σ being the pickup cross-section of the droplet, n the number density of the dopant in the pickup zone and l the length of the pickup region through which the helium droplet beam passes. Assuming a monodispersed size distribution of helium droplets, the pickup probabilities can be calculated, as illustrated in Fig. 10.7. Clearly, as the average pickup number increases, the relative size distribution narrows. The relative size distributions $\Delta n/<n>$, where Δn is the full width at half maximum, are 130%, 70%, 24% and 7% for $<n> = 3$, 10, 100 and 1,000, respectively. This shows the wide size distribution when small clusters are formed in helium droplets, but a clear narrowing as the mean cluster size approaches the nanoscale. This suggests that while helium droplets can never be used to form clusters and nanoparticles with a specific size, the relative spread of sizes can be quite narrow if large particles are formed. Of course, this ignores the size distribution of the helium droplets.

Finally, we discuss the possibility of picking up more than one type of dopant. This can be achieved by adding more than one dopant to a single a pickup cell or by using more than one pickup cell in series (see Fig. 10.3). For the latter arrangement, it is useful to note that the helium droplets travel in a vacuum chamber with a typical velocity from 200 to 300 m/s, depending on the source conditions. The timescales for migration and cooling of the dopants within the droplet are estimated to be ≤ 10 μs [55], which is much shorter than the time required for helium droplets to travel from one pickup cell to the next (typically a few hundred μs). Consequently, dopants picked up in the first cell are at the ambient temperature of helium droplets before they reach the second pickup cell. A consequence of this, which will also be discussed again later, is that helium droplets offer a way of obtaining exquisite control of core-shell structures, simply by using two or more pickup cells and through appropriate control of the order in which dopants are added [56, 57].

Unique Features

Superfluid helium has an ultra-high thermal conductivity (about 30 times higher than copper [20]). Combined with the low equilibrium temperature of 0.38 K, it is unsurprising that helium droplets can cool dopants rapidly to the ambient temperature of helium droplets when they are captured, with a cooling rate that may be as high as 10^{16} K/s [58]. The cooling is achieved by evaporative loss of helium atoms, each removing 5 cm^{-1} energy until all of the excess energy is removed.

The low temperature of helium droplets is important for the investigation of molecules and molecular clusters under cryogenic conditions. At 0.38 K, most molecules are in their ground states. In addition, since helium droplets are isolated nano-matrices with distinct boundaries, molecules and atoms, once trapped, have to locate themselves somewhere inside or on the surface of the droplet and will be cooled to the ambient temperature of 0.38 K. When more than one dopant is added to the droplet the long-range interaction between them is able to drive the aggregation of dopants to form clusters. Here the superfluidity of helium droplets plays a key role in the formation of clusters because atoms and/or molecules can translate almost freely so that their aggregation is not hindered. This is distinct from other potential matrices, such as large Ar clusters, on which the motion of molecules and atoms is very much restricted. Also, much stronger interactions between molecules and solid clusters occur when compared with superfluid helium.

The combination of the low temperature and the superfluidity of helium droplets offers a unique environment for forming clusters. In contrast to rigid rare gas matrices, such as cryogenic argon matrices, the liquidity of helium droplets, perhaps aided by the superfluidity, enables dopant particles to move relatively freely while at a temperature close to absolute zero. An illustration of this was shown earlier, in Fig. 10.6. This shows a clearly rotationally resolved infrared spectrum of HCN–Mg, which indicates that free rotational motion of this particular complex occurs within the helium droplets [53].

Given their unique properties, helium droplets offer extraordinary possibilities for the growth of nanoparticles with an almost unlimited combination of materials. At 0.38 K all materials, apart from helium, will solidify. The sequential addition of materials to the ultra-cold superfluid helium droplets should therefore deliver core-shell clusters, with the core material frozen into place before the shell material is subsequently added. Furthermore, the low temperature will practically eliminate diffusion between the core and the outer layer, something that would be hard to avoid when core-shell particles are formed at much higher temperatures.

Finally, we point out that not all of the advantages mentioned above are met by bulk liquid helium. This is because helium is literally the worst possible solvent, and thus any material added would prefer to locate at the walls of any liquid helium container, rather than reside within the liquid. By way of contrast, there is no wall for a helium droplet and, even given the poor solvation characteristics of liquid helium, the lowest energy location for almost all dopants is at the centre of the droplets [59]. For this reason one can grow and observe materials in helium droplets that could not be prepared in bulk liquid helium.

Quantized Vortices

Vortices are thought to be commonplace in bulk superfluid helium. Quantized vortices are narrow vortex lines around which helium circulates. In 1949 Onsager tried to deduce the wave function describing a superfluid and perceived that a container of rotating superfluid helium should show a distribution of vortices in which the circulation is quantized in units of Planck's constant, h [60]. Experimentally, quantized vortices in superfluid helium were first observed by Vinen in 1961 [61], and quantized vortex rings were discovered by Rayfield and Reif [62] in 1964. Such a vortex can be characterised by a macroscopic wave function and quantized velocity circulation in units of $\kappa = h/M$, where M is the mass of the ^{4}He atom [63, 64].

Rayfield and Reif applied the trapping of electrons and ions in the vortex rings of bulk liquid helium as a probe for quantum vortices. By drawing an analogy with this approach, it has been suggested that atoms and molecules located along the vortices could be used as a possible means to detect quantized vortices in helium droplets [65]. However, to date there has been no experimental evidence for the existence of quantized vortices in helium droplets, althou gh spectroscopic measurements have indicated the occurrence of superfluidity, as discussed earlier. However, theoretical work by Pitaevskii and Stringari has been carried out to ascertain superfluid effects in rotating helium clusters and found that an angular velocity of the helium droplets is necessary for the formation of vortices in the droplets [66]. Calculations have also been carried out by Bauer et al. to determine the stable configurations of a vortex in a freely rotating superfluid droplet, which is constrained to rotate at a fixed angular momentum [67]. The existence of quantized vertices in helium droplets might influence the way dopants agglomerate in helium droplets, and hence they might affect the morphology of nanoparticles.

The first indication of quantized vortices in superfluid helium droplets was reported recently. Gomez et al. formed large helium droplets and added silver atoms to the droplets by oven evaporation [68]. A remarkable observation is that at different helium droplet sizes, i.e. with initial diameters of 100, 300 and 1,000 nm, silver clusters deposited on substrates by collision of the droplets with the substrate show distinct morphologies. At a droplet diameter of 100 nm, roughly spherical silver nanoparticles are seen on the substrate. For droplets with a diameter of 300 nm, the silver particles become more rod-like, and for the largest helium droplets, with diameters of 1,000 nm (ca. 1.7×10^{10} helium atoms), long silver cluster chains a few hundred nanometres length were observed (see Fig. 10.8). To interpret this switch from spherical to rod- and chain-like structures, the authors have suggested that superfluid helium droplets of diameters larger than 300 nm

Fig. 10.8 TEM micrographs obtained upon deposition of Ag-doped droplets of initial average diameter (**A**) 100, (**B**) 300 and (**C**) 1,000 nm and average number of He atoms $<N_{He}> = 10^7$, 3×10^8 and 1.7×10^{10}, respectively (reproduced with permission from [68]; Copyright (2012) by The American Physical Society)

contain quantized vortices, and that the silver atoms are funnelled into these vortices, although it has to be said that these are early days and there is no other evidence currently available to support this claim.

If the interpretation of Gomez et al. is correct, then one would expect that in a large helium droplet, say composed of over 10^7 helium atoms, nanoparticle chains will be commonplace. This may therefore make it difficult to form large and near spherical nanoparticles using large helium droplets. The jury is still out on this issue, but it is clearly possible that the synthesis of nanoparticles using helium droplets will face a severe size limit, i.e. only relatively small compact particles can be formed.

Soft Landing of Particles

As hinted at above, it is possible to remove nanoparticles from helium droplets and deposit them on solid surfaces. The soft landing of clusters and nanoparticles grown in the gas phase onto solid targets has long been a major aim of the gas-phase cluster community for the controlled synthesis and characterisation of nano-structured metal and metal oxide catalysts [69, 70]. However, clusters formed in the gas phase will tend to strike solid surfaces at velocities that can lead to subsequent damage: this is particularly true for mass-selected cluster ions, which are difficult to slow to velocities where some damage, either to the cluster or the surface, is avoided.

In principle, helium droplets can help to deliver a soft landing for a nanoparticle located inside the droplet. The reasoning here derives from the easy removal of helium atoms, which may be able to rapidly dissipate the kinetic energy as the doped droplet collides with a solid surface, thus avoiding depositing any energy into the nanoparticle. Thus, in effect, the helium acts as a cushion that soft-lands the particle onto the solid surface.

The possibility of soft landing using helium droplets has received some theoretical and experimental interest. Aguirre et al. have recently modelled the He–TiO_2(110) interaction using density functional calculations and applied this to the collision of a helium droplet [71]. Their model predicts that the helium droplet spreads on the TiO_2 surface as it collides and leads to the formation of a thin film above the substrate, indicating that the helium layer can act as a cushion for the soft landing of nanoparticles embedded within the droplet. Experimental investigation of the soft landing of nanoparticles formed in helium droplets has been reported by Loginov et al., who explored silver clusters [72]. Assuming that the kinetic energy of the clusters is determined by the velocity of the helium droplets (in the range of 200–300 m/s), the kinetic energy per impacting Ag atom was estimated as 0.034 eV, which is far lower than the binding energy between a silver atom and a silver cluster (3 eV) and the binding to an amorphous carbon surface (~1 eV). Compared with these binding energies, the kinetic energy of a particle embedded within a helium droplet is so low that it should have a negligible impact on both the surface and the particles. Hence the deposition of silver nanoparticles embedded within helium droplets would seem to fall well inside the regime of soft landing [73, 74], which is defined by K_i ~0.1 eV, where K_i is the kinetic energy per atom. The authors concluded that the estimated collision energy per atom with the substrate is less than the lowest yet reported in the literature, 0.05 eV, for Sb_N^+ ions ($N = 90$–2,200) [75].

10.3 Clusters in Helium Droplets

When more than one dopant particle is picked up by the helium droplet, agglomeration will occur leading to the formation of a cluster. The agglomeration of dopants is mainly driven by the intermolecular forces between the foreign species, while the superfluidity of the helium droplet allows almost unhindered migration. Helium droplets have provided a unique environment for the growth of clusters, so they have distinct features when compared with other means for the formation of clusters. In this section we will highlight a few illustrative studies to demonstrate the characteristics of helium droplets in the formation and investigation of atomic and molecular clusters.

10.3.1 High-Spin Metal Clusters

Alkali atoms are known to reside on the surface of helium droplets rather than entering inside. The preference for a surface location arises because the attractive part of the alkali–helium interatomic potential is overwhelmed by the short-range Pauli repulsion. The same tendency occurs for small alkali clusters, but here the atoms can combine in two ways: the spins can either combine in an antiparallel way, leading to the formation of strong metal–metal bonds, or they can be parallel, in which case weakly bound van der Waals clusters result. Higgins et al. found that when alkali dimers are formed on the surface of helium droplets, triplet states are detected in greater abundance than their singlet counterparts using laser spectroscopy, in contrast to the situation found in a heat pipe [76]. Later on, similar observations were made for the trimers Na_3 and K_3, which form high-spin quartet

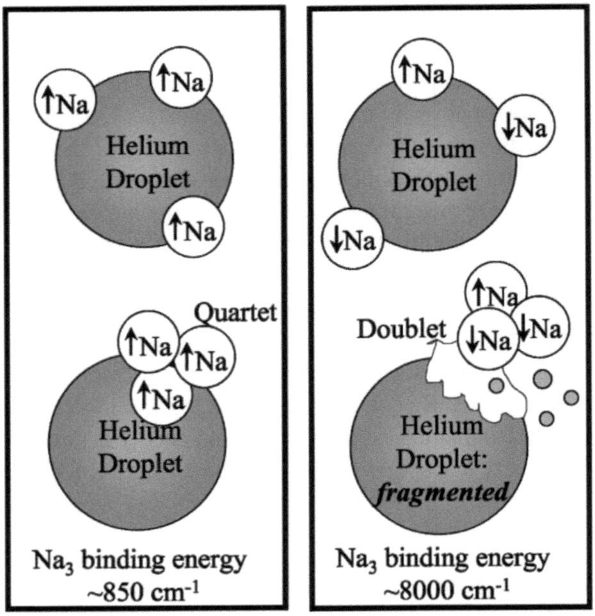

Fig. 10.9 Schematic representation of the formation of sodium quartet and doublet trimers on the surface of a helium droplet (reproduced with permission from [77])

states rather than the more strongly bound doublet states [77]. The mechanism was explained in the context of binding energies of the high-spin and low-spin states, which has to be dissipated by evaporation of helium atoms from the helium droplets. Taking Na_3 as the example (see Fig. 10.9), the formation of a quartet state releases ~850 cm^{-1} energy, which can be dissipated by removal of 170 helium atoms. In contrast, the formation of a doublet state will lead to the evaporation of 1,600 helium atoms. In the latter case the removal of energy now becomes a kinetic bottleneck, and desorption of the alkali clusters competes effectively with this helium evaporation route. In a crude sense, one can imagine that the additional energy release in forming the doublet state leads to the alkali trimer being blasted off the helium droplet surface. In this sense helium droplets offer a unique opportunity to form and study alkali metal clusters in unusual high-spin states.

For larger clusters it has been suggested that the formation of spin-polarised configurations is of low probability, so these clusters will most likely be in their electronic ground states [78]. Thus one might therefore expect that the chance of forming large alkali clusters on helium droplets is remote, given the large energy release when forming low-spin alkali clusters of large size. However, when the metal cluster becomes larger, other factors, such as the more favourable interaction potential between the clusters and the helium, come into play. Theoretical modelling has predicted that the dispersion force increases faster than the repulsive energy, and at some critical size the alkali cluster may become submerged in the helium rather than adopt a surface location [79]. For example, Na_n clusters are predicted to preferentially enter ^{4}He nanodroplets once $n \geq 21$, whereas K_n clusters require approximately 78 atoms to become fully solvated.

These theoretical predictions were recently given strong experimental support by An der Lan et al. [80], who have observed the submersion of alkali metal clusters for the first time using mass spectrometry. Using controlled electron impact energies to construct ion yield curves, a clear surface-to-interior transition was observed at a sufficiently large cluster size. Although it remains unclear what the minimal cluster size is for the submersion to take place, the mass spectra imply that Na_n clusters with $n \geq 21$ submerge, which is in remarkably good agreement with the theoretical prediction by Stark and Kresin [79].

10.3.2 Dipole-Aligned Self-Assembled Clusters

For nonpolar molecules and atomic dopants, their attractive interactions are governed by the dispersion force. This force is weak and so the dopants will, presumably, undergo migration towards each other at random orientations in helium droplets. However for polar molecules long-range attraction can be strong, e.g. due to dipole–dipole interactions, which can influence the relative orientation of these molecules and thus the way they aggregate. In this section we will focus on how molecular clusters are formed in helium droplets when they possess substantial electric dipole moments.

At room temperature, the dipole–dipole interaction is small compared with the rotational energies of molecules, and rotational averaging changes the distance dependence from $1/R^3$ to $1/R^6$ [81]. Consequently, dipole–dipole interactions play only a minor role in determining the structures of molecular clusters at room temperature. However, inside helium droplets the very low rotational temperature, which should be close to 0.4 K, means that there is no rotational averaging and thus the dipole–dipole interaction can be a strong steering force as two molecules approach each other. In particular, the dipole–dipole interaction is maximised when the dipoles align in a head-to-tail fashion, so this would be expected to be the preferred orientation when the cluster forms.

A remarkable illustration of this point was provided by Nauta and Miller, who formed long-chain $(HCN)_n$ clusters ($n = 2$–7, $\mu_{HCN} = 2.98$ D) in helium droplets and confirmed their structures by laser spectroscopy (see Fig. 10.10) [82]. Unlike in a standard seeded supersonic expansion, where HCN molecules tend to form more stable structures such as a cyclic trimer, in helium droplets higher energy structures, i.e. the head-to-tail aligned linear clusters, are formed and are then stabilised by the ultra-cold helium droplets. In other words, the clusters are steered into position by the dipole–dipole forces and become trapped in a shallow potential energy minimum, even though a lower energy minimum is available. This remarkable finding is a consequence of three properties of helium droplets: the very low temperature, the rapid cooling of dopant molecules and the ability of molecules to move freely within the liquid environment.

Similar behaviour was observed in another experiment by Miller and co-workers, this time on formic acid ($\mu_{HCOOH} = 1.41$ D) [83]. Using infrared spectroscopy,

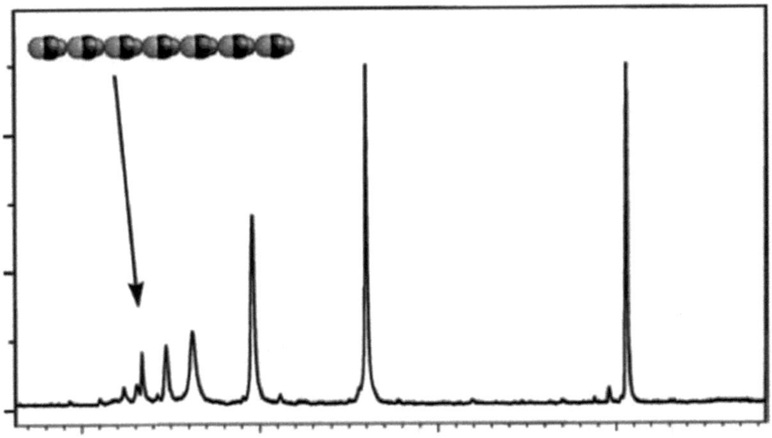

Fig. 10.10 Spectrum of the free C–H stretching region of HCN polymer chains, showing clusters up to at least the heptamer. A linear heptamer chain is shown as an inset. This spectrum was obtained in the presence of a large electric field used to orient the polar chains within the helium droplet, thus collapsing the rotational band contours into a single vibrational peak for each cluster size (reproduced with permission from [82])

formic acid dimers in helium nanodroplets were observed in the 'free' OH and CH-stretching regions. The experimental results show a polar acyclic structure for the dimer, rather than the much more stable cyclic isomer with a pair of hydrogen bonds. Again, the mechanism is the same as that in the case of HCN: when two formic acid molecules are added to a helium nanodroplet, a dipole-aligned single hydrogen bond geometry is formed, instead of the more stable cyclic dimer. We therefore expect that for molecules possessing strong interactions between each other, the likelihood is that aligned clusters will form rather than randomly distributed geometries.

Finally, we briefly discuss the migration of molecules and atoms in a much broader but relevant sense beyond dipole–dipole interactions, in particular the charge transfer process initiated by electron impact ionization. When an energetic electron hits a helium droplet, it has little chance of ionizing the dopant directly; it is more likely to strike a helium atom [84], forming a He^+ cation. The He^+ then undergoes resonant hopping for some 10 hops before becoming self-trapped by forming $He_n{}^+$ ions [85]. The latter is an irreversible process, but if the positive charge reaches a dopant before the 10 hop limit is reached, it can then transfer the charge to the dopant and allow it to be observed in a mass spectrum. Theoretical modelling has shown that the migration of the He^+ ion in helium droplets seems to be steered by following the gradient of the potential well created by the charge–dopant interaction [84, 85]. The interactions between He^+ and the dopant are typically charge–neutral interactions and/or charge–dipole interactions. In this sense the migration of the positive charge is similar to the self-alignment of HCN and HCOOH inside helium droplets.

10.3.3 Core-Shell Molecular Clusters

The earliest work on clusters within helium droplets began with the mass spectrometry of atomic clusters containing one species only, such as Ar [49], Ne [86], Kr [87] and Xe [88]. Since then this work has expanded to include molecular clusters [49, 89–93] (for more references see the review articles in [16, 20–22] and references therein).

Compared with a single type of dopant species, droplets containing two different types of dopants have attracted much less attention. This is somewhat surprising given the potential that helium droplets offer for the formation and the investigation of unusual molecular cluster combinations at very low temperature. One example is ion–molecule reactions at low temperature, which can be initiated in a helium droplet by electron impact. The first specific attempt to study ion–molecule reactions in helium droplets involving two different molecular constituents was carried out by Fárnik and Toennies, who characterised the reactions $N^+ + D_2$, $CH_4{}^+ + D_2$ and $CH_3{}^+ + D_2$ [94]. More recently, in Scheier's group, a number of binary clusters were formed in helium droplets involving C_{60} and another molecular species, such as NH_3 [95] and H_2O [96, 97], and were investigated via mass spectrometry. The major finding was that the C_{60} molecules added to the binary clusters significantly enhance the unprotonated channels of NH_3 and H_2O clusters bounded to C_{60}, i.e. $C_{60}(H_2O)_n{}^+$ and $C_{60}(NH_3)_n{}^+$ ions are prominent when compared with the protonated $C_{60}(H_2O)_nH^+$ and $C_{60}(NH_3)_nH^+$ ions. With a slightly different emphasis, Ren and Kresin have co-embedded water molecules with fragile organic molecules in helium droplets and have shown that small water clusters can have a major effect in softening the ionization process, thus drastically reducing the degree of fragmentation of the organic ions [98].

The sequential addition of different types of materials to helium droplets clearly offers the opportunity to form core-shell structures. Experimental evidence for the formation of core-shell molecular clusters inside helium droplets was recently obtained in Leicester by Liu et al. [56]. Core-shell particles with water clusters at the core and surrounded by an atomic or molecular shell were synthesised by adding water and a co-dopant sequentially to helium nanodroplets using two pickup cells. This work employed mass spectrometry of binary clusters, such as O_2/H_2O, N_2/H_2O and CO_2/H_2O [56]. By reversing the order in which the dopants were added to the helium droplets, changes in the mass spectra were observed which were attributed to core-shell formation. An illustration of the mass spectra recorded is given in Fig. 10.11, where the branching ratio of water clusters is biased toward the unprotonated $(H_2O)_n{}^+$ ions when water is added prior to O_2.

10.3.4 Very Large Molecular Clusters

Helium droplets have wide scope for the study of molecular clusters. However, so far most research in this field has focused on small clusters composed of, at most, a

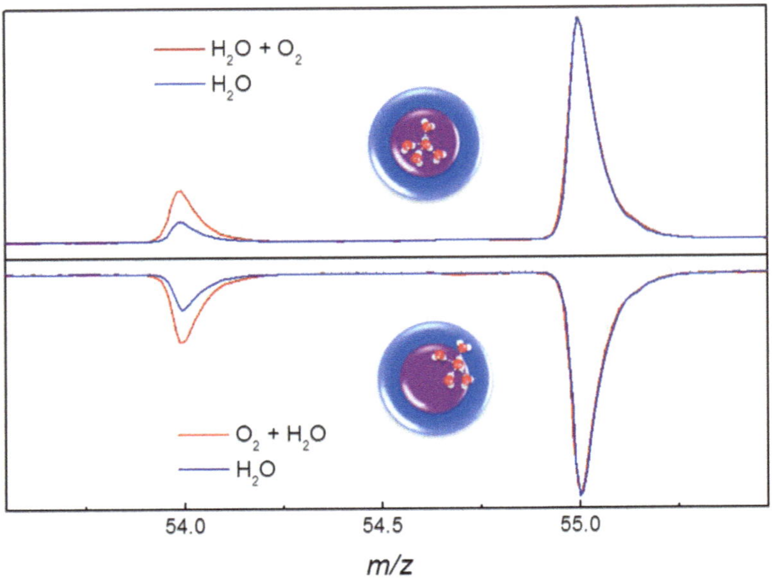

Fig. 10.11 Comparison of the yields of $(H_2O)_3^+$ relative to $(H_2O)_3H^+$ for different pickup orders of H_2O and O_2. The *red line* in the *upper plot* is for addition of H_2O in the first pickup cell and O_2 in the second, while the reverse pickup order applies for the *red plot* in the *lower half.* The *blue lines* are for water addition only (reproduced with permission from the PCCP Owner Societies)

few tens of molecules, and generally much smaller clusters have been targeted such as dimers and trimers. One of the major reasons for the focus on small clusters is that these studies have involved spectroscopy and unravelling the spectroscopic details is only really possible for the smallest clusters. In addition, there are complications from overlapping features of different sized clusters in the spectra (the spectroscopy is not species selective), and this adds to the complication in interpreting the spectra.

Nevertheless, in a few recent experiments, large molecular clusters in helium droplets have been investigated. For example, Mozhayskiy et al. have investigated the transportation of large molecular clusters using helium droplets as the carrier [99]. With helium droplets composed of 10^4–10^7 helium atoms, propyne clusters $(C_3H_4)_n$ with ($n = 10$–10^4) were formed whose size was estimated by the ratio of the maximum attainable flux of the propyne molecules transported by helium droplets to the flux of helium droplets. With a slightly different emphasis, Slipchenko et al. have formed large ammonia clusters in helium droplets and have investigated the size-dependent evolution of the infrared spectroscopy in the N–H stretching region from a single molecule to very large cluster sizes [100]. The largest $(NH_3)_n$ clusters in this study were estimated to have a mean size of 10^4 molecules, which is assumed to be large enough to be regarded essentially as 'bulk' material. The spectra indicate that ammonia clusters in He droplets adopt a hydrogen-bonded structure similar to that in the crystalline form of ammonia. This ability to form crystalline ammonia ice, instead of amorphous ammonia, was ascribed to the directionality of hydrogen bond formation, which guides each added NH_3 molecule at relatively long range into the correct position for crystal formation.

10.4 Metal Clusters and Nanoparticles

10.4.1 Metal Clusters

A wide range of small metal clusters have been investigated in and on helium droplets, including clusters of the alkali metals Na, K, Cs and Rb [50, 51, 76, 77, 79], the coinage metal Ag, the group 13 metal In [101] and the transition metal Cr clusters [102]. Only in a few cases have sizeable metal clusters in helium droplets been studied, such as Ag_n ($n > 40$) [103], Eu_n clusters ($n \sim 20$) [103], Na_n [80] and K_n ($n \sim 100$) [104]. Recently, larger metallic clusters composed of over 1,000 atoms have also been produced in helium droplets, such as Cd_n [50, 105], Zn_n [50, 105] and Mg_n [105].

10.4.2 Deposition of Metal Nanoparticles

Although large metal clusters, such as those of Cd and Mg, have been formed in helium droplets [50, 105], it is only recently that nanoparticles have been synthesised in helium droplets and deposited on substrates. A typical experimental setup is shown in Fig. 10.12, where the major difference from other helium droplet experiments is the inclusion of a deposition system of some sort. With this setup, very large silver clusters composed of over 6,000 Ag atoms have recently been synthesised by Vilesov's group and have been investigated by both transmission electron microscopy [72] and laser spectroscopy [106]. Images of the clusters on

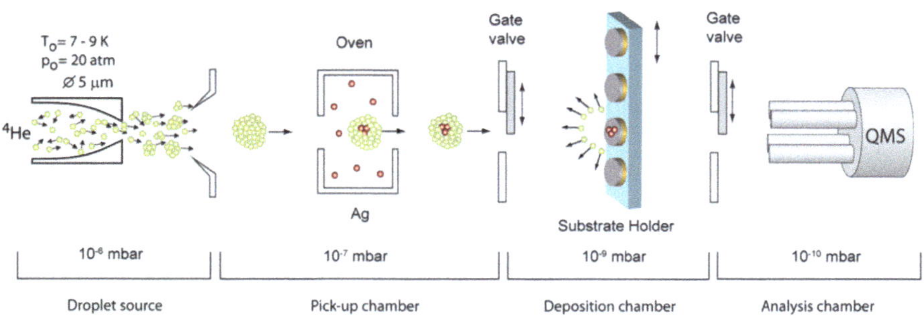

Fig. 10.12 Experimental arrangement for the surface deposition of metal clusters formed in helium droplets. The typical pressure in each vacuum chamber, with the He beam off, is shown (reproduced with permission from [72])

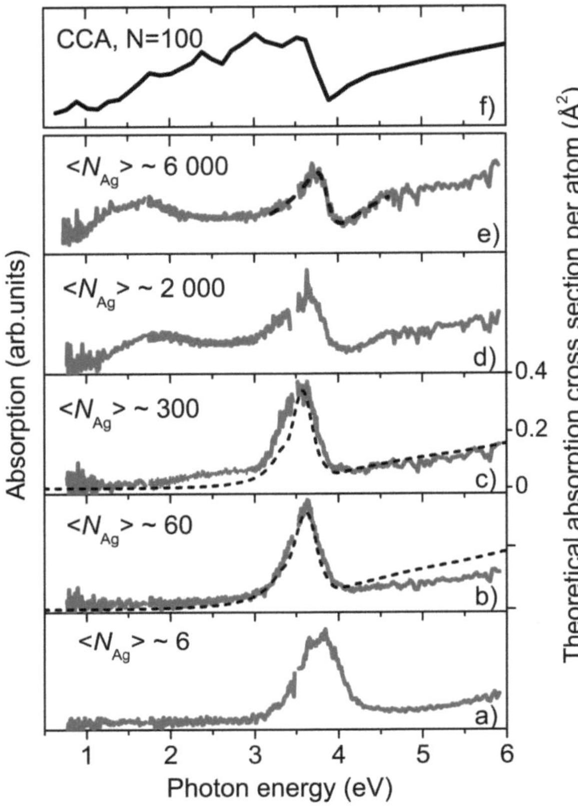

Fig. 10.13 Normalised photoabsorption spectra for Ag particles of different average sizes assembled by pickup in helium droplets (reproduced with permission from [106])

amorphous carbon substrates obtained at short deposition times have shown a broad size distribution of the metal nanoparticles, i.e. for $<n> = 6{,}400$, $\Delta n = 5{,}000$ and for $<n> = 11{,}800$, $\Delta n = 11{,}400$; and generally the $\Delta n/<n> \approx 0.8$. The average sizes of the deposited clusters are in good agreement with an energy balance-based estimate of Ag_n cluster growth in helium droplets. Measurements of the deposition rate indicate that upon impact with the surface, the silver nanoparticles are attached with high probability and seem to be soft-landed intact on the deposition surface.

Spectroscopic investigations of silver nanoparticles inside helium droplets revealed unexpected features in the growth of nanoparticles using helium droplets (see Fig. 10.13) [106]. For small silver nanoparticles ($<n> \sim 300$), the photoabsorption spectra resemble that of the surface plasmon resonance of silver clusters, which falls in the UV region. On the other hand, for $<n> \sim 2{,}000$ and $\sim 6{,}000$ nanoparticles absorption in the near infrared was observed, which can only be accounted for by coupling between small silver nanoparticles, which are formed by multi-centre growth within the helium droplets. This is because silver clusters can grow at different places within the droplets and then subsequently aggregate near the centre of the helium droplet when large helium droplets ($\gg 10^6$ helium atoms) and a high Ag doping rate is used. This essentially produces grain

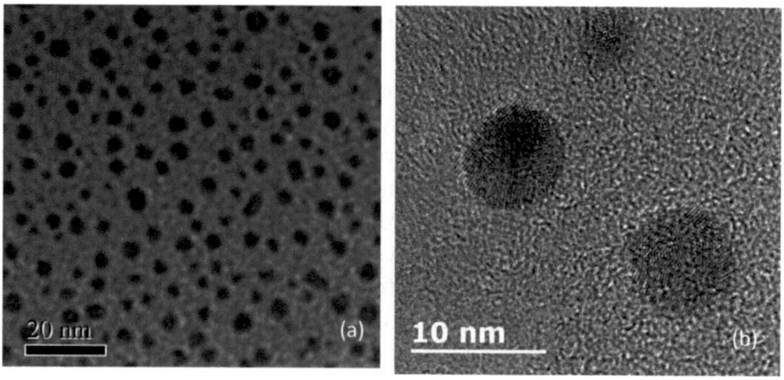

Fig. 10.14 TEM images of nanoparticles. (**a**) Ag nanoparticles at a resolution of 20 nm; (**b**) Au nanoparticles at a resolution of 10 nm

boundaries within the agglomerated nanoparticle, giving rise to the interparticle coupling.

In another experiment, the formation of core-shell nanoparticles by the sequential addition of metal atoms to helium droplets was also attempted by Vilesov's group, who reported the formation of core-shell Ag–Au nanoparticles composed of ~500 metal atoms [99]. However, while conditions were adopted that could lead to the formation of these core-shell nanoparticles, there is no direct evidence, such as from microscopy, to confirm that an actual core-shell structure was formed.

In our laboratory an ultra-high vacuum helium droplet apparatus has recently been constructed which is designed for the synthesis of metallic nanoparticles [57]. Several types of metal nanoparticles, including pure Ag and Ni nanoparticles, as well as the bimetallic systems Ag/Au and Ni/Au, have been synthesised and deposited on substrates for ex situ investigations using high-resolution transmission electron microscopy (TEM) and X-ray photoelectron spectroscopy (TEM). The XPS investigation on Ni/Au core-shell nanoparticles has recently provided the *first* direct evidence for the formation of core-shell nanoparticles using superfluid helium droplets [57].

Some TEM images of nanoparticles deposited using the Leicester apparatus are shown in Fig. 10.14. At a 20 nm resolution the TEM image of silver nanoparticles shows a clear bimodal size distribution, i.e. larger particles with an average diameter of 4 nm are accompanied by relatively smaller ones with diameters <2 nm, with a dearth of particles between these two sizes. The reason for the bimodal distribution is not fully established, but it is likely due to the aggregation of free metal atoms on the deposition targets.

Figure 10.14(b) shows clear evidence of a regular arrangement of atoms into a crystalline structure. This is the first time that the crystalline structure of a metal cluster grown in a helium droplet has been established, thus showing that metal atoms can self-assemble into a regular structure inside superfluid helium.

10.4.3 Challenges and Opportunities

One of the major limitations of using helium droplets as a means to synthesise nanoparticles is the deposition rate. The typical flux of helium droplets that can be used to synthesise nanoparticles is in the range of 10^9–10^{12}/s, which sets the upper limit for the flux of nanoparticles impinging on a substrate surface. On these grounds it would be challenging to produce more than 100 μg of nanoparticles per day, and the likely quantities will be much lower. For fundamental research this may be sufficient in many cases, e.g. for the investigation of fundamental nanoscience, such as the optical and magnetic properties of the particles. It may also be sufficient for applications where a surface coating of the material is required, e.g. for construction of sensors, for heterogeneous catalysis and for data storage based on arrays of nanoparticles. However, if much larger quantities of the nanoparticles are required, as might be the case for some biological or medical applications, there is a major challenge in scaling up the process.

Another challenge is the incorporation of certain metals as dopants, such as the alkali metals, which as we have already seen tend to reside on the surface of helium droplets. Hence, new strategies are needed in order to inject alkali metal clusters into helium droplets. A possible solution is to add alkali metal clusters to helium droplets rather than single atoms, since we saw earlier that sufficiently large clusters of alkali metals will submerge into a helium droplet.

Another challenge is the size limit of nanoparticles that can be produced in helium droplets. To grow the largest possible nanoparticles, one needs the largest possible helium droplets. Very large helium droplets can be formed by passing liquid helium, rather than gaseous helium, through the pinhole nozzle and into the vacuum system. When the temperature is sufficiently low, the helium departs the nozzle as a microjet, which eventually fissures into droplets which have diameters on the order of the microjet beam diameter. These droplets should be called helium microdroplets and may consist of somewhere in the region of 10^{10} or more helium atoms [26]. Inside these ultra-large helium droplets, it is possible in principle to grow nanoparticles with diameters up to several hundred nanometres. However, as we saw earlier even for much smaller droplets, multi-centre growth will come into play, so many small particles are likely to be formed in one large helium droplet rather than a single large particle. So far the largest nanoparticles produced in helium droplets have a diameter less than 10 nm [57] and it has not been proved possible to synthesise larger particles, although these are early days for this technology.

Although there are obvious limitations in using helium droplets as a tool for synthesising nanoparticles, they also offer some very exciting possibilities in nanoscience and nanotechnology. To close this section we summarise the key advantages of helium droplets.

1. Due to the very low temperature, any added materials will attach to and condense within liquid helium, providing it can be delivered initially as a gas or vapour.
2. Helium droplets can sequentially pick up dopants in the gas phase when passing through the pickup region, and therefore they offer the opportunity to form core-shell and even core-multiple shell nanoparticles.

3. The low-temperature environment for the synthesis of nanoparticles minimises the diffusion between layers, and hence the expectation is that nanoparticles can be produced with sharp boundaries between layers.
4. A good degree of control of the size of nanoparticles and the amount of materials in the core and each layer can be achieved by choosing the appropriate pickup conditions, i.e. the droplet sizes and the partial pressures of the dopants in the respective pickup cells.
5. Helium droplets offer the possibility of soft landing the embedded nanoparticles on a solid surface.

10.5 Conclusions

Helium droplets offer a somewhat exotic and cold environment for the formation and investigation of individual molecules and collections of molecules. In this chapter we have attempted to give succinct descriptions of the properties of these droplets, how they can be formed and doped and how their properties can be used for the formation and investigation of clusters. For clusters, the key factors that influence their formation have been addressed and special characteristics of helium droplets that can lead to form unusual cluster structures, such as dipole-aligned molecular chains and surface-located high-spin alkali metal clusters have been described. The properties of helium droplets were then discussed in the context of nanoparticle formation, which opens up new possibilities for using helium droplets as a powerful tool in nanoscience and nanotechnology. By addressing this, we hope to encourage the scientific community to consider the many exciting possibilities offered by nanoparticle growth in helium droplets and to identify clear advantages of this technique over more established techniques for nanoparticle formation.

Acknowledgements SY and AME wish to thank the UK Engineering and Physical Science Research Council (EPSRC), the Leverhulme Trust and the Royal Society for grants in support of their helium droplet research. We also appreciate the Advanced Microscopy Centre at the University of Leicester for providing TEM facilities.

References

1. Ramsey, N.R.: Molecular Beams, Oxford University Press, London and New York (1956)
2. Scoles, G.: Atomic and Molecular Beam Methods, vol. 1, 2. Oxford University Press, New York, NY (1988)
3. Pauly, H.: Atom, Molecule and Cluster Beams, vol. 1, 2. Springer, Berlin (2000)
4. Gough, T.E., Knight, D.G., Scoles, G.: Chem. Phys. Lett. **97**, 155 (1983)
5. Gough, T.E., Mengel, M., Rowntree, P.A., Scoles, G.: J. Chem. Phys. **83**, 4958 (1985)
6. Dunkin, I.R.: Matrix Isolation Techniques: A Practical Approach. Oxford University Press, New York, NY (1998)
7. Himmel, H.-J., Downs, A.J., Greene, T.M.: Chem. Rev. **102**, 4191 (2002)

8. Whittle, E., Dows, D.A., Pimentel, G.C.: J. Chem. Phys. **22**, 1943 (1954)
9. Andrews, L., Moskovits, M.: Chemistry and Physics of Matrix Isolated Species, p. 430. North Holland, Amsterdam (1989)
10. Jacox, M.E.: J. Phys. Chem. Ref. Data **13**, 945 (1984)
11. Jacox, M.E.: J. Phys. Chem. Ref. Data **17**, 269 (1988)
12. Jacox, M.E.: Chem. Phys. **189**, 149 (1994)
13. Jacox, M.E.: J. Phys. Chem. Ref. Data **27**, 115 (1998)
14. Khriachtchev, L., Pettersson, M., Runeberg, N., Lundell, J., Räsänen, M.: Nature **460**, 874 (2000)
15. Wang, X., Andrews, L., Willmann, K., Brosi, F., Riedel, S.: Angew. Chem. Int. Ed. **51**, 10628 (2012)
16. Toennies, J.P., Vilesov, A.F.: Annu. Rev. Phys. Chem. **49**, 1 (1998)
17. Toennies, J.P., Vilesov, A.F., Whaley, K.B.: Phys. Today **2**, 31 (2001)
18. Callegari, C., Jäger, W., Stienkemeier, F.: Handbook of Nanophysics: Nanoparticles and Quantum Dots, p. 4. CRC (2011)
19. Kuyanov-Pozument, K., Skvortsov, D., Slipchenko, M.N., Sartakov, B.G., Vilesov, A.F.: Physics and Chemistry at Low Temperatures, p. 203. Pan Stanford (2011)
20. Toennies, J.P., Vilesov, A.F.: Angew. Chem. Int. Ed. **43**, 2622 (2004)
21. Callegari, C., Ernst, W.E. In: Quack, M., Merkt, F. (eds.) Handbook of High-Resolution Spectroscopy, p. 1551. Wiley, Chichester (2011)
22. Yang, S., Ellis, A.M.: Chem. Soc. Rev. (2013). doi:10.1039/C2CS35277J
23. Pobell, F.: Matter and Methods at Low Temperatures. Springer-Verlag, Berlin (1996)
24. Wilks, J., Betts, D.S.: An Introduction to Liquid Helium. Clarendon, Oxford (1987)
25. Tilley, D.R., Tilley, J.: Superfluidity and Superconductivity. Institute of Physics, Bristol (1990)
26. Grisenti, R.E., Toennies, J.P.: Phys. Rev. Lett. **90**, 234501 (2003)
27. Brink, D.M., Stringari, S.: Z. Phys. D **15**, 257 (1990)
28. Guirao, A., Pi, M., Barranco, M.: Z. Phys. D **21**, 185 (1991)
29. Grebenev, S., Hartmann, M., Havenith, M., Sartakov, B., Toennies, J.P., Vilesov, A.F.: J. Chem. Phys. **112**, 4485 (2000)
30. Grebenev, S., Toennies, J.P., Vilesov, A.F.: Science **279**, 2083 (1998)
31. Kim, H., Seo, K., Tabbert, B., Williams, G.A.: Europhys. Lett. **58**, 395 (2002)
32. Weilert, M.A., Whitaker, D.L., Maris, H.J., Seidel, G.M.: J. Low Temp. Phys. **98**, 17 (1995)
33. Weilert, M.A., Whitaker, D.L., Maris, H.J., Seidel, G.M.: Phys. Rev. Lett. **77**, 4840 (1996)
34. Weilert, M.A., Whitaker, D.L., Maris, H.J., Seidel, G.M.: J. Low Temp. Phys. **106**, 101 (1997)
35. Harms, J., Toennies, J.P., Knuth, E.L.: J. Chem. Phys. **106**, 3348 (1997)
36. Lewerenz, M., Schilling, B., Toennies, J.P.: Chem. Phys. Lett. **206**, 381 (1993)
37. Harms, J., Toennies, J.P., Dalfovo, F.: Phys. Rev. B **58**, 3341 (1998)
38. Harms, J., Toennies, J.P., Knuth, E.L.: J. Chem. Phys. **106**, 3348 (1996)
39. Knuth, E.L., Henne, U.: J. Chem. Phys. **110**, 2664 (1999)
40. Hagena, F.: Phys. Fluids **17**, 894 (1974); Surf. Sci. **106**, 101 (1981); Z. Phys. D At., Mol. Clusters **4**, 291 (1987)
41. Kanaev, V., Museur, L., Laarmann, T., Monticone, S., Castex, M.C., von Haeften, K., Möller, T.: J. Chem. Phys. **115**, 10248 (2001)
42. Gomez, L.F., Loginov, E., Sliter, R., Vilesov, A.F.: J. Chem. Phys. **135**, 154201 (2011)
43. Henne, U., Toennies, J.P.: J. Chem. Phys. **108**, 9327 (1998)
44. Ghazarian, V., Eloranta, J., Apkarian, V.A.: Rev. Sci. Instrum. **73**, 3606 (2002)
45. Slipchenko, M.N., Kuma, S., Momose, T., Vilesov, A.F.: Rev. Sci. Instrum. **73**, 3600 (2002)
46. Yang, S., Brereton, S., Ellis, A.M.: Rev. Sci. Instrum. **76**, 104102 (2005)
47. Yang, S., Ellis, A.M.: Rev. Sci. Instrum. **79**, 016106 (2008)
48. Pentlehner, D., Riechers, R., Dick, B., Slenczka, A., Even, U., Lavie, N., Brown, R., Luria, K.: Rev. Sci. Instrum. **80**, 043302 (2009)
49. Lewerenz, M., Schilling, B., Toennies, J.P.: J. Chem. Phys. **102**, 8191 (1995)
50. Tiggesbäumker, J., Stienkemeier, F.: Phys. Chem. Chem. Phys. **9**, 4748 (2007)

51. Bünermann, O., Droppelmann, G., Hernando, A., Mayo, R., Stienkemeier, F.: J. Phys. Chem. A **111**, 12684 (2007)
52. Ancilotto, F., Lerner, P.B., Cole, M.W.: J. Low Temp. Phys. **101**, 1123 (1995)
53. Douberly, G.E., Miller, R.E.: J. Phys. Chem. A **111**, 7292 (2007)
54. Krückeberg, S., Dietrich, G., Lützenkirchen, K., Schweikhard, L., Walther, C., Ziegler, J.: J. Chem. Phys. **110**, 7216 (1999)
55. Gutberlet, A., Schwaab, G., Birer, Ö., Masia, M., Kaczmarek, A., Forbert, H., Havenith, M., Marx, D.: Science **324**, 1545 (2009)
56. Liu, J., Shepperson, B., Ellis, A.M., Yang, S.: Phys. Chem. Chem. Phys. **13**, 13920 (2011)
57. Boatwright, A., Cheng, F., Spence, D., Latimer, E., Binns, C., Ellis, A.M., Yang, S.: Faraday Disc. **162** (2013, in press)
58. Lewis, W.L., Applegate, B.E., Sztáray, J., Sztáray, B., Baer, T., Bemish, R.J., Miller, R.E.: J. Am. Chem. Soc. **126**, 11283 (2004)
59. Lehmann, K.K.: Mol. Phys. **97**, 645 (1999)
60. Onsager, L.: Nuovo Cimento **6**(Suppl 2), 249 (1949)
61. Vinen, W.F.: Proc. R. Soc. (Lond.) **A260**, 218 (1961)
62. Rayfield, C.W., Reif, F.: Phys. Rev. **136**, A1194 (1964)
63. Donnelly, R.J.: Quantized Vortices in Helium II. Cambridge University Press, Cambridge (1991)
64. Pitaevskii, L., Stringari, S.: Bose-Einstein Condensation. Clarendon, Oxford (2003)
65. Ancilotto, F., Barranco, M., Pi, M.: Phys. Rev. Lett. **91**, 105302 (2003)
66. Pitaevskii, L., Stringari, S.: Z. Phys. D **16**, 299 (1990)
67. Bauer, G.H., Donnelly, R.J., Vinen, W.F.: J. Low Temp. Phys. **98**, 47 (1995)
68. Gomez, L.F., Loginov, E., Vilesov, A.F.: Phys. Rev. Lett. **108**, 155302 (2012)
69. Tong, X., Benz, L., Kemper, P., Metiu, H., Bowers, M.T., Buratto, S.K.: J. Am. Chem. Soc. **127**, 12516 (2005)
70. Kaden, W.E., Wu, T., Kunkel, W.A., Anderson, S.L.: Science **326**, 826 (2009)
71. Aguirre, N.F., Mateo, D., Mitrushchenkov, A.O., Pi, M., de Lara-Castells, M.P.: J. Chem. Phys. **136**, 124703 (2012)
72. Loginov, E., Gomez, L.F., Vilesov, A.F.: J. Phys. Chem. A **115**, 7199 (2011)
73. Meiwes-Broer, K.-H. (ed.): Metal Clusters at Surfaces, Structure, Quantum Properties, Physical Chemistry. Springer, Berlin (2000)
74. Blackman, J. (ed.): Metallic Nanoparticles, Handbook of Metal Physics. Elsevier, Oxford (2009)
75. Bréchignac, C., Cahuzac, P., Carlier, F., de Frutos, M., Masson, A., Mory, C., Colliex, C., Yoon, B.: Phys. Rev. B **57**, R2084 (1998)
76. Higgins, J., Callegari, C., Reho, J., Stienkemeier, F., Ernst, W.E., Lehmann, K.K., Gutowski, M., Scoles, G.: J. Phys. Chem. A **102**, 4952 (1998)
77. Reho, J.H., Higgins, J., Nooijen, M., Lehmann, K.K., Scoles, G., Gutowski, M.: J. Chem. Phys. **115**, 10265 (2001)
78. Vongehr, S., Kresin, V.V.: J. Chem. Phys. **119**, 11124 (2003)
79. Stark, C., Kresin, V.V.: Phys. Rev. B **81**, 085401 (2010)
80. An der Lan, L., Bartl, P., Leidlmair, C., Schöbel, H., Jochum, R., Denifl, S., Märk, T.D., Ellis, A.M., Scheier, P.: J. Chem. Phys. **135**, 044309 (2011)
81. Israelachvili, J.N.: Intermolecular and Surface Forces. Academic, London (1992)
82. Nauta, K., Miller, R.E.: Science **285**, 1895 (1999)
83. Madeja, F., Havenith, M., Nauta, K., Miller, R.E., Chocholoušová, J., Hobza, P.: J. Chem. Phys. **120**, 10554 (2004)
84. Ellis, A.M., Yang, S.: Phys. Rev. A **76**, 032714 (2007)
85. Lewis, W.K., Lindsay, M., Bemish, R.J., Miller, R.E.: J. Am. Chem. Soc. **127**, 7235 (2005)
86. Callicoatt, B.E., Förde, K., Ruchti, T., Jung, L.F., Janda, K.C., Halberstadt, N.: J. Chem. Phys. **108**, 9371 (1998)
87. Ruchti, T., Förde, K., Callicoatt, B.E., Ludwigs, H., Janda, K.C.: J. Chem. Phys. **109**, 10679 (1998)

88. Ruchti, T., Callicoatt, B.E., Janda, K.C.: Phys. Chem. Chem. Phys. **2**, 4075 (2000)
89. Higgins, J., Ernst, W.E., Callegari, C., Reho, J., Lehmann, K.K., Scoles, G., Gutowski, M.: Phys. Rev. Lett. **77**, 4532 (1996)
90. Higgins, J., Callegari, C., Reho, J., Stienkemeier, F., Ernst, W.E., Lehmann, K.K., Gutowski, M., Scoles, G.: Science **273**, 629 (1996)
91. Fröchtenicht, R., Kaloudis, M., Koch, M., Huisken, F.: J. Chem. Phys. **105**, 6128 (1996)
92. Yang, S., Brereton, S., Ellis, A.M.: Int. J. Mass Spectrom. **253**, 79 (2006)
93. Leidlmair, C., Bartl, P., Schöbel, H., Denifl, S., Märk, T.D., Yang, S., Ellis, A.M., Scheier, P.: Chem. Phys. Chem. **13**, 469 (2012)
94. Fárník, M., Toennies, J.P.: J. Chem. Phys. **122**, 014307 (2005)
95. Schöbel, H., Leidlmair, C., Bartl, P., Aleem, A., Hager, M., Echt, O., Märk, T.D., Scheier, P.: Phys. Chem. Chem. Phys. **13**, 1092 (2011)
96. Denifl, S., Zappa, F., Mähr, I., Ferreira da Silva, F., Aleem, A., Mauracher, A., Probst, M., Urban, J., Mach, P., Bacher, A., Echt, O., Märk, T.D., Scheier, P.: Angew. Chem. Int. Ed. **48**, 8940 (2009)
97. Denifl, S., Zappa, F., Mähr, I., Mauracher, A., Probst, M., Urban, J., Mach, P., Bacher, A., Bohme, D.K., Echt, O., Märk, T.D., Scheier, P.: J. Chem. Phys. **132**, 234307 (2010)
98. Ren, Y., Kresin, V.V.: J. Chem. Phys. **128**, 074303 (2008)
99. Mozhayskiy, V., Slipchenko, M.N., Adamchuk, V.K., Vilesov, A.F.: J. Chem. Phys. **127**, 094701 (2007)
100. Slipchenko, M.N., Sartakov, B.G., Vilesov, A.F.: J. Chem. Phys. **128**, 134509 (2008)
101. Bartelt, A., Close, J.D., Federmann, F., Quaas, N., Toennies, J.P.: Phys. Rev. Lett. **77**, 3525 (1996)
102. Ratschek, M., Koch, M., Ernst, W.E.: J. Chem. Phys. **136**, 104201 (2012)
103. Close, J.D., Federmann, F., Hoffmann, K., Quaas, N.: J. Low Temp. Phys. **111**, 661 (1998)
104. An der Lan, L., Bartl, P., Leidlmair, C., Schöbel, H., Jochum, R., Denifl, S., Märk, T.D., Ellis, A.M., Scheier, P.: Phys. Rev. B **85**, 115414 (2012)
105. Diederich, T., Döppner, T., Fennel, T., Tiggesbäumker, J., Meiwes-Broer, K.-H.: Phys. Rev. A **72**, 23203 (2005)
106. Loginov, E., Gomez, L.F., Chiang, N., Halder, A., Guggemos, N., Kresin, V.V., Vilesov, A.F.: Phys. Rev. Lett. **106**, 233401 (2011)

Chapter 11
Reactive Dynamics in Confined Water by Reversed Micelles

Minako Kondo, Ismael A. Heisler, and Stephen R. Meech

Abstract The excited state reactive dynamics of the fluorescence dye molecule, Auramine O, were studied in the confined water environment in reversed micelles formed by ionic and nonionic surfactants. The fluorescence decays were measured by the fluorescence up-conversion method with a time resolution of <70 fs. The time-resolved fluorescence spectra were recreated and analysed using a one-dimensional generalised Smoluchowski equation assuming a time-dependent diffusion coefficient. The fluorescence decay times measured showed a dependence on water droplet sizes, and the reaction time was significantly slowed down in the smallest reversed micelles by both ionic AOT and nonionic surfactant. The reactive friction estimated from the Smoluchowski analysis was enhanced in the confined media which shows good agreement with the reaction times. Therefore, we found out that the interfacial charges are not required for the suppression of the reaction. Interestingly, the slower reaction dynamics were measured in nonionic surfactant reversed micelles than that in reversed micelles by AOT, even when Auramine O is in a similar size of water droplet.

11.1 Introduction

Water plays an essential role as a solvent in biology and chemistry. For this reason the physical chemistry of water has been studied in detail [1–3]. The detailed picture of liquid water dynamics, such as molecular rotation, H-bond breaking and

M. Kondo (✉)
Molecular Photoscience Research Center, Kobe University, Rokkoudai-chyou 1-1, Kobe 657-8501, Japan
e-mail: mkondo@sapphire.kobe-u.ac.jp

I.A. Heisler • S.R. Meech
School of Chemistry, University of East Anglia, Norwich NR4 7TJ, UK
e-mail: i.heisler@uea.ac.uk; s.meech@uea.ac.uk

Z.M. Wang (ed.), *Nanodroplets*, Lecture Notes in Nanoscale Science and Technology 18, DOI 10.1007/978-1-4614-9472-0_11,

rearrangement, is being revealed using vibrational spectroscopy [4] and theoretical calculations [5]. However, in many important systems water molecules are not present as bulk water at all, but as a limited number of water molecules in a spatially constrained environment. For example, confined water plays a critical role in the micro-heterogeneous environment of proteins and membranes. A number of proteins hold a large cavity filled with water molecules [6]. The structure and dynamical properties of water inside the cavity may govern the primary functional events in the protein. Protein-protein interactions and protein folding are also both critically dependent on a confined water layer [7]. Nanometre-sized reverse micelles containing water appear in membrane lipid bilayers, such as cell membranes, and are involved in an array of cellular processes [8]. For membranes themselves interfacial water is essential for their equilibrium structure and biological function.

There are other important applications of reactions in constrained water environments. Water dispersed by reversed micelles in nonpolar solvents could be used for syntheses of the ultrasmall semiconductor CdS nanoparticles or quantum dots. The reversed micelle could be used for controlling nanoparticle size. Therefore there is an increasing application of reversed micelles as a nano-reactor [9–12]. Enzymes could also dissolve in the water droplet within reversed micelles in which they can retain their activity and stability. The catalysed reaction showed slower reaction rate, and the enzyme presents significantly higher activity in water pool of reversed micelles compared to bulk liquids. During the reaction, the existence of counterions within the water pool is also known to change the stability of intermediate states [13]. Therefore, the study of the effect on chemical reactions of the change in water structure and dynamics that occur on confinement has gained much attention [14–18].

11.2 AOT Reversed Micelle System

In this section we will briefly introduce the study of fast dynamics in confined water, focussing on inverse micelle media. The next section will give details on the reactive probe which we selected to study in the inverse micelle media. Section 11.4 shows the sample preparation. The experimental methods, especially ultrafast fluorescence, we have employed in these studies will be explained in Sects. 11.5 and 11.6. Section 11.7 will include a detailed review of our studies of the reactive probe molecule AO in inverse micelles formed by ionic surfactants [19–21]. Finally in Sect. 11.8 we will give a detailed description of the analysis of the time-resolved fluorescence data, giving as an example the dynamics of AO in water nanodroplets confined in nonionic micelles [22].

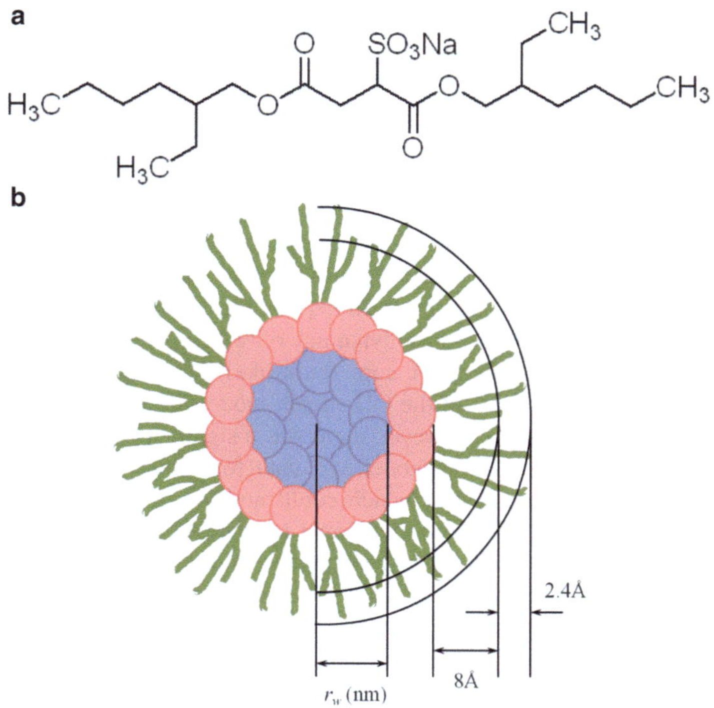

Fig. 11.1 (**a**) The structure of AOT surfactant and (**b**) the reversed micelles [1]

11.2.1 Water Behaviour in Reversed Micelles and Core-Shell Model

One of the most common and widely used models for a constrained aqueous environment is the reversed micelle [23]. Water nanodroplets are formed and stabilised by surfactants in reversed micelles. The size of the water nanodroplet is controllable in the range of 1–10 nm by mixing water, surfactant and nonpolar solvents at appropriate compositions [24]. The most studied reversed micelle system is that formed by Sodium bis(2-ethylhexyl) sulfosuccinate known as AOT (Fig. 11.1). AOT is an anionic surfactant, and the structure and properties of its reversed micelle phase have been studied in detail in both experimental (e.g., small-angle neutron scattering (SANS) [23], vibrational spectroscopy [25–27]) and theoretical studies (e.g., molecular dynamics simulation [28]). For the water/Na-AOT/heptane reversed micelles, the water nanodroplet radius was shown to be determined by the molar ratio of water to Na-AOT concentrations:

$$r_{\rm w} \approx 0.18 w_0, \tag{11.1}$$

where $w_0 = [H_2O]/[AOT]$. This relation was determined by small-angle neutron scattering measurements [29].

The dynamical properties of confined water are found to be considerably different to those of water in the bulk phase. The FTIR absorption spectra of the OD stretch mode of 5% HOD in H_2O in different r_w AOT reverse micelles showed a blue shift from that in the bulk. It is known that weaker hydrogen bonds lead to a water hydroxyl stretch absorption on the blue side compared to bulk water, thus suggesting the weakening of hydrogen bonds in the micelle [30]. Furthermore, the restriction of the structural dynamics of water in AOT reversed micelles was observed in measurements of the water orientational relaxation, which showed a drastic slowing in the reversed micelle compared to the bulk phase [27].

A 'core-shell' model was suggested to explain these effects [14, 27, 31]. This model proposes that the nature and dynamics of water in the reversed micelles are expressed as a combination of core (bulk-like) water and shell (interface-bounded) water. The blue shift of the OH stretch mode was accurately reproduced by this core-shell model [27]. However, there was a significant difference between the predictions of the core-shell model and data for orientational relaxation in small ($w_0 < 10$) reversed micelles, which could not be represented by a sum of two contributions [27]. Since the orientational relaxation depends strongly on the global hydrogen bond network rearrangement, this failure shows that the hydrogen bond network in small r_w micelles cannot be simply separated into core and shell character [32]. In very small droplets (formed with w_0 less than five, $r_w < 1$ nm), the perturbation from interfacial water apparently extends into the core water. Therefore, the core water dynamics in such very small reversed micelles are still slower than in large reversed micelles. When the water droplet size increases ($w_0 = 10$), the water properties can then be separated into two components [33]. Thus, on the basis of studies of H-bond spectra, the core-shell model operates at $w_0 > 16.5$.

11.2.2 Rotation and Solvation Dynamics

There have been a number of studies of the rotation and solvation dynamics of fluorescent probes. In this case the probe is nonreactive, but water dynamics are probed by time-dependent fluorescence anisotropy decay and time-resolved Stokes shift method, respectively [34–37]. Levinger et al. studied the rotational dynamics of Coumarin 343(C343) in water droplets of AOT reverse micelles. Fluorescence depolarisation timescale of C343 was found to be within 200 ps in bulk solution in the presence of ions (since in small micelles the concentration of the counterions may be very high, up to 10 mol dm^{-3}solution). The concentrated electrolyte solution can itself change the bulk water structure and hydrogen bonding strength by the presence of ions [38, 39]. The formation of ion–water interaction influences the bulk water–water H-bonding showing significant dependence on electrolyte concentrations, in particularly highly concentrated electrolyte solutions. The rotational correlation time of C343 in reversed micelles was measured to have much longer timescale about 1.3 ns in $w_0 = 1.7$. This time constant decreases with increasing r_w which suggested that C343 experienced decreasingly restrictive environment in larger water droplets. However the rotational motion of C343 was still significantly slower in the large water droplet of $w_0 = 40$ than in the bulk electrolyte. The result suggested that the C343 presented near the micellar interface and that the dynamics were affected by the AOT interface.

Levinger et al. also studied the solvation dynamics of C343 in the same reversed micelle system. The solvation time was measured to be slower in the reversed micelles than in bulk water, and the slower solvation dynamics was observed in smaller r_w. In addition, the impact of varying the counterion on solvation dynamics in AOT reversed micelles has been investigated using Na^+, K^+, Ca^{2+} and NH_4^+ AOT [25, 40]. Steady-state absorption, emission and time-resolved fluorescence spectra of C343 in these different AOT reversed micelles were measured and compared with the results in bulk aqueous solution. The results suggested that the mobility of water inside AOT reversed micelles is suppressed in all cases, while it was hardly altered in the ionic solutions. Consequently, it was concluded that the immobilisation of water in reversed micellar environments is principally due to the restricted environment rather than to specific water ion interactions. However, water dynamics in the different micelles were influenced differently by different counterions. Specifically, the fastest solvation dynamics were observed in ammonium AOT reversed micelles and assigned to weaker interactions of water with the cation. Correspondingly, water in calcium AOT reversed micelles was considerably restricted compared with the potassium and sodium AOT reversed micelles, reflecting relatively stronger interactions between water and Ca^{2+} compared with the singly charged alkali metal ions. Thus counterion is a potential variable for tuning the properties of the water nanodroplet.

11.3 Reactive Probes for Femtosecond Dynamics Study

The rapid development of systems that produce ultrashort laser pulses has enabled experimental investigations of many ultrafast molecular dynamical processes in real time. Excited state proton transfer (ESPT), intramolecular charge transfer (ICT) and photoisomerisation are some of the ultrafast chemical reactions that have been studied [15]. Some of these processes have been studied in reverse micelles [35]. In this work we will be particularly concerned with chemical reactions involving larger scale structural change. Examples include the Z-to-E photoisomerisation of stilbenes and azobenzenes, which occurs on the timescales of a few hundred femtoseconds to tens of picoseconds. Solvent effects have been observed for the isomerisation of stilbene. In particular it is well established that the reaction rate is significantly affected by the solvent friction (or viscosity) and that the isomerisation is suppressed in viscous solvents [41]. Triphenylmethane (TPM) dyes such as Malachite Green (MG) also reveal a strong viscosity dependence of the fluorescence quantum yield with a decay time of the electronic excited state occurring in the picosecond range. The relaxation dynamics were investigated through nonlinear

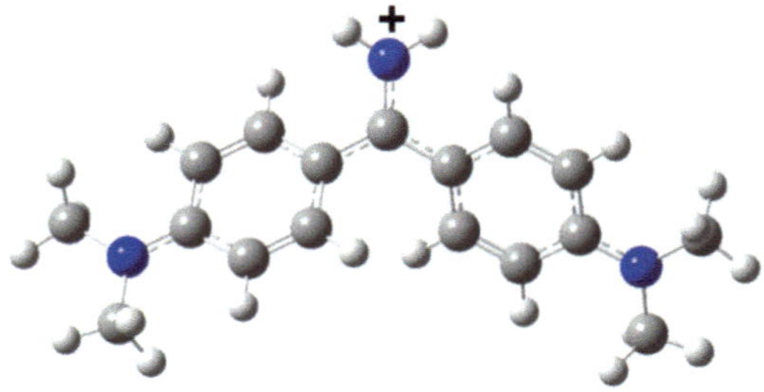

Fig. 11.2 The diagram of Auramine O (*blue*—nitrogen, *grey*—carbon, *white*—hydrogen)

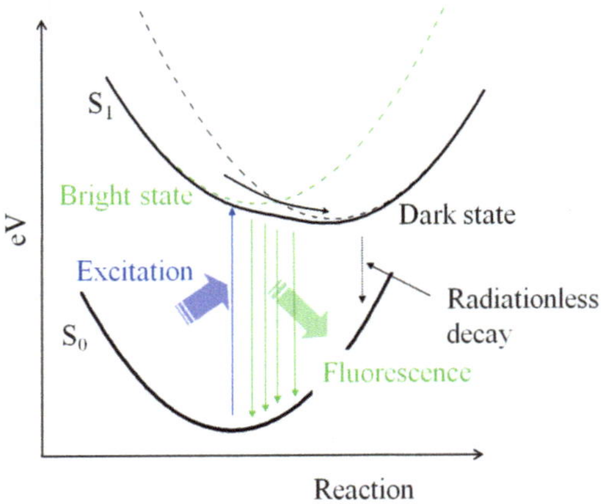

Fig. 11.3 The scheme of potential energy curves illustrating the excited state processes [2]

spectroscopic techniques with picosecond time resolution [42, 43]. The relaxation rate is strongly dependent not only on the solvent viscosity but also on substituents on the phenyl ring. This dependence has been attributed to the radiationless decay promoted by diffusive rotational motions of the phenyl groups in the excited state, leading to internal conversion [44].

The diphenylmethane dye Auramine O (AO, Fig. 11.2) presents a solvent viscosity dependence of its fluorescence quantum yield, being weakly fluorescent in low viscosity solvents and highly fluorescent in very viscous solvents [45]. The dynamics associated with the fluorescence decay in AO are ascribed to an internal conversion process promoted via the rotational diffusion of phenyl groups, analogues to MG. The relaxation dynamics of AO and MG have been studied mainly in bulk solvents. As a result of recent picosecond transient absorption and femtosecond time-resolved fluorescence studies, it was proposed that the initial photoexcited state relaxation is via a barrierless (or nearly barrierless) diffusive internal motion to form a nonradiative state (Fig. 11.3) [46, 47]. Bagchi, Fleming and Oxtoby (BFO) developed a theoretical model for radiationless processes with no internal barrier [48]. In the BFO model the radiationless decay is represented by a coordinate-dependent sink centred on the excited state minimum and a radiative decay rate along the potential surface which is independent of coordinate. They employed a modified Smoluchowski equation to regulate the diffusive motion along the excited state potential surface; models for barrierless diffusive dynamics are described further in Sect. 11.6.

11.4 Sample Preparation

Auramine O (AO, Fig. 11.2), the reactive probe molecule, was purchased from Sigma-Aldrich as dye content $\gg$80%; it was verified by recrystallisation that the other <20% impurities do not affect the resultant fluorescence and absorption spectra.

The model confined aqueous environments to be studied were realised by forming a reversed micelle system where water droplets are stabilised by surfactants in nonpolar solvents. The nonionic surfactant Igepal CO 520, Mn ~ 441 (IG, Fig. 11.4) was also purchased from Sigma-Aldrich. Pure water was obtained as an analytical reagent grade from Fisher Scientific.

The IG reversed micelles have been characterised by small-angle neutron scattering by Lipgens et al. [49]. The radius of the inner and outer water, R_i and R_a, can be calculated through the following Eqs. (11.2) and (11.3), respectively:

$$R_i = 0.19\, w_0 + 0.70, \tag{11.2}$$

$$R_a = 0.22\, w_0 + 2.05, \tag{11.3}$$

where $w_0 = [H_2O]/[IG]$.

In these expressions, R_i represents the core water radius which forms once the solvation shell has been completed by solvating the oxyethylene chain region with water. R_a is the radius including the oxyethylene chain region. For the experiments, $R_i = r_w$ was controlled in the range 2–4 nm, the limits of microemulsion stability. The AO concentration for the IG systems was determined to be less than one AO molecule per one reversed micelle.

11.5 Time-Resolved Fluorescence Spectroscopy

11.5.1 Femtosecond Fluorescence Up-Conversion Setup

Fluorescence decays were measured by the fluorescence up-conversion spectrometer [19]. The laser pulses centred around 820 nm from a Ti:Sapphire oscillator (Coherent Micra 10) were compressed to 20 fs and focused onto a BBO crystal to generate the second harmonic pump beam. The second harmonic beam was

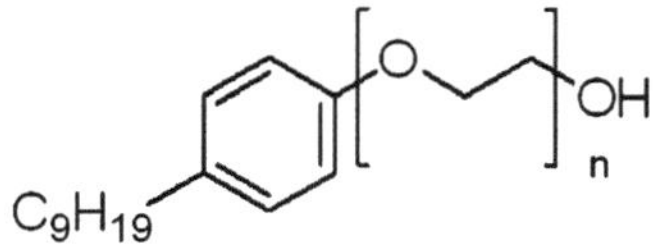

Fig. 11.4 The structure of Igepal CO 520

recompressed with a pair of chirped mirrors and focused onto a 1 mm sample cuvette. The sample fluorescence was collected and focussed on to a BBO crystal with a reflective microscope objective. Fundamental beam pulses at 820 nm travelled via a delay stage and were recompressed using a pair of chirped mirrors. The fundamental beam was used as a gate pulse and mixed in the BBO crystal with sample fluorescence. The up-converted signal amplitudes were detected by a photomultiplier tube and a photon counter. Time resolution of the setup was determined to be 70 fs by up-converting Raman scattering from pure solvent and fitting the result to a Gaussian function.

11.5.2 Decay Curve Fitting and Time-Resolved Fluorescence Spectra

Fluorescence decays were fitted by mathematical functions, $F(t)$, a sum of exponentials convoluted with the instrumental response function (IRF) from the up-converted Raman:

$$F(t) = \sum_{i=1}^{n} A_i \exp(-t/\tau_i)\,. \tag{11.4}$$

The averaged fluorescence decay time was obtained from

$$<\tau>= \frac{\sum_i A_i \tau_i}{\sum_i A_i}. \tag{11.5}$$

From the measurement of fluorescence decays at several fluorescence wavelengths, we can create the time-dependent fluorescence spectra [50]. The fluorescence spectrum at a given time, $S(\lambda, t)$, was obtained from deconvoluted fits to the decay data measured at different wavelengths spanning the emission spectrum, $F(\lambda, t)$. These are weighted according to the intensity in the steady-state spectrum, $S_0(\lambda)$,

$$S(\lambda,t) = \frac{F(\lambda,t)\, S_0(\lambda)}{\int_0^{\infty} F(\lambda,t)\, dt}. \tag{11.6}$$

Usually fluorescence spectra are recorded as a function of wavelength, λ, in nanometre scale. However, from a physical interpretation point of view, it is better to represent the data in terms of energy, using the wave number, $\overline{\nu}\,\left[\mathrm{cm}^{-1}\right]$, scale. As a result, the 3-dimensional fluorescence surfaces were created as a function of both time and wave number.

The time-resolved fluorescence spectra are obtained by slicing the 3D fluorescence surface at different times. These spectra are fit with a log-normal function

$$F\left(\overline{\nu}\right) = h\left\{ \begin{array}{rr} \exp\left[-\ln(2)\{\ln\left(1+\alpha\right)/\gamma\}^2\right] & \alpha > -1 \\ 0 & \alpha \leq -1 \end{array} \right\}, \tag{11.7}$$

where $\alpha \equiv 2\gamma\left[\overline{\nu} - \overline{\nu}_{\mathrm{p}}\right]\Gamma$.

This function is extremely helpful in describing structureless asymmetric spectra, where h represents the height, γ the asymmetry parameter or skewness, Γ the width parameter and $\overline{\nu}_{\mathrm{p}}$ the peak wave number. Recording these parameters as a function of the delay time characterises the temporal evolution of the spectral shape. In addition, the first moment of the emission band, ν_{FM}, was calculated from

$$\overline{\nu}_{\mathrm{FM}} = \frac{\displaystyle\int \overline{\nu}\, I\left(\overline{\nu}\right) d\nu}{\displaystyle\int I\left(\overline{\nu}\right) d\nu}. \tag{11.8}$$

11.6 Generalised Smoluchowski Model

The excited state decay of AO was modelled in terms of the diffusive phenyl rotational motion on a barrierless excited state potential energy surface leading to a curve crossing with a non-emissive charge transfer state. The evolution of the excited state population on the surface was reproduced by solving the Smoluchowski equation. This model was developed for structural dynamics by Glasbeek and co-workers [46, 51–53] and successfully applied to AO in several solvents and AOT stabilised water and also to other excited state reactions [19].

In this model the evolution of the excited state population, $\rho(z,t)$, along a reaction coordinate z is determined by solving a generalised Smoluchowski equation, for the motion of a particle in a harmonic potential as described in the BFO model [48] and by Okumura and Oxtoby [54]:

$$\frac{\partial}{\partial t}\rho\left(z,t\right) = D(t)\frac{\partial}{\partial z}\left(\frac{\partial}{\partial z} + \frac{1}{k_{\mathrm{B}}T}\frac{\partial}{\partial z}S(z)\right)\rho\left(z,t\right) - \kappa\Gamma(z)\rho\left(z,t\right), \tag{11.9}$$

where the first term is the diffusive motion on the excited state potential surface $S_r(z)$ and the second term accounts for decay back to the ground state. In this expression, k_{B} is the Boltzmann constant 0.695 cm^{-1}/K where 1 $\mathrm{cm}^{-1}\mathrm{hc} = 1.986 \times 10^{-12}$J, and T is temperature at 293 K. $D(t)$ is the time-dependent diffusion coefficient, and $\Gamma(z)$ is a Gaussian sink function

$$\Gamma(z) = \frac{1}{\sigma\sqrt{\pi}}\exp\left(\frac{-\left(z - z_0\right)^2}{\sigma^2}\right) \tag{11.10}$$

in which σ represents the width and z_0 is the centre of the Gaussian, and finally κ is the nonradiative decay rate to the ground state. This allows for decay to the ground state with different probabilities along the reaction coordinate. The initial population on the excited state, $\rho(z,0)$, is assumed to be asymmetric and given by a log-normal shape:

$$\rho\left(z,0\right)=\exp\left(-\ln(2)\left(\frac{\ln\left(1+2\gamma z/\left(2\sqrt{k_{\mathrm{B}}T/k}\right)\right)}{\gamma}\right)^{2}\right) \tag{11.11}$$

if $2\gamma z/\left(2\sqrt{k_{\mathrm{B}}T/k}\right) > -1$ and elsewhere $\rho(z,0)=0$; γ represents an asymmetry parameter.

The excited state potential energy surface, $S_r(z)$, is obtained by applying an adiabatic coupling model in which the excited state potential surface was assumed to arise from coupling between a locally excited emissive state, $S_{\mathrm{e}}(z)$, and a dark (charge transfer) state, $S_{\mathrm{d}}(z)$ [55, 56]

$$S_r(z)=\frac{1}{2}\left(S_{\mathrm{e}}(z)-S_{\mathrm{d}}(z)\right)-\frac{1}{2}\sqrt{\left(S_{\mathrm{e}}(z)+S_{\mathrm{d}}(z)\right)^{2}+4C^{2}}, \tag{11.12}$$

where C is the coupling strength parameter. During the evolution from the fluorescent to the dark state, the normalised coordinate-dependent transition moment, $M^2(z)$, can be calculated from

$$M^{2}(z)=\frac{C^{2}}{C^{2}+\left[S_r(z)-S_{\mathrm{e}}(z)\right]^{2}}, \tag{11.13}$$

which is a decreasing function with increasing motion along z and decreases more sharply for weaker coupling.

To compare the model with the experimental data, it is necessary to calculate the time-dependent spectra, given by [55, 56]:

$$I_{\mathrm{fl}}\propto\int dz\ g\left(\nu_0(z),\nu(z)-\nu_0(z)\right)M^{2}(z)\rho\left(z,t\right)\nu^{3}, \tag{11.14}$$

in which $g(\nu_0(z), \nu(z)-\nu_0(z))$ is a line shape function (a log-normal function was used here) which describes the Franck-Condon factor, where $\nu_0(z)$ is the torsional-angle-dependent energy gap between the excited and ground states, that is, $\nu_0(z)=(S_r(z)-G(z))/h$.

In the Stokes–Einstein–Debye equation, the diffusion coefficient is obtained as $D_r=k_{\mathrm{B}}T/\zeta$ which is independent of time. Here, ζ is defined as a friction coefficient. An example of $\rho(z,t)$ calculated by solution of Eq. (11.9) is shown in Fig. 11.5. The distribution broadens and becomes asymmetric with time as it moves down and along the reactive potential energy surface.

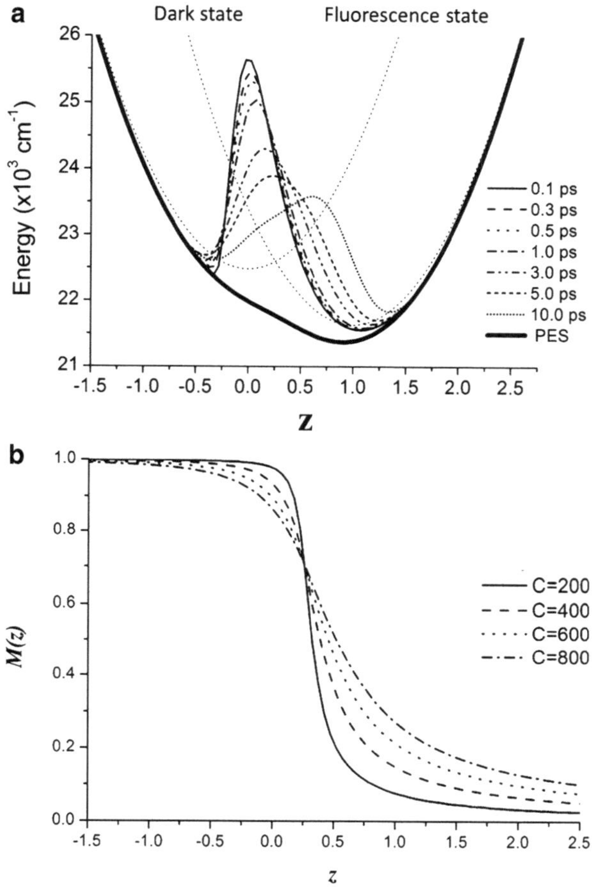

Fig. 11.5 (**a**) The representative excited state potential surface and the evolution of the excited state population $\rho(z,t)$ of AO in decanol. (**b**) The transition dipole moment as a function of the normalised twisting coordinate, z, with different coupling constant values

In this study, the data could not be well fit with a time-independent diffusion coefficient so a time-dependent diffusion coefficient is employed, $D(t)$. Hynes and co-workers [57] examined the influence of solvent dynamics on adiabatic electron transfer (ET) reactions through MD simulations. In their paper, the friction, $\zeta(t)$, on the solvent coordinate was found to be time dependent and showed a steep increase in the early (sub-picosecond) timescale. They stressed that the ET transition reflects the shorter time friction and solvation dynamics. Essentially, the slow part of the friction cannot influence the rate of a fast reaction.

The excited state structure changes in AO are known to lead a photoinduced intramolecular charge transfer state (the dark state) [47]. Hence the diffusion coefficient associated with the AO excited state dynamics is likely to be time dependent. According to Hynes and co-workers [57], the time-dependent diffusion coefficient is given by

$$D(t) = -\left\langle (\delta z)^2 \right\rangle \frac{\dot{\Delta}}{\Delta}, \tag{11.15}$$

where

$$\Delta(t) = \frac{\langle \delta z \delta z(t) \rangle}{\left\langle (\delta z)^2 \right\rangle} \tag{11.16}$$

and $\delta z = z - z_{eq}$ ($z_{eq} = 0$ and 1 for the fluorescence and the dark state, respectively), $\left\langle (\delta z)^2 \right\rangle = \left(\frac{m_L \omega_L}{k_B T} \right)^{-1}$, where m_L and ω_L are the longitudinal solvent polarisation mass and frequency governing oscillation in wells, respectively. $\Delta(t)$ is the normalised reaction coordinate time correlation function. For the case of a medium responding to a CT reaction, this function can be estimated from the time-dependent fluorescence Stokes shift associated with dipole stabilisation (solvation dynamics), in which case z is the solvation coordinate, so we can write

$$\Delta(t) = C(t), \tag{11.17}$$

where $C(t)$ is the solvation time correlation function, defined by

$$C(t) = \frac{\nu(t) - \nu(\infty)}{\nu(0) - \nu(\infty)}. \tag{11.18}$$

In solvation dynamics experiments, this function can generally be well fit by a biexponential function [55]:

$$C(t) = A_1 \exp(-t/\tau_1) + A_2 \exp(-t/\tau_2). \tag{11.19}$$

We will assume as an initial guess for our analyses that $\Delta(t) = C(t)$ is valid for the AO excited state reaction, since motion along the reaction coordinate involves formation and stabilisation of a CT state.

11.7 Reactive Dynamics in Confined Media Formed by AOT Reversed Micelles

11.7.1 Auramine O in Nano-constrained Water in AOT Reversed Micelle

The influence of confinement on the excited state twisting reaction of AO in water was studied through femtosecond fluorescence up-conversion. The confined water system was a reversed micelle system with ionic surfactant, AOT. The size of water droplet, r_w, was controlled between 1 and 10 nm by controlling the ratio

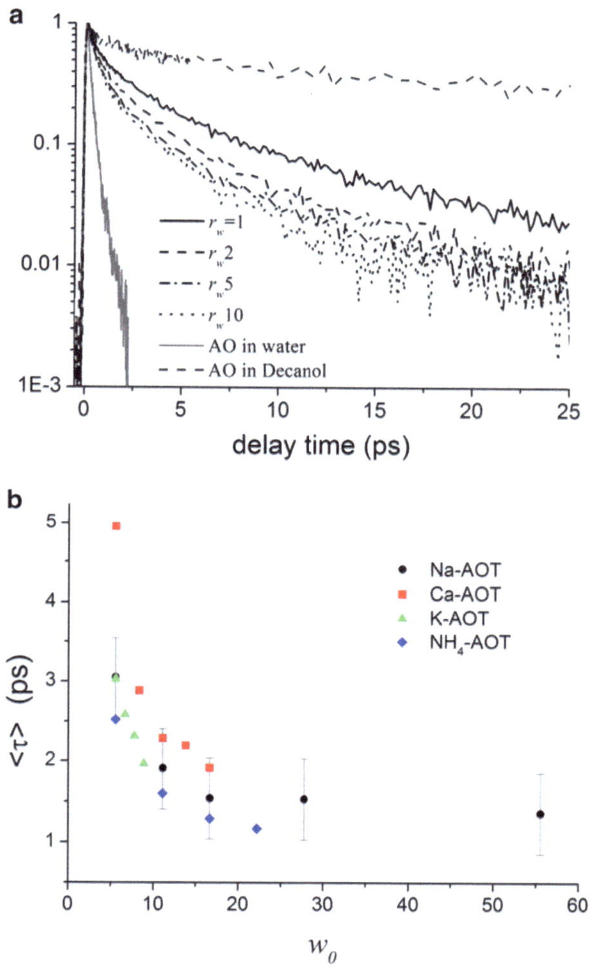

Fig. 11.6 (**a**) Fluorescence decay of AO in reversed micelles and bulk solvents. (**b**) Averaged fluorescence lifetime of AO in AOT reversed micelle changing counterions

of water and surfactant [19]. Time-dependent fluorescence spectra were constructed and analysed through the Smoluchowski model (Sect. 11.6). The comparison of fluorescence decay time of AO in bulk solvents and AOT reversed micelles shows that significantly slower dynamics of the excited state twisting motion of the phenyl ring of AO was observed in confined water in AOT reversed micelles in Fig. 11.6a. Both the fluorescence decay time and quantum yield of AO depend on the reversed micelle sizes, and both were increased with decreasing r_{W}. The friction recovered from the Smoluchowski model (Sect. 11.6) also suggested the restricted twisting dynamics associated with intramolecular charge redistribution in confined media, which is strongly related to the solvation dynamics measured in the AOT reversed micelles.

To understand the effect of changing counterion on the reaction dynamics, the counterion of AOT was changed to Ca^{2+}, K^+ and $NH_4{}^+$ as well as Na^+ as shown in

Fig. 11.6b [20]. The effect of changing the counterions had a minor influence on the reactive dynamics as can be seen in Fig. 11.6b. For the doubly charged counterion Ca^{2+}, a slightly slower reaction reflecting a larger reactive friction was obtained than for singly charged counterions. The same result was observed in the reactive friction obtained by the quantitative Smoluchowski analysis. The result was synchronised with the slower solvation dynamics in Ca^{2+} AOT reversed micelles [20]. On the other hand, the fluorescence decay times in the singly charged counterion were measured to be very similar to one another, although the difference of solvation dynamics with different counterions was more significant. The effect of counterions on solvation dynamics appeared largely perhaps indicating the different dynamics involved—water reorientation for solvation compared to both water reorientation and structural change for AO.

11.7.2 Auramine O in Water and Adsorbed on Regular Micelles

The fluorescence decay of AO was studied extensively in a series of alcohols and as a function of temperature [21]. It was found that viscosity has a major influence on the excited state decay, at least in a series of similar solvents. However, a considerably faster fluorescence decay was measured in bulk water than was expected on the basis of its viscosity. This was explained by the fast solvation dynamics of water. It was concluded that sufficiently fast solvation dynamics can promote a facile barrierless formation of the CT state. This result suggests that both viscosity and solvation dynamics may play a role in the reaction of AO.

The fluorescence decay of AO was also studied in aqueous but unconfined media. These solutions were AO in 1 M Na_2SO_4 salt and adsorbed at the charged interface formed by sodium dodecyl sulphate (SDS). For comparison, a solution of AO in the presence of positively charged cetyltrimethylammonium bromide (CTAB) micelles was also studied. This series allows us to probe the effect of ions and charged interfaces on AO reactivity in the absence of confinement. The AO absorption peak in SDS solutions showed a significant red shift from that in bulk water, which was not seen in salt and CTAB solutions. In addition, the fluorescence intensity and decay time of AO were enhanced in the presence of SDS surfactants, and they could reflect the formation of micelle (Fig. 11.7a). The critical micelle concentration (CMC) of SDS was sensed to be 7.3×10^{-3} mol dm^{-3} which is close to the value of the SDS CMC [58]. However the fluorescence decay of AO could not be changed in the presence of ions or CTAB in Fig. 11.7b. This suggests that the rate of AO reaction was suppressed by the presence of the SDS micelle interface. It suggests in turn that there is an electrostatic interaction between negative interface and positively charged AO molecule. However, there is a further slowdown

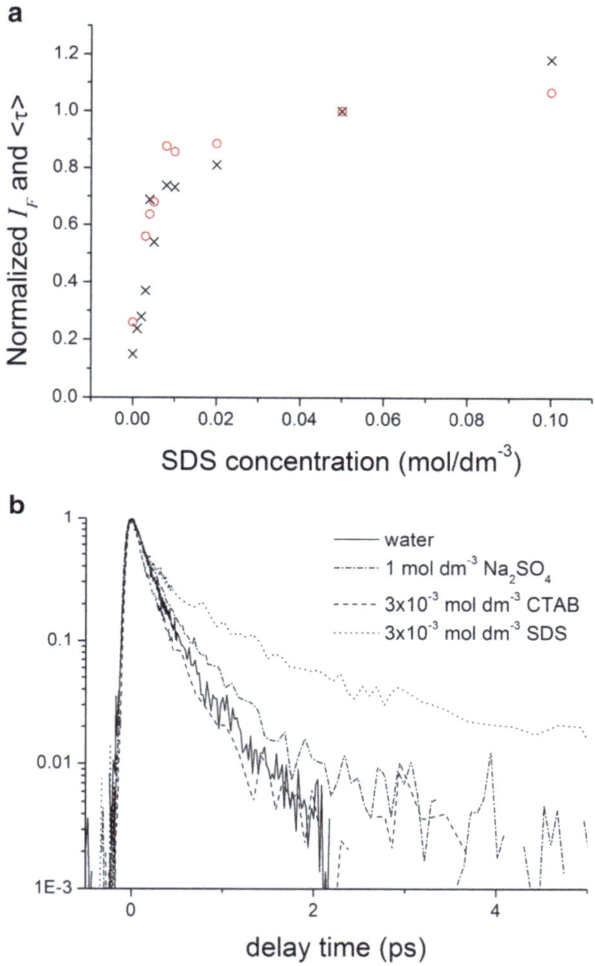

Fig. 11.7 (**a**) Normalised fluorescence yield (cross) and averaged fluorescence decay time (*circle*) as function of SDS concentrations. (**b**) Fluorescence decay of AO in water with salt and micelle interfaces

of AO excited state reaction when adsorbed at the confined interface of the AOT reversed micelle systems. This shows a significant effect of geometrical confinement on the reaction. The change in H-bonding structure between the SDS micelle and AOT reversed micelle–water interfaces may influence the AO excited state reaction. Molecular dynamics simulation points to significant differences in the two interfaces. The sodium ion density was much higher at the AOT-nanoconfined water interface than at the interface between water and SDS. This results in increasing the H-bonding between surfactant and water in the first solvation shell at the AOT interface [59, 60]. This perturbs the water structure strongly. Hence it may be that the suppressed excited state AO reaction arises from the perturbation to the H-bonded network leading to slower aqueous solvation and thus a slower decay.

11.8 Reactive Dynamics of Auramine O in Confined Media: Interfacial Charge Effect

Steady-state and time-resolved fluorescence studies of AO in nanoscale water droplets confined by AOT were described in previous studies as noted in Sect. 11.7 [19]. The effect of modifying the interface by changing the counterion was investigated and found to be small [20]. In addition, it was concluded that although the presence of a charged interface itself leads to slower dynamics of the twisting reaction of AO than in bulk water, the geometrical confinement leads to a significantly larger slowdown of the reaction dynamics [21]. An alternative reversed micelle system in which the water nanodroplet is confined by a neutral nonionic surfactant can be produced. Water dynamics in a nonionic surfactant-stabilised inverse micelle which is similar in structure to IG were studied by MD simulation and by ultrafast infrared pump-probe spectroscopy, and the results were compared to water dynamics observed in Na-AOT reversed micelles [61–66]. This system will allow us to assess further the role of interfacial charge in determining the reaction rate in AO. Thus, our studies of the reactive dynamics of AO in AOT reversed micelle systems (Sect. 11.7) are extended to uncharged IG micelles using the fluorescence up-conversion measurement and Smoluchowski analysis. The results are compared with the measurements in 'simple' solutions of water/polyethylene glycol which serves as a mimic for the IG polar head group [22].

11.8.1 Steady-State and Time-Resolved Fluorescence

Steady-state absorption spectra of AO shows the increase of absorption about 350 nm in the smallest water droplet, r_w =2 nm. The increase can be assigned to the formation of the deprotonated neutral form of AO by comparison with the acid and base pentanol solutions of AO. The result suggests that the oxyethylene head group of IG is incompletely hydrated in the smallest reversed micelles, and this causes the formation of deprotonated AO. The deprotonated AO is not excited at 415 nm excitation wavelength, and so the time-dependent measurements were not affected by the neutral AO.

The fluorescence decays of AO were measured with changing r_w of IG reversed micelles. The fluorescence decay times were dependent on r_w. That is, the decay time increased with decreasing water droplet size which is consistent with the AOT reversed micelle system. However, there is a significant enhancement of the decay time in IG reversed micelles compared to that in AOT reversed micelle system. For example, the averaged decay time of AO in IG is 35 times larger than that in the AOT system, when the $r_w = 2$. Also the water size dependence of the decay time is weak in the IG system, since the decay times of AO in IG systems are sharply decreased from the $r_w = 2$ to 3, but they are almost independent in the range of $r_w = 3$–4. In contrast, the decay times of AO showed a size dependence from $r_w = 1$ to 10 nm in the AOT stabilised reverse micelles.

Table 11.1 The parameters used to fit the log-normal fitting data of AO by Smoluchowski model

			$\Delta(t)$/ps^{-1}					
Medium	S_e/cm^{-1}	S_d/cm^{-1}	a_1	a_2	τ_1/ps	τ_2/ps	$<\tau>$/ps	κ/ps^{-1}
IG $r_w = 2$ nm	22,280	21,430	0.83	0.17	0.33	3.2	0.82	0.04
IG $r_w = 4$ nm	22,230	21,380	0.88	0.12	0.23	1.9	0.43	0.05
AOT $r_w = 1$ nm	22,640	21,790	0.83	0.17	0.32	1.6	0.54	0.14
AOT $r_w = 5$ nm	22,230	21,380	0.94	0.06	0.33	1.2	0.38	0.26
Water	22,480	20,500	0.68	0.32	0.23	0.52	0.32	5

The sink parameter were $z_0 = 0.8$, $\sigma = 0.2$ (for micelle system) and $\sigma = 0.5$ (for bulk water)

11.8.2 Analysis of Time-Dependent Emission Spectra with Smoluchowski Model

Fluorescence up-conversion data of AO in IG micelles $r_w = 2$ and 4 nm were measured at several fluorescence wavelengths across the emission spectrum. Time-resolved emission spectra were recreated by the procedure explained in Sect. 11.5.2. The time-dependent spectra obtained were analysed with the Smoluchowski model. The parameters obtained from the best fit of experimental result are listed in Table 11.1 for the IG and AOT systems and bulk water. The quality of the best fit is illustrated in Fig. 11.8. The normalised reaction coordinate time correlation function, $\Delta(t)$, which correlated to the time-dependent diffusion coefficient showed a good agreement with the qualitative results. That is, the larger fluorescence decay time in the IG system than in bulk water reflects the increased friction experienced by the reaction coordinate in the micelle. Also the decay coefficient from dark state to ground state, κ, significantly increased in bulk water, but became smaller in IG. Both $\Delta(t)$ and κ values suggested slower reactive dynamics and slower radiationless decay of AO in IG than in the AOT system.

11.8.3 Origin of the Friction Increase in IG Reversed Micelles

The significant increase of friction experienced by the reaction coordinate and decrease of decay coefficient from excited to ground state in IG was outlined in Sect. 11.8.2. Both of these contribute to the differences with AO in solution phase and in AOT micelles. One possible reason for these changes is that the head group of IG has a five monomer oxyethylene chain and which may trap the ionic probe molecule. To investigate whether the trapping effect arises from the head group,

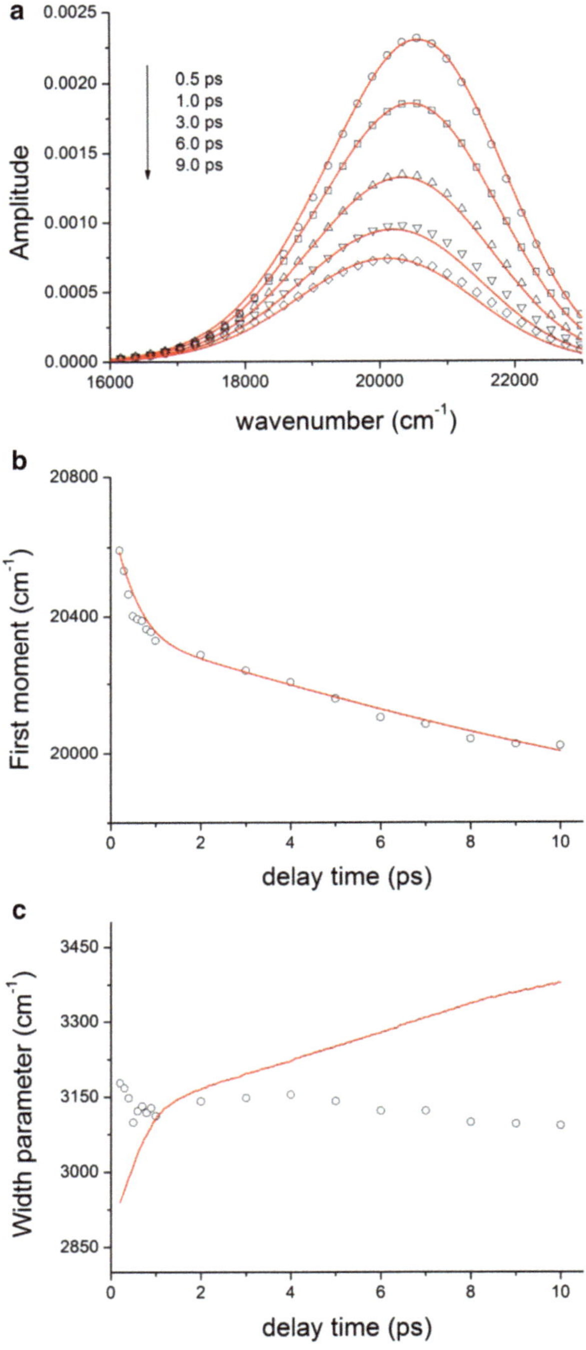

Fig. 11.8 (**a**) Time-resolved spectra, (**b**) first moment and (**c**) width parameter for AO in IG system, $r_{\mathrm{w}} = 2$ nm with the result of Smoluchowski model

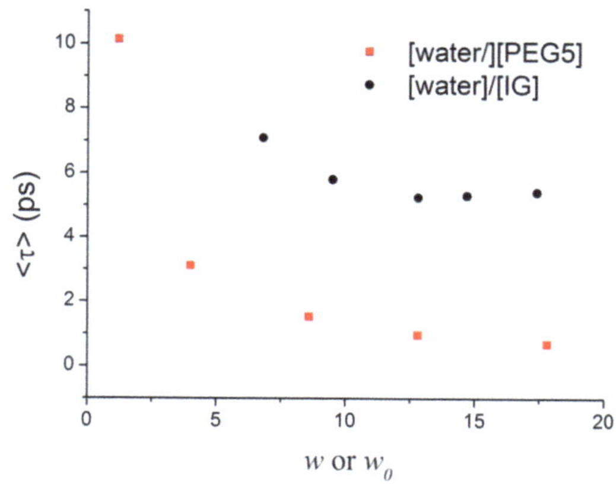

Fig. 11.9 Averaged fluorescence lifetime in PEG5 and water mixtures with $<\tau>$ of AO in IG data

we contrast the observation for AO in IG reversed micelles with the measurements for AO absorption and fluorescence in mixture of water and poly(ethylene glycol), (Mn = 190–200, $n = 4$–5), PEG5 while changing the ratio $w = $ [water]/[PEG5].

The absorption spectra and fluorescence decay of AO in the mixture of water and PEG were measured. Figure 11.9 shows the averaged fluorescence decay time, $<\tau>$, as a function of w_0 or w. The AO lifetime is a strong function of w in the PEG5-rich region. Particularly at $w < 3$ the $<\tau>$ was increased sharply. The result can be due to the increase in media viscosity and the decreased water environment which leads to the suppression of the quenching mechanism specifically found in water [21]. It can be interpreted that the AO preferentially locates in the micelle head group region. However, this argument is not consistent with the result of absorption spectra measurements. The absorption spectra of AO in mixed solvents and in water droplet in IG reversed micelle are compared in Fig. 11.10a. The peak wavelength was red shifted at $w < 5$ and subsequently blue shifted with increasing w. The blue shift is ascribed to the increase of water–AO interactions stabilising the ground electronic state. The blue shift at low w is less readily explained, but perhaps it indicates the solvation of AO by PEG5. The important point of Fig. 11.10 is that the peak position of AO in IG reversed micelle is well matched to that in bulk water rather than in the mixture. Therefore this result suggests that AO presents within the water droplet rather than in the head group region.

We speculate that the orientation of AO phenyl rings relative to oxyethylene chains may be different in AOT and IG reversed micelles, leading to a stronger effect of the interface on the AO twisting reaction. In the AOT reversed micelles, the electrostatic interaction between AO and charged interface was suggested [21]. The positively charged imine group of AO may thus orient on average towards the anionic head group of AOT. In contrast, the AO phenyl rings may preferentially point toward the oxyethylene chains in the IG head group (Fig. 11.11) to allow hydration of the imine group of AO by bulk-like water. Because the rate of twisting of the phenyl rings determines fluorescence lifetime of AO, a bigger effect on the rate may be observed depending on the phenyl rings' direction into the surfactant

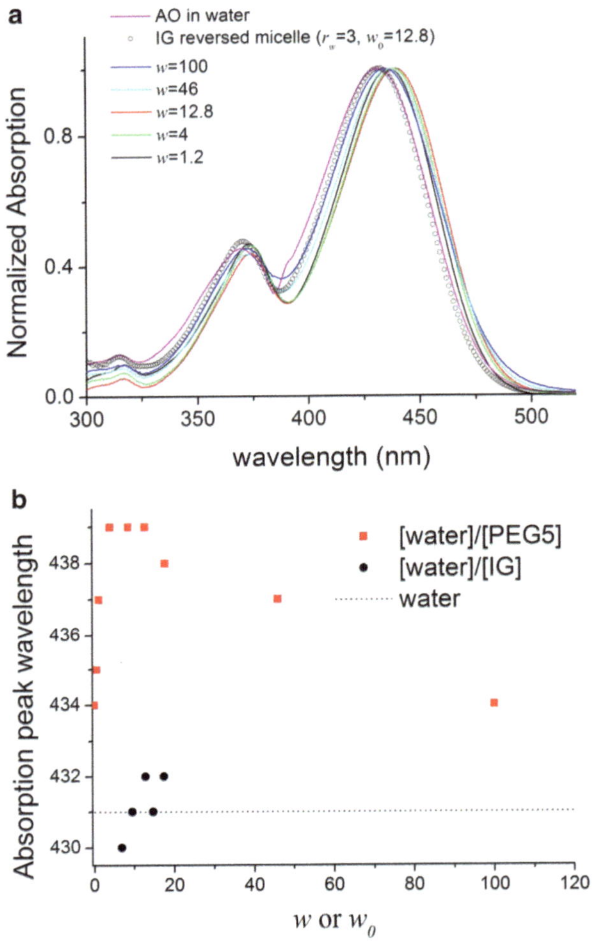

Fig. 11.10 (**a**) Absorption spectra of AO in the water/PEG5 mixed solvent as a function of w. (**b**) Peak wavelength as a function of w compared with w_0(micelle) value. The value for bulk water is shown as a *dashed line*

head group or into the interfacial water in which the slower water dynamics were reported than into bulk-like water [21].

11.9 Conclusions

The excited state reaction of AO has been studied in aqueous nanodroplets in reverse micelles stabilised by the nonionic surfactant IG, and the results were compared with various charged AOT surfactants. Both the experimental methods and the analysis of the barrierless reaction have been outlined. In both IG and AOT stabilised water nanodroplets, the rate of the excited state reaction of AO is dramatically decreased compared with bulk water. Interestingly the averaged fluorescence decay

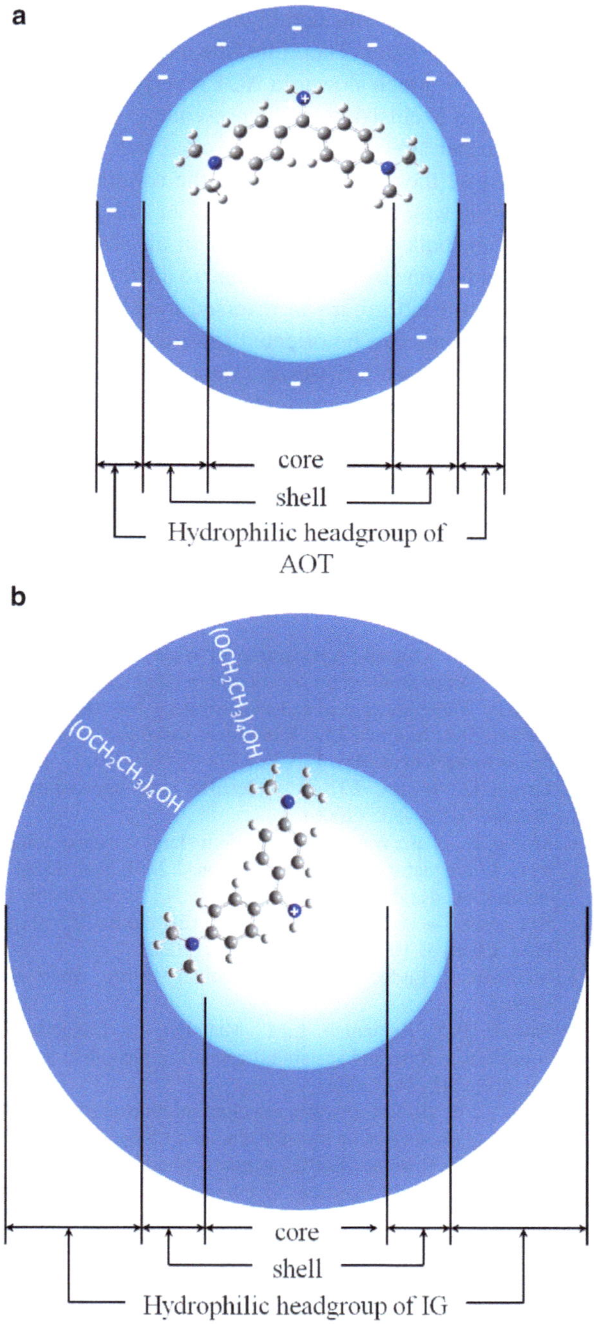

Fig. 11.11 The diagrams of AO in (**a**) AOT and (**b**) IG reversed micelles where the outer blue region expresses the hydrophilic head group of surfactants and the inner sphere is water nanodroplet

time of AO became considerably longer in the IG stabilised water droplet than in the AOT stabilised water. Therefore, the excited state AO twisting reaction which was observed to be much slower in AOT reversed micelles than in bulk aqueous solution is not caused exclusively by the presence of interfacial charge in reversed micelles.

The further slowdown of AO excited state reaction in IG stabilised water could arise because the AO is trapped in the oxyethylene chain region of IG surfactant. However, it was concluded that AO is fully hydrated in nanodroplet water and therefore in bulk-like water region in all sizes of the IG reversed micelles. The friction undergone by AO can be due to the orientation of phenyl rings of AO relative to surfactant head groups in reversed micelles. In water droplets stabilised by IG, the phenyl rings can be oriented towards the IG head group, i.e. in the interfacial water region which has slower water dynamics than in bulk-like core water. This may result in more hindered AO excited state twisting reactions in IG than in AOT reversed micelles. Importantly, the confinement effect has a major impact on the AO reactive dynamics, independent of interface charge.

References

1. Ball, P.: Life's Matrix: A Biography of Water. Farrar, Straus and Grioux, New York, NY (1999)
2. Ball, P.: Water as an active constituent in cell biology. Chem. Rev. **108**(1), 74–108 (2008)
3. Ball, P.: Water: water – an enduring mystery. Nature **452**(7185), 291–292 (2008)
4. Bakker, H.J., Skinner, J.L.: Vibrational spectroscopy as a probe of structure and dynamics in liquid water. Chem. Rev. **110**(3), 1498–1517 (2010)
5. Laage, D., Hynes, J.T.: A molecular jump mechanism of water reorientation. Science **311**(5762), 832–835 (2006)
6. Modig, K., et al.: Water dynamics in the large cavity of three lipid-binding proteins monitored by O-17 magnetic relaxation dispersion. J. Mol. Biol. **332**(4), 965–977 (2003)
7. Mallick, A., Haldar, B., Chattopadhyay, N.: Spectroscopic investigation on the interaction of ICT probe 3-acetyl-4-oxo-6,7-dihydro-12H indolo-[2,3-a] quinolizine with serum albumins. J. Phys. Chem. B **109**(30), 14683–14690 (2005)
8. Mellman, I.: Endocytosis and molecular sorting. Annu. Rev. Cell Dev. Biol. **12**, 575–625 (1996)
9. Sawant, P.D., Ramaniah, L.M., Manohar, C.: Capacity of nano-reactors of AOT micro-emulsions to form and sustain ultra small semiconductor quantum dots. J. Nanosci. Nanotechnol. **6**(1), 241–247 (2006)
10. Petit, C., et al.: Oil-in-water micellar solution used to synthesize CdS particles – structural study and photoelectron transfer-reaction. Langmuir **10**(12), 4446–4450 (1994)
11. Pileni, M.P.: Reverse micelles as microreactors. J. Phys. Chem. **97**(27), 6961–6973 (1993)
12. Wang, W.Z., et al.: Synthesis of US nanoparticles by a novel and simple one-step, solid-state reaction in the presence of a nonionic surfactant. Mater. Lett. **57**(18), 2755–2760 (2003)
13. Oldfield, C., Freedman, R.B., Robinson, B.H.: Enzyme hyperactivity in AOT water-in-oil microemulsions is induced by 'lone' sodium counterions in the water-pool. Faraday Discuss. **129**, 247–263 (2005)
14. Cringus, D., et al.: Ultrafast energy transfer in water-AOT reverse micelles. J. Phys. Chem. B **111**(51), 14193–14207 (2007)
15. Mallick, A., Purkayastha, P., Chattopadhyay, N.: Photoprocesses of excited molecules in confined liquid environments: an overview. J. Photochem. Photobiol. C Photochem. Rev. **8**(3), 109–127 (2007)

16. Bhattacharyya, K.: Solvation dynamics and proton transfer in supramolecular assemblies. Acc. Chem. Res. **36**(2), 95–101 (2003)
17. Bhattacharyya, K., Bagchi, B.: Slow dynamics of constrained water in complex geometries. J. Phys. Chem. A **104**(46), 10603–10613 (2000)
18. Persson, E., Halle, B.: Cell water dynamics on multiple time scales. Proc. Natl. Acad. Sci. USA **105**(17), 6266–6271 (2008)
19. Heisler, I.A., Kondo, M., Meech, S.R.: Reactive dynamics in confined liquids: ultrafast torsional dynamics of Auramine O in nanoconfined water in aerosol OT reverse micelles. J. Phys. Chem. B **113**(6), 1623–1631 (2009)
20. Kondo, M., Heisler, I.A., Meech, S.R.: Ultrafast reaction dynamics in nanoscale water droplets confined by ionic surfactants. Faraday Discuss. **145**, 185–203 (2010)
21. Kondo, M., Heisler, I.A., Meech, S.R.: Reactive dynamics in micelles: Auramine O in solution and adsorbed on regular micelles. J. Phys. Chem. B **114**(40), 12859–12865 (2010)
22. Kondo, M., et al.: Reactive dynamics in confined liquids: interfacial charge effects on ultrafast torsional dynamics in water nanodroplets. J. Phys. Chem. B **113**(6), 1632–1639 (2009)
23. Chen, S.H.: Small-angle neutron-scattering studies of the structure and interaction in micellar and microemulsion systems. Annu. Rev. Phys. Chem. **37**, 351–399 (1986)
24. Clint, J.H.: Surfactant Aggregation (1992)
25. Riter, R.E., Undiks, E.P., Levinger, N.E.: Impact of counterion on water motion in aerosol OT reverse micelles. J. Am. Chem. Soc. **120**(24), 6062–6067 (1998)
26. Piletic, I.R., Tan, H.S., Fayer, M.D.: Dynamics of nanoscopic water: vibrational echo and infrared pump-probe studies of reverse micelles. J. Phys. Chem. B **109**(45), 21273–21284 (2005)
27. Piletic, I.R., et al.: Testing the core/shell model of nanoconfined water in reverse micelles using linear and nonlinear IR spectroscopy. J. Phys. Chem. A **110**(15), 4985–4999 (2006)
28. Harpham, M.R., Ladanyi, B.M., Levinger, N.E.: The effect of the counterion on water mobility in reverse micelles studied by molecular dynamics simulations. J. Phys. Chem. B **109**(35), 16891–16900 (2005)
29. Rees, G.D., Robinson, B.H.: Microemulsions and organogels – properties and novel applications. Adv. Mater. **5**(9), 608–619 (1993)
30. Rey, R., Moller, K.B., Hynes, J.T.: Hydrogen bond dynamics in water and ultrafast infrared spectroscopy. J. Phys. Chem. A **106**(50), 11993–11996 (2002)
31. Onori, G., Santucci, A.: IR investigations of water-structure in aerosol OT reverse micellar aggregates. J. Phys. Chem. **97**(20), 5430–5434 (1993)
32. Faeder, J., Ladanyi, B.M.: Molecular dynamics simulations of the interior of aqueous reverse micelles. J. Phys. Chem. B **104**(5), 1033–1046 (2000)
33. Moilanen, D.E., et al.: Water dynamics in large and small reverse micelles: from two ensembles to collective behavior. J. Chem. Phys. **131**(1), 014704 (2009)
34. Riter, R.E., Willard, D.M., Levinger, N.E.: Water immobilization at surfactant interfaces in reverse micelles. J. Phys. Chem. B **102**(15), 2705–2714 (1998)
35. Datta, A., et al.: Intramolecular charge transfer processes in confined systems. Nile red in reverse micelles. J. Phys. Chem. B **101**(49), 10221–10225 (1997)
36. Levinger, N.E.: Ultrafast dynamics in reverse micelles, microemulsions, and vesicles. Curr. Opin. Colloid Interface Sci. **5**(1–2), 118–124 (2000)
37. Satoh, T., et al.: Excitation wavelength dependence of solvation dynamics in a water pool of a reversed micelle. Chem. Lett. **33**(9), 1090–1091 (2004)
38. Heisler, I.A., Mazur, K., Meech, S.R.: Low-frequency modes of aqueous alkali halide solutions: an ultrafast optical Kerr effect study. J. Phys. Chem. B **115**(8), 1863–1873 (2011)
39. Heisler, I.A., Meech, S.R.: Low-frequency modes of aqueous alkali halide solutions: glimpsing the hydrogen bonding vibration. Science **327**(5967), 857–860 (2010)
40. Pant, D., Riter, R.E., Levinger, N.E.: Influence of restricted environment and ionic interactions on water solvation dynamics. J. Chem. Phys. **109**(22), 9995–10003 (1998)
41. Waldeck, D.H.: Photoisomerization dynamics of stilbenes. Chem. Rev. **91**(3), 415–436 (1991)

42. Saikan, S., Sei, J.: Investigation of the conformational change in triphenylmethane dyes via polarization spectroscopy. J. Chem. Phys. **79**(9), 4154–4158 (1983)
43. Benamotz, D., Jeanloz, R., Harris, C.B.: Torsional dynamics of molecules on barrierless potentials in liquids. 3. Pressure dependent picosecond studies of triphenyl-methane dye solutions in a diamond anvil cell. J. Chem. Phys. **86**(11), 6119–6127 (1987)
44. Yoshizawa, M., et al.: Femtosecond study of S-2 fluorescence in malachite green in solutions. Chem. Phys. Lett. **290**(1–3), 43–48 (1998)
45. Oster, G., Nishijima, Y.: Fluorescence and internal rotation – their dependence on viscosity of the medium. J. Am. Chem. Soc. **78**(8), 1581–1584 (1956)
46. van der Meer, M.J., Zhang, H., Glasbeek, M.: Femtosecond fluorescence upconversion studies of barrierless bond twisting of auramine in solution. J. Chem. Phys. **112**(6), 2878–2887 (2000)
47. Martin, M.M., et al.: Investigation of excited-state charge transfer with structural change in compounds containing anilino subunits by subpicosecond spectroscopy. J. Photochem. Photobiol. A Chem. **105**(2–3), 197–204 (1997)
48. Bagchi, B., Fleming, G.R., Oxtoby, D.W.: Theory of electronic relaxation in solution in the absence of an activation barrier. J. Chem. Phys. **78**(12), 7375–7385 (1983)
49. Lipgens, S., et al.: Percolation in nonionic water-in-oil-microemulsion systems: a small angle neutron scattering study. Langmuir **14**(5), 1041–1049 (1998)
50. Smith, N.A., et al.: LDS-750 as a probe of solvation dynamics: a femtosecond time-resolved fluorescence study in liquid aniline. Chem. Phys. Lett. **303**(1–2), 209–217 (1999)
51. Changenet, P., et al.: Ultrafast twisting dynamics of photoexcited auramine in solution. J. Phys. Chem. A **102**(34), 6716–6721 (1998)
52. Changenet, P., et al.: Fluorescence quenching of auramine in fluid solutions: a femtosecond spectroscopy study. J. Fluoresc. **10**(2), 155–160 (2000)
53. Glasbeek, M., Zhang, H., van der Meer, M.J.: Femtosecond studies of twisting dynamics of auramine in solution. J. Mol. Liq. **86**(1–3), 123–126 (2000)
54. Okuyama, S., Oxtoby, D.W.: Non-Markovian dynamics and barrier crossing rates at high-viscosity. J. Chem. Phys. **84**(10), 5830–5835 (1986)
55. Tominaga, K., et al.: Ultrafast charge separation in adma – experiment, simulation, and theoretical issues. J. Phys. Chem. **95**(25), 10475–10485 (1991)
56. Tominaga, K., et al.: Reaction-rates in the phenomenological adiabatic excited-state electron-transfer theory. J. Phys. Chem. **95**(25), 10485–10492 (1991)
57. Smith, B.B., Staib, A., Hynes, J.T.: Well and barrier dynamics and electron-transfer rates – a molecular-dynamics study. Chem. Phys. **176**(2–3), 521–537 (1993)
58. Rharbi, Y., Chen, L.S., Winnik, M.A.: Exchange mechanisms for sodium dodecyl sulfate micelles: high salt concentration. J. Am. Chem. Soc. **126**(19), 6025–6034 (2004)
59. Chowdhary, J., Ladanyi, B.M.: Molecular dynamics simulation of aerosol-OT reverse micelles. J. Phys. Chem. B **113**(45), 15029–15039 (2009)
60. Bruce, C.D., et al.: Molecular dynamics simulations of sodium dodecyl sulfate micelle in water: the behavior of water. J. Phys. Chem. B **106**(42), 10902–10907 (2002)
61. Abel, S., et al.: Effect of surfactant conformation on the structures of small size nonionic reverse micelles: a molecular dynamics simulation study. Langmuir **22**(22), 9112–9120 (2006)
62. Moilanen, D.E., et al.: Confinement or the nature of the interface? Dynamics of nanoscopic water. J. Am. Chem. Soc. **129**(46), 14311–14318 (2007)
63. Park, S., Moilanen, D.E., Fayer, M.D.: Water dynamics – the effects of ions and nanoconfinement. J. Phys. Chem. B **112**(17), 5279–5290 (2008)
64. Fenn, E.E., Wong, D.B., Fayer, M.D.: Water dynamics at neutral and ionic interfaces. Proc. Natl. Acad. Sci. USA **106**(36), 15243–15248 (2009)
65. Abel, S., et al.: Molecular modeling and simulations of AOT-water reverse micelles in isooctane: structural and dynamic properties. J. Phys. Chem. B **108**(50), 19458–19466 (2004)
66. Sedgwick, M.A., Crans, D.C., Levinger, N.E.: What is inside a nonionic reverse micelle? Probing the interior of Igepal reverse micelles using decavanadate. Langmuir **25**(10), 5496–5503 (2009)

Chapter 12
Brownian Deposition of Nanodroplets and Nanofiber Growth via "Vapor–Liquid–Solid" Route

Sergey P. Fisenko and Dmitry A. Takopulo

Abstract Simulation results of nanodroplet deposition on the wall of cylindrical reactor are presented. Additionally qualitative analytical estimations are given. Contributions of the Brownian diffusion of nanodroplets and thermophoresis are discussed. Application of deposited nanodroplets for the formation carbon nanofiber via "vapor–liquid–solid" route is briefly described.

12.1 Introduction

Many novel products of nanotechnology, among them carbon and silicon nanowires (nanofibers), are obtained by means of modernized CVD reactors. In particular for the formation of carbon nanofibers, a flow of hydrocarbon gases is organized in a CVD reactor. But at first tiny pieces of catalytic metals should be deposited on the substrate. Usually such metals as nickel, cobalt, and iron are used; the nature of substrate is important also for nanofiber formation [1]. The temperature of these tiny particles should be about 700°C and their characteristic size is about 10 nm. Therefore there are a lot reasons to consider such particles as metallic nanodroplets. The principal role of these tiny metal nanodroplets is a catalytic decay of hydrocarbon vapors and then to serve as a stock of solute carbon atoms. After the cooling of the substrate, supersaturated solid solution of carbon atoms is formed in deposited nanodroplets. If the supersaturation is high enough, a carbon cluster is formed and grows. The coalescence of these carbon clusters in the nanodroplet leads to formation of the nanofiber. It is surprising but the external diameter of nanofiber is always equal to the diameter of nanodroplet. Briefly described above, the set of physical–chemical transformation is called "vapor– liquid–solid" (VLS) route.

S.P. Fisenko (✉) • D.A. Takopulo
A.V. Luikov Heat & Mass Transfer of National Academy of Sciences of Belarus,
15, P. Brovka, 220072 Minsk, Republic Belarus
e-mail: fsp@hmti.ac.by; takopulo@mail.ru

Z.M. Wang (ed.), *Nanodroplets*, Lecture Notes in Nanoscale Science and Technology 18,
DOI 10.1007/978-1-4614-9472-0_12,

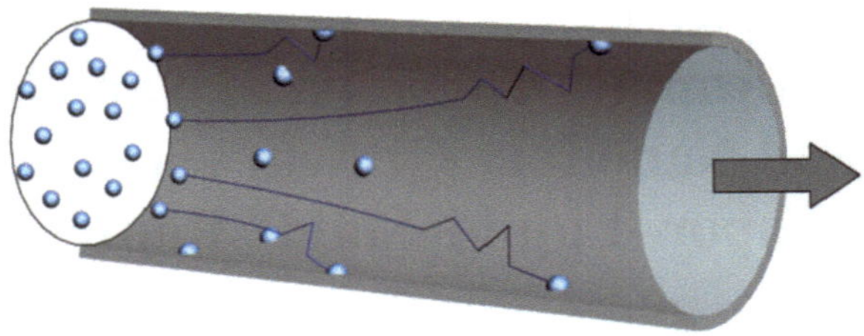

Fig. 12.1 The sketch of CVD reactor

It is worth to mention that the productivity of the CVD reactor for nanofiber growth strongly depends on how many nanodroplets are deposited per unit of wall square of the reactor. The sketch of such reactor is shown in Fig. 12.1. In order to emphasize the nature of Brownian deposition, several stochastic trajectories of metal nanodroplets are drawn.

This chapter is organized at the following way: At first we consider the substantial features of the Brownian deposition of nanodroplets on the wall of cylindrical CVD reactor. In particular we give analysis of a non-isothermal deposition of nanodroplets. It is well known now [2] that practically every metallic nanodroplet at VLS route serves as source of carbon atoms for formation of carbon nanofibers and their growth. We consider the nucleation of carbon clusters on the interfacial of substrate and metallic nanodroplet in the next section. Then the growth of carbon nanofiber, which is formed due to coalescence of carbon clusters, is considered. Main results are summarized in "Conclusions."

12.2 Brownian Deposition of Nanodroplets on the Wall

At first we present our recent results of the simulation of the Brownian deposition of catalytic metallic nanodroplets on the reactor wall. Gas flow is a laminar one, but it is worthy to note that interaction of nanodroplets with gas flow is in the free molecular regime. The key reason is that the mean free path of gas molecules (typically about 200 nm at atmospheric pressure) is much larger than a nanodroplet diameter. For description of the development of the number density of nanodroplets $n(z, r)$ along the reactor, we have the convective diffusion equation

$$\frac{\partial}{\partial z}\left(u(r,z)\,n\right)+\frac{1}{r}\frac{\partial}{\partial r}\left(v_{\mathrm{th}}rn\right)=\frac{1}{r}\frac{\partial}{\partial r}\left[rD\frac{\partial n}{\partial r}\right], \qquad (12.1)$$

where u is the gas velocity profile and v_{th} is the thermophoretic velocity of nanodroplets, which is directly proportional to the gradient of gas temperature in the reactor [3]. Cylindrical system of coordinates is used here due to the symmetry reactor and gas flow in it.

For the free molecular regime of interaction nanodroplet with molecules of gas flow, the coefficient of the Brownian diffusion of spherical nanodroplet D is [4]

$$D = \frac{3kT}{16\pi R^2 p}\sqrt{\frac{2\pi kT}{m}},$$

where R is the nanodroplet radius, k is the Boltzmann's constant, T is the gas temperature, p is pressure, and m is the mass of gas molecule. Below we consider only the steady-state solution of Eq. (12.1).

The boundary conditions for Eq. (12.1) are the following:

the symmetry condition of nanodroplet distribution on the axes of the reactor

$$\frac{\partial n(0,z)}{\partial r} = 0, \tag{12.2}$$

the second boundary condition reflects our assumption that nanodroplet will not return to gas phase if they reach the wall:

$$n(R_r, 0) = 0, \tag{12.3}$$

where R_r is the reactor radius.

For isothermal deposition of nanodroplets, gas temperature is equal to the wall one; therefore thermophoretic velocity is equal to zero. Following [4] we can semi-qualitatively solve Eq. (12.1) and write the expression for the number density of nanodroplets in any point of the reactor

$$n(r,z) \sim n_0 \exp\left[-z/l_B\right] J_0\left(br/R_r\right), \tag{12.4}$$

where n_0 is the initial number density of nanodroplets, J_0 is the Bessel function of the zeroth order, and b is the least root of the equation

$$J_0(b) = 0.$$

The characteristic length of Brownian deposition l_B in the expression (12.4) was obtained in [4]:

$$l_B = \frac{0.27 u_0 R_r^2}{D}.$$

The flux of nanodroplet J on the reactor wall is given by the expression

$$J = -D\nabla n, \tag{12.5}$$

and using Eq. (12.4) we have more detailed expression for J:

$$J \approx \frac{Dbn_0}{R_r}\exp\left[-z/l_B\right]. \tag{12.6}$$

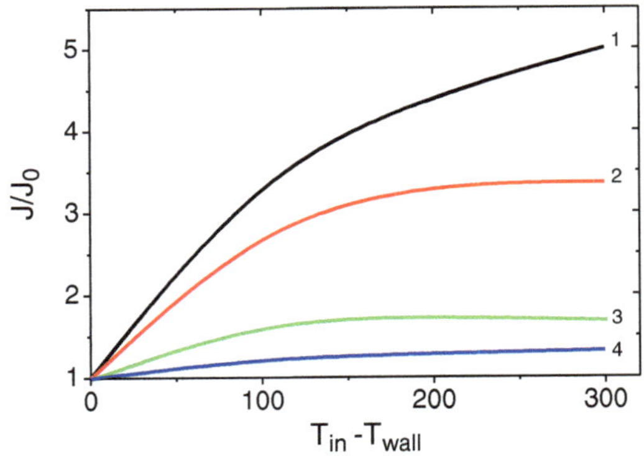

Fig. 12.2 Normalized flow of nanoparticles versus temperature difference. Curve 1 is $z = 5\ R_r$, 2 is for 10 R_r, 3 is for 20 R_r, and 4 is for 40 R_r

If we would like to deposit the N nanodroplets per unit wall square, then the question arises: what time will it take? Deposition time t_d can be estimated by means of Eq. (12.6). We obtain that t_d is

$$t_d = \frac{N}{J} \sim \frac{NR_r}{Dn_0} \exp\left[z/l_B\right]. \tag{12.7}$$

Let us make numerical estimations. For $R_r = 1$ mm and the mean gas (argon) velocity 0.1 m/s, nanodroplet radius $R = 10$ nm, the wall temperature $T = 1{,}000$ K, and characteristic length of Brownian deposition l_B is about 0.3 m. If the reactor length is smaller than l_B, then it is possible to neglect the exponential dependence in Exp. (12.7). Correspondingly, for $N = 10^{11}$ nanodroplets/m^2 and the number density in gas flow $n_0 = 10^{12}$ nanodroplets/m^3, we have t_d is about 10^3 s.

For acceleration of the Brownian deposition of nanodroplets or nanoparticles, usually the thermophoresis is used by experimentalists or engineers. It exists due to temperature difference between wall and inlet gas temperature. It is clear that for deposition process the wall temperature has to be lower than inlet gas temperature, T_{in}. For non-isothermal deposition we have additionally to solve the equation of convective heat conductivity. It has practically the same mathematical form as Eq. (12.1); only the term with thermophoretic velocity is absent. This equation with temperature-dependent thermophysical coefficients was incorporated in our mathematical model. For illustration some numerical results are given in Fig. 12.2.

It can be seen in Fig. 12.2 that temperature difference about 100 K doubles deposition rate of nanodroplets. It is worthy to note that at our notation l_t is the characteristic length of heat transfer process in the reactor. For laminar flow in the cylindrical reactor, l_t was obtained at [5, 6]:

$$l_t = \frac{0.26 u_0 R_r^2 \rho c}{\lambda\left(T_w\right)},$$

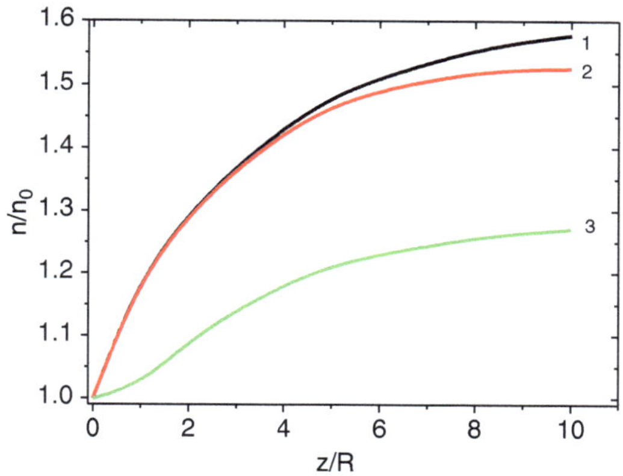

Fig. 12.3 Normalized number density of nanoparticles for $r/R_r = 0.85$. Curve 1 is for cooling effect only; 2 is the joint influence of the cooling and the Brownian diffusion; 3 is for joint influence of all three processes: cooling, Brownian diffusion, and thermophoresis of nanoparticles

where ρ and c are, correspondingly, mass density and heat capacity of gas flow and λ is the heat conductivity of gas at wall temperature. Also it was shown that the temperature difference between gas temperature $T(z)$ and T_w has an exponential decreasing versus coordinate z:

$$T(z) - T_w \sim (T_{\text{in}} - T_w) \exp\left[-z/l_{\text{t}}\right].$$

Thus we can neglect the influence of thermophoresis on deposition rate of nanodroplets for $z > 3l_{\text{t}}$; see some details in [5]. Nevertheless the thermophoretic force drastically enhances the deposition rate of nanodroplets on the reactor wall at the inlet zone.

For non-isothermal deposition of nanodroplets, three phenomena, namely, the flow cooling, the Brownian diffusion, and thermophoresis, affect on the rate of Brownian deposition. It is obvious that the higher the number density of nanodroplets near the wall, the higher the deposition rate. For the position $r/R_r = 0.9$, results of our simulation of the influence of all three factors are shown in Fig. 12.3. At first, due to cooling of gas flow, there is decreasing of velocity of gas flow and, due to the conservation law, increasing of the number density of nanodroplet in the gas flow. Curve 1 in Fig. 12.3 clearly demonstrates this well-known effect. The second effect under consideration is the Brownian diffusion of nanoparticles. Curve 2 shows that the Brownian diffusion plus cooling effect significantly reduces the number density of nanoparticles near the wall and, subsequently, deposits them on the wall. For simulation of gas flow parameters, we use the modified Poiseuille profile [7]. Curve 3 is the result of joint contributions: cooling gas, the Brownian diffusion, and thermophoretic force which affect only on inlet zone of the reactor. In the long run the single player responsible for deposition of nanodroplets is the Brownian diffusion.

For large distance in comparison with l_{t}, the thermophoresis influence is not a so profound one.

12.3 Nucleation of Carbon Clusters in a Metallic Nanodroplet

It was shown in our previous publications [2, 8] that the basic route for carbon nanofiber formation is “vapor–liquid–solid” (VLS) route. This route is important for silicon nanofiber formation as well. It is interesting to note that gold nanodroplets are used for obtaining a silicon nanofiber.

Below we consider the most important physical processes related with this route inside the metallic nanodroplet deposited on the reactor wall. Released from hydrocarbon molecules by catalytic pyrolysis, carbon atoms diffuse into nanodroplet and in some time create the equilibrium solution of carbon atoms in nickel nanodroplet. We don't discuss vapor pyrolysis problem here, but metallic nanodroplets successfully serve as the tool for catalytic decomposition of hydrocarbon vapors.

Characteristic diffusion time τ to reach the equilibrium solution of carbon atoms in the nanodroplet is expressed by formula

$$\tau = R^2/\pi^2 D_c$$

where D_c is the carbon diffusion coefficient in nickel. For example, at 873 K equilibrium, number density of carbon atoms n_e in nickel is about 1.9×20^{26} atoms/m^3. It is useful to compare this value with the number density of nickel atoms in condensed material, which approximately equals 9×10^{28} atoms/m^3.

For “VLS route” a very crucial step is to form a set of carbon clusters in nanodroplet. For this goal the supersaturated solid solution of carbon has to be created. It is well known that the solubility of carbon atoms in nickel increases if temperature of nickel increases. Thus the most effective and practical way to create supersaturated solid solution is at first obtain the equilibrium solution at higher temperature and then cooling the material to the desirable temperature. The supersaturation of solid solution is defined as

$$S = n/n_e(T),$$

where n is the actual number density of carbon atoms in nanodroplet. The influence of the drop temperature on the supersaturation value of carbon solution in nickel is shown in Fig. 12.4. It can be seen that the drop of temperature about 70 K creates high enough supersaturation of the solid carbon solution in metallic nanodroplet for the beginning of nucleation of carbon clusters.

Let us consider the free energy of cluster formation on interface of the nanodroplet and the substrate. For simplicity we assume also that all clusters have the semispherical form (see Fig. 12.5).

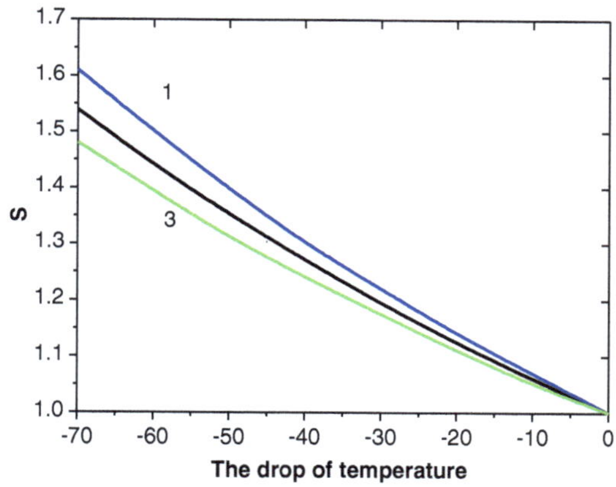

Fig. 12.4 Supersaturation versus the drop of nanodroplet temperature. Curve 1 is for initial temperature 700°C, 2 is for 750°C, and 3 is for 800°C

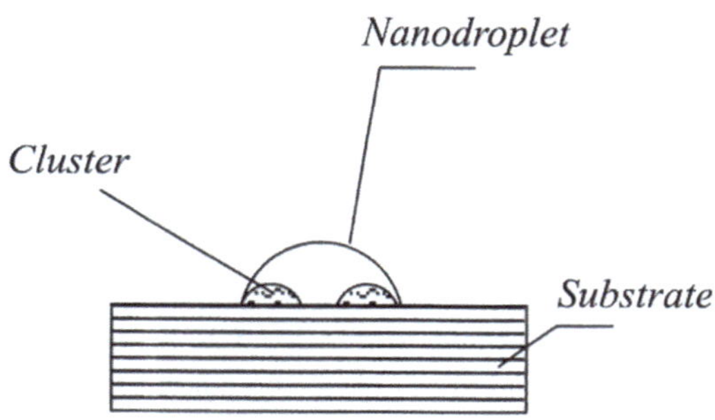

Fig. 12.5 Carbon cluster in catalytic nanodroplet

Then, the free energy $\Delta\Phi(g)$ of the carbon cluster consisting of g atoms at the droplet–substrate interface is

$$\Delta\Phi(g) = -gkT\ln(S) + 2\pi r^2 \ \sigma_{ab} + \pi r^2 (\sigma_{as} - \sigma_{bs}), \tag{12.8}$$

where r is the radius of the carbon cluster, σ_{ab} is the surface tension between the cluster and the material of nanodroplet, and σ_{bs} is the one between nanodroplet and substrate. On the right side of the expression (12.8), the last term, which takes into account the difference between the surface tension coefficients, under some conditions substantially decreases the free energy of carbon clustering. For example, if the substrate is made of carbon material, then $\sigma_{as} = 0$ and the free energy of clustering on the substrate substantially decreases. The free energy of cluster formation inside of nanodroplet is substantially higher, and we can neglect this process as less thermodynamically feasible one.

The critical cluster consisting of g^* atoms is determined from the condition of the minimum of the free energy of cluster formation $\Delta\Phi\ (g)$

$$\frac{\partial \Delta\Phi(g)}{\partial g} = 0. \tag{12.9}$$

For the number of carbon atoms g^* in a critical cluster, we have the expression

$$g^* = \frac{2\pi v_A^2}{3(kT)^3(\ln(S))^2}\left[2\sigma_{ab} + (\sigma_{as} - \sigma_{bs})\right]^3. \qquad (12.10)$$

Obviously the formula (12.10) is the generalization of the well-known Gibbs formula [9]. Substituting Eq. (12.10) into Eq. (12.8) yields the formula for the free energy $\Delta\Phi^*$ of the critical cluster

$$\Delta\Phi^* = \frac{g^* kT \ln(S)}{2}.$$

The basic idea of the classical theory of nucleation [10] is that nucleation is the Brownian diffusion of clusters over the thermodynamic barrier (Eq. 12.8) in the cluster size space. The free energy of cluster formation serves as the thermodynamic barrier. The barrier maximum is equal to $\Delta\Phi^*$. Therefore we can write the expression for nucleation rate I of critical clusters as

$$I = n^2\, 2\pi D_c r^* \exp\left[-\Delta\Phi^*/kT\right], \qquad (12.11)$$

where r^* is the radius of critical cluster. Typically r^* is about 1 nm. For the "vapor–liquid–solid" route to persist, it is necessary that a sufficiently large number of critical clusters arise inside the nanodroplet for a short time. In other words, it is essential to have a high nucleation rate of carbon clusters. The nucleation of carbon clusters is the crucial step in the formation of nanofiber, but there is also a very important step related with the diffusion interaction of growing carbon clusters in the nanodroplet. The coalescence of these carbon clusters gives the beginning of carbon nanofiber; see details on the consideration of this nonlinear process in [11]. Also it is worthy to note that huge release of the latent heat of carbon leads to the melting of metallic nanodroplet during some first milliseconds of nanofiber growth [12]. The melting nanodroplet obviously changes their shape and partially penetrates into the nanofiber. There are a lot of experimental evidences of this penetration.

There is one practical application of developed physical picture. As r^* is about 1 nm, it means that that radius nanodroplets deposited on substrate should be at least several nanometers.

12.4 Growth of Carbon Nanofiber from Deposited Nanodroplet

A sketch of the growing nanofiber is shown in Fig. 12.6. The equation for description of nanofiber growth was obtained in [2]. It has the form of the ordinary differential equation

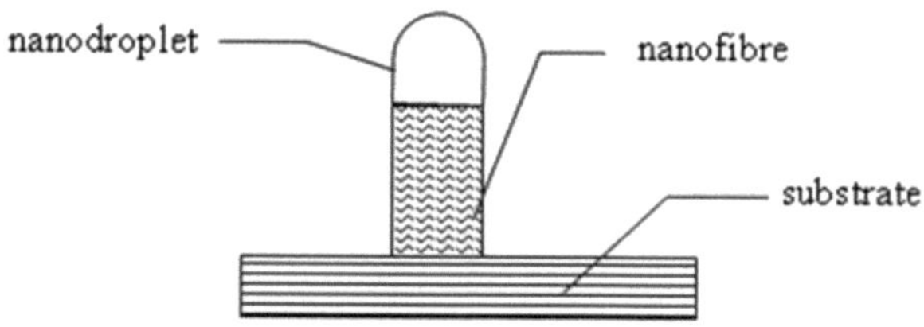

Fig. 12.6 Sketch of nanodroplet and nanofiber

$$\frac{dH}{dt} = D\frac{n - n_e}{Rn_s} \tag{12.12}$$

where t is time, H is a length of nanofiber, and n_s is the number density of carbon atoms in the carbon nanofiber. The interrelationship immediately follows from Eq. (12.12) between the most important parameters:

$$H \sim \frac{Dt}{R}.$$

Thus we have a conclusion that the larger nanodroplet diameter, the smaller the growth rate of the nanofiber. It is worthy to note that this conclusion can be applied to any nanofiber, e.g., silicon nanofiber.

As it was shown in [11], the after nucleation of carbon cluster coalescence plays the crucial role for nanofiber formation. It means that at vast majority of cases, nanofiber has an empty core and can be called tubular nanofiber. After a straightforward modification, the equation of nanofiber growth is

$$\frac{dH}{dt} = DR\frac{n - n_l}{\left(R^2 - R_{\mathrm{i}}^2\right) n_s}, \tag{12.13}$$

where R_{i} is the internal radius of nanofiber. It is obvious that the tubular nanofiber growth rate is higher than for a continuous nanofiber. Again, as r^* is about 1 nm, it follows from coalescence kinetics that the difference R–R_{i} is about several nanometers at least.

12.5 Conclusions

For formation of any kind nanofibers, usually catalytic nanodroplets on substrate are used. Main features of this so-called "vapor–liquid–solid" route are considered in this chapter.

At first we considered the Brownian deposition of spherical nanodroplets on the wall of the cylindrical reactor. For simulation we used the free molecular approximation of interaction nanodroplet with gas flow. Some numerical results are given which are based on solution of the convective diffusion equation for the number density of nanodroplets. The deposition time t_d was estimated by means of semi-qualitative analytical estimations. Also it was shown that the rate of Brownian deposition of nanodroplets can be significantly enhanced by thermophoresis. Thermophoresis is due to temperature difference between inlet temperature of gas flow and wall temperature. In particular temperature difference about 200 K doubles the deposition rate.

Also the creation of a supersaturated solid solution of carbon in deposited nanodroplet is considered. This solution is usually arisen due to the cooling of nanodroplet or the reactor wall. After the diffusion process, it was shown that a cooling nickel nanodroplet on 50–70 K leads to formation of supersaturated solid solution.

A metastable state of supersaturated solid solution of carbon has decay via nucleation of carbon clusters. The expression for nucleation rate of carbon clusters is analyzed. It is worthy to emphasize that the most preferable place for nucleation of carbon clusters is the interfacial between nanodroplet and the substrate (reactor wall). It is shown that the minimal free energy of the critical cluster formation is for cases when materials of nanodroplet and wall coincide.

Finally features of growth of nanofiber from deposited nanodroplet are considered. In particular, it is shown that nanofiber growth rate is inversely proportional to nanodroplet radius and directly proportional to carbon diffusion coefficient in nanodroplet.

Acknowledgments Authors are grateful to Dr. Stankevich Yu. A. for the support and valuable discussions.

References

1. Labunov, V., Shulitski, B., Prudnikava, A., Basaev, A.: Multi-level composite nanostructures based on the arrays of vertically aligned carbon nanotubes and planar graphite layers A. Phys. Status Solidi A **208**, 453–458 (2011)
2. Fisenko, S.P., Borovik, F.N.: Nucleation in a catalytic nanodroplet and growth of nanowires. Tech. Phys. **54**, 246–252 (2009)
3. Talbot, L., Cheng, R.K., Schaefer, R.W., Willis, D.R.: Thermophoresis of particles in a heated boundary layer. J. Fluid. Mech. **101**, 737–758 (1980)
4. Brin, A.A., Fisenko, S.P., Shnip, A.I.: Brownian deposition of nanoparticles from a laminar gas flow through a channel. Tech. Phys. **53**, 1141–1145 (2008)
5. Fisenko, S.P., Khodyko, J.A.: Thermophoresis and the Brownian diffusion of nanoparticles in a flow reactor. Tech. Phys. **57**, 336–343 (2012)
6. Fisenko, S.P., Brin, A.A.: Heat and mass transfer and condensation interference in a laminar flow diffusion chamber. Int. J. Heat Mass Transf. **49**, 1004–1014 (2006)
7. Stankevich, Y.A., Fisenko, S.P.: Reorganization of the Poiseuille profile in nonisothermal flows in the reactor. J. Eng. Phys. Thermophys. **84**, 1318–1321 (2011)
8. Takopulo, D.A., Fisenko, S.P.: On formation of carbon clusters on a substrate in plasma pyrolysis of hydrocarbon gases. J. Eng. Phys. Thermophys. **84**, 1083–1086 (2011)
9. Landau, L.D., Lifshitz, E.M.: Statistical Physics. Pergamon Press, Oxford (1980)
10. Pitaevskii, L.P., Lifshitz, E.M.: Physical Kinetics. Pergamon Press, New York, NY (1981)

11. Fisenko, S.P., Martynenko, V.V., Shabunya, S.I., Zhdanok, S.A.: Coalescence of carbon clusters in a catalytic nanodroplet and formation of carbon nanofibre. Smart Nanocomposites **1**, 162–172 (2011)
12. Fisenko, S.P., Shabunya, S.I., Martynenko, V.V., Zhdanok, S.A.: Surprising thermal effects during growth of carbon nanofibres via "vapor–liquid-solid" route. In: Borisenko, V.E., Gaponenko, S.V., Gurin, V.S., Kam, C.H. (eds.) Physics, Chemistry and Application of Nanostructures, pp. 81–84. World Scientific, Singapore (2011)

Chapter 13
Water Nanodroplets: Molecular Drag and Self-assembly

J. Russell, B. Wang, N. Patra, and P. Král

Abstract Directed transport and self-assembly of nanomaterials can potentially be facilitated by water nanodroplets, which could carry reactants or serve as a selective catalyst. We show by molecular dynamics simulations that water nanodroplets can be transported along and around the surfaces of vibrated carbon nanotubes. We show a second transport method where ions intercalated in carbon and boron-nitride nanotubes can be solvated at distance in polarizable nanodroplets adsorbed on their surfaces. When the ions are driven in the nanotubes by electric fields, the adsorbed droplets are dragged together with them. Finally, we demonstrate that water nanodroplets can activate and guide the folding of planar graphene nanostructures.

J. Russell
Materials Science Division, Argonne National Laboratory, 9700 Cass Ave Bldg 200 Argonne, IL 60546, USA
e-mail: jrussell@anl.gov

B. Wang
Institute of Chemistry, Chinese Academy of Science, Beijing, China
e-mail: wangby@itp.ac.cn

N. Patra
Department of Chemistry, University of Illinois at Chicago, 845 W. Taylor St. Rm 4500 SES, Chicago, IL 60546, USA
e-mail: npatra2@uic.edu

P. Král (✉)
Department of Chemistry, University of Illinois at Chicago, 845 W. Taylor St. Rm 4500 SES, Chicago, IL 60546, USA

Department of Physics, University of Illinois at Chicago, 845 W. Taylor St. Rm 4500 SES, Chicago, IL 60546, USA
e-mail: pkral@uic.edu

Z.M. Wang (ed.), *Nanodroplets*, Lecture Notes in Nanoscale Science and Technology 18,
DOI 10.1007/978-1-4614-9472-0_13,

13.1 Introduction

Molecular transport and self-assembly of nanomaterials have important applications in medicine, nanofluidics, and molecular motors. Nanocarbons, such as carbon nanotubes (CNTs) and graphene, have many useful properties which make them attractive supports for directed molecular transport. CNTs [1] can serve as nanoscale railroads for transport of materials, due to their linear structure, mechanical strength, slippery surfaces, and chemical stability [2]. For example, electric currents passing through CNTs can drag atoms/molecules intercalated/adsorbed on CNTs [3–5]. Polar molecules and ions adsorbed on CNT surfaces can also be dragged by ionic solutions passing through the tubes [6–8]. Recently, nanoparticles [9, 10] and nanodroplets [11, 12] have been dragged along CNTs by hot phonons in thermal gradients. Analogously, breathing [13] and torsional [14] coherent phonons can pump fluids inside CNTs. In this chapter, we explore drag phenomena of nanodroplets on CNTs by vibrations [15] and by coupling to distantly solvated ions [16]. We also investigate droplet-driven self-assembly where nanodroplets guide the folding of planar graphene nanostructures [17].

13.2 Methods

We simulate water nanodroplets on graphene and CNT supports with classical molecular dynamics simulations, using NAMD [18], with the CHARMM27 force field [19], and VMD [20] for visualization and analysis. We estimate parameters of atoms in aliphatic groups and the graphitic support from similar atom types or calculate them ab initio [21], and add them to the force field. The nanodroplets couple to the CNT or graphene support by van der Waals (vdW) forces, described in CHARMM with the Lennard-Jones potential energy [22]

$$V_{\mathrm{LJ}}(r_{ij}) = \varepsilon_{ij}\left[\left(\frac{R_{\mathrm{min},ij}}{r_{ij}}\right)^{12} - 2\left(\frac{R_{\mathrm{min},ij}}{r_{ij}}\right)^{6}\right]. \qquad (13.1)$$

Here, $\varepsilon_{ij} = \sqrt{\varepsilon_i \varepsilon_j}$ is the depth of the potential well, $R_{\mathrm{min},ij} = \frac{1}{2}(R_{\mathrm{min},i} + R_{\mathrm{min},j})$ is the equilibrium vdW distance, and r_{ij} is the distance between a CNT or graphene atom and a water atom.

In order to make sure our CNT and graphene models can be quantitatively matched to experiments, we calculate the flexural rigidity D of our graphene sheets and compare it with theoretical results [23–25]. We simulate a graphene sheet with the size of $a \times b = 3.7 \times 4.0\,\mathrm{nm}^2$, which is rolled on to a cylinder with the radius of $R = a/2\pi$; we use the CHARMM27 force field parameters $k_{\mathrm{bond}} = 322.55$ kcal/a^2, $k_{\mathrm{angle}} = 53.35$ kcal/mol-rad^2 and $k_{\mathrm{dihedral}} = 3.15$ kcal/mol. From the simulations, we calculate the energy associated with the cylindrical deformation of the graphene sheet, to obtain its strain energy density σ_{ela}. This allows us to calculate the flexural rigidity D, by using the formula $\sigma_{ela} = \frac{1}{2} D \kappa^2$, where $\kappa = 1/R$ is valid in the linear elastic regime [26]. The obtained value of $D = 0.194$ nN nm (27.9 kcal/mol) is in close agreement with ab initio results, $D_1 = 0.238$ nN nm [24], and other model studies, giving $D_2 = 0.11$ nN nm [25] and $D_3 = 0.225$ nN nm [26]. Therefore, our simulations should be reasonably close to potential experiments.

13.3 Nanodroplet Transport on Vibrated Nanotubes

Materials adsorbed on macroscopic solid-state surfaces can be transported by surface acoustic waves (SAW) [27]. This method has many practical applications, such as conveyor belt technologies [28], ultrasonic levitation of fragile materials [29], slipping of materials on tilted surfaces [30, 31], threading of cables inside tubes [32], and droplet delivery in microfluidics [33–37].

In this section, we examine the possibility of using SAW at the nanoscale. We use classical molecular dynamics (MD) simulations to model dragging of water nanodroplets on the surface of CNTs by coherent acoustic waves. Such coherent vibrations might be generated by piezo-electric generators [38, 39]. In analogy to coherent control of molecules by light [40], specialized pulses of coherent phonons might also be used in precise manipulation of materials.

13.3.1 Model System

Our model systems are formed by nanodroplets consisting of a number, N_w, of water molecules adsorbed at $T = 300$ K on the (10,0) CNT, and transported along/around its surface by coupling to coherent transversal acoustic (TA) phonon waves, as shown in Fig. 13.1.

The edge atoms at one of the ends of the 450 nm long CNT are fixed. At this end the tube is also oscillated. To prevent the CNT translation, four dummy atoms are placed in its interior at both ends. Otherwise, the tube is left free. A small Langevin damping [41] of 0.01 ps^{-1} is applied to the system to continuously thermalize it, while minimizing the unphysical loss of momenta [6]; the time step is 2 fs. At the two CNT ends, two regions with high damping of 10 ps^{-1} are established to absorb the vibrational waves. One region (35 nm long) is close to the generation point, and the other (180 nm long) is at the other CNT end. We model the systems in a NVT ensemble (periodic cell of $15 \times 15 \times 470$ nm^3).

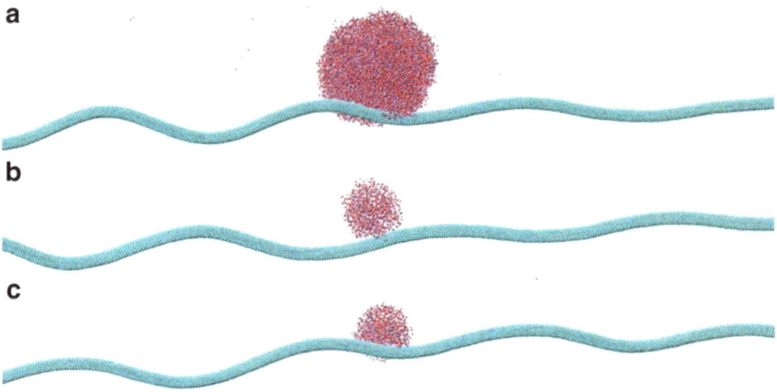

Fig. 13.1 Nanodroplets of (**a**) $N_{\mathrm{w}} = 10{,}000$ and (**b**) $N_{\mathrm{w}} = 1{,}000$ waters adsorbed on the (10,0) CNT and dragged by the linearly polarized TA vibrational wave with the amplitude of $A = 1.2$ ($T = 300$ K). (**c**) A nanodroplet of $N_{\mathrm{w}} = 1{,}000$ adsorbed on the same CNT is dragged by a circularly polarized vibrational wave of $A = 0.75$ nm

13.3.2 Nanodroplet Transport by Linearly Polarized Waves

The vibrational waves are generated at one CNT end by applying a periodic force (orthogonal to its axis), $F = F_0 \sin(\omega t)$, on the carbon atoms separated 35–40 nm from the CNT end. This generates a linearly polarized TA vibration wave, $A_y(t) = A\sin(\omega t)$, where $\omega \approx 208$ GHz, $k = 2\pi/\lambda \approx 0.157\,\mathrm{nm}^{-1}$, and $A \approx 0.3-2.1$ nm for $F_0 = 0.6948 - 5.558$ pN/atom. The TA waves propagate along the nanotube with the velocity of $v_{vib} = \omega/k \approx 1{,}324$ nm/ns, scatter with the nanodroplet, and become absorbed at the tube ends. In our simulations, we let the wave pass around the droplet for a while and then evaluate its average steady-state translational, v, and angular, ω_{d}, velocities.

In Fig. 13.2, we show the (linear) velocities of nanodroplets with $N_{\mathrm{w}} = 1{,}000$ and 10,000 water molecules in dependence on the vibrational amplitude, A. The data are obtained by averaging the droplet motion over trajectories of the length of $t \approx 7.2$ ns. We can see that the 10-times smaller droplet moves about 15-times faster for the same driving conditions. At small amplitudes, $A < 1.2$ nm, the velocities roughly depend quadratically on the driving amplitude. At larger amplitudes, $A > 1.2$ nm, they gain a linear dependence.

The nanodroplet is transported by absorbing a momentum from the vibrational wave. Its steady-state motion is stabilized by frictional dissipation of the gained momentum with the nanotube, which carries it away through the highly damped and fixed atoms. In the first approximation, the droplet motion might be described by the Boltzmann equation. In the steady state, obtained when a wave of a constant amplitude is passed through the CNT, the momentum of the droplet averaged over a short time (50 ps) is constant. Then, the constant driving force acting on the droplet, $\dot{P}_{\mathrm{drive}}$, is equal to the friction force, $\dot{P}_{\mathrm{friction}}$, between the droplet and the CNT [22] (linear motion—vectors omitted),

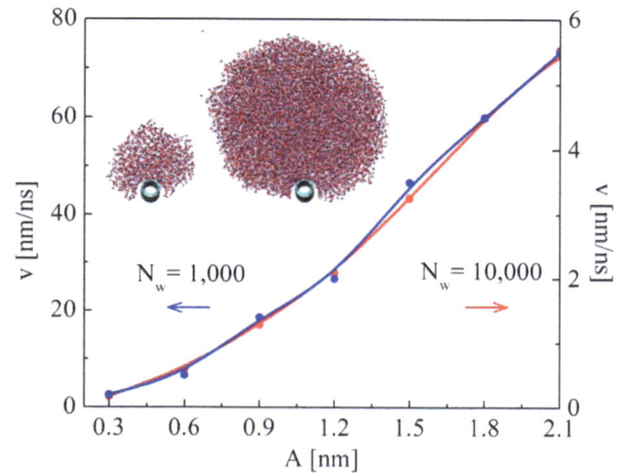

Fig. 13.2 The drag velocity of nanodroplets with $N_w = 1{,}000$ and 10,000 waters in dependence on the amplitude, A, of the linearly polarized wave, with the frequency of $\omega = 208$ GHz. (*inset*) The adsorbed nanodroplets viewed in the CNT axis

$$\dot{P}_{\text{drive}} = \int pF(r)\frac{\partial f}{\partial p}drdp$$
$$= \int p\left(\frac{\partial f}{\partial t}\right)_{\text{coll}} drdp = \dot{P}_{\text{friction}}. \tag{13.2}$$

Here, $f(r,p)$ is the position and momentum distribution function of the waters in the nanodroplet (normalized to N_w) and $F(r)$ is the force acting on each of the waters at r. The collision term $(\frac{\partial f}{\partial t})_{\text{coll}}$ describes scattering of waters with each other and the CNT, where the last option causes the droplet to relax its momentum [42]. In the approximation of the momentum relaxation time [43], τ_p, the damping term can be described as,

$$\int p\left(\frac{\partial f}{\partial t}\right)_{\text{coll}} drdp = \int p\frac{f - f_0}{\tau_p}drdp \approx \frac{P_{\text{droplet}}}{\tau_p}, \tag{13.3}$$

where P_{droplet} is the steady-state average momentum of the nanodroplet [6, 7].

As the acoustic wave propagates along the CNT, it carries the momentum density [44],

$$g(t,x) = \mu\,\omega k A^2\left[1 + \cos(2kx - 2\omega t)\right], \tag{13.4}$$

where μ is the CNT mass per unit length and the other symbols are the same as before. Assuming, for simplicity, that the momentum density of the wave is fully passed to the droplet (only approximately true, as seen in Fig. 13.1), we obtain

$$\dot{P}_{\text{drive}} = \frac{1}{2}\mu\,\omega^2 A^2 = \frac{P_{\text{droplet}}}{\tau_p}. \tag{13.5}$$

Equation 13.5 shows that the droplet velocity scales as

$$v_t = \frac{P_{\text{droplet}}}{m} \approx \frac{A^2 \tau_p}{m}. \tag{13.6}$$

Moreover, the momentum relaxation time, $\tau_p \approx S^{-1}$, can be assumed to scale inversely with the contact area, S, between the droplet and the CNT, due to friction. The 15 times larger velocity of the 10 times smaller droplet with a smaller contact area matches our expectations from (13.6). The quadratic dependence of the droplet velocities on the driving amplitude, A, shown in Fig. 13.2, also roughly agrees with (13.6).

13.3.3 Nanodroplet Transport by Circularly Polarized Waves

Next, we simulate transport of nanodroplets with $N_{\text{w}} = 1{,}000$ and 2,000 waters, adsorbed on the (10,0) CNT, by circularly polarized TA waves. Application of the force of $\mathbf{F}(t) = (F_x, F_y) = F_0\,(\sin(\omega t), \cos(\omega t))$, $F_0 = 0.4864 - 2.084$ pn/atom, on the same C atoms as before generates a circularly polarized wave, $\mathbf{A}(t) = (A_x, A_y) = A\,(\sin(\omega t), \cos(\omega t))$, where $A \approx 0.21 - 0.75$ nm, $\omega \approx 208$ GHz, and $k \approx 0.157\,\text{nm}^{-1}$. The circularly polarized TA waves carry both linear and angular momenta and pass them to the nanodroplets, which are transported along the CNT and rotated around it.

In Fig. 13.3, we plot the translational, v, and the angular, ω_{d}, velocities of the nanodroplets in dependence on the wave amplitude, A. For $N_{\text{w}} = 1{,}000$, ω_{d} rapidly grows with A till $\omega_{\text{d,max}} \approx 50.5$ rad/ns, where the droplet is ejected from the CNT surface due to large centrifugal forces. The larger droplet rotates with $\approx 30 - 40\%$ smaller angular velocity, in analogy to the situation in a linear transport. At $A = 0.4$–0.6 nm, both the linear and angular velocities show certain resonant features for both droplets. At these amplitudes of the circular waves the coupling to the droplets can be dramatically altered, since the wave amplitudes are similar to the droplet sizes. Interestingly, the translational velocities, v, are very similar for both droplets. This might be due to better transfer of linear momentum to the larger droplet from circularly polarized waves.

We can perform similar analysis of the angular momentum passage from the circular wave to the droplet and back to the CNT, like we did for the linear momentum in (13.2)–(13.6). In a steady state, obtained when a circularly polarized wave of a constant amplitude is passed through the CNT, the average angular momentum of the droplet around the (equilibrium position of) CNT axis is constant. The driving momentum of force, $\dot{L}_{\text{drive}}$, acting on the droplet is equal to its friction counterpart, $\dot{L}_{\text{friction}}$, acting between the droplet and the CNT [45]. Assuming that

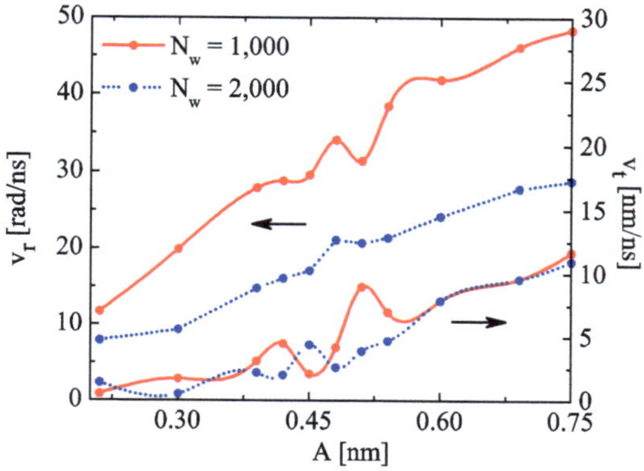

Fig. 13.3 The average translational v and angular ω_d velocities of water nanodroplets with $N_w =$ 1,000 and 2,000, in dependence on the wave amplitude, A, when driven by circularly polarized waves

the whole angular momentum density of the wave is passed to the droplet and using the approximation of the angular momentum relaxation time, we find

$$\dot{L}_{\text{drive}} = f(\mu,\omega,A) = \frac{L_{\text{droplet}}}{\tau_L} = \frac{I\omega_{\text{d}}}{\tau_L} = C, \tag{13.7}$$

where $f(\mu,\omega,A)$ is the angular momentum density (size) of the circularly polarized wave, I is the droplet moment of inertia with respect to the (equilibrium) CNT axis, and τ_L is the angular momentum relaxation time. In the steady state, the average rates of driving and damping are constant, as shown by C. The moment of inertia is $I = \sum_{i=1}^{n} m_{\text{w}} r_i^2 \propto N_{\text{w}}$, where m_{w} is the mass of a water molecule, and r_i is the distance of each water molecule from the (equilibrium) CNT axis. Using this I in (13.7), we find that $\omega_{\text{d}} \propto N_{\text{w}}^{-1}$, in rough agreement with Fig. 13.3.

In order to better understand the droplet-CNT dynamics, we present in Fig. 13.4 the time-dependent motion of the nanodroplet with $N_{\text{w}} = 1{,}000$ transported by circularly polarized waves. The droplet and CNT form a coupled system where the CNT vibrates around its axis and the droplet rotates around it. We describe the droplet rotation around the *actual* position of the CNT by the angle θ of the vector pointing from the center of mass of a CNT segment local to the droplet to the actual droplet center of mass. The CNT segment is defined as a 2 nm section of the CNT bisected by the droplet. The time dependence of θ for different amplitudes A is in Fig. 13.4 (top), the accompanied translation of the droplet along the CNT is in Fig. 13.4 (middle), and the radial distance of the droplet from the CNT axis is in Fig. 13.4 (bottom).

The droplet motion on the circularly polarized waves resembles *surfing*, where the droplet is sometimes grabbed better by the waves and for a while moves fast forward. At small waves, surfers cannot ride waves and neither can the droplet. This happens at $A = 0.21$ nm, where the vertical and longitudinal displacements of the

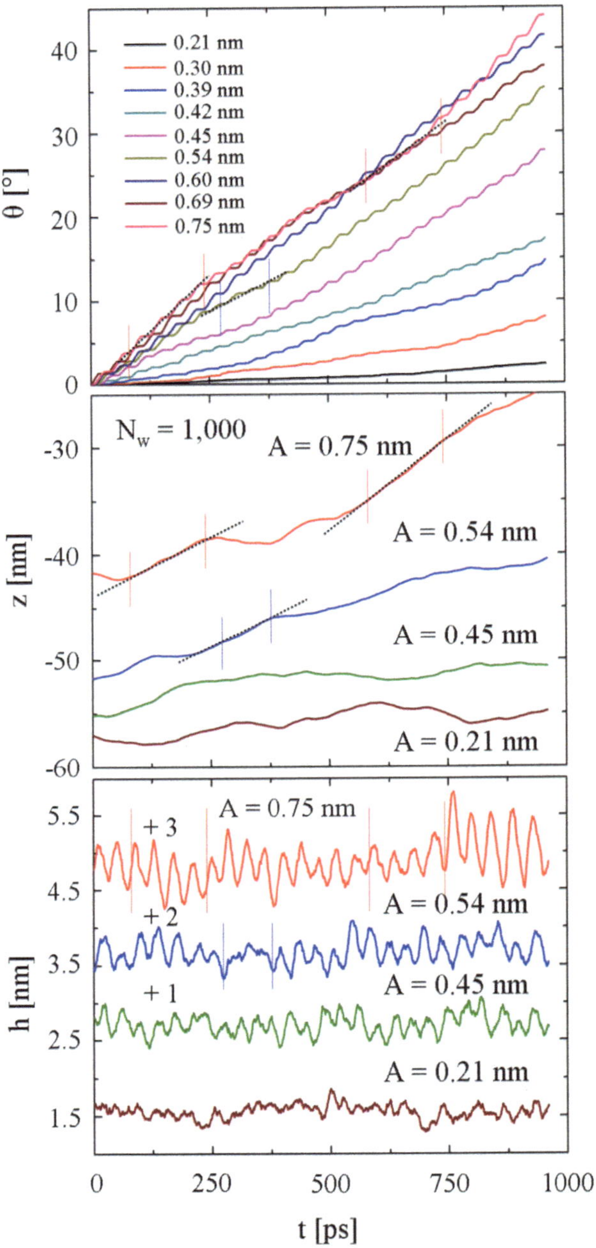

Fig. 13.4 Time-dependent angle of rotation (*top*), position (*middle*), and height (*bottom*) of the $N_W = 1{,}000$ droplet center of mass above the local CNT center of mass as the CNT is driven by circularly polarized waves of $\omega = 208$ GHZ at amplitudes ranging from $A = 0.21$ to $A = 0.75$ nm. *Tangents* between *vertical lines* indicate regions of surfing where the nanodroplet slides down the CNT surface

droplet on the CNT are very small, as shown in Fig. 13.4 (bottom) and (middle), respectively. Therefore, a small momentum is transferred to the water droplet which almost "bobs" in place like a "buoy," much like a surfer waiting for a wave. At larger amplitudes, the droplet can catch some of the waves and glide on them. We can see that at $A > 0.45$ nm, the droplet sometimes ($t \approx 125$ and 700 ps) starts to progress forward quickly. The same is seen even better at the larger amplitudes, $A = 0.54$ and $A = 0.75$ nm, as denoted by the dotted tangential lines. In real surfing, the gravitational force of fixed spatial orientation accelerates the surfer on the traveling tilted wave. On the CNT, the gravitational force is replaced by the inertia forces acting on the nanodroplet surfing of the circularly polarized wave.

13.3.4 Summary of Vibration-Driven Transport Simulations

We have demonstrated that TA vibrational waves on CNTs can drag nanodroplets adsorbed on their surfaces. The droplets perform translational and rotational motions, in dependence on the wave amplitude, polarization, and the droplet size. This material dragging, which complements other transport methods at the nanoscale, could be applied also on planar surfaces, such as graphene. It has potential applications in molecular delivery [46], fabrications of nanostructures [47, 48], and nanofluidics [49].

13.4 Dragging of Polarizable Nanodroplets by Distantly Solvated Ions

Recent studies have demonstrated efficient dragging of molecules adsorbed on CNTs by their Coulombic scattering with electrons passing through the nanotubes [3–5]. Detection of molecular flows around CNTs by related means has also been proposed [50], tested [51–53], and applied in nanofluidic devices [54, 55]. It is of a fundamental and practical interest to design techniques that could also manipulate large molecules and molecular assemblies [56, 57].

13.4.1 Ion Charge Screening in Nanotubes

To follow this goal, we investigate if ions intercalated inside carbon or boron-nitride nanotubes (BNT) can be "solvated at distance" in polarizable molecular assemblies adsorbed on their surfaces [58]. In highly polar solvents, such as water, the strength and long-range order of the charge-dipole Coulombic coupling is significant even if the ion and the solvent are separated by nanometer distances. If this space is filled

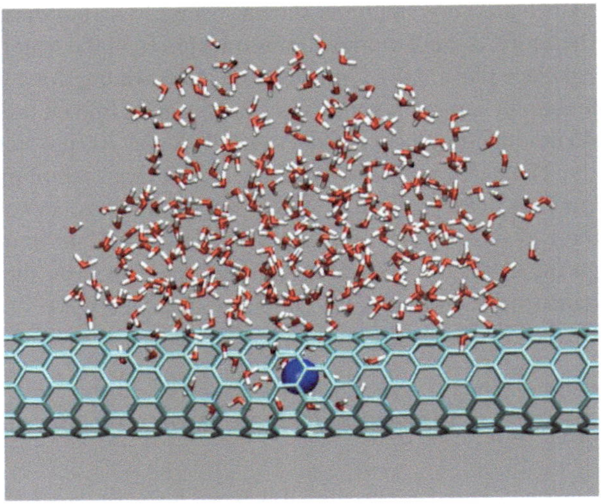

Fig. 13.5 Nanodroplet with $N_\mathrm{W} = 400$ water molecules dragged on the surface of unpolarized (10,0) CNT by a single intercalated Na^+ cation. The ion is driven by the electric field of $\mathscr{E} = 0.1$ V/nm applied along the tube z axis

by a dielectric material with a relatively small dielectric constant, such as the wall of a nanotube with a large band gap, the coupling should be preserved at room temperatures. Therefore, polarizable nanodroplets on the nanotube surfaces might get locked to the intercalated ions and dragged by them.

We first calculate ab initio the electrostatic potential φ generated above a (4,3) CNT (band gap of 1.28 eV) [59] by a Li^+ ion located in its center and φ generated above a (5,5) BNT (band gap of 5.5 eV) [60] by a Na^+ ion. In the calculations, the 5 nm long tubes are kept neutral and frozen, since their structure is rather rigid. The potentials φ are obtained from NBO atomic charges, using the B3LYP density functional and the 3–21g basis set in Gaussian03 [21]. The potential φ of Li^+ is decreased by $\approx$25% due to screening, while that of Na^+ is decreased by $\approx$10%. The same results are obtained in the presence of the electric field of $\mathscr{E} = 0.1$ V/nm applied along the tube axis. When the ions are shifted along the axis by a small distance d, the total energy of both systems changes by $\approx \mathscr{E} d$, as if the nanotubes are absent. These results show that the screening of the ionic field is small in selected CNTs and BNTs, where the ion can be also driven by electric fields. With this in mind, we model for simplicity the ion-droplet dynamics in some typical nanotubes and consider them to be non-polarizable.

13.4.2 Model System

In Fig. 13.5, we display a Na^+ ion intercalated inside an unpolarized (10,0) CNT that is distantly solvated in $N_\mathrm{W} = 400$ water molecules adsorbed on the CNT surface. The system is relaxed in the box of $\approx 1000\ \mathrm{nm}^3$, fitting the coexistence of gas and liquid

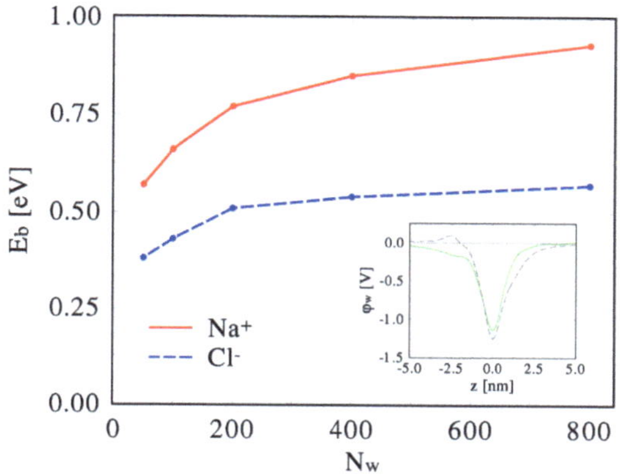

Fig. 13.6 Binding energy E_b between the Na^+ and Cl^- ions and water nanodroplets containing about N_w molecules (molecules that left to the gas phase are neglected). (*inset*) Snapshots of the electrostatic potential φ_w along the axis of the nanotube with the intercalated Na^+ ion

phases at the temperature of $T = 300$ K. The water droplet is spontaneously formed on the CNT even in the absence of the ion, similarly like on other fibers [61].

We model this hybrid system by molecular dynamics with the NAMD software package implementing the CHARMM27 force field as previously described. In this system we model electrostatics with the particle-mesh Ewald method [62]. We simulate the system with periodic boundary conditions in an NVT ensemble. The time step is 1 fs, and a small Langevin damping coefficient of $0.01\,ps^{-1}$ is chosen to minimize the unphysical loss of momenta to the reservoirs [6]. The tube is aligned along the z axis, it is blocked from shifting and left free to vibrate.

13.4.3 Ion-Water Nanodroplet Coupling

We start by exploring the strength of the ion coupling to the distant solvent, characterized by the ion-droplet binding energy E_b. In Fig. 13.6, we show the obtained E_b that is averaged over 10,000 frames, separated by 500 fs intervals. For nanodroplets with $N_w = 100-800$ waters, the binding energies E_b of the Na^+ and Cl^- ions saturate to values that are several times smaller than their solvation energies in bulk water, $E_{solv} = 7.92$ and 6.91 eV [58], respectively. For nanodroplets with $N_w < 100$, the binding energies are smaller, and at $N_w \approx 5-15$ they become comparable to kT.

We can also estimate the coupling energy E_b analytically by assuming that the ion is located at a distance of $d \approx 0.35$ nm above a flat surface of water with the dielectric constant $\varepsilon_w \approx 80$. This gives $E_b \approx -\frac{e^2}{4d}\left(\frac{\varepsilon_w-1}{\varepsilon_w+1}\right) \approx -1$ eV, in a good agreement with Fig. 13.6. The fact that in the simulations the Na^+ ion binds twice as strongly to the nanodroplet than the Cl^- ion is caused by the character of the water polarization:

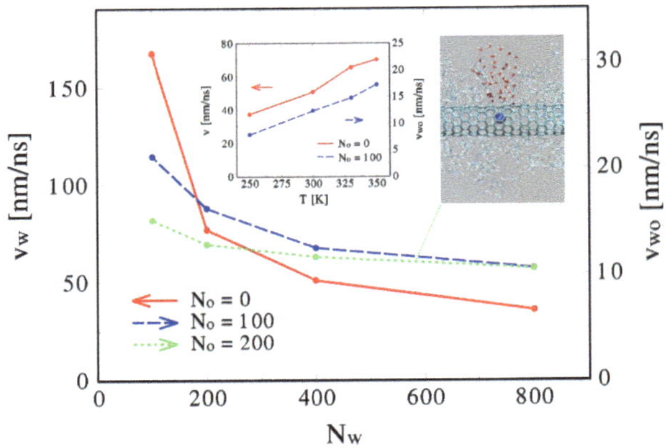

Fig. 13.7 Dependence of the nanodroplet velocity on the number of molecules N_W. Dragging of the droplet in the presence of Octane molecules is considered as well. (*right inset*) Visualization of the nanodroplet with $N_W = 50$ in $N_o = 200$ oil molecules. (*left inset*) Temperature dependence of the droplet speed for $N_W = 400$

the Na^+ ion attracts from each water the O atom that is twice more charged than the H atoms, while Cl^- attracts just one of these two H atoms. The large difference between E_b and the bulk solvation energies is caused by the fact that ions solvated in bulk water are surrounded by two to three times more water molecules that are about twice closer to them. These binding energies are proportionally decreased if the nanotube polarization is included, as discussed above.

In the inset of Fig. 13.6, we also show two snapshots of the 1D electrostatic potential generated by the water droplet ($N_W = 400$) along the axis of the nanotube with the intercalated Na^+ ion. The 1 eV deep well with a potential gradient of 0.5–1 V/nm should be large enough to block the moving ion from leaving the droplet even in the presence of electric fields of $\mathscr{E} \approx 0.1-0.2$ V/nm.

13.4.4 Nanodroplet Dragging with an Ion in an Electric Field

We continue our study by dragging the droplets with the ions in the presence of electric fields aligned along the nanotubes (see the movie [63]). The field acts on the whole system, except on the nanotube that is treated as non-polarizable. In Fig. 13.7 (left axis), we show the velocity, v_W, of droplets with N_W water molecules. The data are calculated from 50 ns simulations ($\approx$ 50 rounds along the CNT) at $\mathscr{E}_0 = 0.1$ V/nm and $T = 300$ K. This statistical averaging is fully sufficient for the presentation of the results [6]. The obtained velocity v_W is proportional to the electric field and it strongly depends on the droplet size. It has practically the same value if the Na^+ or Cl^- ions are used for the dragging. In order to test the scalability of this dragging phenomenon, we also model a droplet with $N_W = 10{,}000$, located between two parallel (10,0) CNTs separated by a 5 nm distance. If one Na^+ ion is placed in each nanotube, and both ions are driven by the field of $\mathscr{E}_0 = 0.1$ V/nm, the droplet moves together with the ion pair at a high speed of $v_w = 6.6$ nm/ns, due to a small contact with the CNTs.

The character of the nanodroplet motion on the nanotubes might be closer to sliding [64] than rolling [65], due to partial wetting of the CNT surface with a large van der Waals binding. In macroscopic systems, the droplet velocity for both mechanisms of motion is controlled by the momentum/energy dissipation of water layers sliding inside the droplet [64]. Both mechanisms give the qualitative dependence of the droplet velocity $v \propto e\mathscr{E}/(r\eta)$, where the droplet radius is $r \propto N_w^{1/3}$ and the water viscosity is $\eta \propto 1/T$ [66]. The results in Fig. 13.6 roughly confirm this dependence even for the motion of nanoscale droplets, but the driving speed scales more steeply with the number of water molecules, $v \propto 1/N_w^{2/3}$. In the inset, we also show for $N_w = 400$ that the temperature dependence is almost linear, $v \propto T$, as expected. To clarify more the character of motion, we test dragging of a nanodroplet by a Na^+ ion that is *directly* solvated in it. The obtained speed of the droplet is about 20% larger than when the ion is inside the tube. This is most likely caused by higher the tendency of the droplet to roll, since the dragging force acts in its center rather than on its periphery at the nanotube surface.

The character of the nanodroplet motion could be dramatically altered, if a monolayer of oil is adsorbed on the nanotube surface. In Fig. 13.7 (right axis), we show that the presence of N_o octane molecules decreases the droplet speed v_{wo} by an order of magnitude, due to friction between water and oil. Smaller droplets, $N_w < 100$, are attached to the ion by a narrow "neck" passing through the oil layer (see inset in Fig. 13.7). Larger droplets, $100 < N_w < 200$, are more or less spherical, significantly submerged inside the oil, and they share a very small surface area with the CNT. In analogy to a water droplet inside a bulk oil, their driving could be described by the Stokes law that is largely valid at the nanoscale [67, 68]. Here, it gives $F = -6\pi r \eta v_{wo}$, where $F = e\mathscr{E}_0 = 16$ pN is the drag force acting on the droplet, r is the droplet radius, and $\eta \approx 0.54$ mPa s is the viscosity of octane at $T = 300$ K. For $N_w = 100$, we find $r \approx 3.5$ Å, so $v_{wo} = 4.5$ m/s. This value is three to four times smaller than that obtained from the simulations, due to the incomplete coverage of the nanodroplet by oil.

If the nanotube is fully submerged in water instead of oil, then, on the contrary, the dynamics of the field-driven ion becomes very different [69]. We simulate this situation in a (10,0) CNT, placed in the periodic water box of $3 \times 3 \times 6.8\,\text{nm}^3$, and present in Table 13.1 the results obtained for the driving field of $\mathscr{E} = 0.1$ V/nm. The velocities of the ions are four to five times larger than those of the ions dragging the water droplet. This is because water molecules around the submerged nanotube just rearrange fast locally when they react to the field-driven ion. At higher temperatures, the ions move faster, since their binding to the water molecules is less stable [70]. The velocities of the Cl^- ion are smaller than those of the Na^+ ion, because Cl^- easily attracts the light H atoms that are not bound (frustrated) in water molecules at the nanotube surface.

Table 13.1 Velocities of ions inside the (10,0) SWNT submerged in water and at $\mathscr{E} = 0.1$ V/nm

Ion (K)	240	270	300
Na^+ (m/s)	464.7	622.5	733.7
Cl^- (m/s)	284.5	401.6	547.9

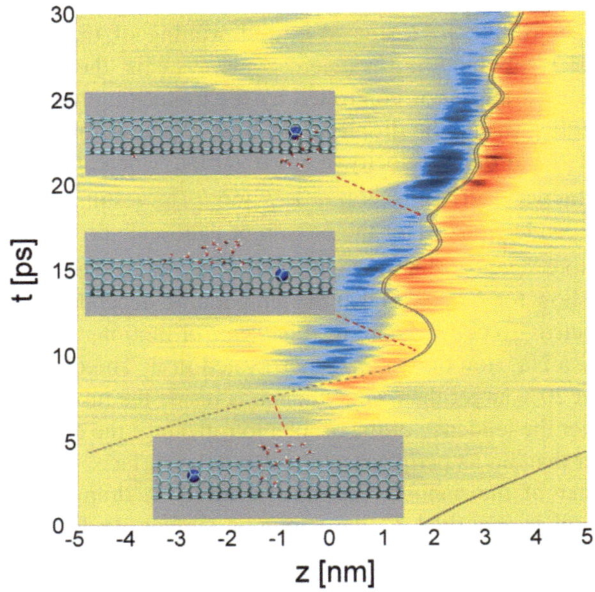

Fig. 13.8 The trajectory of the Na^+ ion that is recaught by a water nanodroplet with $N_w = 20$. The axial position and the time of the ion motion are shown on the *horizontal* and *vertical axis*, respectively. The electric field generated by the water molecules along the CNT axis is plotted by contours

We also discuss the *dynamical stability* of the coupled ion-droplet pair. In larger electric fields or when the nanodroplet is small, the ion might get released, and, in the model with periodic boundary conditions, it might get later recaught by the nanodroplet. In Fig. 13.8, we plot for $\mathscr{E} = 0.02$ V/nm the trajectory of the ion that left the droplet of $N_w = 20$ and was recaught in the next run around it. The ion's trajectory is shown by the dark line, and the time-frame separation on the vertical axis is 100 fs. The electric field along the CNT axis created by the ion-polarized droplet is plotted by contours. The positive/negative regions that lock the ion are obtained from the derivative taken at the sides of the potential well (see inset of Fig. 13.6).

The ion released from the droplet goes once around the tube and reapproaches the droplet with the velocity of $v_{\text{ini}} \approx 1{,}400$ m/s (bottom). After it gets closer to the droplet, the water molecules become fast polarized (inset at $z = -1.3$ nm, $t = 7$ ps). The ion first passes around the droplet, just to be attracted back by several chained molecules protruding from the droplet ($z = 1.7$ nm, $t = 10.0$ ps). This is possible, since a chain of five hydrogen-bonded and aligned water molecules generate at the distance of 1 nm the field of ≈ 0.17 V/nm, which is opposite to and almost an order of magnitude larger than the external field. The deceleration of the ion by this large induced field causes the coupled system to gain high Coulombic potential energy. Thus the ion position oscillates three to four times, before it is fully seized back by the droplet ($z = 2$ nm, $t = 18$ ps). The two start to move together at a much smaller velocity $v_{end} \approx 130$ m/s, while the waters are already interconnected. If the droplet does not catch the ion within several of its runs around the periodic system, the ion might heat and temporally evaporate the droplet. The transient oscillations observed in this ion catching closely resemble quasi-particle formation in condensed matter systems [71].

13.4.5 Summary of Nanodroplet Dragging by Distantly Solvated Ions

The Coulombic dragging of molecules on the surfaces of nanotubes by ions and ionic solutions flowing through them complements the passive transport of gases [72, 73] and liquids [74–77] through CNTs. These phenomena might be used in molecular delivery, separation, desalination, and manipulation of nanoparticles at the nanoscale. When integrated into modern lab-on-a-chip techniques, the methodology could lead to a number of important nanofluidic applications [46].

13.5 Nanodroplet Activated and Guided Folding of Graphene Nanostructures

Recently, graphene monolayers have been prepared and intensively studied [78–81]. Graphene nanoribbons have also been synthesized [82–84], and etched by using lithography [85, 86] and catalytic [87, 88] methods. *Graphene flakes* with strong interlayer vdW binding can self-assemble into larger structures [89–91]. Individual flakes with high elasticity [92–94] could also fold into a variety of 3D structures, such as carbon nanoscrolls [95, 96]. These nanoscrolls could be even prepared from single graphene sheets, when assisted by certain gases or alcohols [97, 98]. In order to reproducibly prepare such stable or metastable structures of different complexity, (1) the potential barriers associated with graphene deformation need to be overcome, (2) the folding processes should be guided, and (3) the final structures need to be well coordinated and stabilized by vdW or other coupling mechanisms.

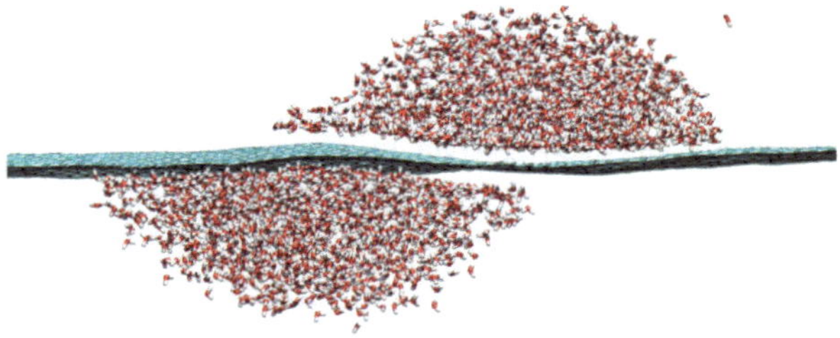

Fig. 13.9 Side view on two water nanodroplets, each with $N_W = 1300$ molecules, adsorbed on the opposite sides of a graphene sheet. The nanodroplets create two shallow holes in the graphene. Eventually, the nanodroplets become adjacent but stay highly mobile. Their dynamical coupling is realized by the minimization of the graphene bending energy associated with the two holes

13.5.1 Nanodroplet Coupling to Graphene

Carbon nanotubes can serve as a railroad for small water droplets [16]. CNTs submerged in water can assemble into micro-rings around bubbles formed by ultrasonic waves [99]. Similar assembly effects might work in 2D graphene-based systems. For example, liquid droplets can induce wrinkles on thin polymer films by strong capillary forces [100]. Droplets can also guide folding of 3D microstructures from polymer (PDMS) sheets [101]. *The question is if nanodroplets (ND) can activate and guide folding of graphene flakes of complex shapes, analogously like chaperones fold proteins* [102].

To answer this question, we study first the interaction of a water nanodroplet, of $N_W = 1{,}300$ waters, with a graphene sheet, of the size of $15 \times 12\,\mathrm{nm}^2$. It turns out that the nanodroplet, equilibrated at $T = 300\,\mathrm{K}$, induces a shallow hole in the graphene sheet, with the curvature radius of $R \approx 5$ nm.[1] The hole formation is driven by vdW coupling, which tends to minimize the surface of the naked droplet but maximize the surface of the graphene-dressed droplet. As shown in Fig. 13.9, two such droplets adsorbed on the opposite sides of the graphene sheet couple to save on the hole formation energy. The two droplets stay together during a correlated diffusion motion on the graphene surface.

[1] In the MD simulations, we apply the Langevin dynamics with 0.01 ps^{-1} damping coefficient, to minimize the unphysical loss of momentum [16], and the time step is 1 fs. The systems are simulated as NVT ensembles inside periodic cells of the following sizes: Fig. 13.9 ($55 \times 35 \times 35\,\mathrm{nm}^3$), Fig. 13.10 (up) ($30 \times 35 \times 35\,\mathrm{nm}^3$), Fig. 13.10 (bottom) ($30 \times 35 \times 35\,\mathrm{nm}^3$), Fig. 13.11 ($15 \times 85 \times 25\,\mathrm{nm}^3$), Fig. 13.12 ($20 \times 120 \times 20\,\mathrm{nm}^3$), and Fig. 13.14 ($20 \times 75 \times 20\,\mathrm{nm}^3$). The graphene–water (or graphene–graphene) binding energies are calculated as the difference of the total vdW energy of the system, when the system components are at the normal binding distance, and when they are separated by 5 nm. Averaging of the energies is done over 100 consecutive frames of the simulation trajectory, with a 1 ps time interval.

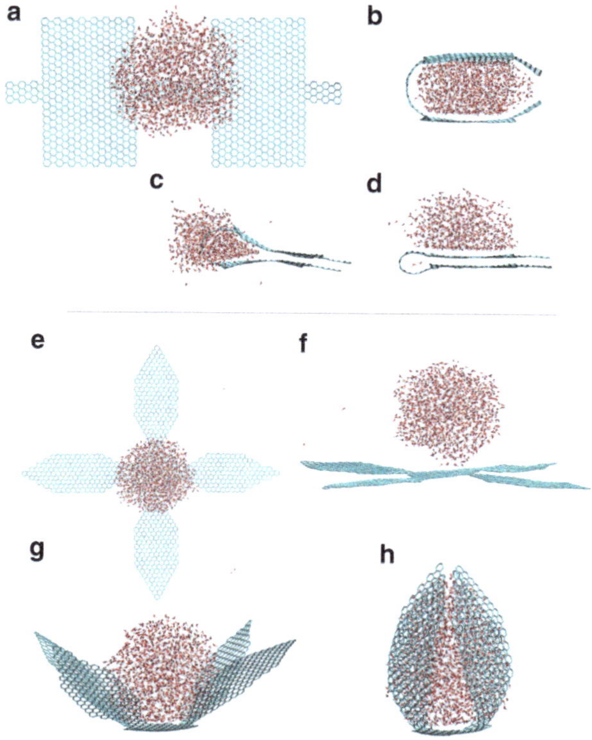

Fig. 13.10 (**a**)–(**d**) Water nanodroplet activated and guided folding of two graphene flakes connected by a narrow bridge. The nanodroplet is squeezed away when the system is heated to $T = 400\,\text{K}$. (**e**)–(**h**) Nanodroplet assisted folding of a star-shaped graphene flake, resembling the action of a "meat-eating flower"

13.5.2 Folding of Flakes

Intrigued by the action of NDs on graphene, we test if they can activate and guide folding of graphene flakes of various shapes. As shown in Fig. 13.10a, we first design a graphene nanoribbon, where two rectangular $3 \times 5\,\text{nm}^2$ flakes are connected by a narrow stripe of $2.5 \times 0.73\,\text{nm}^2$. In the simulations, we fix a few stripes of benzene rings on the right flake, which could be realized if the graphene is partly fixed at some substrate, and position a water nanodroplet ($N_{\text{w}} = 1{,}300$) above the center of the two flakes ($T = 300\,\text{K}$). Figure 13.10b after $t \approx 250\,\text{ps}$, both flakes bind with the droplet and bend the connecting bridge to form a metastable sandwich structure. Figure 13.10c when the temperature is raised to $T = 400\,\text{K}$, the droplet becomes more mobile and fluctuating. Within $t \approx 50\,\text{ps}$, the two flakes start to bind each other, and the water droplet is squeezed away. Figure 13.10d after another $t \approx 60\,\text{ps}$, the flakes join each other into a double layer, and the droplet stays on the top of one flake. When a smaller water droplet with $N_{\text{w}} = 800$ is placed on the nanoribbon, it induces its folding in a similar way and becomes squeezed out even faster in Fig. 13.10c, d.

Similarly, we study the folding of a star-shaped graphene nanoribbon with four blades connected to a central flake, as shown in Fig. 13.10e–f. At $T = 300$ K, a water droplet ($N_{\mathrm{w}} = 1{,}300$) is initially positioned at the height of $h \approx 0.5$ nm above the central flake. Figure 13.10g the droplet binds by vdW coupling with the central flake and induces bending of the four blades. Figure 13.10h after $t \approx 1$ ns, the four blades fold into a closed structure, with waters filling its interior. This effect resembles the action of a "meat-eating flower" [103], where the graphene capsule can store and protect the liquid content in various environments. Notice that slight asymmetry of the flake does not change the character of the assembly process. In real systems, other molecules might also be adsorbed on the graphene flakes. Although these molecules are not considered here, they might coalesce with the water droplets and modify their properties in the assembly process. Experimentally, the droplets could be deposited by Dip-Pen nanolithography [104] or AFM [105]. This deposition can also cause side effects not considered here, such as passage additional momentum to the folding sheet.

Folding of the two flakes into the sandwich, shown in Fig. 13.10a, b, is driven by the decrease of the water–graphene binding energy, $E_{\mathrm{g-w}} = -\sigma_{\mathrm{g-w}} A_{\mathrm{g-w}}$. Here, we estimate the density of the binding energy from our MD simulations (see Footnote 1) to be $\sigma_{\mathrm{g-w}} \approx 20.8$ kcal/(mol nm^2). The water–graphene binding area of the narrow stripe (initial area) and the two flakes (final area) are $A^{\mathrm{ini}}_{\mathrm{g-w}} = 2.5 \times 0.7 = 1.75\,\mathrm{nm}^2$ and $A^{\mathrm{end}}_{\mathrm{g-w}} = 3 \times 5 \times (2) = 30\,\mathrm{nm}^2$, respectively. The elastic bending energy of the connecting stripe is $E_{\mathrm{ela}} = \sigma_{\mathrm{ela}} A_{\mathrm{ela}}$, where $A_{\mathrm{ela}} \approx A^{\mathrm{ini}}_{\mathrm{g-w}}$ is the bending area, and $\sigma_{\mathrm{ela}} = \frac{1}{2} D \kappa^2$ is calculated for $D = 27.9$ kcal/mol and $\kappa = 1/R_{\mathrm{g}}$, where R_{g} is the graphene ribbon radius. In this case, $R_{\mathrm{g}} \approx 1$ nm, so $\sigma_{\mathrm{g-w}} > \sigma_{\mathrm{ela}} \approx 14$ kcal/(mol nm^2). This, together with $A_{\mathrm{g-w}} \approx A_{\mathrm{ela}}$, valid at the beginning of the folding process, means that $E_{\mathrm{g-w}} + E_{\mathrm{ela}} < 0$. During the folding process, the sandwich configuration becomes further stabilized, since $E_{\mathrm{g-w}}$ decreases by an order of magnitude, due to $A_{\mathrm{g-w}} = A^{\mathrm{end}}_{\mathrm{g-w}}$.

The final squeezing of the nanodroplet out of the sandwich, shown in Fig. 13.10c, d, means that graphene–graphene vdW binding is preferable to graphene–water vdW binding, for the force field parameters used in CHARMM27.

13.5.3 *Folding of Ribbons*

We now test if NDs can induce folding of graphene nanoribbons. As shown in Fig. 13.11a, we use a $30 \times 2\,\mathrm{nm}^2$ graphene nanoribbon, with one end fixed. At $T = 300$ K, a ND with $N_{\mathrm{w}} = 1{,}300$ waters is positioned above the free end of the ribbon. Figure 13.11b the free end starts to fold fast around the droplet. Figure 13.11c after $t = 0.6$ ns, the free end folds into a knot structure, touches the ribbon surface and starts to slide fast on it, due to strong vdW binding. Figure 13.11d while the knot is sliding on the ribbon, the droplet is deformed into a droplet-like shape that slips and rolls inside the knot [16]. After sliding over $l = 20$ nm, the water filled knot gains

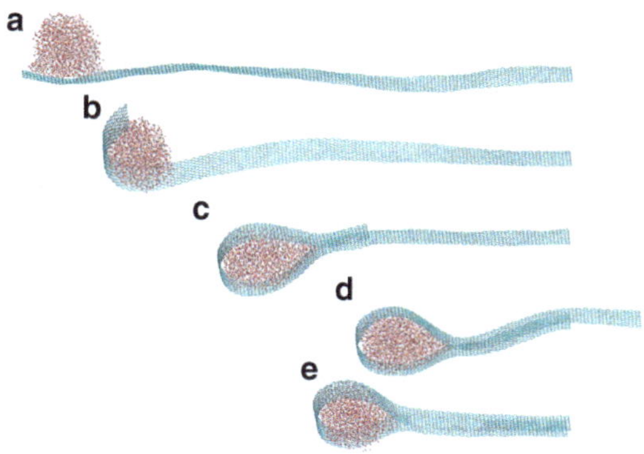

Fig. 13.11 Folding and sliding of a graphene ribbon with the size of 30×2 nm, which is activated and guided by a nanodroplet with $N_w = 1{,}300$ waters and the radius of $R_d \approx 2.1$ nm. (**a**)–(**c**) The free ribbon end folds around the droplet into a knot structure that slides on the ribbon surface (**d**) and (**e**)

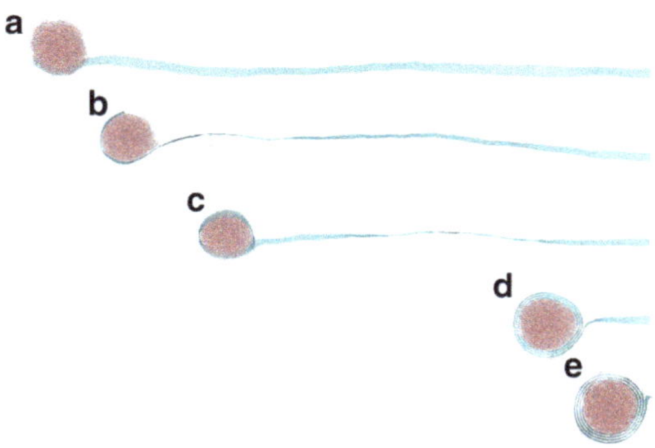

Fig. 13.12 Folding and rolling of a graphene ribbon with the size of 90×2 nm, which is activated and guided by a nanodroplet with $N_w = 10{,}000$ waters and the radius of $R_d \approx 4.2$ nm. (**a**) and (**b**) The ribbon tip folds around the water droplet into a wrapped cylinder, and (**c**)–(**e**) the wrapped cylinder is induced to roll on the ribbon surface (**c**)–(**e**)

the velocity of $v_w \approx 100$ m/s. This velocity is controlled by the rate of releasing potential energy, due to binding but reduced by bending, into the kinetic degrees of freedom, damped by friction. Figure 13.11d the sliding ribbon reaches the fixed end and overstretches into the space, due to its large momentum. Figure 13.11e it oscillates back and forth two to three times before the translational kinetic energy is dissipated.

The existence of CNTs raises the question if we could also "roll" graphene ribbons. This might happen when the droplet is larger and thus when it controls more the ribbon dynamics. In Fig. 13.12a, we simulate the folding of a $90 \times 2\,\text{nm}^2$ graphene ribbon (one end is fixed) at $T = 300$ K, when a droplet of $N_w = 10{,}000$ is initially positioned above the tip of the ribbon. Figure 13.12b, c as before, within $t = 2$ ps, the ribbon tip folds around the spherical droplet into a closed circular cylinder. Figure 13.12d this time, the approaching free end touches the surface at a larger angle and forms a cylinder around the droplet that starts to roll fast on the ribbon surface, like a contracting tongue of a chameleon. After rolling over $l = 60$ nm, the translational and rolling velocities are $v_t \approx 50$ m/s and $\omega_r \approx 12$ rad/ns, respectively. Figure 13.12e the cylinder rolls until the fixed end of the ribbon, where the rolling kinetic energy is eventually dissipated. The folded ribbon forms a multilayered ring structure, similar to multiwall nanotube, which is filled by water.

Let us analyze the conditions under which a graphene ribbon folds. Analogously to the folding of flakes, shown in Fig. 13.10, the folding of ribbon is driven by the competition between the graphene–water binding energy, $E_{\mathrm{g-w}} = -\sigma_{\mathrm{g-w}} A_{\mathrm{g-w}}$, and the graphene bending energy, $E_{\mathrm{ela}} = \sigma_{\mathrm{ela}} A_{\mathrm{ela}}$. From the energy condition, $E_{\mathrm{g-w}} + E_{\mathrm{ela}} < 0$, we obtain

$$\frac{A_{\mathrm{g-w}}}{A_{\mathrm{ela}}} > \frac{\sigma_{\mathrm{ela}}}{\sigma_{\mathrm{g-w}}}. \tag{13.8}$$

In Fig. 13.11, the water droplet has the radius $R_{\mathrm{d}} \approx 2.1$ nm. Assuming that $R_{\mathrm{g}} \approx R_{\mathrm{d}}$, we obtain from the above formula for σ_{ela} that $\sigma_{\mathrm{ela}} = 3.2$ kcal/ (mol nm^2) and $\sigma_{\mathrm{ela}}/\sigma_{\mathrm{g-w}} \approx 0.15$. Since, $A_{\mathrm{g-w}}/A_{\mathrm{ela}} \approx 1$, we see that this case easily fulfills the condition in (13.8). The graphene ribbon slides on itself, and this situation can be called the "sliding phase".

The ratio $A_{\mathrm{g-w}}/A_{\mathrm{ela}}$ and indirectly also σ_{ela} depend on the ratio of the ribbon width w to the droplet radius R_{d}, which thus controls the character of the folding process. When $w < 2R_{\mathrm{d}}$, the droplet can bind, in principle, on the whole width of the ribbon, so $A_{\mathrm{g-w}} \approx A_{\mathrm{ela}}$. For even larger droplets, we eventually get $w < 0.5 R_{\mathrm{d}}$, where our simulations show that the ribbon binds fully to the droplet surface. In this limit, we observe that after folding once around the droplet circumference the ribbon approaches itself practically at the wetting angle, and gains the dynamics characteristic for the "rolling phase", shown in Fig. 13.12.

When $w > 2R_{\mathrm{d}}$, it becomes very difficult for the small droplet to induce folding of the wide ribbon. Then, the droplet binds to the ribbon at an approximately circular area, with a radius $\approx R_{\mathrm{d}}$, because the water contact angle on graphene is about 90° and the droplet has almost the shape of a half-sphere [106]. If we assume that $R_{\mathrm{g}} \approx R_{\mathrm{d}}$, we have $A_{\mathrm{g-w}} \approx \pi R_{\mathrm{d}}^2$ and $A_{\mathrm{ela}} \approx 2R_{\mathrm{d}} w$. From (13.8) and $\sigma_{\mathrm{ela}} = \frac{1}{2} D \kappa^2 = \frac{1}{2} D \frac{1}{R_{\mathrm{d}}^2}$, we then obtain the condition for the ribbon folding

$$R_{\mathrm{d}}^3 > \frac{Dw}{\pi \sigma_{\mathrm{g-w}}}. \tag{13.9}$$

On the other hand, this means that the ribbon does not fold when $w \geq C R_{\mathrm{d}}^3$ ($C = \pi \sigma_{\mathrm{g-w}}/D \approx 4$ nm^{-2}), and this situation can be called the "nonfolding phase".

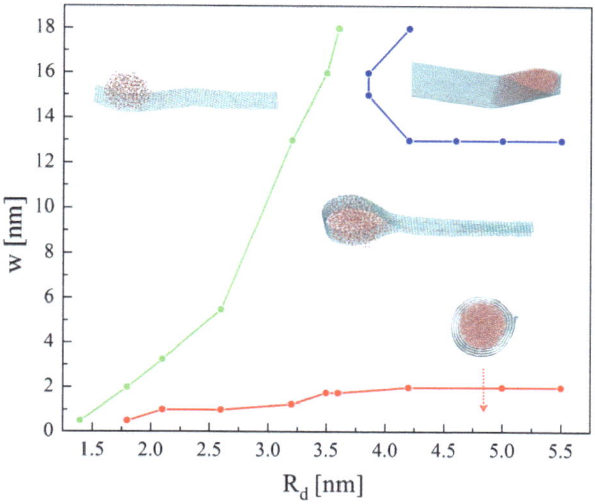

Fig. 13.13 The phase diagram of a nanodroplet and graphene nanoribbon with different folding dynamics. We display the nonfolding, sliding, rolling, and zipping phases

13.5.4 Nanodroplet-Graphene Phase Diagram

In Fig. 13.13, we summarize the results of our simulations in a phase diagram. We display four "phases" characterizing the ribbon dynamics, separated by phase boundary lines. They are called nonfolding, sliding, rolling, and zipping, where the first three were described and briefly analyzed above. The nonfolding phase, where the ribbon end does not fold around the droplet, is characterized by the cubic boundary derived above and shown in Fig. 13.13 (left). In the simulations, we obtain the value $C \approx 2.8\,\text{nm}^{-2}$, in close agreement with the above prediction. The nonfolding phase is adjacent with the sliding phase, which is separated from the rolling phase by the boundary line $w \approx \frac{1}{2} R_{\text{d}}$.

When the graphene ribbon becomes several times wider than the droplet diameter, it may fold around it in the orthogonal direction. Then, the folding dynamics of the graphene ribbon has a character of zipping. This situation corresponds to the "zipping phase", shown in the right top corner of the phase diagram in Fig. 13.13 and explained in detail in Fig. 13.14. Figure 13.14a we place a droplet of $N_{\text{w}} = 17{,}000$ at the free end of the ribbon of the size of $60 \times 16\,\text{nm}^2$. Figure 13.14b the ribbon folds from the two sides of the droplet within $t \approx 250\,\text{ps}$. Figure 13.14c at $t \approx 450\,\text{ps}$, the ribbon starts to "zip," where its two sides touch each other. Figure 13.14d the zipping process continues, and the droplet is transported along the ribbon. After zipping over $l \approx 40\,\text{nm}$, the droplet gains a translational velocity of $v_{\text{t}} \approx 63\,\text{m/s}$. In the zipped region, a chain of water molecules resides inside the turning line of the zipped ribbon. This region can be used like an artificial channel, similarly like CNTs.

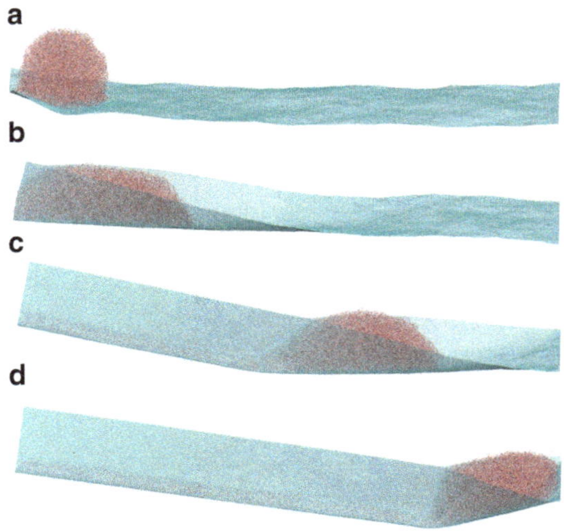

Fig. 13.14 Folding and zipping of a graphene ribbon with the size of $60 \times 16\,\text{nm}$, which is activated and guided by a nanodroplet with $N_{\text{w}} = 17{,}000$ waters and the radius of $R_{\text{d}} \approx 5\,\text{nm}$. (**a**) and (**b**) The ribbon end folds around the droplet. (**c**) and (**d**) The zipping propagates along the ribbon until the fixed end, while water channel is formed in the graphene sleeve

13.6 Conclusions

In summary, we have demonstrated that water nanodroplets can activate and guide folding of graphene nanostructures. The folding can be realized by different types of motions, such as bending, sliding, rolling, or zipping that lead to stable or metastable structures, such as sandwiches, capsules, knots, and rings. These structures can be the building blocks of functional nanodevices, with unique mechanical, electrical, or optical properties [47]. We have also shown how water nanodroplets could be used to carry materials on graphitic nanocarbon supports. We explored drag phenomena of nanodroplets on CNTs by vibrations and by coupling to distantly solvated ions. These studies can lead to future applications in molecular delivery, fabrications of nanostructures, and nanofluidics.

References

1. Iijima, S.: Helical microtubules of graphitic carbon. Nature **354**, 56–58 (1991)
2. Saito, R., Dresselhaus, G., Dresselhaus, M.S.: Physical Properties of Carbon Nanotubes. World Scientific, London (1998)
3. Král, P., Tománek, D.: Laser-driven atomic pump. Phys. Rev. Lett. **83**, 5373–5376 (1999)
4. Regan, B.C., Aloni, S., Ritchie, R.O., Dahmen, U., Zettl, A.: Carbon nanotubes as nanoscale mass conveyors. Nature **428**, 924–927 (2004)

5. Svensson, K., Olin, H., Olsson, E.: Nanopipettes for metal transport. Phys. Rev. Lett. **93**, 145901 (2004)
6. Wang, B., Král, P.: Coulombic dragging of molecules on surfaces induced by separately flowing liquids. J. Am. Chem. Soc. **128**, 15984 (2006)
7. Wang, B., Král, P.: Dragging of polarizable nanodroplets by distantly solvated ions. Phys. Rev. Lett. **101**, 046103 (2008)
8. Zhao, Y.C., Song, L., Deng, K., Liu, Z., Zhang, Z.X., Yang, Y.L., Wang, C., Yang, H.F., Jin, A.Z., Luo, Q., Gu, C.Z., Xie, S.S., Sun, L.F.: Individual water-filled single-walled carbon nanotubes as hydroelectric power converters. Adv. Mater. **20**, 1772–1776 (2008)
9. Schoen, P.A.E., Walther, J.H., Arcidiacono, S., Poulikakos, D., Koumoutsakos, P.: Nanoparticle traffic on helical tracks: thermophoretic mass transport through carbon nanotubes. Nano Lett. **6**, 1910–1917 (2006)
10. Barreiro, A., Rurali, R., Hernandez, E.R., Moser, J., Pichler, T., Forro, L., Bachtold, A.: Subnanometer motion of cargoes driven by thermal gradients along carbon nanotubes. Science **320**, 775–778 (2008)
11. Zambrano, H.A., Walther, J.H., Koumoutsakos, P., Sbalzarini, I.F.: Thermophoretic motion of water nanodroplets confined inside carbon nanotubes. Nano Lett. **9**, 66–71 (2006)
12. Shiomi, J., Maruyama, S.: Water transport inside a single-walled carbon nanotube driven by a temperature gradient. Nanotechnology **20**, 055708 (2009)
13. Insepov, Z., Wolf, D., Hassanein, A.: Nanopumping using carbon nanotubes. Nano Lett. **6**, 1893–1895 (2006)
14. Wang, Q.: Atomic transportation via carbon nanotubes. Nano Lett. **9**, 245–249 (2009)
15. Russell, J.T., Wang, B., Král, P.: Nanodroplet transport on vibrated nanotubes. Phys. Chem. Lett. **3**, 313–357 (2012)
16. Wang, B., Král, P.: Dragging of polarizable nanodroplets by distantly solvated ions. Phys. Rev. Lett. **101**, 046103 (2008)
17. Patra, N., Wang, B., Král, P.: Nanodroplet activated and guided folding of graphene nanostructures. Nano Lett. **9**, 3766–3771 (2009)
18. Phillips, J.C., Braun, R., Wang, W., Gumbart, J., Tajkhorshid, E., Villa, E., Chipot, C., Skeel, R.D., Kale, L., Schulten, K.: Scalable molecular dynamics with NAMD. J. Comput. Chem. **26**, 1781–1802 (2005)
19. MacKerell, A.D. Jr., Bashford, D., Bellot, M., Dunbrack, R.L., Evanseck, J.D., Field, M.J., Fischer, S., Gao, J., Guo, H., Ha, S., Joseph-McCarthy, D., Kuchnir, L., Kuczera, K., Lau, F.T.K., Mattos, C., Michnick, S., Ngo, T., Nguyen, D.T., Prodhom, B., Reiher, W.E. III, Roux, B., Schlenkrich, M., Smith, J.C., Stote, R., Straub, J., Watanabe, M., Wiorkiewicz-Kuczera, J., Yin, D., Karplus, M.: All-atom empirical potential for molecular modeling and dynamics studies of proteins. J. Phys. Chem. B **102**, 3586–3617 (1998)
20. Humphrey, W., Dalke, A., Schulten, K.: VMD: visual molecular dynamics. J. Mol. Graph. **14**, 33–38 (1996)
21. Frisch, M.J., et al: Gaussian 03, Revision C.02. Gaussian, Wallingford (2004)
22. Vukovic, L., Král, P.: Coulombically driven rolling of nanorods on water Phys. Rev. Lett. **103**, 246103 (2009)
23. Tersoff, J.: Energies of fullerenes. J. Phys. Rev. B **46**, 15546–15549 (1992)
24. Kudin, K.N., Scuseria, E.G.: C_2F, BN, and C nanoshell elasticity from ab initio computations. Phys. Rev. B **64**, 235406–234515 (2001)
25. Arroyo, M., Belytschko, T.: Finite crystal elasticity of carbon nanotubes based on the exponential Cauchy-Born rule. Phys. Rev. B **69**, 115415–115425 (2004)
26. Lu, Q., Arroyo, M., Huang, R.: Elastic bending modulus of monolayer graphene. J. Phys. D Appl. Phys. **42**, 102002–102007 (2009)
27. Wang, Z., Zhe, J.: Recent advances in particle and droplet manipulation for lab-on-a-chip devices based on surface acoustic waves. Lab Chip **11**, 1280–1285 (2011)
28. Hashimoto, H., Koike, Y., Ueha, S.: Transporting objects without contact using flexural travelling waves. J. Acoust. Soc. Am. **103**, 3230–3233 (1998)

29. Kim, G.H., Park, J.W., Jeong, S.H.: Analysis of dynamic characteristics for vibration of flexural beam in ultrasonic transport system. J. Mech. Sci. Technol. **23**, 1428–1434 (2009)
30. Miranda, E.C., Thomsen, J.J.: Vibration induced sliding: theory and experiment for a beam with a spring-loaded mass. Nonlinear Dyn. **19**, 167–186 (1998)
31. Vielsack, P., Spiess, H.: Sliding of a mass on an inclined driven plane with randomly varying coefficient of friction. J. Appl. Mech. **67**, 112–116 (2000)
32. Long, Y.G., Nagaya, K., Niwa, H.: Vibration conveyance in spatial-curved tubes. J. Vib. Acoust. **116**, 38–46 (1994)
33. Biwersi, S., Manceau, J.-F., Bastien, F.: Displacement of droplets and deformation of thin liquid layers using flexural vibrations of structures. influence of acoustic radiation pressure. J. Acoust. Soc. Am. **107**, 661–664 (2000)
34. Wixforth, A., Strobl, C., Gauer, C., Toegl, A., Scriba, J., Guttenberg, Z.V.: Acoustic manipulation of small droplets. Anal. Bioanal. Chem. **379**, 982–991 (2004)
35. Alzuaga, S., Manceau, J.-F., Bastien, F.: Motion of droplets on solid surface using acoustic radiation pressure. J. Sound Vib. **282**, 151–162 (2005)
36. Bennes, J., Alzuaga, S., Chabe, P., Morain, G., Cheroux, F, Manceau, J.-F., Bastien, F.: Action of low frequency vibration on liquid droplets and particles. Ultrasonics **44**, 497–502 (2004)
37. Jiao, Z.J., Huang, X.Y., Nguyen, N.-T.: Scattering and attenuation of surface acoustic waves in droplet actuation. J. Phys. A Math. Theor. **41**, 355502 (2008)
38. Mele, E.J., Král, P.: Electric polarization of heteropolar nanotubes as a geometric phase. Phys. Rev. Lett. **88**, 056803 (2002)
39. Michalski, P.J., Sai, N.A., Mele, E.J.: Continuum theory for nanotube piezoelectricity. Phys. Rev. Lett. **95**, 116803 (2005)
40. Král, P., Thanopulos, I., Shapiro, M.: Coherently controlled adiabatic passage. Rev. Mod. Phys. **79**, 53–77 (2007)
41. Schneider, T., Stoll, E.: Molecular-dynamics study of three-dimensions one-component model for distortive phase transitions. Phys. Rev. B **17**, 1302–1322 (1978)
42. Matsumoto, M., Kunisawa, T., Xiao, P.: Relaxation of phonons in classical MD simulation. J. Therm. Sci. Technol. **3**, 159–166 (2008)
43. Král, P.: Linearized quantum transport equations: AC conductance of a quantum wire with an electron-phonon interaction. Phys. Rev. B **53**, 11034–11050 (1996)
44. Benumof, R.: Momentum propagation by traveling waves on a string. Am. J. Phys. **50**, 20–25 (1982)
45. Segal, D., Král, P., Shapiro, M.: Ultraslow phonon-assisted collapse of tubular image states. Surf. Sci. **577**, 86–92 (2005)
46. Craighead, H.G.: Future lab-on-a-chip technologies for interrogating individual molecules. Nature **442**, 387–393 (2006)
47. Baughman, R.H., Cui, C.X., Zakhidov, A.A., Iqbal, Z., Barisci, J.N., Spinks, G.M., Wallace, G.G., Mazzoldi, A., De Rossi, D., Rinzler, A.G., Jaschinski, O., Roth, S., Kertesz, M.: Carbon nanotube actuators. Science **284**, 1340–1344 (1999)
48. Spinks, G.M., Wallace, G.G., Fifield, L.S., Dalton, L.R., Mazzoldi, A., De Rossi, D., Khayrullin, I.I., Baughman, R.H., Pneumatic carbon nanotube actuators. Adv. Mater. **14**, 1728–1732 (2002)
49. Schoch, R.B., Han, J.Y., Renaud, P., Transport phenomena in nanofluidics. Rev. Mod. Phys. **80**, 839–883 (2008)
50. Král, P., Shapiro, M.: Nanotube electron drag in flowing liquids. Phys. Rev. Lett. **86**, 131 (2001)
51. Ghosh, S., Sood, A.K., Kumar, N.: Carbon nanotube flow sensors. Science **299**, 1042 (2003). US Patent No. 6718834 B1
52. Sood, A.K., Ghosh, S.: Direct generation of a voltage and current by gas flow over carbon nanotubes and semiconductors. Phys. Rev. Lett. **93**, 086601 (2004)
53. Subramaniam, C., Pradeep, T., Chakrabarti, J.: Flow-induced transverse electrical potential across an assembly of gold nanoparticles. Phys. Rev. Lett. **95**, 164501 (2005)
54. Bourlon, B., Wong, J., Miko, C., Forro, L., Bockrath, M.: A nanoscale probe for fluidic and ionic transport. Nat. Nanotech. **2**, 104 (2006)

55. Fu, J.P., Schoch, R.B., Stevens, A.L., Tannenbaum, S.R., Han, J.Y.: A patterned anisotropic nanofluidic sieving structure for continuous-flow separation of DNA and proteins. Nat. Nanotech. **2**, 121 (2007)
56. Linke, H., et al.: Self-propelled Leidenfrost droplets. Phys. Rev. Lett. **96**, 154502 (2006)
57. Akin, D., et al.: Bacteria-mediated delivery of nanoparticles and cargo into cells. Nat. Nanotech. **2**, 441–449 (2007)
58. Hitoshi, O., Radnai, T.: Structure and dynamics of hydrated ions. Chem. Rev. **93**, 1157 (1993)
59. Cabria, I., Mintmire, J.W., White, C.T.: Metallic and semiconducting narrow carbon nanotubes. Phys. Rev. B **67**, 121406(R) (2003)
60. Blasé, X., Rubio, A., Louie, S.G., Cohen, M.L.: Stability and band gap constancy of boron nitride nanotubes. Europhys. Lett. **28**, 335 (1994)
61. McHale, G., Newton, M.I., Carroll, B.J.: The shape and stability of small liquid drops on fibers. Oil Gas Sci. Technol. Rev. IFP **56**, 47 (2006)
62. Darden, T., York, D., Pedersen, L.: Particle mesh Ewald: An Nlog(N) method for Ewald sums in large systems. J. Chem. Phys. **98**, 10089 (1993)
63. See EPAPS Document No. E-PRLTAO-101-06380 for three movies. For more information on EPAPS, (2008) see http://www.aip.org/pubservs/epaps.html
64. Kim, H.Y., Lee, H.J., Kang, B.H.: Sliding of liquid drops down an inclined solid surface. J. Coll. Int. Sci. **247**, 372 (2002)
65. Mahadevan, L., Pomeau, Y.: Rolling droplets. Phys. Fluids **11**, 2449 (1999)
66. Brancker, A.V.: Viscosity-temperature dependence. Nature **166**, 905 (1950)
67. Vergeles, M., Keblinski, P., Koplik, J., Banavar, J.R.: Stokes drag and lubrication flows: a molecular dynamics study. Phys. Rev. E **53**, 4852 (1996)
68. Squires, T.M., Quake, S.R.: Microfluidics: fluid physics at the nanoliter scale. Rev. Mod. Phys. **77**, 977 (2005)
69. Mills, D.L.: Image force on a moving charge. Phys. Rev. B **15**, 763 (1977)
70. Laage, D., Hynes, J.T.: A molecular jump mechanism of water reorientation. Science **311**, 832 (2006)
71. Král, P., Jauho, A.P.: Resonant tunneling in a pulsed phonon field. Phys. Rev. B **59**, 7656 (1999)
72. Skoulidas, A.I., Ackerman, D.M., Johnson, J.K., Sholl, D.S.: Rapid transport of gases in carbon nanotubes. Phys. Rev. Lett. **89**, 185901 (2002)
73. Holt, J.K., et al.: Fast mass transport through sub-2-nanometer carbon nanotubes. Science **312**, 1034 (2006)
74. Majumder, M., Chopra, N., Andrews, R., Hinds, B.J.: Nanoscale hydrodynamics: enhanced flow in carbon nanotubes. Nature **483**, 44 (2005)
75. Wang, Z., Ci, L., Chen, L., Nayak, S., Ajayan, P.M., Koratkar, N.: Polarity-dependent electrochemically controlled transport of water through carbon nanotube membranes. Nano Lett. **7**, 697 (2007)
76. Zhou, J.J., Noca, F., Gharib, M.: Flow conveying and diagnosis with carbon nanotube arrays. Nanotechnology **17**, 4845 (2006)
77. Whitby, M., Quirke, N.: Fluid flow in carbon nanotubes and nanopipes. Nat. Nanotechnol. **2**, 87 (2007)
78. Novoselov, K.S., Geim, A.K., Morozov, S.V., Jiang, D., Zhang, Y., Dubonos, S.V., Grigorieva, I.V., Firsov, A.A.: Electric field effect in atomically thin carbon films. Science **306**, 666–669 (2004)
79. Geim, A.K., Novoselov, K.S.: The rise of graphene. Nat. Mater. **6**, 183–191 (2007)
80. Berner, S., Corso, M., Widmer, R., Groening, O., Laskowski, R., Blaha, P., Schwarz, K., Goriachko, A., Over, H., Gsell, S., Schreck, M., Sachdev, H., Greber, T., Osterwalder, J.: Boron nitride nanomesh: functionality from a corrugated monolayer. Angew. Chem. Int. Ed. **46**, 5115–5119 (2007)
81. Laskowski, R., Blaha, P., Gallauner, T., Schwarz, K.: Single-layer model of the hexagonal boron nitride nanomesh on the Rh(111) surface. Phys. Rev. Lett. **98**, 106802 (2007)

82. Li, X., Wang, X., Zhang, L., Lee, S., Dai, H.: Chemically derived, ultrasmooth graphene nanoribbon semiconductors. Science **319**, 1229–1232 (2008)
83. Jiao, L., Zhang, L., Wang, X., Diankov, G., Dai, H.: Narrow graphene nanoribbons from carbon nanotubes. Nature **458**, 877–880 (2009)
84. Kosynkin, D.V., Higginbotham, A.L., Sinitskii, A., Lomeda, J.R., Dimiev, A., Price, B.K., Tour, J.M.: Longitudinal unzipping of carbon nanotubes to form graphene nanoribbons. Nature **458**, 872–876 (2009)
85. Tapasztó, L., Dobrik, G., Lambin, P., Biró, L.P.: Tailoring the atomic structure of graphene nanoribbons by scanning tunnelling microscope lithography. Nat. Nanotechnol. **3**, 397–401 (2008)
86. Stampfer, C., Guttinger, J., Hellmuller, S., Molitor, F., Ensslin, K., Ihn, T.: Energy gaps in etched graphene nanoribbons. Phys. Rev. Lett. **102**, 056403 (2009)
87. Ci, L., Xu, Z., Wang, L., Gao, W., Ding, F., Kelly, K.F., Yakobson, B.I., Ajayan, P.M.: Controlled nanocutting of graphene. Nano Res. **1**, 116–122 (2008)
88. Campos, L.C., Manfrinato, R.V., Sanchez-Yamagishi, J.D., Kong, J., Jarillo-Herrero, P.: Anisotropic etching and nanoribbon formation in single-layer graphene. Nano Lett. **9**, 2600–2604 (2009)
89. Zhu, Z.P., Su, D.S., Weinberg, G., Schlogl, R.: Supermolecular self-assembly of graphene sheets: formation of tube-in-tube nanostructures. Nano Lett. **4**, 2255–2259 (2004)
90. Jin, W., Fukushima, T., Niki, M., Kosaka, A., Ishii, N., Aida, T.: Self-assembled graphitic nanotubes with one-handed helical arrays of a chiral amphiphilic molecular graphene. Proc. Natl. Acad. Sci. USA **102**, 10801–10806 (2005)
91. Chen, Q., Chen, T., Pan, G.-B., Yan, H.-J., Song, W.-G., Wan, L.-J., Li, Z.-T., Wang, Z.-H., Shang, B., Yuan, L.-F., Yang, J.-L.: Structural selection of graphene supramolecular assembly oriented by molecular conformation and alkyl chain. Proc. Natl. Acad. Sci. USA **105**, 16849–16854 (2008)
92. Lee, C., Wei, X., Kysar, J.W., Hone, J.: Measurement of the elastic properties and intrinsic strength of monolayer graphene. Science **321**, 385–388 (2008)
93. Bunch, J.S., Verbridge, S.S., Alden, J.S., van der Zande, A.M., Parpia, J.M., Craighead, H.G., McEuen, P.L.: Impermeable atomic membranes from graphene sheets. Nano Lett. **8**, 2458–2462 (2008)
94. Gómez-Navarro, C., Burghard, M., Kern, K.: Elastic properties of chemically derived single graphene sheets. Nano Lett. **8**, 2045–2049 (2008)
95. Viculis, L.M., Mack, J.J., Kaner, R.B.: A chemical route to carbon nanoscrolls. Science **299**, 1361 (2003)
96. Braga, S.F., Coluci, V.R., Legoas, S.B., Giro, R., Galvao, D.S., Baughman, R.H.: Structure and dynamics of carbon nanoscrolls. Nano Lett. **4**, 881–884 (2004)
97. Yu, D., Liu, F.: Synthesis of carbon nanotubes by rolling up patterned graphene nanoribbons using selective atomic adsorption. Nano Lett. **7**, 3046–3050 (2007)
98. Sidorov, A., Mudd, D., Sumanasekera, G, Ouseph, P.J., Jayanthi, C.S., Wu, S.-Y.: Electrostatic deposition of graphene in a gaseous environment: a deterministic route for synthesizing rolled graphenes? Nanotechnology **20**, 055611 (2009)
99. Martel, R., Shea, R.H., Avouris, P.: Ring formation in single-wall carbon nanotubes. J. Phys. Chem. B **103**, 7551–7556 (1999)
100. Huang, J., Juszkiewicz, M., de Jeu, W.H., Cerda, E., Emrick, T., Menon, N., Russell, T.P.: Capillary wrinkling of floating thin polymer films. Science **317**, 650–653 (2007)
101. Py, C., Reverdy, P., Doppler, L., Bico, J., Roman B., Baroud, C.N.: Capillary origami: spontaneous wrapping of a droplet with an elastic sheet. Phys. Rev. Lett. **98**, 156103 (2007)
102. Ellis, R.J., Vandervies, S.M.: Molecular chaperones. Annu. Rev. Biochem. **60**, 321–347 (1991)
103. Juniper, B.E., Robins, R.J., Joel, D.M.: The Carnivorous Plants. Academic, London (1989). ISBN 0-1239-2170-8
104. Lee, K.B., Park, S.J., Mirkin, C.A., Smith, J.C., Mrksich, M.: Protein nanoarrays generated by dip-pen nanolithography. Science **295**, 1702–1705 (2002)

105. Duwez, A.-S., Cuenot, S., Jérôme, C., Gabriel, S., Jérôme, R., Rapino, S., Zerbetto, F.: Mechanochemistry: targeted delivery of single molecules. Nat. Nanotechnol. **1** , 122–125 (2006)
106. Walther, J.H., Werder, T., Jaffe, R.L., Gonnet, P., Bergdorf, M., Zimmerli, U., Koumoutsakos, P.: Water–carbon interactions III: the influence of surface and fluid impurities. Phys. Chem. Chem. Phys. **6**, 1988–1995 (2004)

Chapter 14
Atomistic Pseudopotential Theory of Droplet Epitaxial GaAs/AlGaAs Quantum Dots

Jun-Wei Luo, Gabriel Bester, and Alex Zunger

Abstract In this chapter, following the introduction to the basic electronic properties of semiconductor quantum dots (QDs), we first briefly introduce our atomistic methodology for multi-million atom nanostructures, which is based on the empirical pseudopotential method for the solution of the single-particle problem combined with the configuration interaction (CI) scheme for the many-body problem which were developed in the solid-state theory group at the National Renewable Energy Laboratory over the past two decades. This methodology, described in Sect. 14.2, can be used to provide quantitative predictions of the electronic and optical properties of colloidal nanostructures containing thousands of atoms as well as epitaxial nanostructures containing several millions of atoms. In Sect. 14.3, we show how the multi-exciton spectra of a droplet epitaxy QD encodes nontrivial structural information that can be uncovered by atomistic many-body pseudopotential calculations. In Sect. 14.4, we investigate the vertical electric field tuning of the fine-structure splitting (FSS) in several InGaAs and GaAs QDs using our atomistic methodology. We reveal the influence of the atomic-scale structure on the exciton FSS in QDs. Finally, a comprehensive and quantitative analysis of the different mechanisms leading to HH–LH mixing in QDs is presented in Sect. 14.5. The novel quantum transmissibility of HH–LH mixing mediated by intermediate states is discovered. The design rules for optimization of the HH–LH mixing in QDs are given in this section.

J.-W. Luo (✉)
National Renewable Energy Laboratory, Golden, CO 80401, USA
e-mail: jun-wei.luo@nrel.gov

G. Bester
Max-Planck-Institut für Festkörperforschung, Heisenbergstrasse 1, 70569 Stuttgart, Germany
e-mail: G.Bester@fkf.mpg.de

A. Zunger
University of Colorado, Boulder, CO 80309, USA
e-mail: alex.Zunger@colorado.edu

Z.M. Wang (ed.), *Nanodroplets*, Lecture Notes in Nanoscale Science and Technology 18,
DOI 10.1007/978-1-4614-9472-0_14,

14.1 Introduction

Until recently, epitaxial quantum dots (QDs) were mostly made by a growth protocol ("Stranski–Krastanov", or SK) [1–3] requiring that the QD material has a significantly different lattice constant (generally larger) than the substrate on which it is grown, e.g., InAs-on-GaAs [1] or InP-on-GaP [3]. Lattice-matched material pairs such as GaAs on AlGaAs or InAs on GaSb were excluded until recently. The advent of the "droplet epitaxy" growth mode [4–6] (involving the growth of cation-element droplets on a substrate and subsequently their crystallization into QDs by incorporation of the anion element) has enabled the epitaxial growth of lattice-matched pairs, thus opening a window to the understanding of the physics of confinement in unstrained semiconductor material such as GaAs. GaAs QDs have recently also been grown using an alternative approach [7–11] where nanoholes are etched on the surface of an AlGaAs layer. The holes are etched by arsenic debt epitaxy (also referred to as local droplet etching) and filled with GaAs. Migration of the GaAs toward the bottom of the holes leads to GaAs QD formation. The QDs are subsequently capped with AlGaAs. The lattice-mismatch-induced strain in In(Ga)As/GaAs QDs represents a main difference from unstrained GaAs/AlGaAs QDs and it markedly modifies the bulk band structure. Figure 14.3 shows that the built-in biaxial strain present in InAs QDs embedded in GaAs [12, 13] lifts the degeneracy of the bulk heavy-hole (HH) and light-hole (LH) bands by as much as 0.18 eV, without considering the quantum confinement effect. The built-in shear strain also couples the HH and LH bands and it appears in the Pikus–Bir Hamiltonian as off-diagonal term [12, 13]. Furthermore, in the droplet case, GaAs represents the QD material, whereas in InAs/GaAs the barrier is GaAs and the QD is InAs. Therefore, the conduction and valence band offsets (confinement potentials) in these two types of QDs are different as shown in Fig. 14.3. Moreover, InAs and GaAs differ in bandgap, electron, and hole effective masses and the relative positions of the conduction band states at Γ, X, and L. It is thus by no means obvious that there will be a similarity in the electronic structure results of GaAs/AlGaAs with InAs/GaAs. Indeed, we find a very different electronic structure in one critical aspect: the order of hole states. In GaAs/AlGaAs the LH-derived S-like state lies between two HH-derived P-like hole states, whereas in InAs/GaAs the LH state is well below the HH-derived P-like hole states.

The symmetry reduction of low-dimensional nanostructures can lead to mixing between electronic states not only from the same bulk band [14] but also from different valleys of the Brillouin zone and different bulk bands [15–18], which are forbidden in their parent bulk semiconductors. Among various possibilities of electronic state mixing, the HH–LH mixing in semiconductor QDs has attracted much attention over the last few years for its profound effects on electronic and optical properties. Specifically, HH–LH mixing is essential to tune the exciton fine-structure splitting (FSS) of an epitaxial grown QD using a vertical electric field [19–21], since it can manipulate FSS only via acting on the bulk $|Z\rangle$ component of the Bloch functions. The QD ground hole state has dominantly bulk HH character [21], whereas the bulk HH band, $|3/2,\pm 3/2\rangle = \mp(|X\rangle \pm i|Y\rangle)|\uparrow,\downarrow\rangle/\sqrt{2}$, contains exclusively $|X\rangle$, $|Y\rangle$ components, and the bulk LH band $|3/2,\pm 1/2\rangle = (1/\sqrt{3})[(|X\rangle \pm i|Y\rangle)|\downarrow,\uparrow\rangle + \sqrt{2}|Z\rangle|\uparrow,\downarrow\rangle]$ contains $|Z\rangle$ component. Thus, mixing LH with HH leads to the control of the FSS via vertical electric field F_z. HH–LH mixing also leads to fast spin decoherence of HH-dominated QD holes [22] by introducing additional efficient spin relaxation channels belong to LH band. In addition, both experimentally and theoretically observed optical polarization anisotropy of neutral excitons (e.g., X^0 and XX^0) and charged trion (e.g., X^{-1} and X^{+1}) radiative recombination is known to arise from HH–LH mixing [23–27].

In the remainder of this chapter, we first briefly introduce our atomistic methodology for multi-million atom nanostructures, which is based on the empirical pseudopotential method [28], combined with the configuration interaction (CI) scheme for the many-body problem developed by solid-state theory group at NREL over the past two decades. This methodology, described in Sect. 14.2, can be used to provide quantitative predictions of the electronic and optical properties of colloidal nanostructures [15, 29–34] containing thousands of atoms as well as epitaxial nanostructures [17, 18, 21, 35–41] containing several millions of atoms. In Sect. 14.3, we then show how the multi-exciton spectra of an unstrained GaAs QD encodes nontrivial structural information that can be uncovered by atomistic many-body pseudopotential calculations. In Sect. 14.4, we investigate the vertical electric field tuning of the FSS in several InGaAs and GaAs QDs and reveal the influence of the atomic-scale structure on the exciton FSS in QDs. Finally, in Sect. 14.5 a comprehensive and quantitative analysis of the different mechanisms leading to HH–LH mixing in QDs is presented. We specifically highlight the discovery of the quantum transmissibility of the HH–LH mixing mediated by QD intermediate states. The design rules for optimization of the HH–LH mixing in QDs are given in this section.

14.2 Atomistic Many-Body Pseudopotential Method for Multi-million Atom Nanostructures

The basis of our methodology, which was reviewed recently by one of us in [42], is divided into four parts, atomic position relaxation, Schrödinger equation for single-particle electronic states, many-body Hamiltonian accounting for Coulomb interaction and correlation effect, and post-processors for optical properties, all feeding into each other. The calculation of the single-particle electronic states requires the input of the geometry and relaxation of the atomic positions to minimize strain. The development of empirical pseudopotentials for each atom type is the prerequisite for the construction of the total crystal potential used subsequently in the Schrödinger equation. The ensuing eigenfunctions are fed into a configuration interaction (CI) treatment to obtain excitations. Finally, from the many-body wavefunctions, observables can be obtained through the use of post-processor tools. These different components will be briefly discussed below and we refer to [28, 42–47] for more detail.

Calculation of Atomistic Strain and Atom Position Relaxation. The first step is to construct a simulation cell (supercell) containing a QD with an assumed shape, size, and composition (gradient) and place the atoms on ideal zinc-blende crystal sites. The atoms within the supercell are then allowed to relax in order to minimize the strain energy using Keating's valence force field (VFF) method [45, 46, 48], including bond stretching, bond bending, and bond bending–bond stretching interactions:

$$
\begin{aligned}
E_{\mathrm{VFF}} = & \sum_i \sum_j^{nn_i} \frac{3}{8} \left[\alpha_{ij}^{(1)} \Delta d_{ij}^2 + \alpha_{ij}^{(2)} \Delta d_{ij}^3 \right] \\
& + \sum_i \sum_{k<j}^{nn_i} \frac{3\beta_{jik}}{8 d_{ij}^0 d_{ik}^0} \left[(\mathbf{R_j} - \mathbf{R_i}) \cdot (\mathbf{R_k} - \mathbf{R_i}) - \cos\theta_{jik}^0 d_{ij}^0 d_{ik}^0 \right] \\
& + \sum_i \sum_{k<j}^{nn_i} \frac{3\sigma_{jik}}{8\sqrt{d_{ij}^0 d_{ik}^0}} \Delta d_{ij} \left[(\mathbf{R_j} - \mathbf{R_i}) \cdot (\mathbf{R_k} - \mathbf{R_i}) - \cos\theta_{jik}^0 d_{ij}^0 d_{ik}^0 \right],
\end{aligned}
\tag{14.1}
$$

where $\Delta d_{ij} = \left[(\mathbf{R_i} - \mathbf{R_j})^2 - (d_{ij}^0)^2 \right] / d_{ij}^0$, $\mathbf{R_i}$ is the coordinate of atom i, d_{ij}^0 is the ideal (unrelaxed) bond length between atoms i and j, and θ_{ijk}^0 is the ideal (unrelaxed) angle of bonds $j - i - k$. $\Sigma_i^{nn_i}$ denotes summation over the nearest neighbors of atom i ($nn_i = 4$ for diamond, zinc-blende, and wurtzite crystal structures). The bond stretching, bond-angle bending, and bond-length–bond-angle interaction coefficients $\alpha_{ij}^{(1)} (\equiv \alpha)$, β_{jik}, and σ_{jik} are directly related to the elastic constants in bulk materials [46]:

$$
\begin{aligned}
C_{11} + 2C_{12} &= \sqrt{\frac{3}{4d^0}} (3\alpha + \beta - 6\sigma) \\
C_{11} - C_{12} &= \sqrt{\frac{3}{d^0}} \beta \\
C_{44} &= \sqrt{\frac{3}{d^0}} \frac{\alpha\beta - \sigma^2}{\alpha + \beta + 2\sigma}.
\end{aligned}
\tag{14.2}
$$

The second-order bond-stretching term is included to correct the pressure dependence of Young's modulus dB/dP, where $B = (C_{11} + 2C_{12})/3$. After the atomic positions are relaxed by minimizing E_{VFF}, the local strain tensor ε at a cation site is calculated by considering a tetrahedron formed by four nearest neighboring anions [45]. The distorted (relaxed) tetrahedron edges ($\mathbf{R_{12}}$, $\mathbf{R_{23}}$, $\mathbf{R_{34}}$) are related to the ideal (unrelaxed) tetrahedron edges ($\mathbf{R_{12}^0}$, $\mathbf{R_{23}^0}$, $\mathbf{R_{34}^0}$) via the local strain tensor ε as illustrated in Fig. 14.1:

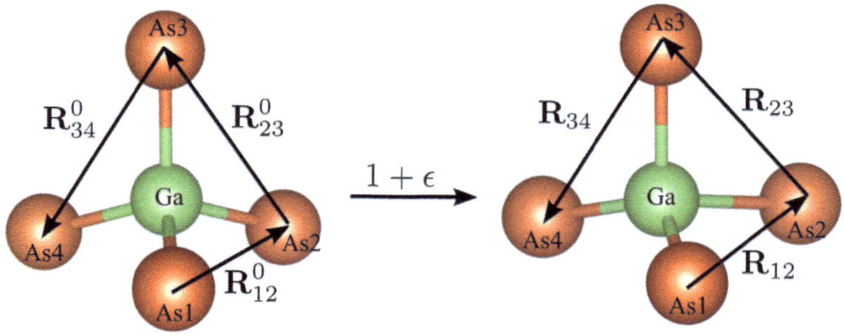

Fig. 14.1 Schematic to illustrate how the local strain is calculated in zinc blende semiconductors. For a cation Ga (or In), three vectors ($\mathbf{R}_{12}$, $\mathbf{R}_{23}$, $\mathbf{R}_{34}$) forming a distorted tetrahedron after atomic relaxation are related to the equivalent vectors ($\mathbf{R}^0_{12}$, $\mathbf{R}^0_{23}$, $\mathbf{R}^0_{34}$) of an ideal tetrahedron via the strain tensor ε

$$(\mathbf{R}_{12},\mathbf{R}_{23},\mathbf{R}_{34}) = (1+\varepsilon)\cdot(\mathbf{R}^0_{12},\mathbf{R}^0_{23},\mathbf{R}^0_{34}). \tag{14.3}$$

Solving the Schrödinger Equation for the Nanostructure. The single-particle QD electronic states are obtained from solving the empirical pseudopotential Schrödinger equation [35, 46, 47],

$$\left(-\frac{\hbar^2}{2m}\nabla^2 + V(\mathbf{r}) + |e|\mathbf{F}\cdot\mathbf{r}\right)\psi_i(\mathbf{r},\sigma) = \varepsilon_i\psi_i(\mathbf{r},\sigma), \tag{14.4}$$

within a basis of linear combination of strained Bloch bands (SLCBB) [43]. Here $\{\varepsilon_i, \psi_i(\mathbf{r},\sigma)\}$ are the eigenvalues and eigenstates of state i with spin σ. The bare electron mass is given by m and $\hbar$ is Planck's constant. An external electric field $\mathbf{F}$ is optionally applied in the supercell [49] for investigating the influence of electric field on QD electronic structure and excitons. The crystal (dot + matrix) potential $V(\mathbf{r})$ is a superposition of overlapping screened atomic (pseudo) potentials centered at the atomic positions:

$$V(\mathbf{r}) = \sum_n\sum_\alpha \hat{v}_\alpha(\mathbf{r}-\mathbf{R}_n-\mathbf{d}_\alpha), \tag{14.5}$$

where $\hat{v}_\alpha(\mathbf{r}-\mathbf{R}_n-\mathbf{d}_\alpha)$ pertains to atom-type α at site $\mathbf{d}_\alpha$ in the nth primary cell $\mathbf{R}_n$ [46, 47]. Thus, it forces upon eigenstates the correct atomically resolved symmetry. The atomic potentials $\hat{v}_\alpha$ were empirically fit to experimental transition energies, spin–orbit splittings, effective masses, deformation potentials of the bulk materials, as well as the band offsets between two materials in a heterostructure [46, 47]. Readers wishing to review the fitting of the GaAs/AlAs and InAs/GaAs pseudopotentials in detail are referred to [47] and [46], respectively. Figure 14.2 shows the calculated square of the single-particle wave functions of the four lowest electron states and the four highest hole states for both strain-free GaAs/AlGaAs and

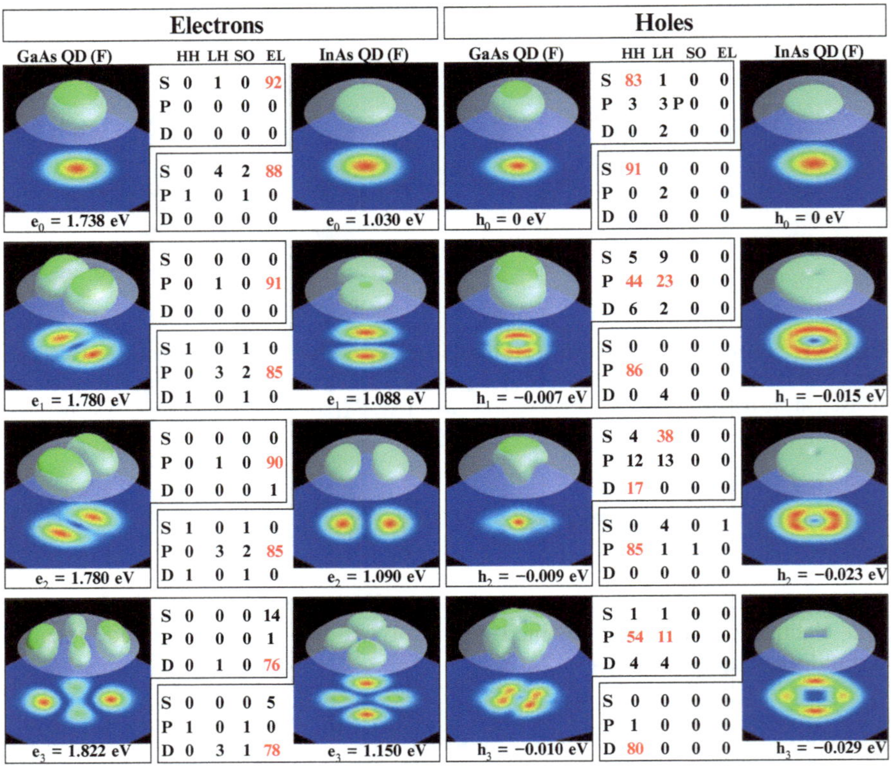

Fig. 14.2 The squared wave functions (3D isosurface and 2D in-plane mapping) of the first four electron and first four hole states for an unstrained GaAs/$Al_{0.3}Ga_{0.7}As$ QD and a strained InAs/GaAs QD, with the same lens shape and the same size (25.2 nm base and 3 nm height). For analysis purposes, we project the wave functions of the QDs on bulk heavy-hole (HH), light-hole (LH), split-off (SO) bands, and the lowest conduction band (CB). The wave functions are further decomposed with respect to their axial angular momentum components (S, P, D)

strained InAs/GaAs QDs with a lens shape. The orbital characters of each state are obtained by decomposing our atomistic electron and hole states with respect to their axial angular momentum components (S, P, D, ...). The bulk Bloch band character of the QD electronic states are gained by projecting them onto bulk Bloch bands at the Γ-point, including HH ($|3/2, \pm 3/2\rangle$), LH ($|3/2, \pm 1/2\rangle$), spin–orbit split (SO) ($|1/2, \pm 1/2\rangle$), bands as well as conduction bands (CB).

Solution of the Many-Body Problem. Once the single particle states of the QD are obtained, the excitonic energies and wave functions, including many-body interactions are calculated in the framework of the CI scheme [44]. In this approach, the excitonic wave functions $\Psi^{(i)}$ are expanded in terms of single-substitution Slater determinants $\Phi_{v,c}$, constructed by promoting an electron from the occupied single-particle state v to the unoccupied single-particle state c:

$$\Psi^{(i)} = \sum_{v=1}^{N_v} \sum_{c=1}^{N_c} C_{v,c}^{(i)} \Phi_{v,c}. \tag{14.6}$$

The coefficients $C_{v,c}^{(i)}$ of the CI expansion are calculated by diagonalizing the CI Hamiltonian for a single exciton:

$$H_{vc,v'c'} \equiv \langle \Psi_{v,c} | H_{\mathrm{CI}} | \Psi_{v',c'} \rangle = (\varepsilon_c - \varepsilon_v)\delta_{v,v'}\delta_{c,c'} - J_{vc,v'c'} + K_{vc,v'c'}, \tag{14.7}$$

where the Coulomb and exchange integrals $J_{vc,v'c'}$ and $K_{vc,v'c'}$ are, respectively, given by

$$J_{vc,v'c'} = e^2 \sum_{\sigma,\sigma'} \int\int \frac{\psi_v^*(\mathbf{r},\sigma)\psi_c^*(\mathbf{r}',\sigma')\psi_{v'}(\mathbf{r},\sigma)\psi_{c'}(\mathbf{r}',\sigma')}{\overline{\varepsilon}(\mathbf{r},\mathbf{r}')|\mathbf{r}-\mathbf{r}'|} d\mathbf{r}d\mathbf{r}' \tag{14.8}$$

$$K_{vc,v'c'} = e^2 \sum_{\sigma,\sigma'} \int\int \frac{\psi_v^*(\mathbf{r},\sigma)\psi_c^*(\mathbf{r}',\sigma')\psi_{c'}(\mathbf{r},\sigma)\psi_{v'}(\mathbf{r}',\sigma')}{\overline{\varepsilon}(\mathbf{r},\mathbf{r}')|\mathbf{r}-\mathbf{r}'|} d\mathbf{r}d\mathbf{r}'. \tag{14.9}$$

The Coulomb potential in the two equations above are screened using a position-dependent and size-dependent screening function $\overline{\varepsilon}(\mathbf{r},\mathbf{r}')$ [44]. The excitonic wavefunctions of Eq. (14.6) are built using 6 valence and 6 conduction band states, including envelope functions with S, P, and D orbital character.

Post-processor Tools. The modification of the potential due to strain can be obtained from the Pikus–Bir Hamiltonian [12] once the atoms within the supercell are relaxed using the VFF method [45, 46, 48] and the strain tensor has been calculated. A comprehensive study of the effect of strain on the band structure has been performed by Bir and Pikus [13]. Here, a simplified Pikus–Bir Hamilton is used to describe the strain-modified confinement potentials, which is, however, not used in our atomistic pseudopotential calculation of the single-particle eigenstates, but serves only as illustration of strain effects. Following [12], in which the model is written in real space, the strain-modified conduction band state is given by

$$E_c(\mathbf{r}) = E_c^0(\mathbf{r}) + a_c(\mathbf{r})Tr[\varepsilon(\mathbf{r})], \tag{14.10}$$

where $E_c^0(\mathbf{r})$ is the conduction band minimum (CBM) of bulk material at $\mathbf{r}$ and a_c is the hydrostatic deformation potential of the CBM, generally at Γ. For valence bands including spin–orbit coupling, the Pikus–Bir Hamiltonian is

$$\begin{aligned} H_v(\mathbf{r}) &= H^{\mathrm{SO}} + a_v Tr[\varepsilon(\mathbf{r})] \\ &\quad - b_v \left[\begin{pmatrix} -2 & 0 & 0 \\ 0 & 1 & 0 \\ 0 & 0 & 1 \end{pmatrix} \varepsilon_{xx} + \begin{pmatrix} 1 & 0 & 0 \\ 0 & -2 & 0 \\ 0 & 0 & 1 \end{pmatrix} \varepsilon_{yy} + \begin{pmatrix} 1 & 0 & 0 \\ 0 & 1 & 0 \\ 0 & 0 & -2 \end{pmatrix} \varepsilon_{zz} \right] \end{aligned} \tag{14.11}$$

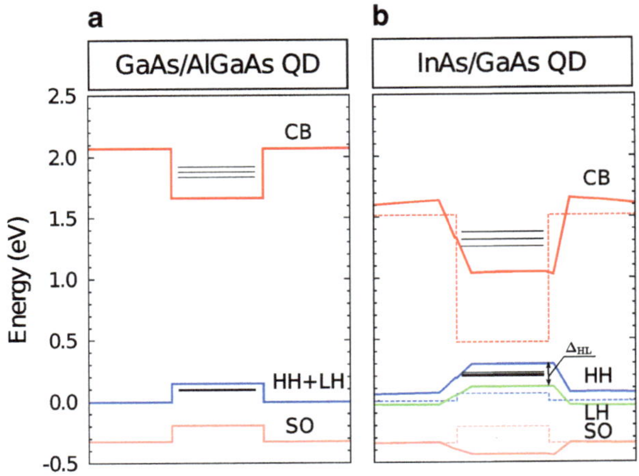

Fig. 14.3 Confinement potential of the lowest conduction band (CB), heavy-hole (HH), light-hole (LH), and split-off (SO) bands for (**a**) a strain-free GaAs/$Al_{0.3}Ga_{0.7}As$ QD and (**b**) a strained InAs/GaAs QD with lens shape of base size 30 and 3 nm height. The *dashed lines* in (**b**) represent the bulk energy levels without considering strain. Δ_{HL} is built-in strain-induced splitting of bulk HH and LH bands

$$-\sqrt{3}d_v\left[\begin{pmatrix}0 & -1 & 0\\ -1 & 0 & 0\\ 0 & 0 & 0\end{pmatrix}\varepsilon_{xy}+\begin{pmatrix}0 & 0 & 0\\ 0 & 0 & -1\\ 0 & -1 & 0\end{pmatrix}\varepsilon_{yz}+\begin{pmatrix}0 & 0 & -1\\ 0 & 0 & 0\\ -1 & 0 & 0\end{pmatrix}\varepsilon_{zx}\right],$$

where H^{SO} is the spin–orbit Hamiltonian [12], a_v is the hydrostatic deformation potential of the VBM, b_v is the biaxial deformation of the valence band maximum (VBM), and d_v is the deformation potential due to shear strain. The value of deformation potentials a_c, a_v, b_v, and d_v are taken from [12, 50]. The calculated strain-modified confinement potentials of a strained InAs/GaAs and an unstrained GaAs/AlGaAs QD using the Pikus–Bir model, as well as the natural band offsets at their equilibrium lattice constants are shown in Fig. 14.3.

After we have calculated the many-body wavefunctions, we have access to observables through the use of post-processor tools. For example, the excitonic optical-absorption spectrum $I(E)$ are calculated with the CI eigenstates of Eq. (14.6) by using Fermi's golden rule [51]:

$$I(E)=\sum_{v}|M_v|^2\exp\left[-\left(\frac{E-E_v}{\sigma}\right)^2\right], \tag{14.12}$$

where $M_v=\sum_{h_i,e_j}C^{(v)}(h_i,e_j)\langle\psi_{h_i}|\hat{p}|\psi_{e_j}\rangle$ is the transition dipole matrix between hole state h_i and electron state e_j, E_v is the exciton energy and the broadening of spectral lines modeled by a chosen σ.

14.3 Geometry of Epitaxial GaAs/(Al,Ga)As QDs as Seen by Excitonic Spectroscopy

Molecular spectroscopy has always been intimately connected with molecular structure and symmetry through fundamental interpretative constructs such as symmetry-mandated selection rules, level degeneracies, and polarization [52]. Yet, the spectroscopy of epitaxial semiconductor QDs—large simple molecules made of 10^3–10^6 atoms such as Si, InAs, or GaAs—has been largely conducted and interpreted without basic knowledge of the underlying structure. Indeed, the extremely rich (10–20 lines), high-resolution ($\sim$10 µeV) single-dot excitonic spectra of such simple "macromolecules" being now measured almost routinely [1, 2, 53–56] has not been accompanied by detailed structural information, other than cross-sectional scanning tunneling microscopy (XSTM) measurements [57, 58] which, however, can produce a range of diverging structures from the same measured relaxation profile on the same QD [39]. A possibly more accurate structure profile of epitaxy QDs can be indirectly obtained from a full three-dimensional electron density map measured by a coherent Bragg rod analysis (COBRA) method [59]. Attempts to bridge the gap between spectra and structure have recently been made in the context of self-assembled (strained) In(Ga)As/GaAs QDs by combining measured excitonic spectra with XSTM structural assessment of the same QD sample, using a quantitative excitonic theory as the bridge. It was found [39] that the calculated excitonic spectra produced by using as input a range of structural models offered by XSTM conflicted with the experimental spectra in a number of crucial aspects. However, a structure derived theoretically by matching the calculated spectra with experiment did agree with the basic data used to derive XSTM structural models (i.e., the measured outer relaxation profile of the cleaved QD). It was concluded that high-resolution excitonic spectra contain significant structural information that can be unearthed using theory as a mining tool.

Recent XSTM measurements [57] suggest that droplet GaAs QDs have Gaussian shape instead of the lens shape often deduced from atomic force microscope (AFM) measurements [56, 60–62] and QD heights of around 14 nm [57, 62]. The exciton band gap measured by optical spectroscopy is about 1.7–1.9 eV [25, 56, 60, 63, 64]. In [41] we discussed the spectra vs. structure link for such QDs. We found that the GaAs QDs grown by droplet epitaxy have indeed a Gaussian-shape, as suggested by the XSTM measurements [57]. However, we found that QDs as seen by optical spectroscopy correspond to QDs with 2–4 nm height rather than the 14 nm determined by XSTM. The fact that XSTM sees tall QDs and spectroscopy sees flat QDs points to the fact that different QDs must have been probed. This was uncovered by theoretical simulations showing that the two experiments could not possibly correspond to the same QD.

Measured Structure. GaAs/GaAlAs QDs grown by droplet epitaxy in Sakoda's group [63] were initially described, on the basis of AFM measurements of uncapped QDs [62], as being lens shaped [56, 60–62] (schematic in left inset of Fig. 14.4),

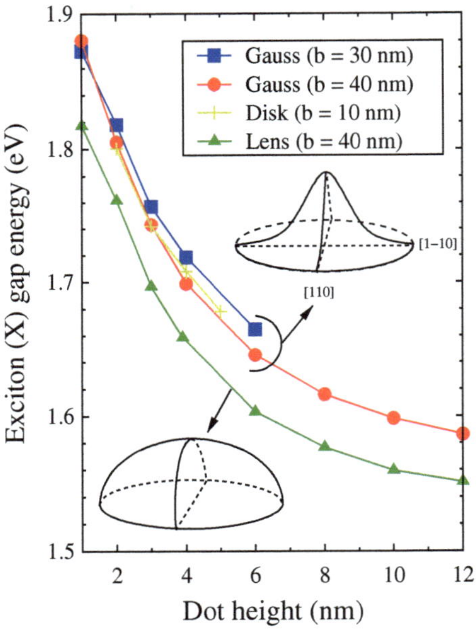

Fig. 14.4 Atomistic many-body pseudopotential calculated exciton emission energy of Gaussian-shaped, lens-shaped, and disk-shaped GaAs/ $Al_{0.3}Ga_{0.7}As$ QDs (base size in diameter given in *parentheses*) as a function of QD height. The base diameter of Gaussian-shaped QD is defined by the largest QD lateral size, which is truncated by the requirement that the QD material must be thicker than 1 ML. Taken from [41]. ©(2011) by the American Physical Society

with averaged $[1\bar{1}0]$-elongated base size of 70×50 nm (spread $\pm 10\%$) and a QD height of 14 nm (spread $\pm 19\%$) [62]. Subsequently, XSTM measurement of these capped QDs were performed by Keizer et al. [57] showing instead a rather different, Gaussian shape (schematic in right inset to Fig. 14.4) with an average base size of 40 nm, height of 14 nm, and a size distribution of 10–20%.

Spectra of Single Exciton. The measured spectroscopy [25, 56, 60, 63, 64] of the QDs grown by Sakoda's group [63] shows that the fundamental exciton emission from many different spectroscopy measurements were in a range of 1.7–1.9 eV.

Calculated Spectra for the Measured Structure Lead to Conflicts with the Assumed Structure. We have calculated the exciton gap energy of lens-shaped, Gaussian-shaped, and disk-shaped strain-free GaAs/AlGaAs QDs using our atomistic many-body pseudopotential method (Fig. 14.4). Notwithstanding the shape, the QDs with calculated exciton energy in the range of the experimental measured exciton energy of 1.7–1.9 eV have a much smaller QD height, of only 1–4 nm compared to the experimentally stated value ($\sim$14 nm) by both AFM [62] and XSTM [57] approaches. This discrepancy, being well outside the measured size distribution in the sample, indicates that the QDs measured by AFM or XSTM are not the same as the QDs seen by optical spectroscopy. We conclude[1] that the QD, the height of which was measured to be 14 nm in [57, 62], is not the same QD that was used in [25, 56, 60, 63, 64] to measure the band gap and exciton fine structure. It is worth mentioning that the QD height decreases from 14 nm when the QDs are grown by droplet epitaxy on a (001)-oriented GaAs substrate [57] to much smaller value of 2.3 ± 0.6 nm when they are grown on a (311)A-oriented GaAs substrate [65]. However, the XSTM [57] and spectroscopy [25, 56, 56, 60, 63] measurements considered here, as well as theory, are all on (001) substrates.

Whereas to first order, the magnitude of the excitonic emission energy reveals information mostly on the QD height, a more detailed measurement can also distinguish different QD shapes. We see from Fig. 14.4 that for the same base size and QD height, the lens-shape QDs have an exciton gap energy that is smaller by as much as ~40 meV than that of Gaussian-shaped QDs and that this is so in a wide range of QD heights of 1–12 nm. If droplet epitaxy grown GaAs/AlGaAs QDs are known to be either lens shaped or Gaussian shaped, this exciton shift is large enough to distinguish the QD shape if the base size, QD height, and exciton energy are accurately measured.

The Shape of the QD as Seen by the Sequence of Multi-exciton Lines. Experimentally, different types of excitons can be created: neutral monoexciton X^0 (1e,1h) or neutral biexciton XX^0 (2e,2h), as well as charged excitons such as positive trion X^+ (1e,2h), negative trion X^- (2e,1h), positive biexciton XX^+ (2e,3h), and negative biexciton XX^- (3e,2h), etc. Figure 14.5 shows the calculated emission spectrum when a single electron–hole pair recombines within such a multi-exciton complex [66]. The spectra consist of a few lines. Specially for XX^+ and XX^- we see several manifolds of four and two lines, respectively, due to various S and P recombination channels and e–h exchange interaction induced FSS of multi-exciton complexes. The emission energy reflects both direct Coulomb interactions and correlation effects between holes and electrons; these interactions ultimately reflect the overlap of the corresponding wave functions which is sensitive to the shape and size of the QD. Such complex and implicit dependences between the sequence of multi-exciton lines ("multi-exciton barcode") and QD structure were used recently to decipher structural features from excitonic features. It was proposed [39] that such barcodes consisting of X, X^+, X^-, XX, XX^+, XX^-, and X^{-2} lines can be correlated with geometrical features of the strained SK-grown InAs/GaAs QDs.

Here we will use this barcoding approach to unearth structural features of another class of QDs based on unstrained, droplet epitaxy grown GaAs/GaAlAs. For this purpose we have calculated the sequence of multi-excitonic lines for a large number of QDs with three different basic shapes (lens shape, Gaussian shape, and disk shape) and many structural parameters within these shapes (height, base size, and shape anisotropy). Using this barcoding method, we can build a link between the structure of strain-free GaAs/AlGaAs QDs and their excitonic emission spectra.

---

[1] We are grateful to Prof. P.M. Koenraad and Dr. M. Takaaki for clarifying to us now that the QDs used in XSTM were different than those used for PL measurements.

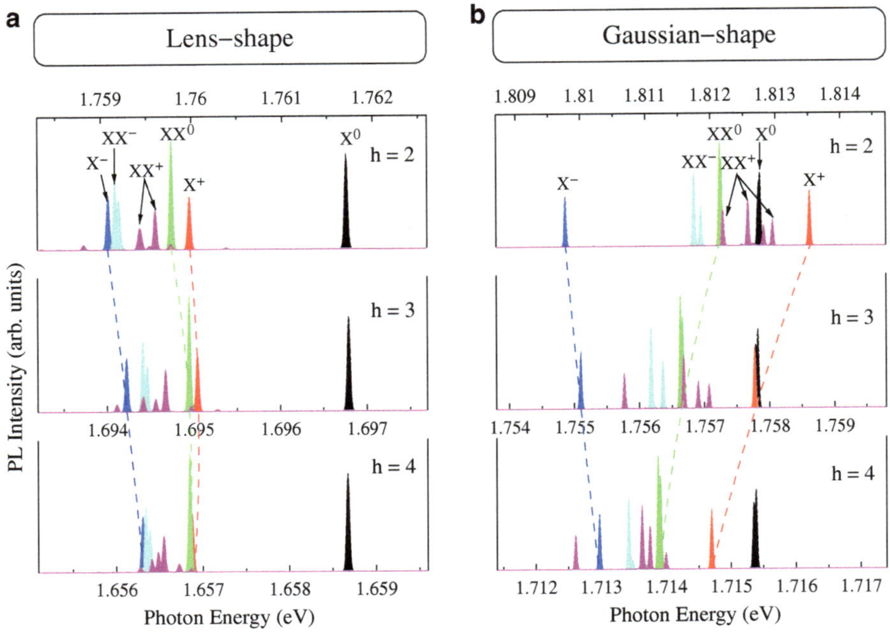

Fig. 14.5 Excitonic emission spectrum of (**a**) symmetric lens-shaped and (**b**) symmetric Gaussian-shaped GaAs/$Al_{0.3}Ga_{0.7}$As QDs with base size of $b = 40 \times 40$ nm and QD height $h = 2$, 3, and 4 nm. Taken from [41]. ©(2011) by the American Physical Society

However, at present, the available experimentally measured excitonic emission spectrum of such QDs includes only neutral monoexciton X^0, positive and negative trions X^+ and X^-, and neutral biexciton XX^0 transitions [25, 56, 56, 60, 63]. Figure 14.5 shows the atomistic calculated emission spectra (where we have aligned the energy of the monoexciton X^0 lines) for lens-shaped (Fig. 14.5a) and Gaussian-shaped (Fig. 14.5b) QDs. In this partial excitonic emission spectrum, we find that the sequence of the following lines always obeys some "hard rules" [39],

$$X^- < XX^- < XX^0 < X^0. \tag{14.13}$$

The hard rules observed in all experimental spectra [25, 56, 60, 63, 64] are that (i) both X^- and XX^0 are red shifted with respect to X^0 (i.e., have positive binding energies) and (ii) the XX line always lies between X and X^-. Hard rule (iii) [39], related to X^{-2}, has not been measured yet for GaAs/AlGaAs QDs. These three hard rules plus the position of the X^0 line will provide the size of the base and the height of the QD [39]. From hard rules (i) and (ii) we estimate that the QDs seen in the optical spectroscopy measurement has a base diameter of 30–40 nm.

Interestingly, we find that the energetic relative position of the positive trion (X^+) is related to the QD shape. Figure 14.5a shows that in lens-shaped QDs the positive trion (X^+) is always redshifted with respect to neutral monoexciton

(X^0). In contrast to lens-shaped QDs, in Gaussian-shaped QDs (Fig. 14.5b) the X^+ has a transition from redshift to blueshift, when the QD height decreases, in agreement with experimental measurements [56]. Furthermore, our calculated transition energy $E_{X^0} = 1.758\,\text{eV}$ also agrees with the experimental value of 1.748 eV [56]. Thus, we conclude that Gaussian shape is more likely in droplet epitaxy-grown GaAs/AlGaAs QDs.

FSS of Mono-exciton vs. QD Shape. The FSS of an exciton [31, 66–68] refers to the splitting of the optical-allowed (bright) exciton states due to both intrinsic crystal asymmetry as well as external shape anisotropy. The role of these two factors has been often misconceived in the literature [56, 60, 63, 69], leading to the misuse of the FSS to infer shape anisotropy: In the Luttinger Hamiltonian representation, the effective mass of the hole is anisotropic in that its value along (100) is different from that along (110). Thus, if one ignores the fact that the QDs under consideration are made of an atomistically discrete material, the symmetry of circular-based QD in this Hamiltonian is C_{4v}. Despite this, numerous papers [25,70] claimed that circular-based lens-shape QDs have D_{2d} symmetry. This is because in a continuum approximation the [110] and [$1\bar{1}0$] directions are equivalent. In such a D_{2d} symmetry, the fourfold degenerate exciton (originating from an electron of $J_z = \pm 1/2$ and a heavy hole of $J_z = \pm 3/2$) splits into doubly degenerate bright state (Γ_5) and two nondegenerate dark states (Γ_1 and Γ_2, respectively). Because Γ_5 is degenerate in this approximation, the FSS is zero for cylindrically symmetric QDs under the continuum point of view. To account for the observed nonzero FSS, the continuum theory assumes that the FSS originates, in its entirety, from deviations from cylindrical symmetry [56, 60, 63, 69]. This shape anisotropy (e.g., elongation in the [$1\bar{1}0$] direction [56, 60, 63]) of the QD lowers the D_{2d} symmetry to C_{2v}. The doubly degenerate bright Γ_5 states splits into two nondegenerate states (Γ_2 and Γ_4 symmetry). The lifting of the degeneracy of the two bright exciton states is referred to as FSS and is used under the continuum point of view to fit the measured FSS into a geometric shape anisotropy. If the base center of the QD does not anchor on a common atom (namely, As atom in GaAs/AlGaAs), then the symmetry of the circular-based QD is C_1, rather than C_{2v}. In the C_1 point group, the two bright exciton states belong to the same irreducible representation and they will couple, if their energy are close enough (namely FSS $< 5\,\mu\text{eV}$ [68]). For QDs with FSS $> 5\,\mu\text{eV}$, the FSS is not sensitive anymore to the choice of the QD base center.

In reality, the [110] and [$1\bar{1}0$] directions are nonequivalent for QDs with zinc-blende crystal lattice. This leads to the fact that a QD with cylindrical shape does not have the commonly thought D_{2d} symmetry, but already has the lower C_{2v} symmetry. Thus, even a shape-symmetric QD has nonzero FSS. Although, this intrinsic crystal anisotropy was pointed out many times in atomistic theories[35, 45, 67], its contribution to the FSS has often been neglected by the community [56, 60, 63, 69]. Figure 14.6 shows the calculated atomistic many-body pseudopotential FSS for symmetric and asymmetric Gaussian-shaped QDs as well as symmetric lens-shaped QDs. In agreement with the atomistic point of view, we see that even the shape-symmetric Gaussian-shaped QD with base size of 30 nm has already a large FSS

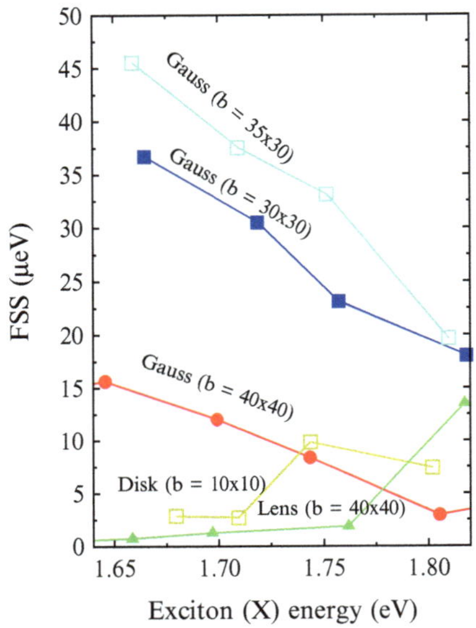

Fig. 14.6 FSS of neutral mono-excitons of GaAs/AlGaAs QDs as a function of exciton emission energy (QD height) for different shapes (symmetric lens shape, symmetric disk shape, and symmetric and asymmetric Gaussian shape) and base sizes (given in *parentheses* in nm). Taken from [41]. ©(2011) by the American Physical Society

(~30 μeV for QDs having an exciton energy of 1.7 eV) and that shape asymmetry additionally adds some (~10 μeV) FSS. Whereas, the increase of base size for the shape-symmetric QDs from 30 to 40 nm reduces the FSS by ~20 μeV. Thus, attributing all of the FSS to shape asymmetry will greatly exaggerate the shape asymmetry.

It is most interesting to note that the slope of the size-dependent FSS, for both symmetric and asymmetric Gaussian-shaped QDs, is opposite to the one seen in lens-shaped and in disk-shaped QDs. Specifically, the FSS of the Gaussian-shaped QDs decreases with increasing exciton emission energy (i.e., decreasing the QD height) in strong contrast to the case of lens-shaped and disk-shaped QDs, where the FSS increases with increasing exciton emission energy. We ascribe these two opposite size-dependent trends of the FSS to two competing effects: (i) FSSs will be enhanced by quantum confinement effect due to increased overlap of electron and hole wave functions; (ii) FSSs will be washed out by random AlGaAs alloy distribution due to an increased wave function leakage with decreasing QD height. Because the wave functions are expected to be more localized inside the QD (in the in-plane direction) in lens-shaped and disk-shaped QDs than in Gaussian-shaped QDs, the item (i) is dominant in lens-shaped and disk-shaped QDs. However, in Gaussian-shaped QDs, item (i) and (ii) are comparable. These factors explain the observed opposite trends. The calculated size-dependent trend of the FSS in Gaussian-shaped QDs is in excellent agreement with experimental measurements [60]. Thus, from the size-dependent trend of the FSS we suggest that droplet epitaxy-grown GaAs/AlGaAs QDs have a Gaussian shape.

In summary, we showed how the multi-exciton spectra of a droplet epitaxy QD encodes nontrivial structural information that can be uncovered by atomistic many-body pseudopotential calculations. We calculated single-particle energy levels, exciton gap, optical emission spectra, and FSSs for a large number of strain-free GaAs/AlGaAs QDs with three different shapes and different structure parameters (base size, height, and shape anisotropy). From such multi-exciton complex emission spectrum ("barcode") and from the size-dependent trends of the FSS, we showed that the droplet epitaxy strain-free GaAs/AlGaAs QDs have a Gaussian shape, in agreement with XSTM measurements, but the previously determined QD height (~14 nm) as "seen" by XSTM [57, 62] could not possibly be consistent with the excitonic signature (1.7–1.9 eV) [25, 56, 56, 60, 63], as the latter must reflect a 1–4 nm tall QD. Both, spectroscopy and XSTM measurements, were done on GaAs QDs capped by an $Al_{0.3}Ga_{0.7}As$ barrier layer. The fact that XSTM sees tall QDs and spectroscopy sees flat QDs suggests that different QDs must have been investigated. Indeed, Koenraad and Takaaki clarified for us that the droplet QDs used in XSTM were different than those used for PL measurements. This approach therefore holds the promise that, with increasing spectral resolution and more multi-excitonic barcode lines, a detailed structural information could be revealed.

14.4 Influence of the Atomic-Scale Structure on the Exciton Fine-Structure Splitting in of QDs in a Vertical Electric Field

One of the leading proposals for the *on-demand* generation of polarization-entangled photons is the utilization of the cascade decay of biexciton–exciton–ground state [71] in semiconductor QDs [72], as illustrated schematically in Fig. 14.7a. A serious impediment to the success of this proposal is the existence of the FSS discussed previously, which must be suppressed below the radiative linewidth (≈1 μeV). The FSS is affected by the atomistic symmetry of the QD confining potentials [35, 73–77] and can be manipulated by strain [78, 79], lateral electric fields [80], vertical electric fields [19, 20, 81], magnetic field [82], and strong coherent lasers [83, 84]. A number of surprising puzzles surround the tuning of the FSS by a vertical electric field. First, it is predicted theoretically [68], and confirmed experimentally [19, 79] and theoretically [85], that for QDs made of random alloys (with symmetry lower than C_{2v}) the two bright components of the excitons undergo an *anticrossing* as a function of field applied along the {100} or {110} directions [68]. Second, since it has been established that the FSS is related to the atomistic in-plane asymmetry between the [110] and [1$\bar{1}$0] crystallographic directions, it would appear that such an intrinsic quantity would not lend itself to tuning via *vertical* field. Nevertheless, it was shown experimentally that the FSS can be tuned rather effectively in In(Ga)As/GaAs QDs by applying an electric field along the growth direction [19]. Third, the role of strain is unclear: while electric field control

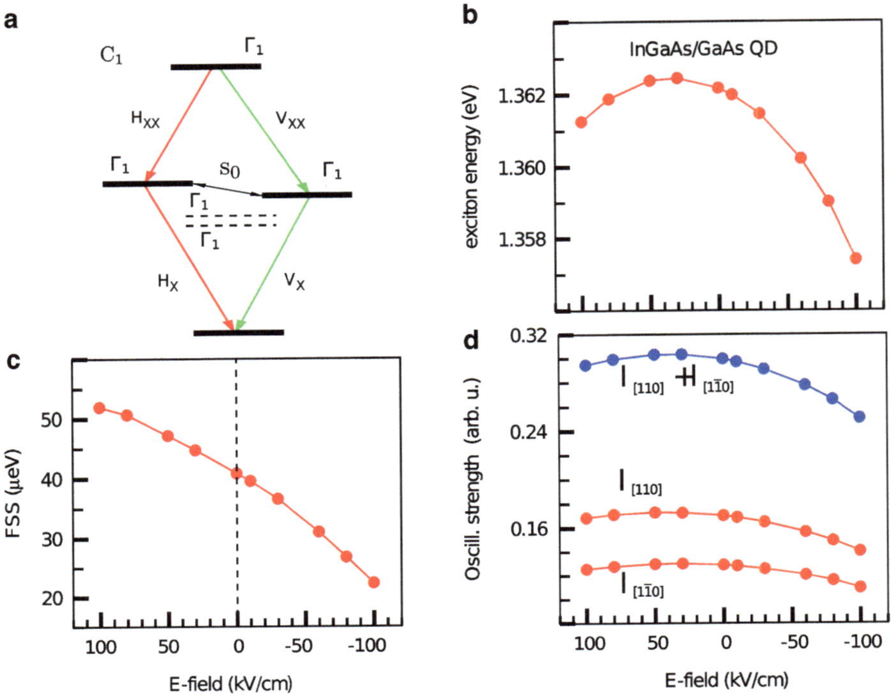

Fig. 14.7 (**a**) Schematic representation of the biexciton → exciton → ground-state cascade. (**b**) Exciton energy, (**c**) FSS, and (**d**) oscillator strength of the bright exciton transitions along the [110] and [1$\bar{1}$0] directions as a function of applied electric field for the strained InGaAs/GaAs QD00 (see Table 14.1). Taken from [21]. ©(2012) by the American Physical Society

was observed in strain-free monolayer thickness fluctuation GaAs QDs [20, 81], investigations of this effect for other strain-free GaAs QDs grown by multistep hierarchical self-assembly [11] or droplet epitaxy [86] have not yet been reported. Finally, even though strain-free GaAs/AlGaAs QDs naturally lack the built-in strain asymmetry that was believed to contribute to FSS in strained InAs/GaAs QDs [36], still significant FSS can exist, casting doubt on our understanding of the role of strain in creating FSS-promoting asymmetries in the potential.

In [21], we clarified the physical process underlying the tuning of the FSS by vertical electric fields by developing a simple mesoscopic model that allows us to analyze our million atom calculations of a large set of InGaAs/GaAs and GaAs/AlGaAs QDs. We found good agreement for InGaAs QDs with existing experiments [19] predicted the FSS in strain-free GaAs/AlGaAs QDs to be tunable well below the radiative linewidth ($\approx$1 μeV). However, we showed that different decorations of the cation lattice in the AlGaAs alloy barrier lead to fluctuations in the minimum FSS in the range of $\pm$1.4 μeV, making a postselection of appropriate QDs necessary. We showed that measurements of the FSS and the polarization angle at zero field suffice to identify appropriate QDs.

Table 14.1 Sizes and compositions of different QDs investigated in this section

QD	Composition	Size (nm) a, b, h	Barrier (% Al) Top	Barrier (% Al) Bottom
Lens shape				
00	$In_{0.8}Ga_{0.2}As$	10, 7.5, 2.5	0	0
01	GaAs	45, 45, 3	35	35
02	GaAs	70, 50, 3	45	45
03	GaAs	70, 50, 3	35	45
04	GaAs	60, 40, 2	35	45
05	GaAs	25, 31, 3.9	35	35
Gaussian shape				
06	GaAs	30, 30, 3	30	30
07	GaAs	30, 30, 4	30	30
08	GaAs	30, 30, 6	30	30
09	GaAs	35, 30, 3	30	30
10	GaAs	35, 30, 4	30	30
11	GaAs	35, 30, 6	30	30
12	$Al_{0.06}Ga_{0.94}As$	30, 30, 3	30	30
13	$Al_{0.06}Ga_{0.94}As$	30, 30, 6	30	30
14	$Al_{0.06}Ga_{0.94}As$	35, 30, 3	30	30
15	$Al_{0.06}Ga_{0.94}As$	35, 30, 6	30	30

The sizes a, b, and h describe the elliptic axis along the [110] and [$1\bar{1}0$] directions and the height, respectively. Taken from [21]. ©(2012) by the American Physical Society

We consider lens-shaped and Gaussian-shaped QDs with the properties (composition and geometry) given in Table 14.1. The atom positions are relaxed using the VFF method [46] and the single particle states are calculated using the atomistic empirical pseudopotential approach [43,46] as outlined in Sect. 14.2. We apply an external electric field following [49]. The direct and exchange Coulomb interactions are calculated from the atomic wave functions, and the correlated excitonic states are calculated by the CI approach as shown in Sect. 14.2 using 12 electron and 12 hole states (spin included), thus accounting for correlations.

Before we present our numerical results, we introduce a mesoscopic simple model where the Hamiltonian is split into different components:

$$H = H_{C2v} + \delta H_{C1} + q_s F z, \tag{14.14}$$

where q_s is the charge of a particle in band s, i.e., $-e(+e)$ for conduction (valence) bands, H_{C2v} is the Hamiltonian of the QD with C_{2v} point group symmetry, which must be supplemented by the deviation from this symmetry by the term δH_{C1}. This latter term represents the random alloy present in the barrier and possible impurities inside the GaAs QD, as well as shape asymmetries. In the space of the two bright states $|1\rangle$ and $|2\rangle$ the Hamiltonian has a simple form:

$$H = \begin{pmatrix} E_1 + \delta E_1 + \gamma_1 F & s_0/2 \\ s_0/2 & E_2 + \delta E_2 + \gamma_2 F \end{pmatrix}. \tag{14.15}$$

The exciton energies of the high symmetric hypothetical structure given by $E_1 = \langle 1|H_{C2v}|1\rangle$ and $E_2 = \langle 2|H_{C2v}|2\rangle$ are different due mainly to strain [36] (nearly vanishing in the case of strain free GaAs QDs). The lowering of the symmetry to C_1 leads to the terms $\delta E_1 = \langle 1|\delta H_{C1}|1\rangle$ and $\delta E_2 = \langle 2|\delta H_{C1}|2\rangle$ and also to $s_0/2 = \langle 1|\delta H_{C1}|2\rangle$ and $\gamma_i = \langle i|q_s z|i\rangle$. Redefining $E_1 + \delta E_1$ as E_0 and $\delta = E_2 - E_1 + \delta E_2 - \delta E_1$ and removing the linear term in the field from $|1\rangle$ leads to

$$H = \begin{pmatrix} E_0 & s_0/2 \\ s_0/2 & E_0 + \delta + (\gamma_2 - \gamma_1)F \end{pmatrix}, \tag{14.16}$$

which corresponds to the anticrossing model used by Bennett et al.[19]:

$$E \begin{pmatrix} \cos\theta \\ \sin\theta \end{pmatrix} = \begin{pmatrix} E_0 & s_0/2 \\ s_0/2 & E_0 - \gamma(F - F_0) \end{pmatrix} \begin{pmatrix} \cos\theta \\ \sin\theta \end{pmatrix}. \tag{14.17}$$

We identify $\gamma = \gamma_1 - \gamma_2$ and $\gamma F_0 = \delta$ from Eqs. (14.15) and (14.17). This simple reformulation clarifies the origin of the terms:

- γ represents the difference in the response of $|1\rangle$ and $|2\rangle$ to the applied field and γF_0 the intrinsic FSS due to the inequivalence of [110] and [1$\bar{1}$0] in C_{2v} (small for a strain-free structure) and the lowering to C_1 symmetry through atomistic alloy effects.
- s_0 is the FSS at the anticrossing and quantifies the coupling between the bright states. In a pure GaAs QD embedded in a pure AlAs matrix, the bright states are expected to cross [68] due to the high C_{2v} symmetry of the structure and $s_0 = 0$. However, the reduction of the QD symmetry due to the alloy fluctuations in the AlGaAs barrier at the QD interface leads to an avoided crossing [68] with $s_0/2 \neq 0$.
- F_0 is the field at the anticrossing. As the field approaches F_0, the exciton eigenstates become a coherent mixture with components $\sin\theta$ and $\cos\theta$, where θ is the angle describing the orientation of the lowest eigenstate relative to the [110] crystal axis.

The solution of Eq. (14.17) yields the eigenvalues ($E_\pm$) and angles [19]:

$$E_\pm = E_0 - \frac{\gamma(F - F_0)}{2} \pm \frac{1}{2}\sqrt{\gamma^2(F - F_0)^2 + s_0^2} \tag{14.18}$$

$$\theta = \pm\tan^{-1}\left[\frac{s_0}{\gamma(F - F_0) \pm (E_- - E_+)}\right]. \tag{14.19}$$

We note at this point that the model of Eq. (14.15) does not include any field dependence of the off-diagonal terms. Such terms would lead to an additional

Table 14.2 Transition energy E_0 and FSS parameters defined in Eq. (14.17) and extracted from our numerical results

QD	E_0 (meV)	s_0 (μeV)	γ (μeV cm/kV)	F_0 (kV/cm)
00	1363	?	0.15	+ 273
01	1644	0.1	0.11	+17
02	1650	0.1	0.08	−48
03	1643	0.1	0.08	−48
04	1742	0.9	0.14	−43
05	1679	0.3	0.33	+29
06	1762 ± 2	0.8 ± 0.3	0.85 ± 0.08	−21 ± 5
07	1718 ± 2	0.4 ± 0.1	0.95 ± 0.06	−26 ± 3
08	1666 ± 1	0.9 ± 0.8	1.06 ± 0.07	−25 ± 2
09	1754	0.9	0.79	−33.5
10	1714	0.4	0.78	−37.4
11	1660	0.7	0.96	−40.5
12	1806 ± 5	1.2 ± 0.7	0.74 ± 0.11	−14 ± 7
13	1727 ± 2	1.2 ± 0.5	0.85 ± 0.09	−15 ± 9
14	1799 ± 2	1.3 ± 1.0	0.73 ± 0.03	−25 ± 6
15	1721 ± 2	1.8 ± 1.4	0.84 ± 0.07	−40 ± 5

The error bars represent the range of parameters we obtain by running five different random alloy realizations (see Fig. 14.9). Taken from [21]. ©(2012) by the American Physical Society

coupling of the two bright states and could be used to tune the FSS through zero (if it would exactly compensate $s_0/2$). In our case of vertical field, this coupling is negligible, but in the case of a field with a component along a low symmetry direction (any direction but [110] or [1$\bar{1}$0]) this term should exist. A future investigation of this effect would be worthwhile.

We first present our results for the strained $In_{0.8}Ga_{0.2}As$ QD00 (see Table 14.1) an emission energy that fits the measured results of Bennett et al. [19] very well. Figure 14.7 shows the Stark shift, FSS, and the oscillator strength as a function of vertical electric field.

We obtain a nearly linear change in the FSS with the E field in agreement with the experimental results [19]. A fit of our numerical results to the model of Eq. (14.17) yields the parameters given in Table 14.2. For the field dependence of the FSS, γ, we obtain a value of 0.15 μeV cm/kV, somewhat lower than the value of 0.28 μeV cm/kV reported by Bennett et al. [19]. The strong shape and size dependence of the slope can explain this discrepancy and will be illustrated below. Our results for the set of strain-free GaAs QDs given in Table 14.1 are shown in Figs. 14.8 and 14.9, where we plotted the Stark shift, the FSSs, the polarization angle θ, and the oscillator strength as a function of the vertical E field. The results of the fit to the model of Eq. (14.17) are given in Table 14.2. We make the following observations.

FSS and Polarization Angle. The anticrossing described by Eq. (14.17) can be seen in Figs. 14.8c and 14.9 as a reduction of the FSS until the value s_0, followed by an increase. The anticrossing is accompanied by a rotation of the polarization angle of

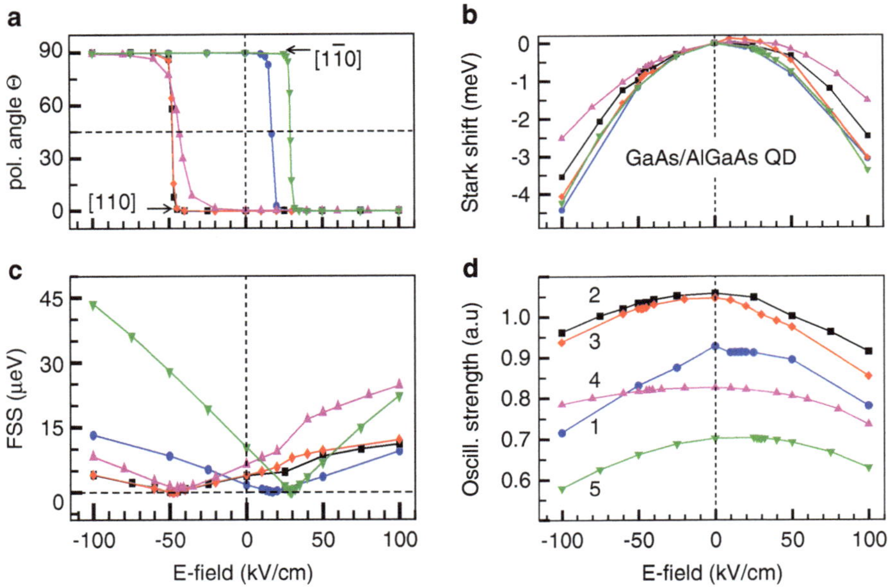

Fig. 14.8 Results for GaAs/AlGaAs strain-free QDs. (**a**) Polarization angle θ, (**b**) stark shift, (**c**) FSS, and (**d**) sum of intensities along the [110] and the [1$\bar{1}$0] crystal directions as a function of applied vertical E field. The *circles*, *squares*, *diamonds*, *up-triangles*, and *down-triangles* are for the QD01, QD02, QD03, QD04, and QD05, respectively. Taken from [21]. ©(2012) by the American Physical Society

the lowest energy exciton state [68], as shown in Fig. 14.8a. At the field F_0, where the anticrossing occurs, the polarization angle changes more rapidly when s_0 is small in agreement with the model.

Shape and Size Effects on the Tunability γ*.* Table 14.2 reveals that γ increases with the height of the QDs: tall QDs are more tunable in the vertical electric field, which correlated with the polarizability of the exciton states. Comparing QD05 and QD07 with similar dimensions but different shapes shows that Gaussian-shaped QDs have a larger γ value.

Shape and Size Effects on s_0*.* From Table 14.2 we conclude that the shape effect on s_0 is rather moderate, while the size effect shows a trend for larger values of s_0 in larger QDs. This latter trend is, however, overshadowed by a very strong random alloy effect (see next).

Random Alloy Effects on s_0 *and* F_0*.* In Fig. 14.9a–c we generated the same QD structure with different random realizations of the barrier material. In Fig. 14.9e–g the QDs have a 6% Al content and these Al atoms are randomly distributed in five different realizations within the QDs. These variations represent fluctuations that should be encountered experimentally. We notice that both s_0 and F_0 are significantly affected by these atomistic effects. For instance, the pure GaAs QD

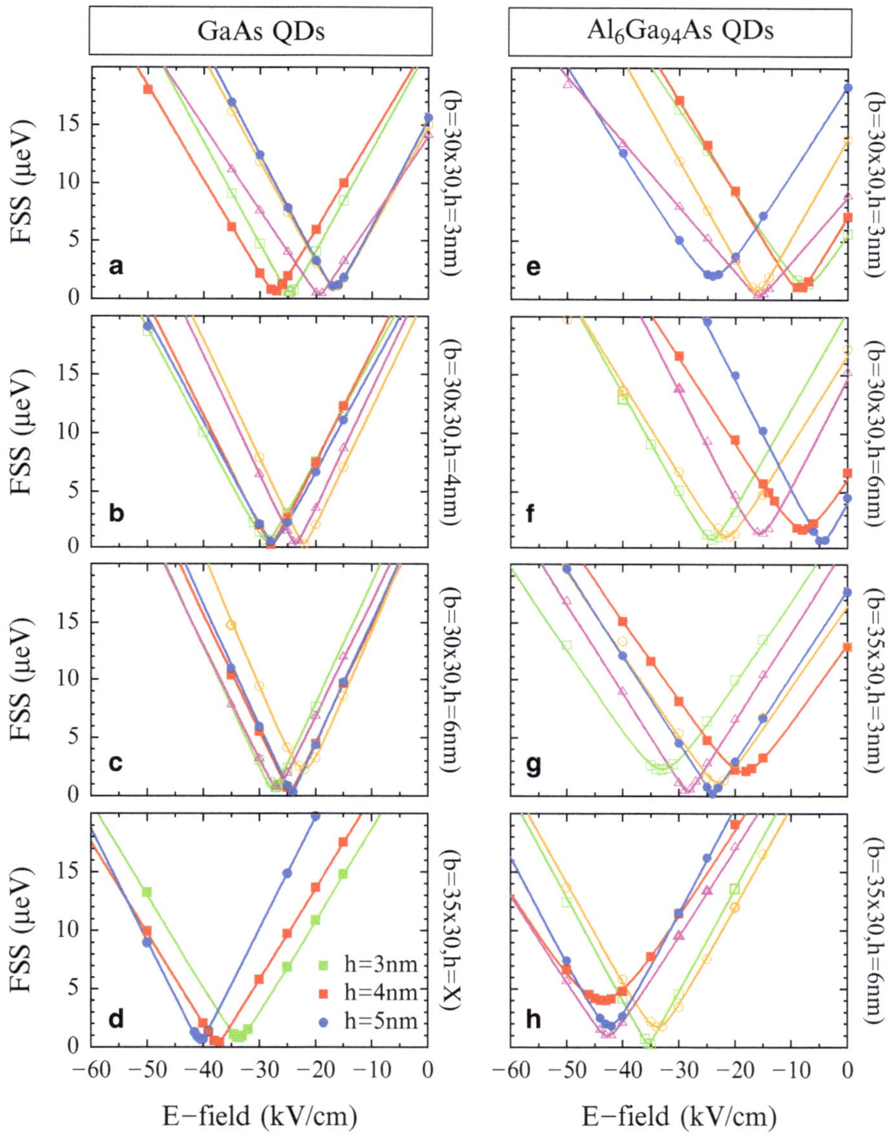

Fig. 14.9 FSS as a function of the electric field for various QDs listed in Table 14.1. Five different alloy configurations of the $Al_{30}Ga_{70}As$ matrix have been used in (**a**)–(**c**) and five different QD alloy configurations in (**e**)–(**h**). Taken from [21]. ©(2012) by the American Physical Society

QD08 can exist with s_0 of 0.1 or 1.7 μeV by merely changing the realization (i.e., the random distribution of the cations) of the barrier material. Furthermore, QD15 can exist with s_0 of 0.4 or 3.2 μeV by changing the random distribution of the 6%Al atoms inside the QD. The sensitivity of s_0 and F_0 on the alloy realization is in agreement with our model (Eq. (14.15)), where these terms have been shown to originate from the lowering of the symmetry (term δH_{C1}).

Random Alloy Effect on γ. The value of γ is only weakly dependent on the details of the alloy realization (see Fig. 14.9) but rather strongly on the QDs geometry, size, and composition (see Fig. 14.8c). Indeed, following our model, γ gives the difference in the response of $|1\rangle$ and $|2\rangle$ to the applied field and is directly related to the light-hole component of the exciton state. For a pure heavy-hole exciton, γ vanishes. The light-hole component does change significantly for different shapes (QD01, QD02, QD03, QD04, and QD05 have 3.5%, 2.4%, 2.6%, 5.0%, and 8.2%, light-hole component, respectively) but remains constant for different alloy realizations.

Oscillator Strength. Figure 14.8d shows a moderate change of the oscillator strength, in the range of 10%, with varying E field in the range of -100 to $+100\,\mathrm{kV/cm}$.

How to Select QDs with Small s_0. Our present work shows not only that GaAs QDs are good candidates to achieve small FSS via vertical electric field but also that rather large fluctuations of s_0 should be expected within one homogenous set of QDs (that differ only by random alloy effects and have the same shape, size and composition). A selection of appropriate QDs (as practiced experimentally [87,88]) will therefore be advantageous, if not necessary. From Eqs. (14.18) and (14.19) at zero field ($F = 0$) we can derive the following expressions:

$$F_0 = \frac{\Delta E \cos 2\theta}{\gamma} \quad ; \quad s_0 = -\Delta E \sin 2\theta, \tag{14.20}$$

where ΔE is the FSS at $F = 0$. Interestingly, s_0 does not depend on the slope γ and only requires a single measurement of the FSS and the corresponding polarization angle θ at zero field. We have used Eq. (14.20) in Table 14.2 to report our value of F_0 for QD00. For the value of s_0, however, if ΔE is large, a small inaccuracy in the measurement, or the calculation in our case, of the angle θ will lead to an inaccurate determination of s_0. With a ΔE of $50\,\mu$eV and an angle accuracy of 2° we obtain s_0 with an error bar of $\pm 3.5\,\mu$eV, which is too large to be useful. However, Eq. (14.20) is very useful for QDs where ΔE is not too large, which represent the QDs that will require weaker E fields to be tuned.

In summary, we showed that the FSS in GaAs/AlGaAs and InGaAs/GaAs self-assembled QDs can be effectively tuned by a vertical electric field. Indeed, the tuning rate for GaAs QDs is between 0.1 and $1\,\mu$eV cm/kV, depending on size and geometry, and is surprisingly similar to the tuning rate obtained with lateral electric fields ($0.15\,\mu$eV cm/kV [89]). Our results for InGaAs QDs are in good agreement with experiment, while the results for GaAs QDs represent predictions. The minimum FSS, s_0, for GaAs QDs, is between 0.1 and $1.8\,\mu$eV, depending on size and geometry. However, alloy fluctuations in the surrounding barrier lead to a variations of s_0 in the range of $\pm 1.4\,\mu$eV calling for a postselection of the "best QDs," for which we suggest a simple experimental procedure requiring only one measurement at zero field.

This dependence of s_0, and also F_0, on the random atomic arrangement is in agreement with the expectations from a simple mesoscopic model that shows these terms to be proportional to the "amount of deviation from C_{2v}" symmetry toward the lower C_1 symmetry. Hence, a QD made of a random alloy (with formally C_1 symmetry) with an atomic decoration of the lattice that will *resemble* the C_{2v} symmetry, will have the smallest s_0. This represents a striking example of an observable, where the conventional treatment of a random alloy through a replacement of the atomic distribution by an average (VCA [90]) or an effective medium (CPA [90]) fails. In this case, the position of each and every atom in a structure made of several thousand atoms is relevant.

As we mentioned above, the FSS can be tuned to zero in the case of vertical electric field plus a component of field along a low symmetry direction (any direction but [110] or [$1\bar{1}0$]). Because the model of Eq. (14.15) under a single vertical field does not include any field dependence of the off-diagonal terms, an additional field giving rise to field-dependent terms to the off-diagonal terms can be used to compensate field-independent terms and then tune the FSS to zero. Indeed, it was demonstrated in a recent experiment [91] in InGaAs/GaAs QDs, where in addition to vertical electric field a field of anisotropic biaxial strain delivered by piezoelectric actuators was applied in the (001) plane. However, our finding of the significant influence of random alloy fluctuations in both QD and barrier on the values of minimum FSS s_0 and F_0 show that tuning several QDs to the optimum properties in the same sample is unlikely. The realization of large arrays of QD entanglement sources [92] would therefore require a receipt to suppress the alloy randomness, which is certainly a formidable challenge. Thus, the alloy randomness will prevent us to integrate large number of entanglement sources into one chip.

14.5 HH–LH Mixing in Semiconductor QDs

HH–LH mixing occurs only when states derived from bulk HH band and from bulk LH band, respectively, belong to the same irreducible representation of the point group in a nanostructure. The point group of self-assembled QDs on (001) substrates are often mis-assigned to be D_{2d} [23, 25, 70, 93], because the QD shape resembles a lens or a truncated cone with a circular base [39] and the in-plane [110] and [$1\bar{1}0$] directions are incorrectly considered to be equivalent. The underlying bulk HH and LH bands at the Γ-point (Γ_8 in bulk T_d symmetry) transform to two *different* representations Γ_7 and Γ_6 in the D_{2d} point group [94], and therefore HH–LH mixing is expected to be forbidden in QDs. The experimentally observed HH–LH mixing in self-assemble In(Ga)As/GaAs [24] and CdTe/ZnTe [23] QDs were thus attributed to strain, which was assumed to deform the QD symmetry from D_{2d} to C_{2v}. In this scenario, HH–LH mixing is allowed since both HH and LH transform to the *same* representation Γ_5 of C_{2v}. HH–LH mixing is proportional to the magnitude of shear strain components (described by the off-diagonal terms of the Pikus–Bir Hamiltonian [12, 13]). One therefore expects HH–LH mixing to vanish in strain-free QDs. However, large HH–LH mixing was recently observed experimentally in strain-free GaAs/AlGaAs QDs grown by molecular droplet epitaxy [4, 86]. The anisotropic QD geometric shape, which also lowers the QD symmetry from D_{2d} to C_{2v} is alone considered as the origin of this unexpected HH–LH mixing [25].

In reality, the atomistic symmetry of ideal circular-based lens-shape or Gaussian-shape QDs is already C_{2v} because the [110] and [1$\bar{1}$0] directions are nonequivalent. This is unlike T_d zinc-blende crystal and D_{2d} symmetric (001) quantum well (QW), where the [110] direction is identical to the [1$\bar{1}$0] direction with S_4 operations i.e., 90° rotation followed with reflection through a (001) mirror plane [94]. *HH–LH mixing is consequently intrinsically allowed in self-assembled QDs even without built-in strain or QD shape anisotropy.* Furthermore, HH–LH mixing was found earlier even for $k = 0$, in D_{2d} QWs [95–97] and assigned to stem from C_{2v} interfaces [98, 99] since HH–LH mixing is allowed under local microscopic point group C_{2v}. The k-linear terms, originating from Rashba and Dresselhaus spin–orbit interactions, existing as off-diagonal terms in bulk valence band Luttinger–Kohn [100] or Kane [101] Hamiltonians, can also induce HH–LH mixing in QDs as a result of quantum confinement [22, 99, 102]. Fischer and Loss [22] and Tanaka et al. [102] studied HH–LH mixing in strain-free QDs taking account of the mechanism of the bulk linear terms alone. We notice that the alloy randomness in the QDs or barrier further lowers the QD symmetry and thus enhances HH–LH mixing. In summary, there are six mechanisms discussed in connection to HH–LH mixing in QD: (i) built-in shear strain [23, 24]; (ii) QD shape anisotropy [25]; (iii) intrinsic nonequivalence of the [110] and [1$\bar{1}$0] directions; (iv) low local microscopic symmetry of the interfaces [49, 98, 99]; (v) bulk k-linear terms [22, 99, 102]; and (vi) alloy randomness in the QD or barrier, which was often neglected. To the best of our knowledge, the relative importance of these six HH–LH mixing mechanisms in QDs has yet to be addressed.

In this section a comprehensive and quantitative analysis of the aforementioned mechanisms is given using the atomistic empirical pseudopotential method [43, 46, 47] for both strain-free GaAs/Al(Ga)As and strained In(Ga)As/GaAs QDs with various shapes and sizes. We find that in strain-free nominal C_{2v} GaAs QDs (the nominal point group defines the QD symmetry, neglecting the symmetry breaking by alloy randomness), mechanisms (iii) (intrinsic nonequivalence of the [110] and [1$\bar{1}$0] directions) and (iv) (interface effects) contribute 60% and 40%, respectively, to the HH–LH mixing. In strained In(Ga)As/GaAs QDs, mechanism (i) (built-in shear strain) provides 80% of the HH–LH mixing and the remaining 20% originate from mechanisms (iii) and (iv) in the ratio 3:2. Most interestingly we discover a quantum chain mediated by QD intermediate states that effectively brings the ground state of bulk LH band closer in energy to the ground state of bulk HH band and then enhances the HH–LH mixing. We refer this phenomena to quantum transmissibility of the HH–LH mixing.

The bulk band character of the QD states is gained by projection onto the bulk HH = $|3/2, \pm 3/2\rangle$, LH = $|3/2, \pm 1/2\rangle$, and spin–orbit split (SO) = $|1/2, \pm 1/2\rangle$ bands. However, all available HH–LH mixing mechanisms blend together in the QD states and prevent them to be accessed individually. We develop a formula, based on perturbation theory, to isolate the impact of the aforementioned mechanisms onto the HH–LH mixing of the QD ground hole state h_0. Specifically, we separate the QD Hamiltonian into two parts: $H = H_0 + \delta V$, where H_0 is the QD Hamiltonian without HH–LH mixing producing unperturbed QD ground states derived from bulk HH and LH bands: $\Psi^0_{\mathrm{HH}} = |u_{\mathrm{HH}};\phi^0_{\mathrm{HH}}\rangle$ and $\Psi^0_{\mathrm{LH}} = |u_{\mathrm{LH}};\phi^0_{\mathrm{LH}}\rangle$, respectively, where $|u_{\mathrm{HH}}\rangle$ and $|u_{\mathrm{LH}}\rangle$ are bulk Bloch functions and ϕ^0_{i} are the envelop functions associated with the bulk band i. The perturbative potential is $\delta V = \delta V_{(\mathrm{i})} + \delta V_{(\mathrm{ii})} + \delta V_{(\mathrm{iii})} + \delta V_{(\mathrm{iv})} + \delta V_{(\mathrm{v})} + \delta V_{(\mathrm{vi})}$, accounting for the six HH–LH mixing mechanisms (i)–(vi), respectively. δV couples unperturbed QD states. The perturbed QD ground state h_0, with dominant bulk HH character, can now be evaluated by a perturbation expansion [18, 22, 23]:

$$|\Psi_{h_0}\rangle = \lambda_{\mathrm{HH}}|\Psi^0_{\mathrm{HH}}\rangle + \lambda_{\mathrm{LH}}|\Psi^0_{\mathrm{LH}}\rangle + \cdots, \tag{14.21}$$

where λ_{LH} is a mixing coefficient given by

$$\lambda_{\mathrm{LH}} = \langle\Psi^0_{\mathrm{LH}}|\delta V|\Psi^0_{\mathrm{HH}}\rangle/\Delta^0_{\mathrm{HL}}, \tag{14.22}$$

λ_{HH} is a renormalization factor and Δ^0_{LH} is the energy splitting between $|\Psi^0_{\mathrm{HH}}\rangle$ and $|\Psi^0_{\mathrm{LH}}\rangle$. The numerator of Eq. (14.22) is the HH–LH coupling matrix element $\delta V_{\mathrm{HL}} = \langle\Psi^0_{\mathrm{LH}}|\delta V|\Psi^0_{\mathrm{HH}}\rangle$ which can be inferred (as shown in Fig. 14.10) by fitting the bulk LH character of h_0 to Eq. (14.22). Δ^0_{HL} is approximated by the energy splitting Δ_{HL} of the highest HH- and highest LH-dominated QD hole states. The relative importance of the HH–LH mixing mechanism γ for a QD is obtained from the ratio of $\delta V_{\gamma,\mathrm{HL}}$ to δV_{HL}, where $\delta V_{\gamma,\mathrm{HL}} = \langle\Psi^0_{\mathrm{LH}}|\delta V_\gamma|\Psi^0_{\mathrm{HH}}\rangle$ is the component of HH–LH coupling matrix originating from mechanism γ. The challenge is how to isolate $\delta V_{\gamma,\mathrm{HL}}$, $\gamma = (\mathrm{i}),\ldots,(\mathrm{vi})$, from the overall δV_{HL}. In the following we extract each $\delta V_{\gamma,\mathrm{HL}}$ using a comparative study approach through the design of different types of QDs.

Mechanism (i) Built-In Nonuniform Strain. The lattice-mismatch-induced strain, which has a significant impact on the electronic properties of strained QDs, is the main difference between strained In(Ga)As/GaAs QDs and strain-free GaAs/AlGaAs QDs. Figure 14.3b shows that the built-in biaxial strain [12, 13] splits the bulk InAs HH and LH bands by as much as 0.18 eV in an In(Ga)As/GaAs QD, otherwise bulk HH and LH bands are degenerate, as shown in Fig. 14.3a for a strain-free GaAs/AlAs QD. As mentioned above, the magnitude of the HH–LH mixing is proportional to shear strain components as they are present in the off-diagonal terms of the Pikus–Bir Hamiltonian [12, 13]. Figure 14.10 shows the LH character λ^2_{LH} of the h_0 state (represented as △) as a function of Δ_{HL} for 24 nominal C_{2v} lens-shape or Gaussian-shape GaAs/Al(Ga)As QDs with height varying from 2 to 7 nm. The corresponding base sizes are listed in Table 14.3. Here, nominal refers to symmetry excluding alloying effect. Despite different shapes, sizes, and barrier compositions, λ^2_{LH} of all QDs, except QD #3, fall on the blue line described by Eq. (14.22) with an overall HH–LH coupling matrix element $\delta V_{\mathrm{HL}} = 2.15$ meV. The single value of δV_{HL} for all GaAs QDs illustrates the negligible contribution of the QD height and QD shape to δV_{HL} for C_{2v} symmetry QDs. The LH character

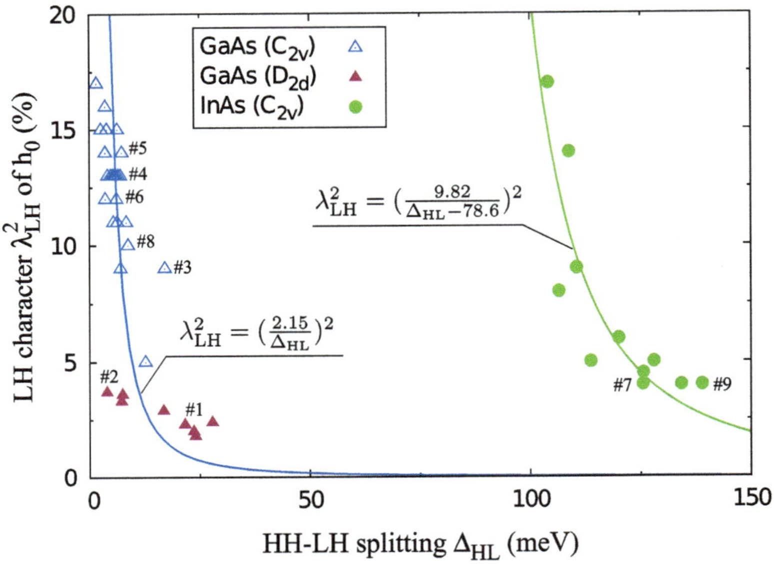

Fig. 14.10 Bulk LH character λ^2_{LH} of QD ground hole state h_0 as a function of HH–LH splitting Δ_{HL} for nominal C_{2v} lens-shape and Gaussian-shape GaAs/Al(Ga)As, lens-shaped In(Ga)As/GaAs, and nominal D_{2d} disk-shaped GaAs/Al(Ga)As QDs. The *numbering* of the QDs corresponds to the numbers in Table 14.3, which gives QD geometry properties and more detailed results, except #8 and #9 which correspond to the QDs shown and described in Fig. 14.2

λ^2_{LH} of 11 nominal C_{2v} lens-shaped In(Ga)As/GaAs QDs with varying QD height from 2 to 6 nm are shown by green dots in Fig. 14.10. In contrast to GaAs QDs, the value of λ^2_{LH} of In(Ga)As QDs can not be described by Eq. (14.22) using a single δV_{HL} value. Surprisingly, if one subtracts $\delta E_{HL} = 78.6$ meV from Δ_{HL}, then all data points lie around a curve (green line in Fig. 14.10) given by Eq. (14.22) with $\delta V_{HL} = 9.82$ meV. The enhancement of the HH–LH coupling matrix element δV_{HL} from 2.15 meV for strain-free GaAs QD to 9.82 meV for strained In(Ga)As QDs exclusively arises from the built-in nonuniform strain. We therefore conclude that the mechanism of built-in strain contributes around 80% to the HH–LH mixing since $\delta V_{(i),HL}/\delta V_{HL} = 0.78$ in strained In(Ga)As/GaAs QDs.

Quantum Transmissibility of HH–LH Mixing in QDs. The "red shift" δE_{HL} of the effective HH–LH splitting ($\Delta_{HL} - \delta E_{HL}$) required to fit the data in strained In(Ga)As/GaAs QDs arises from the effect of intermediate QD states energetically located between the HH and LH two ground states $|\Psi^0_{HH}\rangle$ and $|\Psi^0_{LH}\rangle$. In strained In(Ga)As/GaAs QDs there are tens of intermediate QD states derived from the bulk HH band between $|\Psi^0_{HH}\rangle$ and $|\Psi^0_{LH}\rangle$. This is a consequence of the large splitting between bulk HH and LH bands (shown in Fig. 14.3) induced by built-in strain. The direct coupling between $|\Psi^0_{HH}\rangle$ and $|\Psi^0_{LH}\rangle$ is expect to be very small since it is inversely proportional to Δ_{HL} (Eq. (14.22)). However, the unperturbed LH ground state $|\Psi^0_{LH}\rangle$ in a QD can strongly mix with its nearest HH-dominated QD states and

Table 14.3 Bulk band (HH, LH, and SO) characters of the highest three hole states h_0, h_1, and h_2 for strain-free GaAs/Al(Ga)As and strained $In_{0.6}Ga_{0.4}As$/GaAs QDs

QD	Shape cylindrical Composition	Point group	HH/LH/SO (%) $E(h_0)$	$E(h_1)$	$E(h_2)$
#1	Cylindrical	D_{2d}	96/2/0	92/5/0	92/5/1
	GaAs/AlAs		0.0	−8.1	−8.6
#2	Gaussian	$C_1(D_{2d})$	95/4/0	42/57/0	15/84/0
	GaAs/AlGaAs		0.0	−2.8	−4.0
#3	Gaussian	C_{2v}	87/9/1	54/42/2	43/52/2
	GaAs/AlAs		0.0	−3.2	−7.1
#4	Gaussian	$C_1(C_{2v})$	85/13/0	34/64/0	47/51/0
	GaAs/AlGaAs		0.0	−7.1	−10.3
#5	Elong. [1$\bar{1}$0]	$C_1(C_{2v})$	85/14/0	37/61/0	49/49/0
	GaAs/AlGaAs		0.0	−7.3	−10.1
#6	Elong. [110]	$C_1(C_{2v})$	87/12/0	34/64/0	44/54/0
	GaAs/AlGaAs		0.0	−6.2	−9.7
#7	Lens	$C_1(C_{2v})$	94/4/0	89/8/1	89/8/1
	InGaAs/GaAs		0.0	−12.1	−17.2

The Ga composition in the AlGaAs alloy barrier is 70%. All QDs have a height of 3 nm. The base size is 25.2 nm for both disk-shape and lens-shaped QDs. The base size is 30 nm and 35 × 30 nm for circular (#3 and #4) and elongated (#5 and #6) Gaussian-shape QDs, respectively. The energy levels (in meV) of h_1 and h_2 with respect to h_0 are also given beneath the bulk band character of each QD. The point group in parentheses is the symmetry of the QD without considering alloy randomness

these states increase the LH character of their own nearest HH-dominated QD states and eventually brings LH character to $|\Psi^0_{\mathrm{HH}}\rangle$. This process forms a transmission chain for HH–LH mixing. We refer to this enhancement of the HH–LH mixing mediated by intermediate QD states as quantum transmissibility of the HH–LH mixing. This quantum transmissibility is further confirmed by the special point (QD #3) of a nominal C_{2v} GaAs QD in Fig. 14.10. In its QD family, the impact of quantum transmissibility of HH–LH mixing is exclusively acting on QD #3 because its $h_0 \approx |\Psi^0_{\mathrm{HH}}\rangle$ and $h_2 \approx |\Psi^0_{\mathrm{LH}}\rangle$ are mediated by a HH-dominated QD state (h_1) (as shown in Table 14.3), whereas in the remaining QDs of this family the state $h_0 \approx |\Psi^0_{\mathrm{HH}}\rangle$ is immediately followed by $h_1 \approx |\Psi^0_{\mathrm{LH}}\rangle$. Surprisingly, the redshift energy δE_{HL} modifying the HH–LH splitting with respective to the bare HH–LH splitting is universal for one QD family, e.g. $\Delta_{\mathrm{HL}} - 78.6$ meV, for a family of nominal C_{2v} In(Ga)As/GaAs QDs. To the best of our knowledge, this novel quantum transmissibility of HH–LH mixing in QDs is discovered and quantitatively analyzed here for the first time.

Mechanism (vi) Effect of Alloy Randomness in the Barrier or Inside the QD. Its negligible contribution to HH–LH mixing in QDs ($\delta V_{(\mathrm{vi}),\mathrm{HL}} \sim 0$) is exhibited by the fact that (shown in Fig. 14.10) both InAs/GaAs and $In_{60}Ga_{40}As$/GaAs QDs and

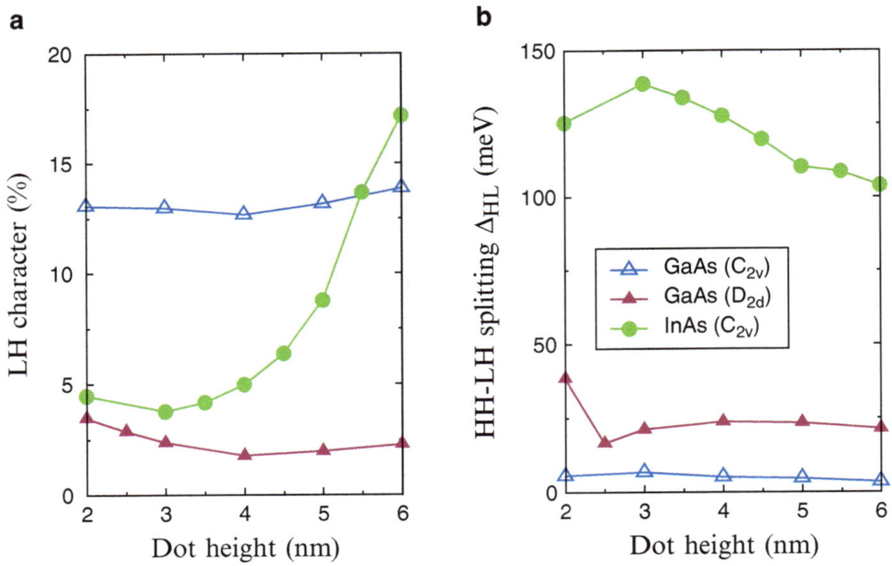

Fig. 14.11 (**a**) LH character of QD ground hole state h_0 and (**b**) HH–LH splitting Δ_{HL} as a function of QD height for nominal D_{2d} and C_{2v} GaAs/Al(Ga)As QDs and nominal C_{2v} In(Ga)As/GaAs QDs

both GaAs/AlAs and GaAs/AlGaAs QDs share an overall HH–LH coupling matrix element $\delta V_{\mathrm{HL}} = 9.82\,\mathrm{meV}$ and 2.15 meV, respectively. A further analysis of alloy effects is done via studying λ_{LH}^2 variation induced by fluctuations in the random atomic configuration. These atomic fluctuations have a significant effect on both FSS [21] and optical polarization [39] in QDs. Five random atomic configurations of the alloy barrier for a Gaussian-shape 3 nm height GaAs/$Al_{30}Ga_{70}As$ QD lead to $\lambda_{\mathrm{LH}}^2 = 13.3, 13.0, 12.9, 13.1, 12.8\%$, with a standard deviation $\sigma = 0.2\%$. Meanwhile, five random atomic configurations of a $In_{60}Ga_{40}As$/GaAs alloy QD with 3 nm height give rise to four $\lambda_{\mathrm{LH}}^2 = 3.6\%$ and one $\lambda_{\mathrm{LH}}^2 = 3.8\%$. Our assertion of negligible alloy effect on HH–LH mixing is well supported by these results.

Mechanism (v) Bulk k-Linear Terms. Fischer and Loss [22] considered this mechanism as the only origin of HH–LH mixing in strain-free QDs and derived $\lambda_{\mathrm{LH}} = \sqrt{3}\beta_{\mathrm{HL}}\gamma_3/(2\sqrt{2}\gamma_2) \times a_z L/(L^2 - a_z^2)$ for a strain-free QD of base size L and QD height a_z, where $\gamma_{2,3}$ are Luttinger parameters and β_{HL} accounts for the difference in effective masses between the bands. For flat GaAs QDs ($a_z \ll L$), $\lambda_{\mathrm{LH}} \simeq 0.53 a_z/L$ [22] is linear proportional to the QD height a_z for fixed base size L. However, using atomistic pseudopotentials we find that λ_{LH}^2 of both D_{2d} disk-shaped and C_{2v} lens-shaped GaAs/Al(Ga)As flat QDs (shown in Fig. 14.11a) to be nonmonotonic with a weak dependence on QD height. This marked difference to the model of Fischer and Loss [22] demonstrates the negligible impact of bulk k linear terms on HH–LH mixing in flat QDs.

Mechanism (iv) Low Local Microscopic Symmetry Interfaces. The impact of local atomic symmetry of C_{2v} interface on HH–LH mixing was first proposed by Ivchenko et al. [98, 99] using phenomenological Hamiltonian to explain unexpected HH–LH mixing observed in D_{2d} GaAs/AlAs QWs [95–97]. The filled triangles in Fig. 14.10 show that the LH character λ^2_{LH} of disk-shaped nominal D_{2d} GaAs QDs embedded in AlAs and $Al_{0.3}Ga_{0.7}As$ barriers can be fitted by the functions $[0.8/(\Delta_{HL}-13)]^2$ and $(0.8/\Delta_{HL})^2$, respectively. Quantum transmissibility of HH–LH mixing plays again an important role in D_{2d} GaAs/AlAs QDs giving rise to a redshift of 13 meV in the effective HH–LH splitting. Table 14.3 shows that the confinement potential of the AlAs barrier is so large that additional QD states exist between the two QD ground hole states of $|\Psi^0_{HH}\rangle$ and $|\Psi^0_{LH}\rangle$ in D_{2d} GaAs/AlAs QDs. No such intermediate states exist in QDs embedded in $Al_{0.3}Ga_{0.7}As$. The low local atomic symmetry interfaces induce a HH–LH coupling $\delta V_{(iv),HL} = 0.8$ meV. Moreover, the barrier composition independence of $\delta V_{(iv),HL} = 0.8$ meV further confirms the negligible contribution of alloy randomness effect onto the HH–LH mixing.

Mechanism (ii) QD Shape Anisotropy. Seven out of the 24 nominal C_{2v} GaAs QDs have an anisotropic shape: 6 QDs are elongation along the $[1\bar{1}0]$ direction and one along the [110] direction, having base size of 35×30 nm and various heights. The LH character λ^2_{LH} of seven irregular QDs, together with 17 circular-based QDs, fall on a single curve (shown in Fig. 14.10) indicating the minor contribution of the QD shape anisotropy to the HH–LH mixing. Specifically, the LH characters are $\lambda^2_{LH} = 14, 13$, and 12% for QDs elongated along the $[1\bar{1}0]$ direction (QD #5), with a circular base (QD #4), and elongated along [110] (QD #6), respectively. Mechanism (ii) induces a HH–LH mixing with magnitude $\delta V_{(ii),HL} \simeq 0.2$ meV which is less than 10% of the overall δV_{HL}. This finding indicates the incorrect link between HH–LH mixing, in strain-free GaAs/AlGaAs QDs, and QD shape anisotropy [25].

Mechanism (iii) Intrinsic Nonequivalence of the [110] and $[1\bar{1}0]$ Directions. The impact of QD shape anisotropy and alloy randomness on HH–LH coupling is fairly small, as discussed above. HH–LH coupling $\delta V_{HL} = 2.15$ meV of nominal C_{2v} GaAs QDs is thus mainly arising from the remaining mechanisms (iii) and (iv), i.e., intrinsic nonequivalence of the [110] and $[1\bar{1}0]$ directions and low symmetry interfaces. Interface-induced HH–LH coupling is found to be $\delta V_{(iv),HL} = 0.8$ meV and intrinsic nonequivalent of the [110] and $[1\bar{1}0]$ directions leads consequently to $\delta V_{(iv),HL} \simeq 1.35$ meV. Therefore, mechanisms (iii) and (iv) contribute 50–60% and 50–40%, respectively, of the HH–LH mixing in nominal C_{2v} GaAs QDs, depending on QD shape.

In conclusion, we comprehensively and quantitatively analyzes the impact of a total of six mechanisms leading to HH–LH mixing in semiconductor QDs. The novel quantum transmissibility of HH–LH mixing meditated by intermediate states is highlighted. We find that the HH–LH mixing in strain-free nominal D_{2d} GaAs QDs majorly arises from the mechanism (iv) low local symmetry interfaces with a HH–LH coupling $\delta V_{HL} = 0.8$ meV. In strain-free nominal C_{2v} GaAs QDs, mechanisms (iii) and (iv) related to atomistic interfaces and intrinsic nonequivalence of the [110] and $[1\bar{1}0]$ directions contribute 50–60% and 50–40%, respectively, to the HH–LH mixing. The mechanism (ii) related to the QD shape anisotropy contributes less than 10%. In strained In(Ga)As/GaAs QDs, mechanism (i) given by built-in shear strain contributes 80% to the HH–LH mixing and the remaining 20% stems from intrinsic nonequivalence of the [110] and $[1\bar{1}0]$ directions and the interface effect in a ratio of 3:2. Most importantly thing the HH–LH mixing in strained QDs is mainly due to the quantum transmissibility of HH–LH mixing meditated by intermediate QD states. Since the HH–LH coupling δV_{HL} is nearly insensitive to the QD morphology for a type of QDs, the HH–LH mixing can only be tuned by designing QDs with specific HH–LH splitting Δ_{HL}. For nominal C_{2v} In(Ga)As/GaAs QDs (as shown in Fig. 14.11b), flatter QDs usually have larger HH–LH splitting and consequently smaller HH–LH mixing. This is a consequence of quantum confinement and built-in strain that tends to be larger in flat structures. Whereas, for nominal C_{2v} GaAs/Al(Ga)As QDs, the QD height is inefficient to tune the HH–LH mixing. The lens-shape GaAs QDs often exhibit larger HH–LH splittings than Gaussian-shape GaAs QDs and thus have weaker HH–LH mixing. Increasing the Ga composition of the barrier for GaAs/Al(Ga)As QDs leads to enhanced HH–LH splitting and to reduced HH–LH mixing.

References

1. Poem, E., Shemesh, J., Marderfeld, I., Galushko, D., Akopian, N., Gershoni, D., Gerardot, B.D., Badolato, A., Petroff, P.M.: Phys. Rev. B **76**, 235304 (2007)
2. Abbarchi, M., Kuroda, T.T., Mano, T., Sakoda, K., Mastrandrea, C., Vinattieri, A., Gurioli, M., Tsuchiya, T.: Phys. Rev. B **3**, 774 (2007)
3. Hatami, F., Masselink, W.T., Schrottke, L., Tomm, J.W., Talalaev, V., Kristukat, C., Goñi, A.R.: Phys. Rev. B **67**, 085306 (2003)
4. Koguchi, N., Ishige, K.: Jpn. J. Appl. Phys. **32**, 2052 (1993)
5. Moison, J.M., Houzay, F., Barthe, F., Leprince, L., André, E., Vatel, O.: Appl. Phys. Lett. **64**, 196 (1994). http://link.aip.org/link/?APL/64/196/1
6. Wang, Z.M., Liang, B.L., Sablon, K.A., Salamo, G.J.: Appl. Phys. Lett. **90**, 113120 (3 pp.) (2007). http://link.aip.org/link/?APL/90/113120/1
7. Lee, J.H., Wang, Z.M., Strom, N.W., Mazur, Y.I., Salamo, G.J.: Appl. Phys. Lett. **89**, 202101 (3 pp.) (2006). http://link.aip.org/link/?APL/89/202101/1
8. Wang, L., Křápek, V., Ding, F., Horton, F., Schliwa, A., Bimberg, D., Rastelli, A., Schmidt, O.G.: Phys. Rev. B **80**, 085309 (2009). http://link.aps.org/doi/10.1103/PhysRevB.80.085309
9. Sonnenberg, D., Graf, A., Paulava, V., Hansen, W., Heyn, C.: Appl. Phys. Lett. **101**, 143106 (4 pp.) (2012). http://link.aip.org/link/?APL/101/143106/1
10. Atkinson, P., Zallo, E., Schmidt, O.G.: J. Appl. Phys. **112**, 054303 (5 pp.) (2012). http://link.aip.org/link/?JAP/112/054303/1
11. Rastelli, A., Stufler, S., Schliwa, A., Songmuang, R., Manzano, C., Costantini, G., Kern, K., Zrenner, A., Bimberg, D., Schmidt, O.G.: Phys. Rev. Lett. **92**, 166104 (2004). http://link.aps.org/doi/10.1103/PhysRevLett.92.166104
12. Wei, S.-H., Zunger, A.: Phys. Rev. B **49**, 14337 (1994)
13. Bir, G., Pikus, G.: Symmetry and Strain-Induced Effects in Semiconductors. Wiley, New York (1974)
14. Li, S.-S., Xia, J.-B.: Phys. Rev. B **55**, 15434 (1997)

15. Luo, J.-W., Franceschetti, A., Zunger, A.: Phys. Rev. B **78**, 035306 (2008)
16. Zhang, L., d'Avezac, M., Luo, J.-W., Zunger, A.: Nano Lett. **12**, 984 (2012)
17. d'Avezac, M., Luo, J.-W., Chanier, T., Zunger, A.: Phys. Rev. Lett. **108**, 027401 (2012)
18. Luo, J.-W., Chantis, A.N., van Schilfgaarde, M., Bester, G., Zunger, A.: Phys. Rev. Lett. **104**, 066405 (2010)
19. Bennett, A.J., Pooley, M.A., Stevenson, R.M., Ward, M.B., Patel, R.B., de la Giroday, A.B., Skold, N., Farrer, I., Nicoll, C.A., Ritchie, D.A., et al.: Nat. Phys. **6**, 947 (2010)
20. Ghali, M., Ohtani, K., Ohno, Y., Ohno, H.: Nat. Commun. **3**, 661 (2012)
21. Luo, J.-W., Singh, R., Zunger, A., Bester, G.: Phys. Rev. B **86**, 161302 (2012)
22. Fischer, J., Loss, D.: Phys. Rev. Lett. **105**, 266603 (2010)
23. Koudinov, A.V., Akimov, I.A., Kusrayev, Y.G., Henneberger, F.: Phys. Rev. B **70**, 241305 (2004)
24. Kowalik, K., Krebs, O., Lemaître, A., Gaj, J.A., Voisin, P.: Phys. Rev. B **77**, 161305 (2008)
25. Belhadj, T., Amand, T., Kunold, A., Simon, C.-M., Kuroda, T., Abbarchi, M., Mano, T., Sakoda, K., Kunz, S., Marie, X., et al.: Appl. Phys. Lett. **97**, 051111 (3 pp.) (2010)
26. Liao, Y.-H., Liao, C.-C., Ku, C.-H., Chang, Y.-C., Cheng, S.-J., Jo, M., Kuroda, T., Mano, T., Abbarchi, M., Sakoda, K.: Phys. Rev. B **86**, 115323 (2012)
27. Tonin, C., Hostein, R., Voliotis, V., Grousson, R., Lemaître, A., Martinez, A.: Phys. Rev. B **85**, 155303 (2012)
28. Wang, L.-W., Zunger, A.: Phys. Rev. B **51**, 17398 (1995)
29. Luo, J.-W., Li, S.-S., Xia, J.-B., Wang, L.-W.: Appl. Phys. Lett. **88**, 143108 (2006)
30. Luo, J.-W., Franceschetti, A., Zunger, A.: Nano Lett. **8**, 3174 (2008)
31. Luo, J.W., Franceschetti, A., Zunger, A.: Phys. Rev. B **79**, 201301 (2009)
32. Luo, J.-W., Franceschetti, A., Zunger, A.: Nano Lett. **9**, 2648 (2009)
33. Zhang, L., Luo, J.-W., Zunger, A., Akopian, N., Zwiller, V., Harmand, J.-C.: Nano Lett. **10**, 4055 (2010)
34. Luo, J.-W., Stradins, P., Zunger, A.: Energy Environ. Sci. **4**, 2546 (2011)
35. Bester, G., Nair, S., Zunger, A.: Phys. Rev. B **67**, 161306 (2003)
36. Bester, G., Zunger, A.: Phys. Rev. B **71**, 045318 (2005)
37. Luo, J.-W., Li, S.-S., Xia, J.-B., Wang, L.-W.: Phys. Rev. B **71**, 245315 (2005)
38. Luo, J.-W., Bester, G., Zunger, A.: New J. Phys. **11**, 123024 (2009)
39. Mlinar, V., Bozkurt, M., Ulloa, J.M., Ediger, M., Bester, G., Badolato, A., Koenraad, P.M., Warburton, R.J., Zunger, A.: Phys. Rev. B **80**, 165425 (2009)
40. Luo, J.-W., Zunger, A.: Phys. Rev. Lett. **105**, 176805 (2010)
41. Luo, J.-W., Zunger, A.: Phys. Rev. B **84**, 235317 (2011)
42. Bester, G.: J. Phys. Condens. Matter **21**, 023202 (2009)
43. Wang, L.-W., Zunger, A.: Phys. Rev. B **59**, 15806 (1999)
44. Franceschetti, A., Fu, H., Wang, L.-W., Zunger, A.: Phys. Rev. B **60**, 1819 (1999)
45. Pryor, C., Kim, J., Wang, L.W., Williamson, A.J., Zunger, A.: J. Appl. Phys. **83**, 2548 (1998)
46. Williamson, A.J., Wang, L.-W., Zunger, A.: Phys. Rev. B **62**, 12963 (2000)
47. Luo, J.-W., Bester, G., Zunger, A.: Phys. Rev. B **79**, 125329 (2009)
48. Keating, P.N.: Phys. Rev. **145**, 637 (1966)
49. Bester, G., Zunger, A.: Phys. Rev. B **72**, 165334 (2005)
50. Vurgaftman, I., Meyer, J.R., Ram-Mohan, L.R.: J. Appl. Phys. **89**, 5815 (2001)
51. Yu, P.Y., Cardona, M.: Fundamentals of Semiconductors: Physics and Materials Properties, 3rd edn. Springer, Berlin (2004)
52. Barrow, G.M.: Introduction to Molecular Spectroscopy, 3rd edn. McGraw-Hill, New York (1962)
53. Krenner, H.J., Clark, E.C., Nakaoka, T., Bichler, M., Scheurer, C., Abstreiter, G., Finley, J.J.: Phys. Rev. Lett. **97**, 076403 (2006)
54. Ester, P., Stufler, S., de Vasconcellos, S.M., Bichler, M., Zrenner, A.: Phys. Status Solidi (c) **3**, 3722 (2006)
55. Sychugov, I., Juhasz, R., Valenta, J., Linnros, J.: Phys. Rev. Lett. **94**, 087405 (2005)

56. Abbarchi, M., Kuroda, T., Mano, T., Sakoda, K., Mastrandrea, C.A., Vinattieri, A., Gurioli, M., Tsuchiya, T.: Phys. Rev. B **82**, 201301 (2010)
57. Keizer, J.G., Bocquel, J., Koenraad, P.M., Mano, T., Noda, T., Sakoda, K.: Appl. Phys. Lett. **96**, 062101 (3 pp.) (2010)
58. Blokland, J.H., Bozkurt, M., Ulloa, J.M., Reuter, D., Wieck, A.D., Koenraad, P.M., Christianen, P.C.M., Maan, J.C.: Appl. Phys. Lett. **94**, 023107 (3 pp.) (2009)
59. Kumah, D.P., Wu, J.H., Husseini, N.S., Dasika, V.D., Goldman, R.S., Yacoby, Y., Clarke, R.: Appl. Phys. Lett. **98**, 021903 (3 pp.) (2011)
60. Abbarchi, M., Mastrandrea, C.A., Kuroda, T., Mano, T., Sakoda, K., Koguchi, N., Sanguinetti, S., Vinattieri, A., Gurioli, M.: Phys. Rev. B **78**, 125321 (2008)
61. Kuroda, K., Kuroda, T., Sakoda, K., Watanabe, K., Koguchi, N., Kido, G.: Appl. Phys. Lett. **88**, 124101 (3 pp.) (2006)
62. Jo, M., Mano, T., Sakoda, K.: J. Appl. Phys. **108**, 083505 (3 pp.) (2010)
63. Mano, T., Abbarchi, M., Kuroda, T., Mastrandrea, C., Vinattieri, A., Sanguinetti, S., Sakoda, K., Gurioli, M.: Nanotechnology **20**, 395601 (2009)
64. Abbarchi, M., Kuroda, T., Mastrandrea, C., Vinattieri, A., Sanguinetti, S., Mano, T., Sakoda, K., Gurioli, M.: Physica E Low Dimens. Syst. Nanostruct. **42**, 881 (2010)
65. Keizer, J.G., Jo, T.M.M., Noda, T., Sakoda, K., Koenraad, P.M.: Appl. Phys. Lett. **98**, 193112 (2011)
66. Mlinar, V., Zunger, A.: Phys. Rev. B **80**, 205311 (2009)
67. Ivchenko, E.L.: Phys. Status Solidi (a) **164**, 487 (1997)
68. Singh, R., Bester, G.: Phys. Rev. Lett. **104**, 196803 (2010)
69. Plumhof, J.D., Křápek, V., Wang, L., Schliwa, A., Bimberg, D., Rastelli, A., Schmidt, O.G.: Phys. Rev. B **81**, 121309 (2010)
70. Bayer, M., Ortner, G., Stern, O., Kuther, A., Gorbunov, A.A., Forchel, A., Hawrylak, P., Fafard, S., Hinzer, K., Reinecke, T.L., et al.: Phys. Rev. B **65**, 195315 (2002)
71. Benson, O., Santori, C., Pelton, M., Yamamoto, Y.: Phys. Rev. Lett. **84**, 2513 (2000)
72. Michler, P. (ed.): Single Semiconductor Quantum Dots Series: NanoScience and Technology. Springer, Berlin (2009)
73. Bayer, M., Kuther, A., Forchel, A., Gorbunov, A., Timofeev, V.B., Schäfer, F., Reithmaier, J.P., Reinecke, T.L., Walck, S.N.: Phys. Rev. Lett. **82**, 1748 (1999)
74. Warburton, R.J., Schaflein, C., Haft, D., Bickel, F., Lorke, A., Karrai, K., Garcia, J.M., Schoenfeld, W., Petroff, P.M.: Nature **405**, 926 (2000)
75. Takagahara, T.: Phys. Rev. B **62**, 16840 (2000)
76. Besombes, L., Kheng, K., Martrou, D.: Phys. Rev. Lett. **85**, 425 (2000)
77. Seguin, R., Schliwa, A., Rodt, S., Potschke, K., Pohl, U.W., Bimberg, D.: Phys. Rev. Lett. **95**, 257402 (2005)
78. Seidl, S., Kroner, M., Högele, A., Karrai, K., Warburton, R., Badolato, A., Petroff, P.: Appl. Phys. Lett. **88** (2006)
79. Plumhof, J.D., Křápek, V., Ding, F., Jöns, K.D., Hafenbrak, R., Klenovský, P., Herklotz, A., Dörr, K., Michler, P., Rastelli, A., et al.: Phys. Rev. B **83**, 121302 (2011)
80. Kowalik, K., Krebs, O., Lemaitre, A., Laurent, S., Senellart, P., Voisin, P., Gaj, J.A.: Appl. Phys. Lett. **86**, 041907 (2005)
81. Marcet, S., Ohtani, K., Ohno, H.: Appl. Phys. Lett. **96**, 101117 (2010)
82. Stevenson, R.M., Young, R.J., See, P., Gevaux, D.G., Cooper, K., Atkinson, P., Farrer, I., Ritchie, D.A., Shields, A.J.: Phys. Rev. B **73**, 033306 (2006)
83. Jundt, G., Robledo, L., Högele, A., Fält, S., Imamoğlu, A.: Phys. Rev. Lett. **100**, 177401 (2008)
84. Muller, A., Fang, W., Lawall, J., Solomon, G.S.: Phys. Rev. Lett. **103**, 217402 (2009)
85. Gong, M., Zhang, W., Guo, G.-C., He, L.: Phys. Rev. Lett. **106**, 227401 (2011)
86. Watanabe, K., Koguchi, N., Gotoh, Y.: Jpn. J. Appl. Phys. **39**, L79 (2000)
87. Young, R.J., Stevenson, R.M., Shields, A.J., Atkinson, P., Cooper, K., Ritchie, D.A., Groom, K.M., Tartakovskii, A.I., Skolnick, M.S.: Phys. Rev. B **72**, 113305 (2005)

88. Stevenson, R.M., Salter, C.L., Nilsson, J., Bennett, A.J., Ward, M.B., Farrer, I., Ritchie, D.A., Shields, A.J.: Phys. Rev. Lett. **108**, 040503 (2012)
89. Vogel, M.M., Ulrich, S.M., Hafenbrak, R., Michler, P., Wang, L., Rastelli, A., Schmidt, O.G.: Appl. Phys. Lett. **91**, 051904 (3 pp.) (2007)
90. Jaros, M.: Rep. Prog. Phys. **48**, 1091 (1985)
91. Trotta, R., Zallo, E., Ortix, C., Atkinson, P., Plumhof, J.D., van den Brink, J., Rastelli, A., Schmidt, O.G.: Phys. Rev. Lett. **109**, 147401 (2012)
92. Ladd, T.D.: Physics **5**, 109 (2012)
93. Li, S.-S., Xia, J.-B., Yuan, Z.L., Xu, Z.Y., Ge, W., Wang, X.R., Wang, Y., Wang, J., Chang, L.L.: Phys. Rev. B **54**, 11575 (1996)
94. Koster, G., Dimmock, J., Wheeler, R., Statz, H.: Properties of the Thirty-Two Point Groups. MIT, Cambridge (1963)
95. Magri, R., Zunger, A.: Phys. Rev. B **62**, 10364 (2000)
96. van Kesteren, H.W., Cosman, E.C., van der Poel, W.A.J.A., Foxon, C.T.: Phys. Rev. B **41**, 5283 (1990)
97. Chang, Y.-C., Schulman, J.N.: Appl. Phys. Lett. **43**, 536 (1983)
98. Ivchenko, E.L., Kaminski, A.Y., Rössler, U.: Phys. Rev. B **54**, 5852 (1996)
99. Winkler, R.: Spin-Orbit Coupling Effects in Two-Dimensional Electron and Hole Systems. Springer Tracts in Modern Physics, vol. 191. Springer, Berlin (2003)
100. Luttinger, J.M., Kohn, W.: Phys. Rev. **97**, 869 (1955)
101. Kane, E.O.: J. Phys. Chem. Solids **1**, 249 (1957)
102. Tanaka, T., Singh, J., Arakawa, Y., Bhattacharya, P.: Appl. Phys. Lett. **62**, 756 (1993)

Chapter 15
Local Droplet Etching: Self-assembled Nanoholes for Quantum Dots and Nanopillars

Christian Heyn, David Sonnenberg, and Wolfgang Hansen

Abstract We review the mechanism and recent applications of the self-organized patterning of semiconductor surfaces by local droplet etching (LDE). LDE is a nanofabrication technique that is applicable in situ during molecular beam epitaxy (MBE) and fully compatible with state-of-the-art MBE systems. Most importantly, as a local etching technique that works with a number of different materials, it adds a new degree of freedom to established self-assembling techniques. During LDE, metallic droplets drill nanoholes into a semiconductor surface with structural parameters adjustable over a wide range by the process conditions. In subsequent overgrowth steps the holes are filled for the formation of nanostructures like, e.g., quantum dots (QDs) or quantum pillars. In comparison to other QD systems, the LDE dots have the key advantages that they are strain-free, highly uniform, and that their size is precisely adjustable. In addition, vertically stacked quantum dot molecules have been realized. Crystalline nanopillars are created by a combination of in situ LDE with ex situ selective etching that are highly attractive for studies of ballistic phonon and electron transport, e.g., in the field of thermoelectrics.

15.1 Introduction

Nanostructured materials are superior compared to the bulk in many aspects. In particular the tunable optoelectronic properties of low-dimensional semiconductor heterostructures are highly attractive [1]. A prominent technique for the highly reproducible creation of such semiconductor nanostructures is the molecular beam epitaxy (MBE) with its precise control of the composition profile inside the device combined with a high crystalline perfection [2]. Utilizing self-assembly

C. Heyn (✉) • D. Sonnenberg • W. Hansen
Institute of Applied Physics, University of Hamburg, Jungiusstr. 11, 20355 Hamburg, Germany
e-mail: heyn@physnet.uni-hamburg.de; dsonnenb@physnet.uni-hamburg.de; hansen@physnet.uni-hamburg.de

Z.M. Wang (ed.), *Nanodroplets*, Lecture Notes in Nanoscale Science and Technology 18, 363
DOI 10.1007/978-1-4614-9472-0_15, © Springer Science+Business Media New York 2013

mechanisms, nanostructures with one- or zero-dimensional confinement for charge carriers can be created without the need of lithographic processing. For conventional semiconductor MBE, crystalline material is deposited on a semiconductor crystalline substrate. As an alternative, the metal droplet-based techniques have attracted an increasing interest during the last years. Nevertheless, the benefit of depositing a liquid metal during epitaxy of crystalline nanostructures is not directly obvious. In this context, one might consider the surface tension of the liquid metals as the key property. The surface tension is the driving force for processes minimizing the free surface area and, thus, causes a localization of the planarly deposited metal into an ensemble of isolated droplets. These droplets can now be functionalized during epitaxy for the self-assembly of semiconductor nanostructures.

A prominent example for the functionalization of a liquid metal is the droplet epitaxy. This method has been established for by now about 20 years to fabricate self-assembled semiconductor quantum dots (QDs) [3–16] and represents an interesting alternative to the strain-driven formation of InAs QDs in Stranski–Krastanov (SK) growth mode [17–19]. In comparison to SK growth, the method of droplet epitaxy is more flexible regarding the choice of the QD material. For example, fabrication of strain-free GaAs QDs [5, 7, 11–13], InGaAs QDs with controlled In content [6,8], InAs QDs [9], and InSb QDs [4] has been demonstrated. A typical droplet epitaxy process is performed in two steps. First, the droplet material is deposited in Volmer–Weber growth mode [20] and, second, the droplets are crystallized into semiconductor nanostructures during post-growth annealing. A sketch illustrating the droplet epitaxy of GaAs nanostructures is shown in Fig. 15.1 with typical values of the process temperature T and the arsenic flux gauge reading F_{As}. By varying the arsenic flux during post-growth droplet crystallization, also the fabrication of quantum ring complexes has been demonstrated [10, 12].

This chapter focuses on a qualitatively different functionalization of the liquid metal droplets, i.e., the local droplet etching (LDE). Here, the droplets act as a local etchant drilling nanoholes into semiconductor surfaces (Fig. 15.1). In contrast to droplet epitaxy, where material is added to the surface, in LDE material is locally removed. We would like to emphasize that LDE is fully compatible with the state-of-the-art MBE fabrication of semiconductor heterostructures. In this sense, LDE extends conventional MBE, which is a bottom up technology, by a new degree of freedom allowing now the combination with self-assembled top-down structuring.

Droplet etching takes place during post-growth annealing at very low arsenic flux and significantly higher temperatures than those applied during droplet epitaxy (Fig. 15.1) [21, 22]. The high temperature is advantageous in view of the reduced incorporation of unwanted background impurities or crystal defects. The method was introduced by Wang et al. [23] for etching of GaAs surfaces with Ga droplets. Later, we have expanded the range of materials and performed etching with Ga [21, 22,24], Al [25–27], and InGa [28,29] droplets on GaAs [22], AlGaAs [21,24,28,29], and AlAs [25–27] surfaces. Examples of AlGaAs surfaces with LDE nanoholes are shown in Fig. 15.2a, b. In subsequent overgrowth steps the holes are filled for the formation of nanostructures like, e.g., quantum dots or quantum pillars.

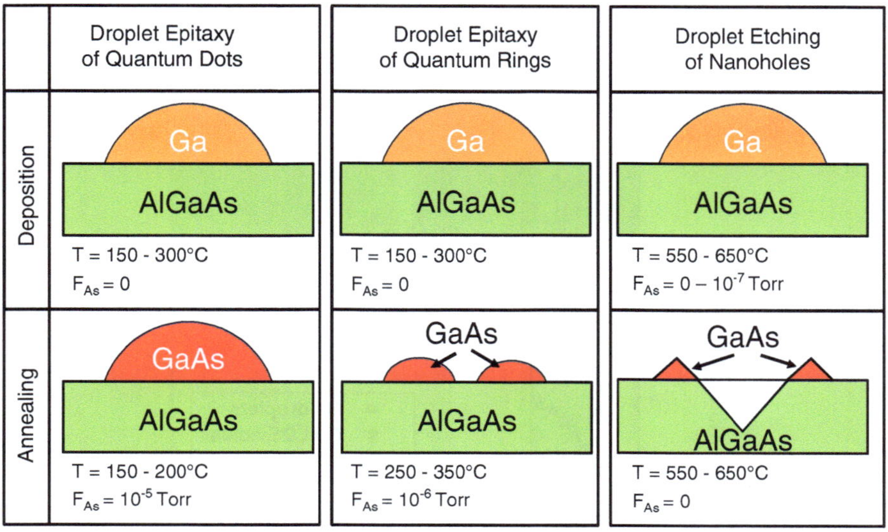

Fig. 15.1 Comparison of various droplet-based self-assembled nanostructuring techniques during semiconductor epitaxy. The processes divide into a metal droplet material deposition step and a post-growth annealing. Typical values of the process temperature T and the arsenic flux gauge reading F_{As} are given. Dependent on F_{As} during annealing, droplet epitaxy yields either quantum dots or rings. For reference, during GaAs homoepitaxy F_{As} of typical 10^{-5} Torr is applied. Droplet etching takes place at significantly higher T and minimized F_{As}, which results in material removal and the formation of nanoholes in the surface

This chapter is organized as follows: in Sect. 15.2, the mechanism behind the droplet etching process and the tunability of the hole structural parameters are addressed. Section 15.3 covers the fabrication of strain-free GaAs quantum dots by nanohole filling as an important application of the LDE technology. Section 15.4 describes the creation of nanopillars, i.e., for thermoelectric applications.

15.2 Local Droplet Etching

Local droplet etching is performed in a conventional MBE growth chamber on atomically flat GaAs, AlAs, or AlGaAs surfaces. A typical LDE process is performed in two steps (Fig. 15.1) [21, 22]. In the first step, Ga, Al, In, or a mixture of these materials is deposited as liquid metallic droplets being nucleated in Volmer–Weber growth mode [20]. During post-growth annealing, the initial droplets transform into nanoholes surrounded by a wall. Examples of AlGaAs surfaces with nanoholes are shown in Fig. 15.2a, b. The relevant process parameters are the temperatures T_1 and T_2 during droplet deposition and annealing, respectively, the droplet material deposition time t_1 and flux F yielding the coverage $\theta = Ft_1$ of deposited droplet material, the arsenic fluxes $F_{As,1}$ and $F_{As,2}$ during droplet deposition and annealing,

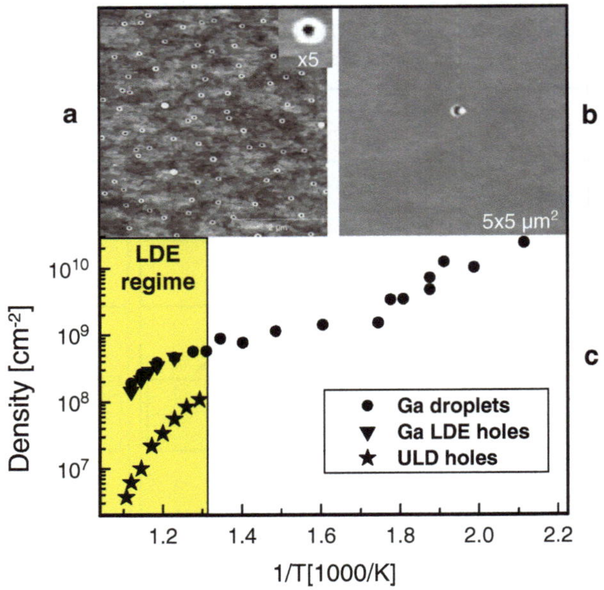

Fig. 15.2 AFM micrographs of AlGaAs surfaces after local droplet etching (LDE) using (**a**) Ga droplets and (**b**) Al droplets in the ultra-low density (ULD) regime and (**c**) density of droplets and droplet etched nanoholes as function of the droplet deposition temperature T. The temperatures T_2 that can be used for droplet etching are marked in *yellow* [13]

and the annealing time t_2. For simplicity, here we discuss processes with constant temperature $T = T_1 = T_2$. The influence of a varied $T_2 > T_1$ has been studied in [21]. The coverage with droplet material is in our experiments usually $\theta = 3.2$ ML (monolayers) and the flux $F = 0.8$ ML/s for Ga and $F = 0.4$ ML/s for Al droplets.

15.2.1 Mechanism and Simulation of Droplet Etching

The deposited droplets perform the droplet etching process and, therefore, crucially determine the properties of the later nanoholes. Figure 15.2c demonstrates that the droplet density can be adjusted very efficiently by the deposition temperature T_1 over several orders of magnitude [13, 22]. However, for droplet etching, the usable range of annealing temperatures T_2 is limited as is indicated by the yellow area in Fig. 15.2c.

A sequence of atomic-force microscopy (AFM) images illustrating the transformation of the droplets into nanoholes is shown in Fig. 15.3a. As an interesting point, the onset of etching is delayed with respect to the end of droplet deposition [21]. Caused by Ostwald ripening [30] during this delay time, the droplet density at the onset of etching is slightly reduced compared to the as-grown density [21]. When etching starts, we assume that every droplet forms a nanohole.

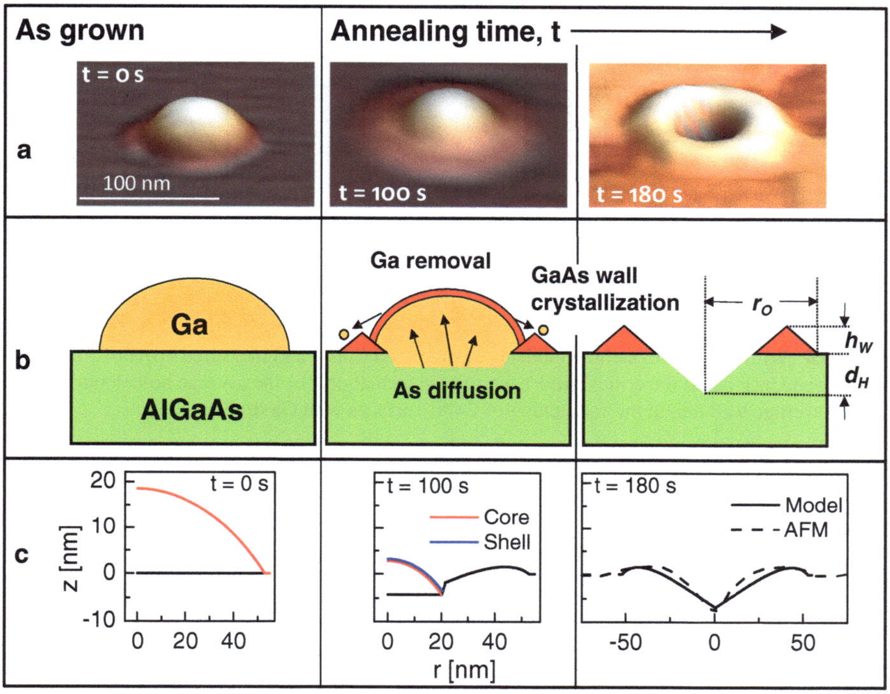

Fig. 15.3 (**a**) AFM images illustrating the transformation of an initial droplet into a nanohole with wall during post-growth annealing. (**b**) Scheme of the relevant processes for etching with Ga droplets. The central processes are diffusion of As from the substrate into the liquid droplet (= etching), removal of Ga from the droplet, assumed formation of a GaAs shell which surrounds the liquid droplet core, and crystallization of the GaAs wall. The characteristic structural parameters, i.e., the wall outer radius r_O, the nanohole depth d_H, and the wall height h_W are indicated. (**c**) Annealing time-dependent results of the droplet etching model discussed in [22]. *Lines* represent snapshots of the calculated droplet core (*red*), shell (*blue*), and substrate surface (*black*) morphologies. *Dashed lines* represent AFM linescans of typical nanoholes fabricated at the process parameters of the calculations

As is visible in Figs. 15.2a, b, and 15.3a, the nanohole openings are surrounded by distinct walls. Photoluminescence studies in [28, 29, 31] establish that LDE with Ga droplets yields crystalline walls composed of GaAs, whereas Al droplets result in optically inactive AlAs walls. Strong PL emission in samples etched with Ga droplets on AlGaAs surfaces show that walls are created that provide a quantum-ring like confinement in the AlGaAs matrix. Accordingly, for LDE with Al droplets no optical emission is detected. Further results indicate that the volume of material stored in a wall is approximately equal to the volume of material removed from a hole, and that these volumes represent approximately 3% of the initial droplet volume [24].

Regarding the transformation of the droplets into nanoholes with walls, the combined hole drilling and wall formation during LDE represents an interesting point. Obviously, the droplet performs two opposite operations: first, the removal of

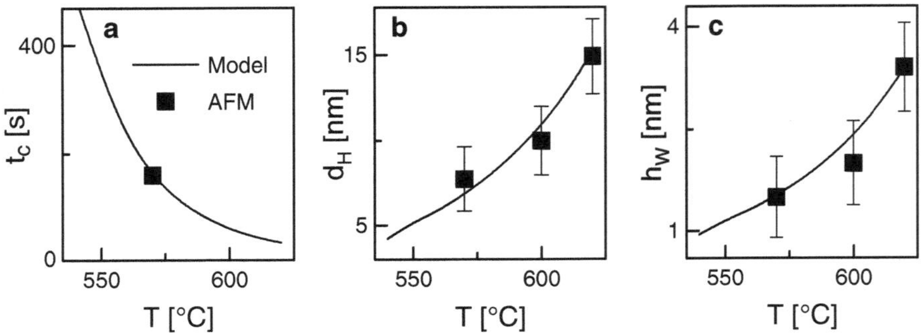

Fig. 15.4 With the droplet etching model [22] simulated (*lines*) and measured (*symbols*) temperature dependence of (**a**) the critical time t_c up to hole formation, (**b**) the average hole depth d_H, and (**c**) the average wall height h_W for etching of GaAs surfaces with Ga droplets [22]

material from the substrate, which results in nanohole formation and, second, the deposition of material on the substrate surrounding the nanohole opening as wall. This indicates a strong inhomogeneity of the droplets during the etching process. To explain this effect, we assume a core–shell geometry of the droplets with a liquid Ga core that etches the substrate and a GaAs precipitation zone at the droplet surface (denoted as GaAs shell) that causes the wall formation (Fig. 15.3b). The investigations of Kumah et al. [32] also report a core–shell configuration for a slightly different material system, i.e., indium droplets on GaAs exposed to antimony.

In order to understand the mechanism behind droplet etching, we have developed a kinetic rate-equations based model of the dynamics during etching under consideration of a core–shell geometry for the droplets [22]. In detail, the starting point of the model is a Ga droplet deposited on a GaAs surface (Fig. 15.3b). Etching of the GaAs substrate below the droplet takes place by diffusion of As into the liquid droplet material driven by the concentration gradient. This leads to an increase of the As concentration inside the droplet. The etching is accompanied by removal of the droplet material. This removal may take place either by desorption of Ga atoms from the droplet surface or more probably due to very recent experiments by Ga detachment and spreading over the substrate surface. In addition, the Ga removal causes a local increase of the As concentration at the droplet boundary, which now exceeds the very small maximum solubility [33] of As in liquid Ga. This results in precipitation of GaAs that is assumed to form a thin GaAs shell surrounding the liquid droplet core. Furthermore, the shell is assumed to crystallize the GaAs wall around the nanohole opening. Simulations using this approach quantitatively reproduce experimental surface morphologies with walls and nanoholes after Ga droplet etching (Fig. 15.3c) [22].

Figure 15.4 shows a comparison of results simulated using the above droplet etching model (Sect. 15.2.1) with experimental data [22]. In Fig. 15.4a, the critical time t_c for etching up to the complete removal of the droplet material is shown as function of the temperature T. We find a faster etching at higher temperatures

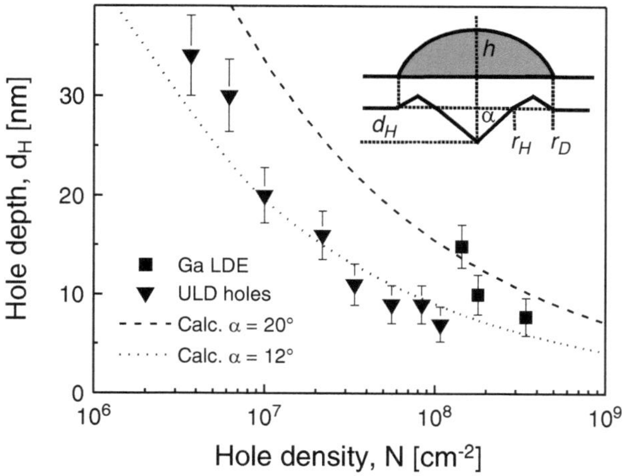

Fig. 15.5 Relation between hole density N and depth d_{H} measured (*symbols*) and estimated (*lines*) using Eq. (15.1). In the experiments, the hole density is varied by the process temperature (see Fig. 15.2). In the calculations, $\alpha = 20°$ is used for Ga LDE holes and $\alpha = 12°$ for ULD holes. The *inset* shows a cross-sectional sketch of the simplified droplet and hole geometries

and a significant increase of the etching time for temperatures below 550 °C. Accordingly, this temperature is indicated in Fig. 15.2c as low-temperature limit of the LDE process. Results on the temperature-dependent hole depth d_{H} are plotted in Fig. 15.4b. Here, we find a depth increasing with temperature from about 5 up to 14 nm. Also the wall height h_{W} increases with temperature from about 1 up to 3 nm. We would like to note that all simulation results agree very well with experimentally obtained values.

15.2.2 Morphology of Droplet-Etched Nanoholes

The above model of the droplet etching mechanism (Sect. 15.2.1) provides no information on the influence of the nanohole morphology on the temperature-dependent droplet density N. We discuss now a simple analytical expression of this relation. As a starting point, we assume that the first deposited Ga monolayer is consumed for the transition of the initial As terminated into a Ga terminated surface [21]. The remaining Ga is stored in the droplets, which yields the average number of atoms inside a single droplet: $v = (\theta - 1)/n$, with the deposited coverage θ of droplet material, the normalized droplet density $n = N/N_0$, and the density $N_0 = 6.25 \times 10^{14}\,\mathrm{cm}^{-2}$ of surface sites. The droplet volume is $V = v(V_{\mathrm{M}}/N_{\mathrm{A}})$, with Avogadro's constant N_{A} and the volume V_{M} of a mol of the droplet material. We assume now that when etching the droplets can be described as a segment of a sphere with volume $V = \pi\kappa^2 r_{\mathrm{D}}^3(1 - \kappa/3)$, where $\kappa = h/r_{\mathrm{D}}$, h is the droplet height, and $r_{\mathrm{D}} = [(\theta - 1)V_{\mathrm{m}}/(nN_{\mathrm{A}}\pi\kappa^2[1 - \kappa/3])]^{1/3}$ is the droplet radius (see inset in Fig. 15.5). It is furthermore assumed that every droplet forms a hole and that the droplet radius at the etching process start is equal to the outer radius of the later wall around the hole opening [21, 22, 28]. Experimental observations from a large number of nanoholes reveal that the radius r_H of the hole opening is approximately half of the wall radius: $r_H \simeq 0.5 r_D$ [28]. This allows to estimate the hole depth from:

$$d_H = 0.5 \tan(\alpha) \left[\frac{(\theta - 1) V_m}{n N_A \pi \kappa^2 (1 - \kappa/3)} \right]^{1/3} \tag{15.1}$$

with the angle α between the hole sidewall and the flat surface. Figure 15.5 shows values of the hole depth as function of their density both measured and estimated using Eq. (15.1) with $\theta = 3.2$ ML, $\kappa = 0.4$ [28], and $\alpha = 20°$ for Ga LDE holes [29] or $\alpha = 12°$ for Al etching of ultra-low density holes [34]. The central result of this estimation is that, for a constant coverage, an increasing hole density is always accompanied by a reduction of the hole depth. To consider, in addition, the process temperature T, the droplet density in Eq. (15.1) can be described by a scaling law: $N = j \exp[E_N/(k_B T)]$, where k_B is Boltzmann's constant, E_N a characteristic energy, and j a constant [21, 35].

After this discussion of the average structural properties of the nanoholes, we now address the distribution of the hole depths. Figure 15.6a shows a typical AlGaAs surface after etching with Ga droplets at a low temperature of $T = 610$ °C. We find two different types of holes, i.e., shallow holes (S) with depth up to 7 nm and holes (D) that are deeper. A profile of a shallow hole is shown in Fig. 15.6b and a deep hole in Fig. 15.6c. The hole depth distribution is plotted in Fig. 15.6d. Clearly visible is that the depth distribution is bimodal-like [36]. The shallow holes with broad depth distribution are very interesting for the fabrication of ensembles of quantum dots that act as white-light emitters [26] (see Sect. 15.3.1). The origin of the bimodal hole depth distribution is not completely clear, so far. Preliminary results indicate that this effect is related to the size distribution of the droplets at the onset of etching.

We have applied two modifications of the LDE process in order to achieve uniform nanoholes with narrow depth distribution. Both modifications account for etching with Al droplets which is relevant for QD fabrication. The first modification uses a higher process temperature of about $T = 650$ °C and an additional AlAs layer for thermal surface stabilization [25, 26]. This process will be described in Sect. 15.3.2. The second modification is based on an optimization of the As flux during droplet deposition and will be described in the following section.

15.2.3 Ultra-Low Density Nanoholes

Figure 15.7 shows that a finite Arsenic flux $F_{As,1}$ during droplet deposition increasing up to 4×10^7 Torr changes the densities of shallow and deep holes only

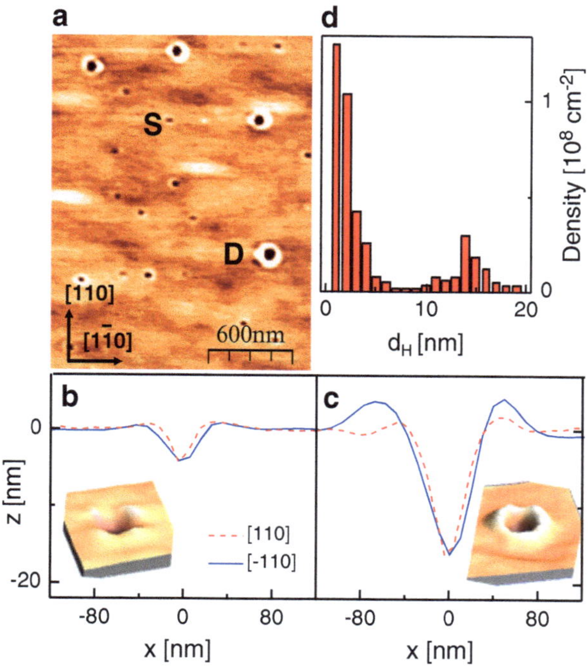

Fig. 15.6 Depth distribution of LDE nanoholes in AlGaAs after etching with Ga droplets at low $T = 610\,°\text{C}$. (**a**) *Top view* AFM image with shallow (S) and deep (D) holes. 3D AFM images and AFM profiles of (**b**) a typical shallow and (**c**) a deep hole. (**d**) Hole depth distribution [27]

slightly. A further increase up to $F_{\text{As},1} \simeq 7 \times 10^7$ Torr leads to a slight reduction of both densities. Importantly, at 7×10^7 Torr $< F_{\text{As},1} < 9 \times 10^7$ Torr, the situation changes qualitatively and now the formation of shallow holes vanishes, whereas still deep holes are formed [34]. These holes have an ultra-low density (ULD) in the $10^6\,\text{cm}^{-2}$ range uniformly over the whole wafer. At $F_{\text{As},1} > 9 \times 10^7$ Torr hole formation is completely suppressed. By an increase of the process temperature, the ULD-hole density is further reduced to less than $10^6\,\text{cm}^{-2}$.

Obviously, the additional As supplied during Al droplet material deposition substantially modifies the droplet nucleation and forces the generation of larger droplets with low density on cost of the smaller ones. During annealing, this droplet size distribution is transformed into a hole depth distribution according to the mechanism described in Sect. 15.2.1 with only deep holes of ULD.

The low density suggests the ULD holes for the fabrication of laterally isolated quantum dots (Sect. 15.3.3) [34]. Furthermore, according to the relation between the hole density and depth (Eq. (15.1)), the ULD holes have depths of 25–30 nm that are significantly deeper than the holes with higher density described above (Fig. 15.5). This opens the possibility to fabricate vertically stacked quantum-dot molecules in a single hole (Sect. 15.3.4).

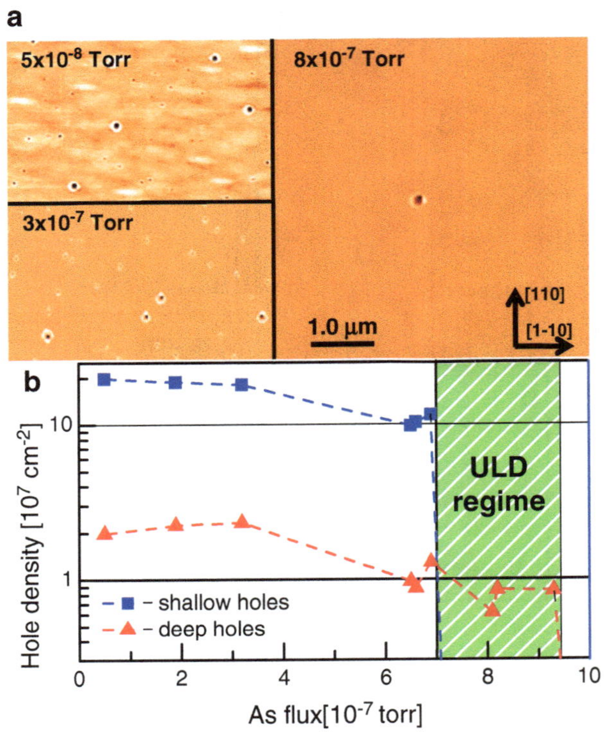

Fig. 15.7 Evolution of the hole density with increasing Arsenic flux during droplet deposition. Here, Al droplets are used for etching of AlGaAs surfaces. During post-growth annealing the As flux is minimized. (**a**) AFM images after droplet etching with indicated As flux. (**b**) Densities of shallow and deep holes as function of the As flux. The ultra-low density (ULD) regime with only deep holes is *shaded* [34]

15.3 GaAs Quantum Dots by Nanohole Filling

Recently, we have demonstrated the creation of a novel type of strain-free GaAs quantum dot by filling of LDE nanoholes in AlGaAs as barrier material [25–27]. For these experiments, etching is performed using Al droplets in order to avoid an additional quantum ring-like confinement induced by the wall (Fig. 15.8a). The nanoholes are filled by deposition of a GaAs layer with thickness d_f in a pulsed mode with 0.5 s deposition and 30 s pause, respectively. We assume that all GaAs that enter the nanohole opening diffuse downwards and fill the nanohole starting from its bottom. Using this approach, the dots are assumed to have an inverted cone-like shape and the QD height h_{QD} can be calculated from the filling level [25]. Finally, the dots are capped by overgrowth with 120 nm AlGaAs barrier material. The nominal Al content in all AlGaAs layers is 0.35. Controlled by the parameters of the LDE process, we distinguish three types of nanohole templates and resulting QDs (types 1–3, schemes are shown in Fig. 15.8) that are described in the following.

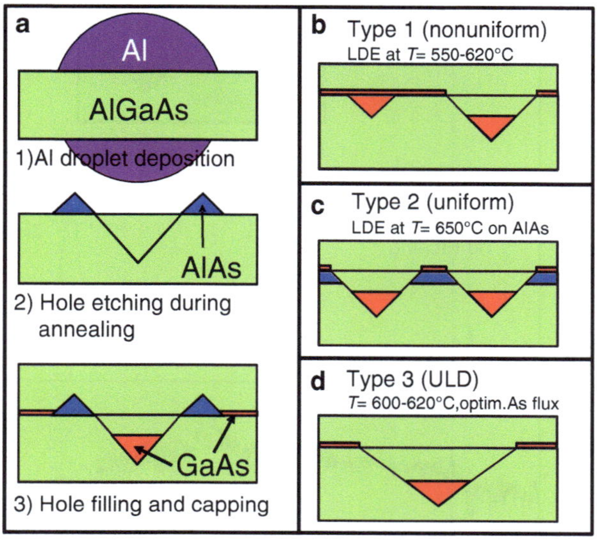

Fig. 15.8 (**a**) Scheme of the fabrication of GaAs quantum dots by self-assembled droplet etching with Al droplets and subsequent nanohole filling. (**b**) Type 1 QDs fabricated by mostly complete filling of shallow holes with broad size distribution. Ensembles of these dots are interesting as white-light emitters. (**c**) Type 2 QDs fabricated by partial filling of deep holes. These dots are highly uniform and their size is precisely adjustable by the filling level. (**d**) Type 3 ultra-low density (ULD) QDs. These dots are interesting for single-dot spectroscopy

15.3.1 Type 1 QDs: White-Light Emitters

The type 1 QDs are fabricated using LDE on AlGaAs at low temperatures of T = 550–620 °C [26]. Using these process conditions, the nanoholes have a bimodal depth distribution with shallow and deep holes (Fig. 15.6). Importantly, during filling with GaAs for QD generation the shallow holes are completely filled whereas the deep holes are only partially filled (Fig. 15.8b). The optical emission from ensembles of type 1 QDs is dominated by the completely filled shallow holes with broad size distribution. Corresponding photoluminescence (PL) measurements are shown in Fig. 15.9. A reference sample without filling shows no optical signal (Fig. 15.9a) and, thus, demonstrates that there is no background emission from the AlGaAs layers. A second reference sample with filling but without etching shows one strong PL peak at energy $E = 1.900\,\text{eV}$ (Fig. 15.9b) that is related to the strongly confined $d_F = 0.65\,\text{nm}$ thick GaAs quantum well formed by the deposited GaAs. PL measurements of samples that contain a layer with shallow QDs show a broadband optical emission over a range of at least 300 meV without pronounced peaks (Fig. 15.9c, d). These results are consistent with the picture of a very broad QD size distribution caused by complete filling of the nonuniform shallow holes. Furthermore, no clear dependence on the thickness d_{f} of the deposited GaAs layer

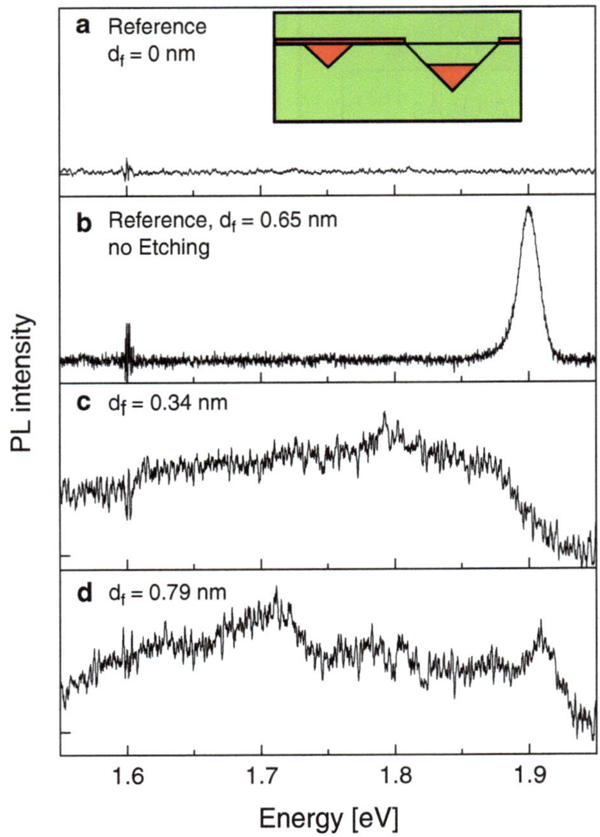

Fig. 15.9 PL emission at $T = 4\,\mathrm{K}$ from ensembles of type 1 GaAs QDs fabricated by complete filling of shallow nanoholes in AlGaAs. (**a**) Reference sample with etching but without filling. (**b**) Reference sample without etching but with filling with $d_f = 0.65\,\mathrm{nm}$. (**c**) Sample with QDs and $d_f = 0.34\,\mathrm{nm}$. (**d**) Sample with QDs and $d_f = 0.79\,\mathrm{nm}$. d_F is the thickness of the GaAs layer deposited for hole filling. The peak at 1.6 eV is an artifact caused by the FIR spectrometer used for ensemble PL measurements [26]

is visible. This shows that already the thin filling layer yields a complete filling of the shallow holes. Such type 1 QDs allow the integration of a single layer with nearly white-light emission into MBE grown heterostructures.

15.3.2 Type 2 QDs: Uniform Dots

In [36] we have observed that an increase of the temperature during LDE leads to a reduction of the density of shallow holes. However, the maximum temperature of GaAs or AlGaAs surfaces is limited to about 630 °C by the onset of As desorption.

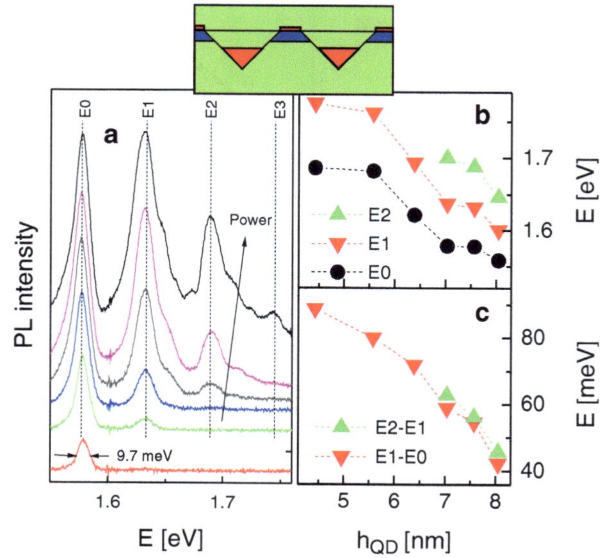

Fig. 15.10 (**a**) Excitation power series (8.5 ...450 W/cm^2) of the PL emission at $T = 4$ K from a highly uniform ensemble of type 2 GaAs QDs ($h_{QD} = 7.6$ nm) fabricated by partially filling (0.57 nm filling level) of deep holes in AlAs/AlGaAs. The spectra are vertically shifted for clarity. *Dashed lines* indicate calculated transition energies assuming a parabolic confinement potential. (**b**) Influence of the dot height h_{QD} on the energy of the $E0$, $E1$, and $E2$ peaks. (**c**) Influence of the dot height h_{QD} on the quantization energies $E1 - E0$ and $E2 - E1$ [25]

To overcome this limitation, we have grown a 5-nm thick AlAs layer before the LDE process for thermal stabilization of the surface. High-temperature LDE ($T = 650$ °C) with Al droplets on an AlAs surface results in the formation of only deep holes with average depth of 14 nm and density of about 4×10^8 cm^{-2} [25]. Filling of such deep holes with GaAs yields type 2 QDs as depicted in Fig. 15.8c. Most importantly, the type 2 dots are only partially filled and the QD size is perfectly controlled by the amount of GaAs deposited for filling. Due to the spatial separation, we assume an only negligible interaction between the deep hole QDs and the GaAs quantum well formed outside the nanoholes by filling. This yields a better defined confinement in comparison to type 1 dots in completely filled shallow holes.

Low-temperature PL measurements from ensembles of type 2 QDs are plotted in Fig. 15.10a. Importantly, at low excitation power, dot ensembles exhibit a very sharp PL ground-state peak E_0 with minimum FWHM as small as 9.7 meV. This sharp peak establishes that only partially filled, deep holes form the highly uniform QD ensemble. We calculate a QD height of 7.6 nm according to [25]. Additional sharp peaks arise at higher energy with increasing excitation power. These are associated with excited states with quantized shell structure that is well described by a parabolic potential with a slightly anisotropic QD base [25].

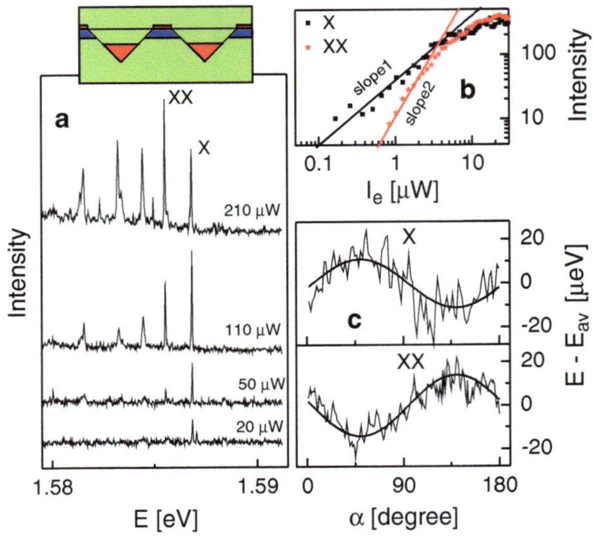

Fig. 15.11 (**a**) Single-dot PL spectra of a type 2 GaAs QD taken at varied excitation power I_e as indicated. The laser energy is 2.33 eV. The spectra are vertically shifted for clarity. The exciton peaks are labeled as X and the biexciton as XX. (**b**) *Symbols*: excitation power dependence of the X and XX peaks. *Lines*: fits with slope = 1 for the X and slope = 2 for the XX peaks. (**c**) Position of the X and XX peak maxima as function of the analyzer polarization angle α together with fits using a sinus function. The neutral exciton fine-structure splitting (FSS) is 22 μeV for the X and 28 μeV for the XX peak, respectively [27]

Figure 15.10b demonstrates that the QD emission energies can be tuned over a wide range, i.e., from 1.55 up to 1.69 eV (ground-state energy E_0), by controlling the hole filling level and with this the dot size. That means the optical emission lies between 733 and 800 nm, which is within the high-sensitive range of typical silicon photo-detectors. The energy level separation is also tunable from 40 up to 90 meV (Fig. 15.10c). Interestingly, QDs with height smaller than 7.5 nm show additional peaks that we attribute to transitions between electrons in the ground-state and holes in the second excited state [25]. Further studies demonstrate room-temperature emission from the QDs and that several layers with LDE QDs can be stacked without peak broadening [37].

The uniform type 2 QDs have typical densities of $4 \times 10^8\,\text{cm}^{-2}$. This allows to address single dots by using a focused laser for excitation. Single-dot PL spectroscopy was performed in a micro PL setup at $T = 4\,\text{K}$ [27]. Typical PL spectra at varied excitation power I_e are plotted in Fig. 15.11a. At low excitation power, the spectra exhibit a sharp line representing the neutral exciton peak X. The measured linewidth of 50 μeV is resolution-limited. With increasing I_e, the biexciton peak XX and higher excitonic complexes arise. We observe no charged excitons (trions) in our samples. The exciton and biexciton peaks are identified on the basis of their excitation-power dependence, with slope of one for X and of two for XX (Fig. 15.11b). In our samples the biexciton is always redshifted by 1.3–1.8 meV compared to the exciton.

Polarization-dependent measurements of the neutral exciton and biexciton peaks reveal a polarization angle α dependent shift of the peak maxima (Fig. 15.11c) that is related to the excitonic fine-structure splitting (FSS) [38]. Values of the FSS between 20 and 40 μeV are determined. A small value of the FSS is very important for the generation of indistinguishable photons for applications in the field of quantum cryptography [39]. The phase-shift between the exciton and biexciton peaks of 180° is in agreement with common observations [40], which supports our identification of these peaks.

Measurements of the excitonic lifetimes of single GaAs QDs are performed under quasi-resonant excitation into the QD d-shell and with a streak camera for data recording [41]. The data are analyzed using a three-level rate model and yield lifetimes of 390 ps for the exciton and 426 ps for the biexciton peak. A discussion of the lifetimes on the basis of Fermi's golden rule with a comparison of lifetimes in other QD systems is given in [41].

15.3.3 Type 3 QDs: Ultra-Low Density Dots

Also the ULD nanoholes described in Sect. 15.2.3 are filled with GaAs for QD generation (type 3 QDs, Fig. 15.8d). The main advantages of type 3 ULD dots in comparison to the above type 2 QDs are the further reduced density and the increased hole depth. As an additional interesting feature, the position of buried ULD dots is visible on the surface as a several nanometer high hillock, even after capping with 200 nm AlGaAs [34]. This surface visibility allows the simple integration of ULD QDs into subsequently prepared device geometries.

Single-dot PL measurements of type 3 ULD dots exhibit sharp excitonic features with a redshifted biexciton [34]. Figure 15.12a demonstrates that the energy of the exciton peak can be controlled very precisely from 1.55 up to 1.68 eV by the hole filling level. This range is close to that of the type 2 dots discussed above. The biexcitons are always redshifted compared to the excitons and the exciton–biexciton splitting is in the range between 1.7 and 2.3 meV.

Due to their low density, the ULD dots can be studied also under non-resonant high-excitation conditions, without the risk to excite neighbored dots by exciton diffusion. An example is shown in Fig. 15.12b. At low excitation power (0.1 μW), the spectrum is dominated by the sharp exciton and biexciton peaks. With increasing excitation power, the peaks broaden significantly and a redshift is observed. The peak broadening is probably related to fluctuating charges induced by the strong laser irradiation. Furthermore, additional peaks arise that represent the shell structure of the quantum dot. At an excitation power of 60 μW at least six quantized shells are observable. The quantization energy of about 25 meV is smaller compared to the value of 43 meV for a type 2 dot of similar height (Fig. 15.10c). This is probably related to the QD shape. For flat QDs the strong confinement in growth direction controls the QD ground-state energy and the weaker confinements in the lateral directions the quantization energy (energy separation of the excited levels).

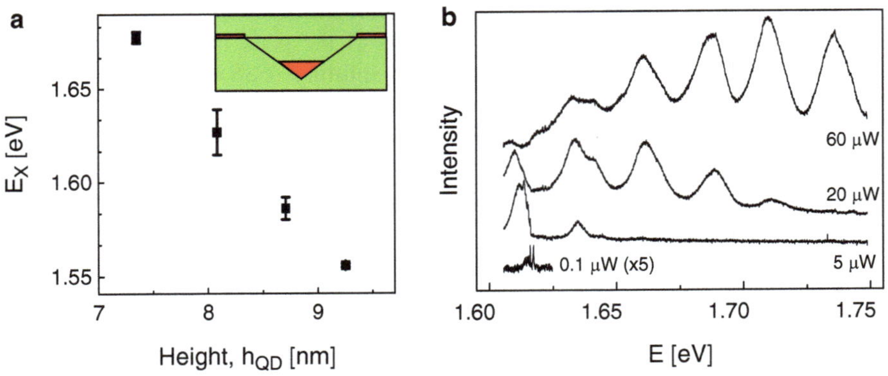

Fig. 15.12 (**a**) Energetic position of the exciton peak from type 3 ULD-QDs with height h_{QD} varied by the hole filling level. Every data point represents the average of at least five dots on one sample. (**b**) PL spectra under high excitation conditions from a single type 3 ULD-QD with $h_{QD} = 8.1$ nm. The respective excitation power is indicated [34]

The ULD holes have an angle $\alpha = 12°$ [34] between the surface and the hole side-facet, that is smaller compared to $\alpha = 20°$ [29] for the holes used for type 2 dots. As a consequence, the ULD dots are broader with smaller lateral confinement, which causes their smaller quantization energy.

15.3.4 Vertically Stacked Quantum Dot Molecules

Quantum dot molecules (QDMs) composed of two closely spaced quantum dots are very interesting objects since they represent the simplest possible interacting system formed by quantum nanostructures. For the fabrication of vertically stacked QDMs we have double filled deep ULD holes. The two QDs inside a hole are separated by an AlGaAs barrier with well-defined thickness. A cross-sectional scheme of a QDM is shown in Fig. 15.13a. A huge advantage of this structure is the possibility to independently adjust the size of both QDs and the barrier thickness.

This is demonstrated by the PL measurements shown in Fig. 15.13b, c. In both spectra two peaks are visible, which represent the ground-state energy of the respective two dots in the vertically stacked QDM. Smaller dots are fabricated in the sample of Fig. 15.13c, which results in a clear increase of the energy of the PL peaks. The goal here is to fabricate QDMs composed of dots with about equal energetic positions in order to study resonant coupling effects inside the QDM [42]. In order to tune the coupling, the Stark effect is utilized which allows to tune the QD optical emission via an electrical field. Figure 15.13d shows an example of a QDM where the PL lines are shifted up to 25 meV by a vertical electrical field. For studies of the coupling inside the QDM, it is furthermore very advantageous that the barrier thickness between the respective dots can be varied in a precisely controlled fashion.

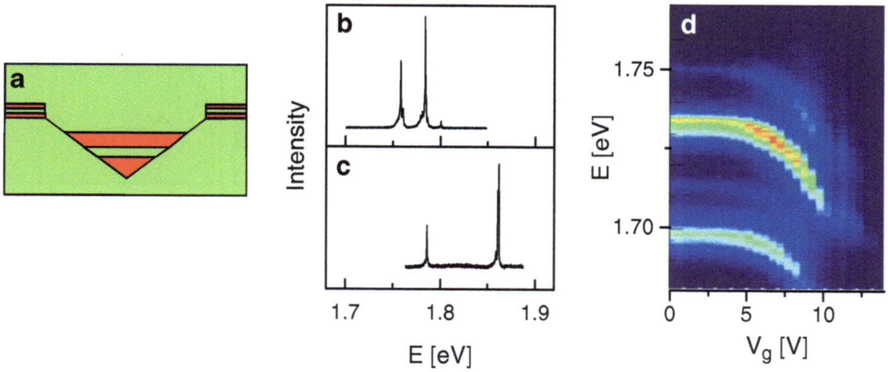

Fig. 15.13 (**a**) Cross-sectional scheme of a vertically stacked quantum dot molecule (QDM) fabricated by double filling of a deep ULD hole. *Green* marks AlGaAs and *red* GaAs. (**b**) PL spectra from a single vertically stacked QDM with respective lines from the two nonresonant dots. (**c**) PL spectra from a single QDM with smaller dots. (**d**) *Color plot* showing the influence of a vertical electrical field V_g on the PL emission from a single vertically stacked QDM. Values of the Stark-shift up to 25 meV are measured

15.4 Nanopillars

This section addresses the fabrication and properties of ultrashort semiconductor nanopillars that are a further interesting application of the droplet etching technique. These crystalline nanopillars represent a new type of semiconductor nanostructure for which many applications can be envisioned. They can, e.g., stabilize an only few nanometer thin gap between two epitaxial semiconductor layers in so-called air-gap heterostructures (AGHs) [43] or they can be embedded in a higher bandgap material forming quantum point contacts for charge carriers between two stacked 2D or 3D reservoirs. As an example, it is suggested to use AGHs for thermoelectrical applications. The efficiency of a thermoelectric generator is defined as the ratio between the generated electrical power and the heat flow through the device. It has been demonstrated that nanostructured materials are often superior to bulk crystals since they can offer a reduced heat flow and, thus, a higher thermoelectric efficiency [44–46]. In this field, we have studied the thermal transport through crystalline nanopillars [47].

15.4.1 Fabrication of the Nanopillars

Schemes of the fabrication procedures for freestanding nanopillars and nanopillars stabilizing two epitaxial layers in AGHs are sketched in Fig. 15.14. The central difference between both procedures is the thickness of the AlAs layer. For the

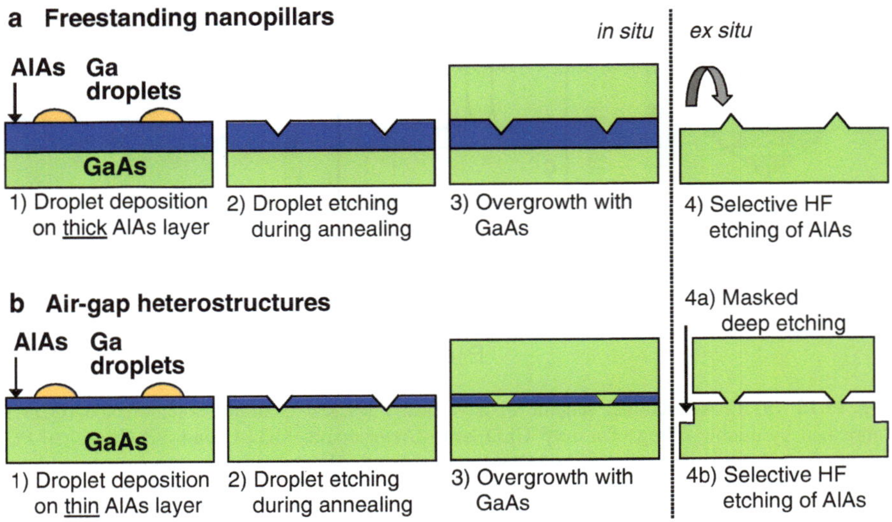

Fig. 15.14 Schemes of the fabrication procedures for freestanding nanopillars and nanopillars stabilizing two epitaxial layers in so-called air-gap heterostructures (AGHs). (**a**) The fabrication of freestanding nanopillars starts with the growth of a thick (75 nm) AlAs layer on a GaAs substrate. For etching, Ga droplets are deposited on the AlAs layer and nanoholes are formed during subsequent annealing. The holes are filled by overgrowth with 500 nm GaAs. After transfer of the sample out of the MBE growth chamber, only the AlAs layer is removed by selective wet-chemical HF etching. (**b**) The fabrication of AGHs requires similar steps. However, here a thinner AlAs layer (2–8 nm) is used, where the holes are deeper than the AlAs layer thickness. The starting edge for selective HF etching is opened by a masked deep etching process. A such created AGH is composed of two epitaxial layers separated only by crystalline nanopillars

creation of freestanding nanopillars, the AlAs layer must be thicker than the depth of the LDE holes (Fig. 15.14a). On the other hand, for the fabrication of AGHs, the AlAs layer must be thinner than the hole depth (Fig. 15.14b) [43].

An example of an ensemble of freestanding nanopillars is shown in Fig. 15.15a. It should be mentioned here that a direct AFM imaging of the initial nanoholes under air is not possible due to the very rapid oxidation of the highly reactive AlAs surface. The pillar density can be varied by the temperature during LDE in the range between 2 and $6 \times 10^8\,\mathrm{cm}^{-2}$ and the diameter at the base is 90–130 nm. The pillar height of 9–13 nm allows AGHs with gaps up to about 8 nm.

Optical microscopy and reflectivity measurements were applied in order to prove the presence of the air gap between the GaAs layers and its stabilization by the nanopillars [43]. Figure 15.15b shows a top-view micrograph on a sample after processing according to the AGH fabrication procedure of Fig. 15.14b. Dominant features are the lithographically defined square deep-etched windows with respective size of $100 \times 100\,\mu\mathrm{m}^2$. The borderline of such a window is marked by the arrows c_1 in Fig. 15.15b. The arrows c_2 mark an additional weaker contrast that we attribute to the borderline up to which AlAs has been removed during the HF underetching step. The origin of the different contrasts is illustrated in Fig. 15.15c. The slightly

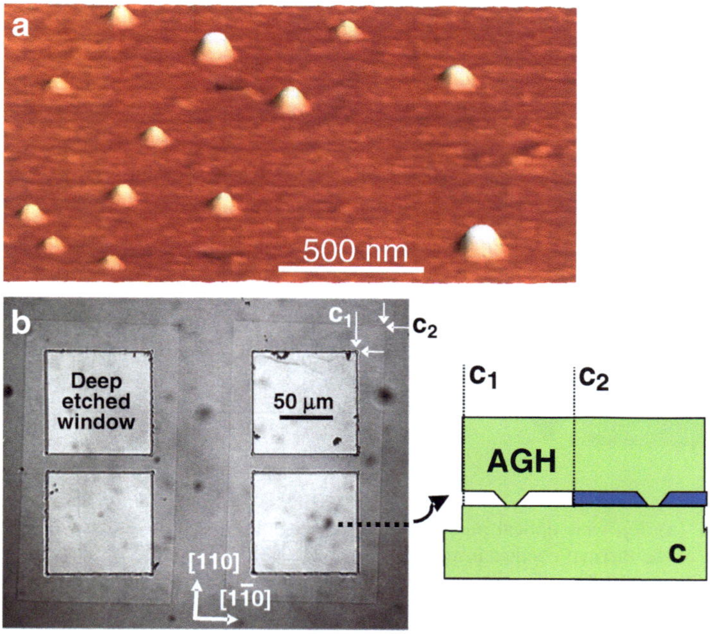

Fig. 15.15 (**a**) Three-dimensional $2 \times 2\,\mu m^2$ AFM image of GaAs nanopillars created by etching with Ga droplets at $T = 600\,°C$. (**b**) Optical micrograph from different features of an AGH sample with AlAs layer thickness of 8 nm. The *arrows* c_1 mark contrasts caused by the lithographically defined $100 \times 100\,\mu m$ deep-etched windows, which open the starting edge for the selecting HF etching. The *arrows* c_2 mark the borderline up to which AlAs have been removed during the selective HF underetching step. (**c**) Cross-sectional scheme along the *dashed line* in (**b**) illustrating the contrasts c_1 and c_2. The AGH area is between both contrasts. Gap and nanopillar size are strongly magnified with respect to the lateral distance. *Green* marks GaAs and *blue* AlAs [43]

misaligned rectangles reveal that the HF etching is anisotropic and that the mask for deep etching has not been perfectly aligned with the crystal symmetry. Note that the horizontal bridges between the deep etched windows are completely underetched. To verify the presence and thickness of the air-gaps, this completely underetched part of the sample was studied with spatially resolved reflectivity measurements. An analysis of the measured reflectivity data by fitting them with calculation results obtained using the transfer-matrix method yields slightly thicker air gaps than expected from the thicknesses of the initial AlAs layers [43].

15.4.2 Thermal Transport Through Nanopillars

For measurements of the thermal conductance through the pillars a four-point probed metallic wire is prepared on the sample surface (Fig. 15.16a). This wire is used as mask for the deep etching step (Fig. 15.14b) and, thus, defines the area of

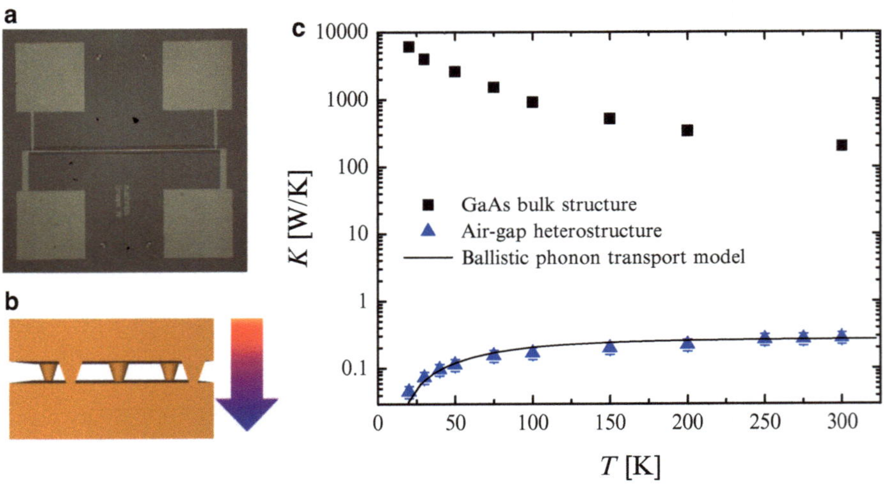

Fig. 15.16 (**a**) *Top view* optical micrograph of a four-point probed metallic wire for 3ω measurements of the thermal conductance. The Au wire is 1.6 mm long, the width is 15 µm, and the thickness 30 nm. (**b**) The thermal conductance is measured perpendicular to the sample surface through the nanopillars. (**c**) Temperature-dependent thermal conductance of a reference GaAs bulk crystal calculated as described in the text (*black symbols*) and a typical AGH (*blue symbols*). The line is calculated under the assumption that the nanopillars can be treated as ballistic point-contacts for phonons [47]

the AGH. Furthermore, the wire is simultaneously used as a heater and thermometer in the 3ω-method to measure the thermal conductance perpendicular to the sample surface (Fig. 15.16b). Details of the 3ω-method are described in [47–49].

Figure 15.16c shows an example for the thermal conductance through an AGH together with calculated reference values of a GaAs bulk structures. The reference values are determined by multiplying the specific thermal conductivity of GaAs bulk material with the quotient of the area beneath the heater wire and the AGH thickness. Obviously, the AGH sample has a strongly reduced thermal conductance in comparison to the bulk. We attribute this reduction mainly to the reduced area for the thermal current flow. Furthermore, the monotonous increase with temperature in contrast to the behavior of bulk GaAs indicates qualitative different phonon transport mechanisms in both systems. The GaAs bulk data are dominated by phonon umklapp scattering. On the other hand, the pillars are much shorter than the mean free path for phonon scattering. So the data obtained for the AGH can be explained assuming that phonon transport through the nanopillars is ballistic. With a simple Boltzmann transport model assuming ballistic phonon transport the temperature dependence of the conductance data can be indeed well described [47, 50].

15.5 Conclusions

Local droplet etching is a new and versatile technique for the self-assembled creation of a variety of semiconductor nanostructures. LDE is fully compatible with state-of-the art MBE technology and takes full advantage of its precise control on the nanostructures structural properties. First applications demonstrate, e.g., the fabrication of GaAs QDs with highly uniform or broadband optical emission. These QDs are defect-free, strain-free, without unintentional intermixing, and the optical emission energy is precisely adjustable. Furthermore, ultra-low density QDs and vertically stacked strain-free QD molecules are realized. Nanopillars separating epitaxial layers in so-called AGHs establish a strong reduction of the thermal conductivity and represent ballistic point-contacts for phonons. These examples demonstrate the flexibility of the LDE method, from which we expect even more interesting applications in the future.

Acknowledgements The authors thank Andrea Stemmann and Jochen Kerbst for MBE growth and AFM characterizations, Andreas Graf, Achim Küster, Christian Strelow, and Tim Köppen for PL measurements, and Thorben Bartsch for measurements of the thermal conductance. Furthermore, we thank Stefano Sanguinetti, Takaaki Mano, Nobuyuki Koguchi, and Richard Warburton for very helpful and stimulating discussions. Finally, we thank the "Deutsche Forschungsgemeinschaft" for financial support via HA 2042/6-1, GrK 1286, and SSP 1386.

References

1. Bimberg, D., Grundmann, M., Ledentsov, N.N.: Quantum Dot Heterostructures. Wiley, Chichester (1999)
2. Herman, M.A., Sitter, H.: Molecular Beam Epitaxy: Fundamentals and Current Status. Springer, Berlin (1989)
3. Chikyow, T., Koguchi, N.: Jpn. J. Appl. Phys. **29**, L2093 (1990)
4. Koguchi, N., Takahashi, S., Chikyow, T.: J. Cryst. Growth **111**, 688 (1991)
5. Koguchi, N., Ishige, K.: Jpn. J. Appl. Phys. **32**, 2052 (1993)
6. Mano, T., Watanabe, K., Tsukamoto, S., Fujioka, H., Oshima, M., Koguchi, N.: Jpn. J. Appl. Phys. **38**, L1009 (1999)
7. Watanabe, K., Koguchi, N., Gotoh, Y.: Jpn. J. Appl. Phys. **39**, L79 (2000)
8. Mano, T., Watanabe, K., Tsukamoto, S., Koguchi, N., Fujioka, H., Oshima, M., Lee, C.D., Leem, J.Y., Lee, H.J., Noh, S.K.: Appl. Phys. Lett. **76**, 3543 (2000)
9. Kim, J.S., Koguchi, N.: Appl. Phys. Lett. **85**, 5893 (2004)
10. Kuroda, T., Mano, T., Ochiai, T., Sanguinetti, S., Sakoda, K., Kido, G., Koguchi, N.: Phys. Rev. B **72**, 205301 (2005)
11. Yamagiwa, M., Mano, T., Kuroda, T., Tateno, T., Sakoda, K., Kido, G., Koguchi, N.: Appl. Phys. Lett. **89**, 113115 (2006)
12. Huang, S., Niu, Z., Fang, Z., Ni, H., Gong, Z., Xia, J.: Appl. Phys. Lett. **89**, 031921 (2006)
13. Heyn, Ch., Stemmann, A., Schramm, A., Welsch, H., Hansen, W., Nemcsics, Á.: Phys. Rev. B **76**, 075317 (2007)
14. Abbarchi, M., Mastrandrea, C.A., Kuroda, T., Mano, T., Sakoda, K., Koguchi, N., Sanguinetti, S., Vinattieri, A., Gurioli, M.: Phys. Rev. B **78**, 125321 (2008)

15. Stock, E., Warming, T., Ostapenko, I., Rodt, S., Schliwa, A., Töfflinger, J.A., Lochmann, A., Toropov, A.I., Moshchenko, S.A., Dmitriev, D.V., Haisler, V.A., Bimberg, D.: Appl. Phys. Lett. **96**, 093112 (2010)
16. Lee, J.H., Wang, Zh.M., Kim, E.S., Kim, N.Y., Park, S.H., Salamo, G.J.: Nanoscale Res. Lett. **5**, 308 (2010)
17. Moison, J.M., Houzay, F., Barthe, F., Lepronce, L., Andre, E., Vatel, O.: Appl. Phys. Lett. **64**, 196 (1994)
18. Madhukar, A., Xie, Q., Chen, P., Konkar, A.: Appl. Phys. Lett. **64**, 2727 (1994)
19. Leonard, D., Krishnamurthy, M., Fafard, S., Merz, J.L., Petroff, P.M.: J. Vac. Sci. Technol. B **12**, 1063 (1994); Leonard, D., Fafard, S., Zhang, Y.H., Merz, J.L., Petroff, P.M.: J. Vac. Sci. Technol. B **12**, 2516 (1994)
20. Volmer, M., Weber, A.: Z. Phys. Chem. **119**, 277 (1926)
21. Heyn, Ch., Stemmann, A., Hansen, W.: Appl. Phys. Lett. **95**, 173110 (2009)
22. Heyn, Ch.: Phys. Rev. B **83**, 165302 (2011)
23. Wang, Zh.M., Liang, B.L., Sablon, K.A., Salamo, G.J.: Appl. Phys. Lett. **90**, 113120 (2007)
24. Heyn, Ch., Stemmann, A., Eiselt, R., Hansen, W.: J. Appl. Phys. **105**, 054316 (2009)
25. Heyn, Ch., Stemmann, A., Köppen, T., Strelow, Ch., Kipp, T., Mendach, S., Hansen, W.: Appl. Phys. Lett. **94**, 183113 (2009)
26. Heyn, Ch., Stemmann, A., Köppen, T., Strelow, Ch., Kipp, T., Grave, M., Mendach, S., Hansen, W.: Nanoscale Res. Lett. **5**, 576 (2010)
27. Heyn, Ch., Klingbeil, M., Strelow, Ch., Stemmann, A., Mendach, S., Hansen, W.: Nanoscale Res. Lett. **5**, 1633 (2010)
28. Stemmann, A., Heyn, Ch., Köppen, T., Kipp, T., Hansen, W.: Appl. Phys. Lett. **93**, 123108 (2008)
29. Stemmann, A., Heyn, Ch., Hansen, W.: J. Appl. Phys. **106**, 064315 (2009)
30. Ostwald, W.: Z. Phys. Chem. **34**, 495 (1900)
31. Heyn, Ch., Stemmann, A., Strelow, Ch., Köppen, T., Sonnenberg, D., Graf, A., Mendach, S., Hansen, W.: J. Nanoelectron. Optoelectron. **6**, 62–67 (2011)
32. Kumah, D.P., Shusterman, S., Paltiel, Y., Yacoby, Y., Clarke, R.: Nat. Nanotechnol. **4**, 835 (2009)
33. Thurmond, C.D.: J. Phys. Chem. Solids **26**, 785 (1965)
34. Sonnenberg, D., Graf, A., Paulava, V., Hansen, W., Heyn, Ch.: Appl. Phys. Lett. **101**, 143106 (2012)
35. Venables, J.A.: Philos. Mag. **27**, 693 (1973); Venables, J.A., Spiller, G.D., Hanbücken, M.: Rep. Prog. Phys. **47**, 399 (1984)
36. Heyn, Ch., Stemmann, A., Hansen, W.: J. Cryst. Growth **311**, 1839 (2009)
37. Polojärvi, V., Schramm, A., Guina, M., Stemmann, A., Heyn, Ch.: Nanotechnology **22**, 105603 (2011)
38. Singh, R., Bester, G.: Phys. Rev. Lett. **104**, 196803 (2010)
39. Benson, O., Santori, C., Pelton, M., Yamamoto, Y.: Phys. Rev. Lett. **84**, 2513 (2000)
40. Gammon, D., Snow, E.S., Shanabrook, B.V., Katzer, D.S., Park, D.: Phys. Rev. Lett. **76**, 3005 (1996)
41. Heyn, Ch., Strelow, Ch., Hansen, W.: New J. Phys. **14**, 053004 (2012)
42. Bayer, M., Hawrylak, P., Hinzer, K., Fafard, S., Korkusinski, M., Wasilewski, Z.R., Stern, O., Forchel, A.: Science **291**, 451 (2001)
43. Heyn, Ch., Schmidt, M., Schwaiger, S., Stemmann, A., Mendach, S., Hansen, W.: Appl. Phys. Lett. **98**, 033105 (2011)
44. Majumdar, W.: Science **303**, 777 (2004)
45. Vineis, C.J., Shakouri, A., Majumdar, A., Kanatzidis, M.G.: Adv. Mater. **22**, 3970 (2010)
46. Nielsch, K., Bachmann, J., Kimling, J., Böttner, H.: Adv. Energy Mater. **1**, 713 (2011)
47. Bartsch, Th., Schmidt, M., Heyn, Ch., Hansen, W.: Phys. Rev. Lett. **108**, 075901 (2012)
48. Cahill, D.G.: Rev. Sci. Instrum. **61**, 802 (1990)
49. Cahill, D.G., Katiyar, M., Abelson, J.R.: Phys. Rev. B **50**, 6077 (1994)
50. Jeong, C., Lundstrom, M.: Appl. Phys. Lett. **100**, 233109 (2012)

Index

Z.M. Wang (ed.), *Nanodroplets*, Lecture Notes in Nanoscale Science and Technology 18, DOI 10.1007/978-1-4614-9472-0, © Springer Science+Business Media New York 2013

MIX
Papier aus verantwortungsvollen Quellen
Paper from responsible sources
FSC® C105338

If you have any concerns about our products,
you can contact us on
ProductSafety@springernature.com

In case Publisher is established outside the EU,
the EU authorized representative is:
Springer Nature Customer Service Center GmbH
Europaplatz 3, 69115 Heidelberg, Germany

Printed by Libri Plureos GmbH
in Hamburg, Germany